Social
Evolution
in Ants

MONOGRAPHS IN BEHAVIOR AND ECOLOGY

Edited by John R. Krebs and Tim Clutton-Brock

Social Evolution in Ants

ANDREW F.G. BOURKE AND

NIGEL R. FRANKS

Princeton University Press

Princeton, New Jersey

Library of Congress Cataloging-in-Publication Data

Bourke, Andrew F.G., 1961–
Social evolution in ants/Andrew F.G. Bourke, Nigel
R. Franks.
p.   cm. — (Monographs in behavior and ecology)
Includes bibliographical references (p.   ) and
indexes.
ISBN 0-691-04427-9 (cl). — ISBN 0-691-04426-0
(pbk.)
1. Ants-Behavior. 2. Social evolution in animals. 3.
Kin selection (Evolution) 4. Sex ratio. I. Franks,
Nigel R. II. Title. III. Series.
QL568.F7B63 1995
595.79'6045248–dc20   95–5959

This book has been composed in 10/12 Times by
Wearset, Boldon, Tyne and Wear

Printed in the United States of America by
Princeton Academic Press

10 9 8 7 6 5 4 3 2 1

10 9 8 7 6 5 4 3 2 1
(Pbk.)

Contents

Preface and
Acknowledgments

Ants have always provoked the amazement of people at large, and the intellectual curiosity of biologists in particular. To Darwin, their sterile castes posed "one special difficulty, which at first appeared to me insuperable, and actually fatal to my whole theory" (Darwin 1859 p. 236). In fact, Darwin found several difficulties in the sterile workers of ants (Cronin 1991 p. 298). The most pressing for modern biologists is the question of how sterility and self-sacrifice can evolve under natural selection. In 1964, W.D. Hamilton's theory of kin selection provided the radical and elegant solution to this problem. Hamilton's work revolutionized our understanding of natural selection and adaptation, particularly in the field of animal behavior (G.C. Williams 1966; Dawkins 1976). In addition, after being largely inspired by social insects, Hamilton's theory repaid the debt by sparking off an explosion in research on kin selection in ants and their relatives. In this, it was aided by the major synthetic works of E.O. Wilson (1971, 1975). Another fillip was provided by the paper of Trivers and Hare (1976), which linked kin selection with sex ratio evolution and the occurrence of conflicts of interest within insect societies. With these studies, social evolution in ants became a topic of concern to evolutionary biologists and behavioral ecologists of all kinds.

An additional facet of social evolution to which ant studies are central is the organization of work within societies. Prompted largely by the pioneering researches of Oster and Wilson (1978), the study of the division of labor in ants has also outgrown its original boundaries to become a topic of interest to many types of biologists. Its modern aspect includes, among other things, the study of self-organization in complex systems.

The inquiries initiated by Darwin, Hamilton, Wilson, Trivers, and others have created a rich and sophisticated body of theory and data on ant social evolution. In this book, we describe its present state. There are many recent books about ants. A sample includes those by Dumpert (1981), Passera (1984), Sudd and Franks (1987), and the magnificent treatise of Hölldobler and Wilson (1990). This book offers something that builds on these works, but is nevertheless different in purpose. Aided by the full panoply of modern adaptationist logic, we explore in depth the fundamental topics in evolutionary biology and behavioral ecology to which ant studies continue to make an important contribution. In full, these include sex ratio evolution, kin conflict, the ecology of social systems, and self-organization. We also discuss the evolution of eusociality, along with the logic of kin selection theory and associated concepts of selection. In addition, we review two topics that have been relatively neglected in ants, namely life history strategies and mating biology. Throughout, we have aimed to make comparisons with other types of social insects, and other organisms altogether, where this seemed appropriate. Inevitably, we have not found space to consider several further, equally important facets of ant sociality. Examples are foraging systems, the evolution of social parasitism, and mutualisms between ants and plants. For these, and for a comprehensive, beautifully illustrated survey of ant biology in general, we refer readers to Hölldobler and Wilson (1990). In this book, we invite all those interested in evolutionary biology and behavioral ecology, including those whose favorite organisms are (for now) birds and mammals, to share in the knowledge and understanding that comes from the study of social evolution in ants.

We thank the many people who helped us produce this book. Laurent Keller and Peter Nonacs each read the whole work in draft, and made dozens of useful comments. The following people read and helped improve various chapters, passages, or groups of chapters: Tim Benton, Koos Boomsma, George Chan, Scott Creel, William Foster, Raghavendra Gadagkar, Charles Godfray, Jürgen Heinze, Ian Owens, Christian Peeters, Andrew Pomiankowski, Francis Ratnieks, and Mary Jane West-Eberhard. We thank them all; we always followed their advice, but from time to time we had second thoughts and changed the passage back to our version again. We particularly thank those colleagues who commented with good grace on passages opposed to their viewpoint. Many additional colleagues kindly sent unpublished or in press work. Simon Fraser and Ana Sendova-Franks prepared some of the figures.

A.F.G.B. thanks Jesus College, Cambridge, and the Zoology Department of the University of Cambridge, for support during the early stages of the writing. A.F.G.B. wrote the bulk of his contribution as a member of the Ecology Group of the Institute of Zoology, Zoological Society of London. Thanks go to Steve Albon and all the other members of the Group, especially Ian Owens, for their backing and encouragement. The following libraries are acknowledged for providing essential facilities: The Balfour Library, Department of Zoology, University of Cambridge; The Library of the Zoological Society of London, Regent's Park, London; and The Bloomsbury Science Library, University College London.

N.R.F. thanks the School of Biology and Biochemistry, University of Bath, and the Wissenshaftskolleg zu Berlin, Institute of Advanced

Study, Berlin, for their support. This book was partly written when N.R.F. was a Fellow at the Wissenschaftskolleg during the academic year 1993–1994. A debt of gratitude is owed to friends and colleagues at both Bath and Berlin for their encouragement and support. In particular, thanks go to Guy Blanchard, Don Braben, Scott Camazine, Jean-Louis Deneubourg, Melanie Hatcher, Mike Mogie, Glenda Orledge, Lucas Partridge, Alan Rayner, Stuart Reynolds, Tom Seeley, Ana Sendova-Franks, Lesley Smart, Tim Stickland, Chris Tofts, and Rüdiger Wehner.

Finally, we thank John Krebs and Tim Clutton-Brock for initiating this project, and for their professional help Emily Wilkinson and Kevin Downing of Princeton University Press, and Fisher Duncan, London. The order of authorship of this book was determined alphabetically.

Andrew Bourke, Nigel Franks
London and Bath, December 1994

Social
Evolution
in Ants

1 Kin Selection

1.1 Introduction

The basis of the modern evolutionary study of animal behavior is Darwin's theory of natural selection. But the more immediate foundations of the study of sociality and altruism in animals come from a development of Darwinian theory, W.D. Hamilton's theory of kin selection. Hamilton's theory is fundamental to the understanding of the social evolution of ants and their relatives. So the main purpose of this chapter is to explain the meaning and power of kin selection theory. Chapter 2 explores the relation between kin selection and other concepts of selection such as group selection and colony-level selection. Chapter 3 concentrates on specific models of social evolution in the Hymenoptera.

This book aims to interpret the social biology of ants from an adaptationist viewpoint, deploying the theory of natural selection as a tool of explanation. It therefore pursues what critics call the "adaptationist programme" (Gould and Lewontin 1979). However, embracing adaptationism need not mean that all evolutionary change must be regarded as adaptive, or all features of organisms as adaptations. Nor does it mean that natural selection, the mechanism of adaptive evolutionary change, is either omnipotent, or unconstrained, or a perfect optimizing agent. Lastly, it does not imply that other facets of evolutionary biology are irrelevant or uninteresting.

Instead, adaptationists hold that the evolution of adaptation by natural selection is a unique and pervasive feature of the living world. Therefore, it is a fruitful exercise, in terms of gaining understanding, to interpret biology in adaptive terms. This may not always be successful. On the other hand, the recent expansion of adaptationist thinking into animal behavior has been very successful. Many features of behavioral biology that previously seemed just natural history curiosities can be explained adaptively (for example, the kin conflicts within societies described in Chapter 7). After all, natural selection, though not omnipotent, is powerful; though not unconstrained, is not totally bound by history; and though not a perfect optimizer, can achieve design good enough to give the illusion of creation. These are the features of natural selection that justify the adaptationist program. For full defences of adaptationism and associated concepts see, for example, Cain (1964), G.C. Williams (1966, 1985, 1992), Dawkins (1982a, 1986), Mayr (1983), Parker and Maynard Smith (1990), Cronin (1991), and Reeve and Sherman (1993).

This chapter also makes the case for a gene-centered approach to understanding natural selection. In other words, it advocates the gene selectionist or "selfish gene" perspective introduced to evolutionary biology by G.C. Williams (1966) and Dawkins (1976, 1982a, 1986). Gene selectionism is controversial (e.g. Wimsatt 1980; Wright 1980; Dawkins 1982a; Gould 1983a, 1992; Brandon and Burian 1984; Sober 1984a,b; Brandon 1985; Dover 1988; Ohta 1992). Nevertheless, a look at today's behavioral ecology textbooks suggests that it is already the dominant mode of evolutionary explanation. But the arguments for gene selectionism in general, and for its use in the study of social insects in particular, are still worth presenting.

First, since natural selection thinking underpins all of this book, what natural selection is taken to mean needs explaining. Second, regarding the main point of this chapter, kin selection theory is best understood as a logical consequence of gene selectionism. So a strong case for kin selection first has to set out its gene selectionist premises. Dawkins (1976, 1979, 1982a) has already clearly explained the gene selectionist logic of kin selection theory. Nevertheless, the necessity of kin selection for understanding social evolution has not been fully appreciated. Kin selection is still mistaken for an elaborate kind of selection on individuals, or suspected of being a mere scientific fashion. So another reason for a full explanation of gene selectionism is to justify kin selection theory from first principles, in the hope that this can dispel the lingering skepticism over the theory. In this we are reinforcing the message conveyed by earlier authors. In particular, a debt to the ideas and writings of Dawkins will be obvious and is acknowledged here.

An alternative to the gene selectionist understanding of natural selection focuses on the hierarchical organization of life and on the different levels at which selection is believed to act. The social insects, especially "advanced" ones like the ants, are often pressed into the service of this "levels-of-selection" perspective. This is because their colonies are held to represent prime examples of higher-level units subject to natural selection in their own right. As a result, the scientific literature on social insects is full of references to "colony-level selection" and "superorganisms." For this reason, the next chapter analyzes the issues surrounding levels-of-selection theory, colony-level selection, and the superorganism concept. It concludes that, with qualifications, these ideas can be accommodated by the gene selectionist perspective. This is the final reason for initially concentrating on gene selectionism.

Therefore, this chapter starts very generally by setting out the case for gene selectionism (Section 1.2). It then explains how this underpins kin selection theory (Sections 1.3, 1.4). Next, Sections 1.5 to 1.9 elaborate on the structure and scope of the theory, covering topics such as inclusive fitness, how kin selection works at high frequencies of "genes for altruism," gene expression in Hamilton's rule, selection on loci other than loci for altruism, and parental manipulation theory. The conclusion is that kin selection theory is a logical corollary of gene-centered natural selection.

1.2 Natural Selection as Gene Selection

The world's living things are characterized by the detailed fit of their form and function to their way of life and environment – that is, by adaptive complexity. Darwin's (1859) theory of natural selection is biology's explanation for adaptation (e.g. Grant 1963; G.C. Williams 1966; Leigh 1971). Darwin framed natural selection theory in terms of individuals, stating the process to be a consequence of three properties. First, individuals vary. Second, they show heredity. And third, individuals with advantageous traits outreproduce those without such traits (Darwin 1859 pp. 80–81). Therefore, Darwin deduced, since the offspring of favorably endowed individuals would inherit their parents' beneficial features, favorable variations would accumulate in successful lineages, leading to the close fit of individuals to their lifestyle and environment that is organismal adaptation. Modern definitions of natural selection that characterize the process in terms of variation, heredity, and fitness differences (e.g. Lewontin 1970; Endler 1986) are essentially Darwin's individual-centered definition in more up-to-date language.

However, Dawkins (1976, 1978, 1982a,b, 1983, 1986) realized that the fundamental element in natural selection is a replicator, and so offered a new way of characterizing the process. A replicator is an entity of which copies are made, or which makes copies of itself (Dawkins 1982a p. 293, b). Replicators have two other important properties. The first is that each heads a potentially never-ending lineage of descendant replicators (Dawkins' [1982a p. 83] "germ-line" criterion). The second is that the copying process is exact to the point that alterations to replicator structure are preserved in successive replications (Dawkins 1978, 1982a pp. 97–99, b, 1989a pp. 273–274). Dawkins' "replicator version" of natural selection theory runs as follows. First, there are replicators (structures that show high-fidelity copying). Second, there is a level of error, or mutation, in the copying process, producing undirected variation in each generation of replicators. Third, the structure and properties of replicators influence their survival and rate of replication. Given these conditions, the accumulation of favorable mutations within replicators will lead to the appearance of adaptive complexity benefiting the replicators.

Adaptation cannot arise in one stage, in what Dawkins (1986) calls single step selection, because this is too improbable. However, adaptation can arise through successive bouts of selection, in each of which new traits are added to – and if favorable preserved in – the existing set, yielding a complex, well-adapted final product. Dawkins (1986) calls this cumulative selection. Only this process can overcome the improbability inherent in adaptation that single step selection fails to address. This reasoning shows why the replicator concept is fundamental to natural selection theory: only replicators can form the basis of a process of cumulative selection, because only when mutations are conserved in replication can favorable ones accumulate. So Dawkins' argument is that adaptive complexity requires cumulative selection, which in turn requires replicators.

According to natural selection theory the world is presently full of units that exist because of their ancestors' success in leaving descendants. Therefore, going back in time, one would expect to find ancestral units that, though lacking foresight, were acting as if concerned to be successful in leaving descendants for the future. But the present time is just an arbitrary point in the continuum of history. So natural selection theory leads biologists to expect to find at all times a world inhabited by "units of self-interest" (Dawkins 1976 p. 12), where this means entities acting to ensure their successful propagation into the future. Such units must then persist over time, because they can only have inherited their "self-interest" from their ancestors if there is a correspondence between their past and present selves. Put another way, if these entities disintegrate, then they have no future to be successfully projected into. So entities undergoing natural selection, as units of self-interest, require what Dawkins (1986) terms durability.

Replicators have durability, but not in the sense of physical persistence. Instead, it follows from the earlier definition of replicators that their internal structure has the capacity to become sufficiently irregular to store information. It also follows that this information can be transmitted down the generations in the form of copies. Therefore, the durability of replicators is persistence in the form of copied messages (G.C. Williams 1985, 1992 p. 11; Dawkins 1986; Gliddon and Gouyon 1989), where the information preserved specifies adaptations. So replicators, with this kind of durability, can be units of self-interest.

In nature, the usual kinds of replicators are genes, lengths of DNA. Genes exist in aggregations embedded in bodies. Under gene selectionism, bodies are "vehicles" for replicators (Dawkins 1982a,b) – entities constructed by collectives of genes as dwelling places and aids to survival. (Hull's [1980] "interactor" broadly corresponds to Dawkins' vehicle. G.C. Williams [1992 p. 10] suggests the additional distinction between the "codical" [informational] and material "domains" inhabited by replicators and vehicles respectively; see also Cronin [1991].) Bodies are not replicators themselves because alterations to bodily structures are not transmitted in reproduction. This is the same as asserting that acquired characters are not inherited, or that inheritance is not Lamarckian. Bodies of course show heredity, but this is because of the genes they contain. Groups also are not replicators, because alterations to group structure are not conserved. Instead, groups are, like bodies, vehicles for replicators, because it may pay genes to instruct bodies to form societies for the genes' benefit. In addition, both bodies and groups lack durability, since no individual or group persists indefinitely, even in "message" form. Therefore, they are not units of self-interest in the sense that replicators are (Dawkins 1976).

Many species engage in sexual reproduction, which generally entails a fair meiosis (every gene in a body has the same 0.5 chance of getting into a gamete) and chromosomal crossing-over. The existence of crossing-over means that in species with sex it is impossible to say exactly what length of DNA constitutes a gene. Instead, an operational definition can be adopted of a gene as a piece of DNA small enough to have durability (persist down the generations in the form of copies), and large enough to encode meaningful information (G.C. Williams 1966, 1992 p. 18; Dawkins 1976, 1978, 1982b). The existence of fair meiosis also means that genes in sexual species are not usually in competition with genes at other loci (position on a chromosome) in the same individual (for exceptions see Section 1.8). Instead, genes in these species compete with each other in a highly structured way. In particular, they compete with their alleles in the population for representation at their shared locus (G.C. Williams 1966 pp. 57–58; Dawkins 1982a p. 283,b). This is made apparent by remembering that a rise in gene frequency, the definition of success for a focal gene, means a rise in the number of copies of a gene at its locus relative to the number of copies of its alleles. However, the selective background of a gene will still depend on genes at other loci in the body. A gene may even be favored on the basis of its success when matched with an allele, as in the case of heterozygote advantage (G.C. Williams 1966 p. 58; Dawkins 1976 p. 91, 1982a p. 52). Therefore gene selectionism needs to be applied in a qualified and sophisticated way to modern sexually reproducing species.

Another important point about modern genes is that some genes persist down the generations in groups within the larger collective (the genome) as linked "gene complexes" (Dawkins 1978). These complexes act like a large replicator. In addition, the entire genome of a sexual organism comes to represent a mutually stable set of replicators. This is because, as just mentioned, part of the selective environment of genes is made of the other genes it is likely to share a body with (Dawkins 1976). Therefore, in modern bodies adaptive complexity stems from the accumulation of favorable mutations both within the structure of individual replicators (by nucleotide substitution in genes, or gene substitution in gene complexes) and within the mutually stable community of replicators represented by the genome (by gene substitution).

Additional reasons exist for why, aside from being essential for cumulative selection, replicators are fundamental to any deep theory of natural selection. The first is that the arrow of causality points from genes to bodies (Dawkins 1983). Although this has been disputed (e.g. Gould 1983b; Sober 1984a), the causal priority of genes is evident from embryology and development, which involve a one-way flow of information from genes to bodies. So, since genes also carry information "vertically" (in heredity), an organismic trait cannot evolve by natural selection unless it is subject both to genetic control and genetic variation. Therefore, to regard gene frequency changes as simply the passive tracking of changes in the frequencies of traits (e.g. Bateson 1982; Gould 1983b, 1992), as "a kind of genetic bookkeeping" (Wimsatt 1980 p. 158), is to reverse the causal structure of natural selection theory. Traits are only worth considering as candidates for natural selection if they aid the survival and replication of genes (Dawkins 1983; G.C. Williams 1985).

The causal primacy of genes also has a temporal element, which becomes clear from considering the origin of life. Replicators occur in all living things (as RNA or DNA) and, as just discussed, control their development. From the earlier arguments, only replicators generate adaptive complexity. And, unlike bodies (which are too complex), replicators could have arisen spontaneously, by Dawkins' (1986) "single step selection." From these points, several authors have concluded that complex life originated as free replicators or "naked genes" (e.g. Dawkins 1982b p. 50), self-replicating molecules that were not embedded in bodies (Dawkins 1976, 1986). Replicators must then have evolved to acquire their complex cellular and bodily vehicles. In short, bodies exist because of genes, and not the other way round (Dawkins 1982a,b, 1983).

The other main reason why the replicator concept is so central to natural selection theory is that it has generality. Adaptive complexity must be based on the natural selection of replicators, but these may be of many kinds (Dawkins 1983). Biologists typically regard nuclear, transcribed DNA as the prime genetic material. But the replicator concept also explains the existence and behavior of other kinds of replicating biological entity, for example (genes in) viral nucleic acids, mitochondrial DNA, "selfish" nuclear DNA, sex-ratio distorters, "killer" chromosomes, and plasmids (e.g. Doolittle and Sapienza 1980; Eberhard 1980, 1990; Orgel and Crick 1980; Cosmides and Tooby 1981; Dawkins 1982a; Werren et al. 1988; Hurst 1991, 1993a; Ebbert 1993).

The argument has now reached an important corollary of replicator-centered natural selection theory. This is the idea that all adaptations, including those seen in vehicles, ultimately exist for the benefit of the replicators responsible for them, and not for the good of the vehicles (Dawkins 1976, 1982a). This follows from the earlier demonstration that only replicators, not vehicles, have the durability to be units of self-interest. Put another way, adaptation arises from the accumulation of favorable mutations and – as argued earlier – this process can only occur in replicators. The idea that adaptations benefit replicators becomes important when the interests of replicators and vehicles, genes and bodies for example, do not coincide. An example comes from Hamilton (1967), who showed that a meiotic drive gene linked to the Y sex chromosome could produce an extreme male-biased sex ratio (its adaptation) that, in an unstructured population, could drive the population extinct. Therefore, recognizing the primacy of the replicator's interests shows that adaptation may not always involve "good design" at the organismic level. It may even fail to benefit the replicator itself in the long term. This merely emphasizes that natural selection is a mechanistic process without foresight.

Another example where the interests of replicators and vehicles do not coincide forms the main subject of this chapter – kin-selected altruism. The view that adaptations primarily benefit replicators suggests that when bodies behave in a self-destructive fashion, it could be for the selective advantage of a gene responsible for the behavior (Dawkins 1976, 1982a). This insight is central to the kin-selectionist explanation of self-sacrificial behavior (Section 1.4).

Lastly, these arguments imply that the traditional categories of individual selection, kin selection, and group selection are all, fundamentally, aspects of gene selection (Dawkins 1978). They do not rule out, however, a parallel approach to natural selection that focuses on the different levels at which selection acts, providing these levels are understood as levels of vehicles. This is the distinction between the individual- and replicator-centered views of natural selection (Chapter 2). In addition, in common with other behavioral ecologists, social insect biologists need to be aware that a gene selectionist, adaptationist approach presupposes a (simple) genetic basis to the traits they study (Grafen's [1984 p. 63] "phenotypic gambit"). Put another way, at some level of analysis all evolutionary hypotheses of adaptation in behavioral ecology have to be consistent with population genetics (Grafen 1984, 1988). To conclude, the arguments in this section show why we seek to understand adaptation in biology, including social insect biology, in terms of natural selection for adaptations serving the interests of genes.

1.3 The Problem of Altruism

The most obvious adaptive feature of ant biology is sociality, the habit of living in groups or colonies. The adaptiveness of sociality is demonstrated by the sophisticated design features shown by insect colonies, and by sociality's undoubted contribution to the ecological success of ants (E.O. Wilson 1987, 1990).

But ants are more than just social. All modern ants are eusocial ("truly social") or are workerless social parasites secondarily derived from eusocial species (Buschinger 1990a; Bourke and Franks 1991). The key trait of eusociality is that members of the society display a reproductive division of labor: some are fertile individuals (sexuals or reproductives, such as queens) and some are either completely sterile or show limited fertility (neuters or workers). The other defining features of eusociality (E.O. Wilson 1971 p. 4) are an overlap of adult generations in the society, and cooperative brood care, which together mean that the workers help raise the young of reproductives in the parental generation. Eusocial animals include the eusocial Hymenoptera (all ants, some bees, some wasps), termites, aphids (Aoki 1987; Itô 1989; Benton and Foster 1992), ambrosia beetles (Kent and Simpson 1992), thrips (Crespi 1992a), and some mole-rats (Sherman et al. 1991; Jarvis et al. 1994). According to how flexibly the definition is interpreted, they also include many other social vertebrates (Brockmann 1990; Gadagkar 1994; Crespi and Yanega 1995; Sherman et al. 1995).

From the definition of eusociality, it is apparent that another important biological trait of ants is altruism. Biological altruism can itself be defined in various ways (e.g. Alexander 1974; Orlove 1975; Crozier 1979; Starr 1979; Uyenoyama and Feldman 1980; Bertram 1982), but most definitions are framed in terms of effects, not motivation or psychology, and in terms of individuals, not genes. (Genetic altruism cannot evolve because a gene suffering a net loss in its representation in future generations cannot, by definition, increase in frequency [Alexander 1974]. The problem of altruism is the problem of how individual-centered altruism evolves.) The definition to be followed here is that *altruism occurs when an individual behaves in such a way that the result is an increase in the survival or offspring production of another individual and a decrease in its own survival or offspring production.* So eusociality is "sociality-with-altruism," because workers give up their own chances of survival and reproduction to promote those of the brood they rear. Because of a focus of interest on the sacrifice in reproduction, this behavior is often specifically termed reproductive altruism (e.g. Trivers 1985 p. 169).

As already described, Darwin formulated natural selection theory in terms of selection acting on the individual. Altruism then emerged as a problem for the theory, which predicted that each individual should behave as if trying to maximize its number of offspring. In the language of the previous section, Darwin saw the individual organism as the "unit of self-interest." Altruism, and in particular extreme reproductive altruism involving sterility, contradicted this prediction, and so seemed to undermine the theory. This is the problem of altruism (see Cronin [1991] for a historical review).

Darwin's solution to this problem in the social insects was that workers could evolve if they were "profitable to the community," by which he meant the colony (Darwin 1859 p. 236). He also discussed, considering them greater problems than the problem of sterility, how workers could evolve to differ morphologically from queens and from one another. These passages (Darwin 1859 pp. 236–242) indicate that, as regards the evolution of sterility, Darwin envisaged colonies with workers benefiting by producing extra queens and males. So sexuals from these colonies would have been favored by selection, given that they would transmit to their own sexual offspring the profitable trait of worker-production. With hindsight, it is apparent that Darwin therefore closely anticipated modern, kin-selectionist explanations of worker sterility (E.O. Wilson 1975 p. 117; Alexander et al. 1991; Seger 1991; see also Cronin 1991 p. 298).

Nowadays, biologists recognize three ways in which altruism can evolve by natural selection acting on genes (e.g. Alexander 1974; West-Eberhard 1975; Ridley and Dawkins 1981; Trivers 1985). These are: (1) by kin selection (Hamilton 1963, 1964a,b, 1970, 1971a, 1972); (2) via "delayed benefits" (e.g. Bertram 1982 p. 257); and (3) by manipulation or social (or brood) parasitism. By "delayed benefits" is meant, for example, that an individual may join a social group as a helper in the hope of later inheriting the position of breeder. This kind of route to altruism is therefore sometimes called a "hopeful reproductive" or "mutualism" route (West-Eberhard 1978a; Seger 1991; Section 3.2). Another important kind of delayed benefits altruism occurs when an individual makes a self-sacrifice in return for a future repayment by the beneficiary. This is termed reciprocal altruism (Trivers 1971; Axelrod and Hamilton 1981; Axelrod 1984; Ligon 1991; Dugatkin et al. 1992; Mesterton-Gibbons and Dugatkin 1992). Since reciprocal altruism may occur between relatives, this route to altruism and the kin selection one are not mutually exclusive (e.g. Nee 1989; Ligon 1991; Dugatkin et al. 1994). Over the time-scale of the individual's lifespan, delayed benefits altruism profits the "altruist" individual itself. It is therefore only altruism in a short-term sense (Alexander 1974), and so is more accurately classified as cooperative (see below) rather than altruistic behavior. Social parasitism typically involves the exploitation of preexisting altruistic behavior (generally evolved by kin selection) in a host of a different species, although it can be intraspecific. Parental manipulation of offspring has also been proposed as a promoter of altruism (Alexander 1974), but Section 1.9 argues that this idea falls within kin selection theory.

Another suggested mechanism for the evolution of altruism is group selection (e.g. Wynne-Edwards 1962). Group selection is controversial, partly because the term has been used to describe different processes (Maynard Smith 1976, 1982a; Wade 1978b; D.S. Wilson 1983; Grafen 1984). In one sense (not Wynne-Edward's), it is legitimate to say that kin selection for altruism involves group selection (e.g. Wade 1980). This topic is discussed in Chapter 2. The following sections consider the evolution of altruism by kin selection, and how kin selection theory solved Darwin's problem of altruism.

1.4 Kin Selection and Hamilton's Rule

Kin selection theory was formulated by Hamilton (1963, 1964a,b) as inclusive fitness theory, and termed kin selection by Maynard Smith (1964). Several earlier authors had appreciated that altruistic behavior could evolve via benefits to relatives (e.g. Darwin 1859; Fisher 1930; Haldane 1932, 1955; Williams and Williams 1957). But Hamilton was the first to develop kin selection as an evolutionary principle of far-reaching and radical importance. *Kin selection is the natural selection of genes for social actions via the sharing of these genes between the performer of the action and its relatives (kin).* For other definitions of kin selection, and reviews, see for example West-Eberhard (1975), Kurland (1980), Boorman and Levitt (1980), Michod (1982), and Trivers (1985).

To explain this definition of kin selection. First, by a gene "for" a trait is meant, as is conventional, the gene that makes the difference between whether the trait is shown or not. So if bearers of a gene G on average show a trait T, and nonbearers do not show it, G is the gene "for" trait T (e.g. Dawkins 1979). Next, a social action occurs when an individual (the actor) behaves so as to increase or decrease the personal fitness (survival or offspring production) of other individuals (often termed neighbors, or recipients of the action). As a result, the actor may alter its own survival or offspring production. So genes for social actions are genes whose bearers, on average, perform social acts. It is evident from these definitions that there can be four mutually exclusive types of social action, according to whether the actor and recipient gain or lose personal fitness. These types – cooperation, selfishness, altruism, and spite – are defined in Table 1.1.

Table 1.1

Types of Social Action

| | | Effect on Recipient's Survival or Offspring Number | |
		Gains	Loses
	Gains	Cooperation	Selfishness
Effect on actor's survival or offspring number			
	Loses	Altruism	Spite

After Hamilton (1964a, 1970), Trivers (1985), Gadagkar (1993).

Kin selection theory applies to the evolution of all four social actions. However, in practice it has been mostly used to explain altruism, because this created the greatest puzzle for individual selection theory. Another important feature of the definition of altruism is that it encompasses parental care (Clutton-Brock 1991), because parents who care reduce their own survival (compared to parents who do not care) while promoting that of their young. Correspondingly, since offspring are relatives, kin selection theory underpins the evolution of parental care, and was indeed formulated with parental care in mind (Hamilton 1963, 1964a). However, again, kin selection theory has not been widely invoked to explain parental care, because this seemed adequately explained by individual selection (Dawkins 1979).

Hamilton derived a condition for the spread of a gene for a social action now known as Hamilton's rule (Charnov 1977). Hamilton's rule is the mathematical embodiment of kin selection theory. It involves three terms: the magnitude of the change in the actor's personal fitness; the magnitude of the change in the personal fitness of the recipient; and

BOX 1.1 THE REGRESSION DEFINITION OF RELATEDNESS

Relatedness is a measure of genetic similarity, but its definition forms a complex subject that has been much debated in population genetics (e.g. Hamilton 1970, 1972; Orlove 1975; Michod and Anderson 1979; Michod and Hamilton 1980; Uyenoyama and Feldman 1981; Michod 1982; Pamilo and Crozier 1982; Grafen 1985, 1986, 1991; Bennett 1987a; Moritz and Southwick 1992 p. 237; Queller 1992a, 1994a; Gayley 1993). In kin selection theory (Section 1.4), relatedness is best defined as a regression coefficient (Hamilton 1970, 1972). Consider an outbreeding population consisting of groups or pairs of potential social interactants (potential donors of a social action, or actors, and potential recipients of the action). Regression relatedness at a locus equals the sign and slope of the regression line obtained when the focal gene's average frequency among the potential recipients within groups is regressed, across all the groups, on its within-group frequency in a random potential actor (e.g. Pamilo and Crozier 1982).

This is illustrated in Figure 1.1. The regression of the gene frequencies across groups is positive, so by definition the interactants within the groups are related. If they were not, then in Figure 1.1 relatedness would be represented by the flat dashed line at p on the Y axis. So a relatedness of zero does not imply that individuals have no genes in common, but instead that they share the focal gene with only random probability (p).

Clearly, positive relatedness means that if a potential actor has the focal gene with high (above-average) frequency, then so do the potential recipients, its group-mates. Say A is a gene for a social action (Section 1.4). Then the frequency of A in potential actors also determines the likelihood of the social action being performed (for example, in the extreme case of the frequency of A in actors being zero, the action will obviously not be performed). So positive relatedness at loci for social actions means that social actions are likely to be directed at individuals (recipients) that share the same gene with above-average frequency. This is why kin selection for altruism works, and why relatedness is central to it (Section 1.4).

Note also that the average frequency of gene A in the potential recipients and potential actors is the same, and equals p, the population average gene frequency (Figure 1.1). This condition forms a basic assumption of kin selection theory (e.g. Grafen 1984; Maynard Smith 1989 p. 173). For, as already described, the concept of relatedness does not involve average gene frequencies, but how individuals with frequencies other than the average associate. In kin selection theory, the roles of potential recipient and potential actor are not determined by gene frequency (except that individuals without the gene for the social action will not be actors), but are determined conditionally. In other words, contrary to some misunderstandings of kin selection, there is no genetic difference between actors and recipients (Section 1.7).

BOX 1.1 CONT.

Another important point is that relatedness and the population average gene frequency are independent of one another. In Figure 1.1, p was arbitrarily set fairly low. But p could be any value without the slope and sign of the regression line being different. This reemphasizes the point that relatedness concerns the chance of gene sharing independently of the average chance (set by p).

What is the link between regression relatedness and kinship? This becomes apparent if groups of potential interactants are families. Clearly, regression relatedness will then measure genetic similarity due to ties of pedigree. If parents have a focal gene, their offspring share it. If parents lack the gene, offspring lack it. So being members of the same family is a way of associating with individuals likely to share focal genes. Therefore kinship guarantees positive relatedness.

Let us now prove that applying the regression definition of relatedness yields the familiar relatedness value of $r = 0.5$ for diploid sibs. The method comes from Pamilo and Crozier (1982). Consider a population of outbred diploids divided into family groups. At a focal (autosomal) locus, the frequency of a gene A equals p, and that of its allele a equals q, where $p + q = 1$. The problem is to find the regression relatedness between sibs for gene A.

Each parent in the population can be of genotype AA, Aa, or aa. Assuming Hardy-Weinberg equilibrium, the frequencies of these genotypes are p^2, $2pq$, and q^2 respectively. So each family in the population will be headed by a parental pair in one of the possible combinations (mating types) shown in column 1 of Table 1.2.

The frequencies of these mating types are then calculated as the product of the parental genotype frequencies, summed over the possible ways of achieving each mating type (column 2 in Table 1.2). In addition, Table 1.2 gives, in each kind of family, the genotypes of the offspring and their ratios of abundance (column 3), the frequency of gene A in each potential actor (each genotype of sib) (column 4), and the frequency of A among potential recipients (all sibs) (column 5).

For calculating relatedness, the numbers in column 5 of Table 1.2 (Y values) are regressed on those in column 4 (X values). Each (X, Y) pair must also be weighted by its frequency (f) in the population, which is given by the product of X's genotype frequency among the offspring and the frequency of the family type. For example, the frequency of the pair ($X = 0.5$, $Y = 0.5$) is $1/2(4p^2q^2) + 2p^2q^2 = 4p^2q^2$. The plot of the X and Y values from Table 1.2, with the corresponding f values, is therefore as shown in Figure 1.2 (compare with Figure 1.1).

The next step is to find the slope of the line through the points in Figure 1.2. This equals the regression coefficient of Y on X, which is defined as

BOX 1.1 CONT.

$$\Sigma f(X - \bar{X})(Y - \bar{Y})/\Sigma f(X - \bar{X})^2$$

where $\bar{X}$ and $\bar{Y}$ are the mean values of X and Y respectively. These mean values must both equal p. This is known because all offspring can be potential actors or recipients, and if there is Hardy-Weinberg equilibrium the average frequency of a gene among offspring equals that among their parents, which was set at p. (This reasoning reflects the assertion above that the average frequencies of a focal gene in potential actors and recipients are the same, and equal the gene's population mean frequency.) Using the regression formula with the X, Y, and f values from Table 1.2 and Figure 1.2, and $\bar{X}$ and $\bar{Y}$ values of p, one finds after a lot of algebra that the regression coefficient equals $(1/4)/(1/2) = 0.5$. So sib-sib regression relatedness in diploids is 0.5. Similar calculations would yield all the familiar between-kin relatedness values.

Four more points about regression relatedness need making.

1. From the regression definition, relatedness can in theory take any value, positive or negative (Grafen 1991). But in most cases relatedness lies between 0 (no relatedness) and 1 (clonality). Negative relatedness means that bearers of a focal gene systematically associate with individuals bearing the gene with below-average frequency. The unusual conditions leading to this are discussed by Grafen (1985).

2. Relatedness can also be understood as the average proportion of genes at a locus in the recipient that is identical with any allele at the locus in the actor (Grafen 1986). This equals the probability that a gene in the recipient is present in the actor. The concept of relatedness as a proportion or probability appears frequently in the sociobiological literature. It is usually qualified by the statement that the gene copies in question must be "identical by descent," meaning "descended from the same copy of the gene in their most recent common ancestor" (Dawkins 1979 p. 191). However, the choice of most recent ancestor in the criterion for identity by descent is arbitrary, since every copy of a particular gene must descend from a common ancestral copy if one traces back through enough generations (Seger 1981; Grafen 1985). So the regression definition of relatedness allows biologists to discard the problematic idea of identity by descent (e.g. Gayley 1993).

3. The regression definition also underlies methods of estimating relatedness from gene frequency data obtained from social groups by, for example, electrophoretic allozyme analysis (Pamilo and Crozier 1982; Queller and Goodnight 1989; Pamilo 1989, 1990a). These measures assume that relatedness at neutral allozyme loci estimates relatedness at the putative loci for social actions (under weak selection). This is reasonable because kinship makes average relatedness between any two

BOX 1.1 CONT.

individuals the same across all their loci (Section 1.8).

4. Lastly, on a point of terminology, the regression definition explains why the relatedness between an actor A and a recipient R should strictly be termed b and phrased as the relatedness of R to A, $b_{R.A}$. This is because, by statistical convention, the regression of a dependent variable Y on an independent variable X is denoted $b_{Y.X}$ (Michod and Anderson 1979; Crozier and Pamilo 1980; Pamilo and Crozier 1982).

Boxes 1.3 and 3.1 discuss relatedness further. Box 1.3 describes Grafen's (1985) "geometric view of relatedness," which is formally equivalent to the regression definition. Box 3.1 deals with relatedness levels

the relatedness between the actor and the recipient. Relatedness is a measure of the genetic similarity between two individuals. In kin selection theory, it is formally defined as a regression coefficient, as Box 1.1 describes.

Why does relatedness feature in the gene selectionist explanation of altruism? Imagine a wildebeest carrying a gene that causes it to eat less

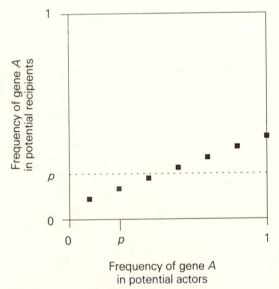

Figure 1.1   The regression relatedness between potential social interactants measured over a number of groups (represented by the black squares) for a gene *A* at a locus. Regression relatedness is given by the sign and slope of the line formed by these points. The average frequency of *A* in the population is *p*.

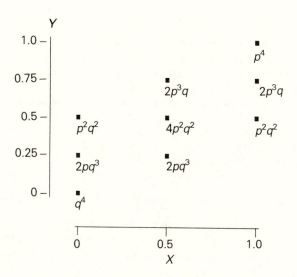

Figure 1.2 The regression of gene frequency in potential recipients (*Y*) on gene frequency in random potential actors (*X*) in a population of families of outbred diploids. The expressions below each point are the weights of the points (*f*).

Table 1.2

Genotype and Gene Frequencies in Families in Randomly Mating Diploids

1 Parents Heading Family	2 Frequency of Family	3 Offspring (sib) Genotypes (with ratios if >1 type)	4 Frequency of A in Random Potential Actor (X)	5 Frequency of A in Potential Recipients (Y)
$AA \times AA$	p^4	AA	1.0	1.0
$AA \times Aa$	$4p^3q$	AA, Aa (1 : 1)	1.0, 0.5	0.75
$AA \times aa$	$2p^2q^2$	Aa	0.5	0.5
$Aa \times Aa$	$4p^2q^2$	AA, Aa, aa (1 : 2 : 1)	1.0, 0.5, 0	0.5
$Aa \times aa$	$4pq^3$	Aa, aa (1 : 1)	0.5, 0	0.25
$aa \times aa$	q^4	aa	0	0

grass. By the earlier definition, this action is altruistic, because food loss decreases the actor's personal fitness and increases that of its neighbors (each remaining wildebeest in the herd has slightly more grass to eat). Will this gene spread through the wildebeest population by kin selection?

The answer is no, because wildebeest herds do not consist of groups of relatives. So the beneficiaries of the abstaining animal are a random section of the population, and share the focal gene with random probability, set by the gene's mean frequency in the population. When these individuals reproduce, their extra offspring will also bear the gene with random probability, because unless other factors intervene the average gene frequencies of offspring equal those of their parents. Therefore, the original animal's action brings about no increase in the frequency of the causative gene.

Suppose, however, that wildebeest lived in stable, sedentary family groups. (West-Eberhard [1975] discusses the influence of herd structure on kin selection.) Then the original animal's action would benefit a non-random section of the population, namely those with an above-average probability of bearing the focal gene via the sharing of genes inherent in kinship. Altruistic restraint in eating could then be positively selected, because the extra young produced as a result would also bear the gene with above-average frequency, resulting in an increase of the gene in the population. This example demonstrates why relatedness is crucial to the action of selection on a gene for altruism (see also Box 1.1). It also foreshadows the later discussion of Grafen's (1985) geometric view of relatedness (Section 1.6; Box 1.3).

In addition, this example illustrates the following point. Let a positive effect of a gene be one that increases its bearer's offspring production relative to the effect of the gene's allele, and a negative effect be one that does the opposite. Then all genes have either a positive effect on their bearer, or a negative effect, or no effect. Similarly, all genes have either a positive effect on the offspring production of a bearer's neighbors, or a negative one, or none. Therefore, excluding genes of null effect, all genes are genes for social actions. Dawkins (1979) made this point with particular reference to genes for selfishness and altruism. So there is nothing "special" about genes for social actions. On the contrary, they are probably very common. The issue is not whether these genes exist, but whether conditions are such that they can spread through populations. Hamilton's rule is the evolutionary principle specifying these conditions. Further, recognizing that genes for social actions potentially encompass many traits suggests that kin selection theory has very broad applicability. In fact, it is relevant whenever conspecific organisms interact in ways that influence personal fitness and are nonrandom with respect to relatedness. This means that kin selection thinking has justifiably entered many fields that are not traditionally part of social biology at all, for example parasitology (Wickler 1976; Frank 1994), population ecology (Charnov and Finerty 1980), botany (Haig 1987; Queller 1989a), microbiology (Shub 1994), and medicine (Westoby 1994).

Box 1.2 gives a simple derivation of Hamilton's rule, especially as it applies to the evolution of altruism. Briefly, Hamilton's rule for altruism is that a gene for altruism spreads if the inequality $rb - c > 0$ is satisfied. In this expression, r is the regression relatedness between altruist and beneficiary (Box 1.1), b is the benefit of altruism in terms of the extra offspring the beneficiary gains, and c is the cost of altruism in terms of the offspring production lost by the altruist (Box 1.2).

In the context of the study of social biology, Hamilton's rule implies that eusocial evolution is influenced both by genetical factors (affecting relatedness) and by ecological and ergonomic ones (affecting benefit and cost). However, the need to take account of both genetics and ecology in the study of eusociality was for a time neglected, despite warnings that it should not be (e.g. West-Eberhard 1975; Evans 1977). This was because, along with kin selection theory, Hamilton (1964b) proposed that haplodiploid sex determination in the Hymenoptera could lead to high relatedness levels especially favorable to eusocial evolution. This "haplodiploidy hypothesis" (Chapter 3) concentrated attention on genetic factors and, as noted by Dawkins (1989a p. 316), led many biologists to believe that kin selection theory required particularly high relatedness to account for eusociality. They therefore doubted the theory whenever such levels were found to be absent. This was a mistake. Kin selection theory states that altruist and beneficiary must show some level of relatedness, as opposed to none at all. In addition, high relatedness will certainly facilitate eusocial evolution. But Hamilton's rule shows that altruism can evolve if relatedness is very low (for example, if r and c are both low but b is very high). The preoccupation with genetic factors also obscured the common ecological factors such as nest-site shortage or a patchy distribution of food that promote eusociality in insects and similar phenomena (helpers at the nest, cooperative breeding) in birds and mammals. However, such ecological parallels are now being increasingly recognized (e.g. Vehrencamp 1979; Andersson 1984; Strassmann and Queller 1989; Brockmann 1990; Pamilo 1991a; Sherman et al. 1991; Crespi 1994).

Hamilton's rule suggests a powerful qualitative way of looking at kin selection. Exactly why does the gene for altruism spread through a population under kin selection? The answer is that the gene causes its bearer to care for individuals having an above-average probability of sharing the same gene, through relatedness. So, in terms of its representation in the next generation, what the gene loses in the sacrifice of the

BOX 1.2 DERIVING HAMILTON'S RULE

This box gives an informal derivation of Hamilton's rule (cf. West-Eberhard 1975; Craig 1979; Section 1.4). Michod (1982) and Grafen (1985) review the large population genetics literature on the rule. Papers formally proving it include Charnov (1977), Wade (1978a), Charlesworth (1980), Aoki (1981), Michod (1982), Queller (1984, 1985, 1992a), Grafen (1985), Goodnight et al. (1992), Gayley (1993), and Sibly (1994). Note that although several proofs deal with a single "gene for altruism," others are general inclusive fitness models or explicitly treat altruism as a quantitative trait influenced by many genes (e.g. Yokoyama and Felsenstein 1978; Boyd and Richerson 1980; Aoki 1982; Cheverud 1985).

Imagine an individual A (the actor) that carries a gene G for a social action. It influences another individual, the recipient R, with the result that A's expected offspring number is changed by m offspring, and R's is changed by n offspring. Let A's young be denoted Ay, and R's young, Ry. In addition, let the regression coefficient of relatedness between two individuals X (actor) and Y (recipient) be $r_{Y.X}$ (Box 1.1).

$$
\begin{array}{ccc}
 & \text{act} & \\
\text{A} & \text{---}\!\!\rightarrow & \text{R} \\
 & & \\
\text{Change by } m & & \text{Change by } n \\
 & & \\
\text{Ay} & & \text{Ry}
\end{array}
$$

The change to gene G's frequency caused by A's social action is $(m \times r_{Ay.A})$ via effects on A's own young, and $(n \times r_{Ry.A})$ via effects on R's young. For example, if m were positive and n negative, m individuals each with a chance of $r_{Ay.A}$ of bearing the gene would be added to the population, and n individuals bearing the gene with a chance of $r_{Ry.A}$ would be lost from it.

By definition, gene G increases in frequency if the total change in its frequency is positive, that is if

$$(m \times r_{Ay.A}) + (n \times r_{Ry.A}) > 0.$$

This is Hamilton's rule in a general form, demonstrating the assertion (Section 1.4) that it applies to the spread of genes for all kinds of social actions. To adapt it for the evolution of altruism, note that in this case, by definition, m is negative (and is conventionally called c for cost), whereas n is positive (and is conventionally called b for benefit). So the rule becomes

$$(-c \times r_{Ay.A}) + (b \times r_{Ry.A}) > 0.$$

BOX 1.2 CONT.

This can be altered as follows. First, divide each side by $r_{Ay.A}$ and rearrange to give $[(r_{Ry.A}/r_{Ay.A}) \times b] - c > 0$. Second, note that, assuming $r_{Ay.A} = r_{Ry.R}$ (parent–offspring relatedness is uniform), then $r_{Ry.A}/r_{Ay.A}$ equals $r_{Ry.A}/r_{Ry.R}$. Finally, note that this last term equals $r_{R.A}$, since the relatedness of A with R ($r_{R.A}$) multiplied by the relatedness of R with R's young ($r_{Ry.R}$) equals the relatedness of A with R's young ($r_{Ry.A}$). Therefore, Hamilton's rule can be rewritten as

$$rb - c > 0$$

where r equals $r_{R.A}$. So, *Hamilton's rule states that a gene for altruism spreads if the genetic profit, calculated as the recipient's extra offspring production multiplied by the relatedness of the altruist and the recipient, minus the genetic loss, which is the fall in the altruist's offspring production, is greater than zero.*

Hamilton's rule is now a well-accepted theorem in population genetics (e.g. Grafen 1985). However, its validity depends on certain assumptions. These include additivity of gene effects (each separate instance of social behavior has the same effect on fitness), weak selection, and outbreeding (e.g. Grafen 1984, 1985; Queller 1984, 1985, 1989a; Maynard Smith 1989 pp. 172–173).

altruistic body it occupies, it can redeem many times over in the enhanced survival or reproduction of the beneficiaries. In effect, then, the gene spreads because it promotes care for copies of itself (Dawkins 1979). This perspective reveals why kin selection theory, as well as explaining the evolution of altruism towards nondescendant kin, also accounts for the evolution of parental care (Hamilton 1963, 1964a; Maynard Smith 1964; Dawkins 1979; Michod 1982). Both cases involve a gene promoting care for copies of itself in related individuals. Both are therefore aspects of the same, gene-selectionist phenomenon, namely kin selection.

Kin selection theory has had a major impact on evolutionary biology as a whole. With the theory, Hamilton solved Darwin's problem of how traits disadvantageous to their bearers in terms of personal fitness could evolve. In other words, he solved the problem that altruism posed for the individual-centered theory of natural selection. Further, Hamilton did this by recasting the problem in gene-level terms, through imagining a gene for altruism and then deriving a population genetics condition for the positive selection of such a gene. This revealed the inadequacy of individualistic natural selection as an explanation of adaptive phenomena compared to the gene-level view. Hamilton's work therefore helped bring about the gene selectionist revolution in evolutionary biology of the past twenty-five years. This has involved the reformulation of natural selection theory in explicitly genetic terms (Section 1.2), most notably by G.C. Williams (1966) and Dawkins (1976). In fact, gene-level thinking was already present in the work of Fisher (1930), Wright (1931), and Haldane (1932), but it has taken modern gene selectionists to make it indispensable, especially in the field of behavior.

Many biologists use gene-centered language in discussing natural selection, and so clearly accept gene selectionism in some sense. But often their phraseology betrays a traditional individual-level perspective. For example, modern texts often make statements like "animals are naturally selected to pass on as many copies of their genes as possible to the next generation." Such statements fail to acknowledge the full radicalism of the gene selectionist revolution (Dawkins 1978, 1979, 1982a,b). This can be illustrated with regard to kin selection theory, where such statements translate into something like "individuals are selected to aid relatives because they share a high proportion of genes with them." This formulation is true in a sense (because a high average fraction of shared genes means that the gene for altruism is shared with high probability). But it fails to capture the essence of kin selection theory, because it distracts the focus from the party for whose benefit altruism has evolved, namely the gene for altruism. Clearly, the individual itself does not benefit from altruism in individual selection terms (by definition). Also, it cannot be taken for granted that all the other genes in the individual apart from the gene for altruism will benefit (Dawkins 1982a; Section 1.8). This means that statements of the above sort are not only theoretically unsatisfactory but also at times misleading. Such statements are admittedly sometimes used as a lazy shorthand, for example "Female workers can propagate their genes more effectively by raising their exceptionally closely related fertile sisters than by producing offspring themselves" (Franks and Bourke 1988 p. 48). But to represent kin selection as an elaborate kind of individual selection hampers understanding by obscuring its gene selectionist logic (Dawkins 1979).

One additional result of this half-hearted acceptance of gene selectionism is that biologists often use individual-level selection as a synonym for gene-level selection. This may not always matter, because in the largely nonsocial, sexual diploid organisms studied by many behavioral ecologists, the predictions of the two perspectives will often coincide. Nevertheless, this synonymizing is undesirable because it again loses the focus on causative genes and because it blurs Dawkins' (1982a,b) important distinction between replicators and vehicles

(Sections 1.2, 2.4). Natural selection is the process by which replicators promote their own survival and replication (Section 1.2). Commonly, this involves genes promoting their survival and replication through the agency of organisms (e.g. Dawkins 1978). This in turn may often involve individuals maximizing their gene transmission, and even their number of offspring, but need not.

Finally, kin selection theory is qualitatively very well supported, in that most animal societies with some degree of altruism are societies of relatives (examples in E.O. Wilson 1975). Certainly this is true in the eusocial insects, including the ants.

1.5 Inclusive Fitness

Now that the basic idea of kin selection has been presented, this and the following four sections discuss a number of issues arising from the theory, beginning with inclusive fitness. Fitness is the measure that evolutionary biologists use to gauge evolutionary success under natural selection within populations. But fitness can be measured in many ways (reviewed by Dawkins 1982a; Endler 1986). For example, field biologists may count the lifetime number of offspring of individuals, whereas population geneticists measure the average offspring production of individuals of given genotypes, relative to some standard. Both these measures refer to offspring number. In devising kin selection theory, Hamilton (1964a,b) recognized that a gene for a social action could be positively or negatively selected due to its effects on relatives other than offspring. Hamilton therefore invented an "inclusive" fitness, so-called because it included not just an individual's genetic representation in offspring, but in other classes of relative as well. Unfortunately, there has been much confusion over the precise definition of inclusive fitness (Grafen 1982, 1984, 1988, 1991; Creel 1990). For example, Grafen (1982) pointed out that inclusive fitness is often incorrectly defined as an individual's number of offspring, plus all the offspring of its relatives, with these being weighted by the appropriate coefficient of relatedness. The problem with this quantity is that it is theoretically almost never-ending (because every member of a population is connected by a pedigree link with a focal individual to some degree). It also fails to predict the direction of gene frequency change, as a measure of fitness should (Grafen 1982).

The correct definition of inclusive fitness is more subtle and complex. However, as Grafen (1984) demonstrated, this need not be worrying, because an inclusive fitness calculation and Hamilton's rule provide two ways of deciding the same thing, namely the direction of selection on a gene for a social action. In inclusive fitness terms, Hamilton's rule is the statement that a gene for a social action spreads if bearers have above-average inclusive fitness (Grafen 1991). But Hamilton's rule has the advantage of a clearer logic (Grafen 1984). There may also be major practical problems in measuring inclusive fitness in the field (Grafen 1982, 1984, 1988; Queller and Strassmann 1989). So this section is designed to clarify inclusive fitness as a concept, not to present a methodology for its measurement.

An exact verbal definition of inclusive fitness can be constructed as follows (Hamilton 1964a; Grafen 1982; Creel 1990). Consider a population of animals. Each will experience what Hamilton (1964a) called the "social environment," which is that part of its environment represented by interactions with conspecific neighbors. These neighbors may increase or decrease a focal individual's offspring production (by help or hindrance), and a focal individual may in turn increase or decrease (by helping or hindering) the offspring production of neighbors.

In this population, one could in principle calculate the total number of extra offspring conferred by help, divided by the total number of individuals (helpers and breeders) in the population. This quantity will be termed the average per capita amount of help. Similarly, one could work out the total number of offspring lost due to hindrance, over the total population size. Let this be the average per capita amount of hindrance. The inclusive fitness of a focal individual then equals the following two-part expression:

(1) the number of offspring produced by the focal individual, minus the average per capita amount of help (if there is helping in the population), plus the average per capita amount of hindrance (if there is hindering in the population); plus

(2) the number of extra offspring the focal individual's help confers on its neighbors (if it does help them), minus the loss in offspring its hindrance causes to its neighbors (if it does hinder them), with these quantities being weighted by the respective coefficients of relatedness between the focal individual and its neighbors, and then summed over all affected neighbors.

Part (1) of this definition corresponds to Hamilton's (1964a p. 8) "production of adult offspring . . . stripped of all components which can be considered as due to the individual's social environment, leaving the fitness which he would express if not exposed to any of the harms or benefits of that environment." Part (2) corresponds to Hamilton's "certain fractions [i.e. coefficients of relatedness] of the quantities of harm and benefit which the individual himself causes to the fitnesses of his neighbours." Note, however, that in Hamilton's verbal definition, the quantity "stripped" from the offspring production of a focal individual receiving help is *all* its offspring due to help, rather than the average per capita amount of help. Creel (1990) noticed that, to achieve consistency with Hamilton's (1964a) mathematical formulation of inclusive fitness, the first part of the verbal definition needed, as above, to be recast in terms of average per capita effects.

To see why this matters, consider a population of breeders and helpers in which the breeders cannot reproduce without help and in which helpers never reproduce. An example would be a population of queens of eusocial Hymenoptera with totally sterile workers. On the original definition, the inclusive fitness of all breeders in this population would be zero (Creel 1990). This is because in calculating part (1) of a breeder's inclusive fitness, all of its offspring would have to be "stripped" away, being entirely due to help received, leaving a part (1) of nothing; and because a breeder gives no help itself, meaning part (2) of its inclusive fitness is also zero. Creel's (1990) correction disposes of this paradoxical result, since now not all offspring due to help received are stripped. Instead, only the average number of extra offspring conferred per head of population is removed.

Creel's (1990) correction also shows why a breeder that produces many reproductive offspring in this kind of population would be fitter than one with few, as seems intuitively reasonable. Part (1) of a breeder's inclusive fitness is the only relevant quantity, since all breeders have a zero part (2). In other words, biologically, the inclusive fitness of breeders will depend on the amount of help they attract. A breeder attracting an above-average amount of help will have a high part (1) fitness when the average per capita amount of help received is stripped away. This means it will have above-average inclusive fitness. On the other hand, a breeder attracting a below-average amount of help will have a low part (1) fitness after the average per capita amount of help is removed, leaving a below-average inclusive fitness. Therefore, Creel's (1990) modified definition correctly gives productive breeders higher inclusive fitness than less productive ones, even though every breeder depends on help for all its offspring production.

Similar considerations apply to the helpers in this population. Now the relevant quantity is part (2) of inclusive fitness, since all sterile helpers have the same part (1) (they have no offspring and the other terms are all average values). If a helper confers a lot of assistance, many relatives bearing the helping gene will be added to the population. If it confers little, the opposite will be the case. Therefore, as well as giving and justifying a definition of inclusive fitness, this section has now also shown that for both breeders and helpers inclusive fitness is a satisfactory measure of fitness, in that it correctly predicts the direction

of gene frequency change. In effect, it has confirmed the earlier assertion that in the social environment an inclusive fitness calculation is
equivalent to applying Hamilton's rule for determining the direction of
selection on genes for social actions.

1.6 Kin Selection Works at All Gene Frequencies

An important question raised by recognizing kin selection to be an
explicitly gene-level theory is the following. How does kin selection for
altruism work when a gene for altruism becomes very common in a
population? This appears to lead to universal, indiscriminate altruism,
and yet phenomena such as nestmate recognition among ants (Section
7.3) show that this does not occur. But the logic described earlier suggests that, in a population with a high frequency of a gene for altruism,
every individual should regard all others as related by a level approach-
ing 1.0 at the altruism locus, and should therefore help them regardless
of genealogical ties. So it seems that relatedness becomes a redundant
concept at high frequencies of the gene for altruism, because all members of the population now share the gene with high probability.

There are two ways to resolve this problem, the first being an argument made by Dawkins (1979). Any gene for helping is likely to operate via a behavioral "rule of thumb," such as – in the social insects –
"rear sexual forms in the nest in which you live." The statistical association between nest-sharing and relatedness means that a rule like this
will usually lead to kin being aided. Therefore, when a gene for helping
becomes very frequent, or even fixed in a population, the rule also
becomes universal. The result is a population in which help is still
directed at nestmates (kin), so help remains discriminating.

The second, more rigorous approach to this question is to demonstrate theoretically that kin selection can operate at all frequencies of
the gene for altruism, and that relatedness remains meaningful at all
frequencies. This was shown to be the case by Hamilton (1964a) and is
implicit in the regression definition of relatedness (Box 1.1). However,
the point is most clearly made using Grafen's (1985) "geometric view of
relatedness", as Box 1.3 explains in full. In brief, regression relatedness
can be regarded as the extra probability, over and above the average
"background" probability (set by the gene's mean frequency in the
population), that kinship adds to two individuals' chances of sharing a
focal gene. Therefore, relatedness remains a valid concept whether the
background level is high or low. From this follows the conclusion that
kin selection can operate at all frequencies of a gene for a social action
(Box 1.3).

BOX 1.3 THE GEOMETRIC VIEW OF RELATEDNESS

Grafen (1985) presented a formal proof of Hamilton's rule as a theorem in population genetics (Boxes 1.1, 1.2). This involved proving the identity between "Hamilton's relatedness" – the r term in Hamilton's rule – and "pedigree relatedness" – the concept of relatedness involving kinship. The geometric view of relatedness is Grafen's (1985) device for proving this identity, and hence the validity of Hamilton's rule. Both concepts of relatedness, as described below, also turn out to be strictly equivalent to regression relatedness (Box 1.1). This box explains Grafen's (1985) argument (see also Maynard Smith 1989 p. 172; Grafen 1991).

Hamilton's rule states that a gene for altruism spreads if $rb - c > 0$ (Box 1.2). (The present box refers specifically to a gene for altruism, but the proof applies to genes for social actions in general.) The relatedness term (r) can be viewed as the fractional, genetic valuation that the actor (altruist) places on one offspring of the recipient, compared to one offspring of its own (e.g. Hamilton 1972). For example, if r is 0.5, the actor would sacrifice one offspring to save at least two offspring of the recipient. So, to the actor, each offspring of the recipient is genetically worth 0.5 of one of its own. Generalizing, Hamilton's rule says that the actor values one offspring of the recipient at an "rth" of one of its own. This "r" is "Hamilton's r" and is the concept of relatedness needed for Hamilton's rule to work.

Hamilton's r can be represented geometrically by plotting on a one-dimensional scale the frequency of a gene for altruism in three classes of individual. The first of these is an actor. Let its gene frequency equal A and be set at $A = 1$. This is because the actor must have the gene for altruism by definition (i.e. $A > 0$), and for simplicity is assumed to be, say, a homozygous diploid (this does not affect the conclusions). The other two kinds of individual are a recipient (gene frequency $= R$), and a random member of the population. The gene frequency of the latter (μ), by definition, equals the average frequency in the population. No assumptions are made about the values of R and μ, except that $\mu < R < A$. Given this, λ is defined as the fraction of the distance between μ and A at which R lies, i.e. $\lambda = (R - \mu)/(A - \mu)$. When λ equals, say, 1/4, the plot is as shown in Figure 1.3.

Now imagine an individual at R which has, say, eight offspring. This will increase the frequency of the altruist gene, since R is higher than the gene's average frequency (μ), and the offspring of individuals at R will also contain the gene with frequency R. This increase can be decomposed into two elements: when $\lambda = 1/4$, an individual at R having eight offspring has the same positive effect on the altruism gene's frequency as if an individual at μ had six offspring and one at A had two offspring. To see why, consider Grafen's (1985) metaphor of the gene frequency scale as a lever

BOX 1.3 CONT.

with its fulcrum at μ. The turning moment is the same whether there are eight bricks at R, or six bricks at μ and two at A. For the six bricks at μ have no effect on the lever's tilt, being directly above the fulcrum. This leaves the equivalency between eight bricks at R and two bricks four times the distance from the fulcrum, at A (Figure 1.4).

Generalizing, an individual at R having N offspring when R is λ of the way from μ to A has the same positive effect on gene frequency as an individual at μ having $(1 - \lambda)N$ offspring, and one at A having the remaining λN offspring. For, as in the lever analogy, reproduction by the notional individual at μ has no effect on gene frequency, because it only involves adding to the population individuals with the population mean frequency. Therefore, production by an individual at R of N offspring increases gene frequency by the same amount as the production by an individual at A of λN offspring, because A is relatively $1/\lambda$ times higher than the average gene frequency than R.

Now set N to equal 1. It follows that if an individual at R has one offspring when it is λ of the way from μ to A, this raises gene frequency by as much as an individual at A having λ offspring. Therefore λ equals the fractional, genetic valuation that an individual at A places on the production of one offspring by an individual at R. So λ equals "Hamilton's r." Or, by the geometric view, relatedness equals $(R - \mu)/(A - \mu)$. In words, *relatedness equals the frequency of a focal gene in one individual (the potential recipient) minus the gene's population mean frequency, divided by the frequency of the gene in another individual (the potential actor) minus the gene's population mean frequency.*

It now needs to be shown that this is equivalent to the pedigree definition of relatedness. Consider a sexual diploid species with random breeding. What is "geometric" parent–offspring relatedness in this species? Say an individual bearing a focal gene with frequency A mates and has offspring. This individual's mate must, on average, have a gene frequency of μ, the population mean, because of the assumption of random breeding. The gene frequency of a typical offspring (call it R) will be the average of the parental frequencies, i.e. $(\mu + A)/2$. This is because in sexual diploids an offspring is genetically half its mother and half its father, since it receives an equal number of chromosomes from each parent. (Note that R will therefore lie between μ and A, matching the earlier assumption.) So geometric parent–offspring relatedness equals $(R - \mu)/(A - \mu)$ $= [(\mu + A)/2 - \mu]/(A - \mu) = 0.5$. This is also the pedigree relatedness between parent and offspring in diploids. So geometric relatedness and pedigree relatedness are the same (Grafen 1985).

To recap, the geometric view of relatedness proves the identity of the relatedness term in Hamilton's rule and the relatedness deduced from

BOX 1.3 CONT.

pedigrees. Furthermore, it is the concept of relatedness that makes Hamilton's rule work, in that a potential altruist should be selected to value a potential beneficiary's reproduction according to their geometric relatedness. So the geometric view also provides a proof of Hamilton's rule.

On top of this, regression relatedness (Box 1.1) and geometric relatedness are the same, as Grafen (1985) formally demonstrated. One way to see this equivalence is to recall from Box 1.1 that regression relatedness also refers to a deviation from an average gene frequency. In addition, Box 1.1 showed that regression relatedness is the same as pedigree relatedness (and hence geometric relatedness). For regression relatedness is obtained by regressing gene frequencies across groups of potential interactants. So if these groups are families, it measures genetic similarity due to kinship. Lastly, regression relatedness equals the genetic valuation an actor places on a recipient's reproduction (and so again is the same as geometric relatedness). To understand this, remember that regression relatedness can be defined as the average proportion of genes at a locus in the recipient that is identical with any allele at the locus in the actor (Grafen 1986; Box 1.1). When they have young, both recipient and actor will pass on a random allele at a locus to each offspring. Therefore, regression relatedness must give the actor's valuation of the recipient as an offspring producer relative to the actor's valuation of itself as an offspring producer (Hamilton 1972; Grafen 1986).

Summing up, regression relatedness, pedigree relatedness, and the concept of relatedness as a genetic exchange rate are all the same thing, and are equivalent to geometric relatedness. A valuable feature of the geometric view of relatedness is to bring these connections to light.

Finally, the geometric view of relatedness shows that Hamilton's rule applies at all frequencies of a gene for altruism (Section 1.6). In deriving the formula for geometric relatedness, no assumption was made about the value of μ, the gene's mean population frequency, other than $\mu < R < A$. Imagine that relatedness equals 0.5, and that μ can freely slide up and down the gene frequency scale from 0 to 1. Then, even if μ is very high (close to A at 1), relatedness can still be represented as half of the (short) way from μ to A. Therefore, it remains a meaningful concept at high gene frequencies. Put more formally, the expression $(R - \mu)/(A - \mu)$ has a positive value for all values of μ provided that $0 < \mu < R < A \leq 1$. So, at all frequencies of a gene for altruism, relatedness guarantees an above-average chance of gene-sharing between potential interactants. Consequently, at high gene frequencies, altruists should still value the reproduction of relatives more highly than that of random members of the population.

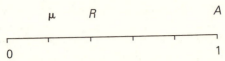

Figure 1.3   A gene frequency plot when the gene frequency in a recipient (*R*) lies a quarter of the way between the average gene frequency (μ) and the gene frequency in an actor (*A*, set at 1). (From Grafen 1985; by permission of Oxford University Press)

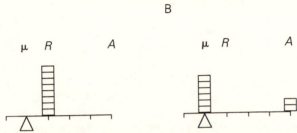

Figure 1.4   A diagram of Grafen's (1985, 1991) metaphor of gene frequency change as a system of weights on a lever. *R* lies a quarter of the way from the fulcrum (μ) to *A*. So placing eight bricks at *R* (Figure 1.4A) has the same effect as placing two bricks at *A* and six at the fulcrum (Figure 1.4B). (From Grafen 1991; by permission of Blackwell Scientific Publications Ltd)

1.7 Gene Expression in Kin Selection Theory

This section continues to explore facets of kin selection by considering the expression (penetrance) of genes for altruism under the theory. The essential point is that carriers of a kin-selected gene for altruism must include *both* helper phenotypes, such as workers, *and* reproducer phenotypes, such as queens. The reason is as follows. Hamilton's rule, as it applies to altruism, is about the spread through a population of a gene causing altruistic helping behavior in the gene bearers. But no gene can spread at all if no carriers of the gene ever have offspring. Therefore, if kin selection is to operate, some bearers of the gene for altruism must be reproducers. In other words, bearers must include both helpers and reproducers. This could come about in various ways. For example, the gene could have facultative expression (some adult bearers help, and others reproduce), or it could have obligate expression but be expressed at different life stages (for example, young bearers help and old ones reproduce). This point has been made by several previous authors (e.g. Orlove 1975; Charlesworth 1978; Crozier 1979; Dawkins 1979; Grafen 1984, 1985; Hamilton 1987a; Parker 1989).

To put it another way, imagine a gene for altruism with obligate expression such that all carriers were always helpers and none were ever reproducers. Then if bearers helped reproductive individuals that they met, they would be directing their aid towards nonbearers, so such a gene could never spread. Kin selection works because bearers of the gene for altruism help reproducers who are also bearers of the gene (with above-average frequency). Such reproducers must therefore bear the gene but not express it, for example because the gene is expressed facultatively.

This reasoning is present in the scenario for the origin of eusociality in ants proposed in Section 3.3. Here one of its implications needs highlighting. If a facultative helping–reproducing gene spreads through a population of solitary nesters to fixation, the result is a population that is entirely social but in which the helper–reproducer dichotomy (worker–queen dichotomy) is nongenetic. In other words, kin selection reasoning implies that sexual-worker caste determination in eusocial insects should be nongenetic (Orlove 1975; Crozier 1979, 1992). This is indeed the case throughout the termites and the social Hymenoptera (e.g. Craig and Crozier 1978), with the possible exceptions of the stingless bees (Michener 1974) and two ant species (Winter and Buschinger 1986; Heinze and Buschinger 1989), where genetic control of caste is presumably secondary (Crozier 1977, 1979). In all other eusocial insects the factors affecting queen–worker caste determination are numerous and complex (Watson et al. 1985; Hölldobler and Wilson 1990), but, as far as is known, exclusively environmental.

These arguments reflect the assumption in kin selection theory that potential actors and recipients do not differ in their average gene frequencies (Box 1.1). They also rebut West-Eberhard's (1988) assertion that nongenetic caste determination in social insects is a contradiction of "allelic" versions of kin selection theory (see also Crozier 1992; West-Eberhard 1992). Lastly, they show that West-Eberhard's (1987a,b, 1988) "epigenetic" theory of the evolution of insect sociality is not unique in invoking the facultative nature of worker behavior. This last point is further discussed in Section 3.4.

1.8 The Gene for Altruism and the Interests of the Rest of the Genome

As has been emphasized, kin selection theory is an explicitly gene-level theory. It concerns the fate of genes for social actions – altruism in the present context. This raises the following question. If a gene for altruism is being positively selected, how should selection act on all the other genes in the genome of the altruist? A gene-level perspective cannot automatically assume that these genes and the altruism gene share the

same interests. So it needs explaining why in fact each individual altruist usually acts as a unit.

There is a reason why there is generally no conflict of interest within
individuals between a gene for altruism and other genes. This is that, on
average, the relatedness values between individuals at all loci on chromosomes sharing the same inheritance system are equal (Dawkins
1982a p. 149; Grafen 1984, 1985; Trivers 1985 p. 128). For example, in
sexual diploids with chromosomal sex determination, all the chromosomes apart from the sex chromosomes (the autosomes) share a common inheritance system. Therefore, on average, the relatedness
between two individuals is equal for all loci on the autosomes. To see
why, imagine the chain of pedigree joining two individuals. The reason
the probability of sharing genes changes at any link is meiosis. Further,
at any meiosis, the probability of gene-sharing between parent and offspring is the same for all autosomal loci – one half. Since all loci in an
individual are connected in the pedigree chain to all corresponding loci
in another individual by the same number of links with the same probability of gene-sharing at each link, average relatedness between all loci
must be equal.

This reasoning means that if at an autosomal locus for altruism the
relatedness between the altruist and recipient is, say, 0.25, then relatedness at every other autosomal locus is also 0.25, on average. So if
Hamilton's condition ($rb - c > 0$) is fulfilled for the gene for altruism, it
is also fulfilled at every other locus. Therefore, complete agreement
exists over the performance of the altruistic behavior between the rest
of the autosomal genome and the gene for altruism. Put another way,
when selection favors the altruism gene, all the other autosomal genes
are indirect beneficiaries, and to the same extent, because all have the
same chance of being present in the recipient. An altruist should therefore behave as a unit, as is observed. Of course, individuals also behave
as units in most behavior, and the reason is the same. All genes on chromosomes having the same inheritance system share an equal probability
(0.5 for autosomes) of being present in an individual's gametes, and
hence in its offspring, because of a fair meiosis (Dawkins 1982a p. 135).
Therefore, gene selectionism predicts that all autosomal genes should
cooperate fully for the survival and reproduction of the body they
inhabit, as long as meiosis remains fair (Section 2.5). (For simplicity,
this argument ignores the possibility of genomic imprinting [Box 4.2].)

The condition that only genes on the same kind of chromosome share
interests becomes important when exceptions are considered. Say a
gene for altruism is on the Y chromosome of a species with human-style
genetics (a diploid with chromosomal sex determination in which XX
individuals are females, XY ones males, with no crossing-over between

the X and Y chromosomes). Then the relatedness between brothers at the altruism locus would be $r = 1.0$ (because each brother inherits the same Y chromosome from the father), but at all the autosomal loci relatedness would be the normal value for diploid sibs of 0.5. Therefore, the gene for altruism would be selected to make bearers more helpful to brothers than would be favored by selection on autosomal genes. For example, if the gene were for suicide to save brothers' lives, it would be selected if it caused its carrier to die on behalf of one or more brothers. But the autosomal genes would require two or more brothers to be saved, and dying for only one would cause more autosomal genes to be lost (1 unit) than gained (0.5 units). Mutant genes ("modifiers") at autosomal loci that rendered the "outlaw" altruism gene ineffective would therefore be selected (Alexander and Borgia 1978; Dawkins 1982a).

These considerations are not simply speculation. Selection for autosomal modifiers may account for the characteristic small size and inertness of the Y chromosome. This was suggested by Alexander and Borgia (1978) and Trivers (1985 p. 136), following Hamilton's argument attributing Y inertness to modifiers for suppressing Y meiotic-drive genes (Hamilton 1967; Leigh 1977; Cosmides and Tooby 1981; Dawkins 1982a p. 140). Whether these particular arguments are correct, the general point is that if relatedness between individuals differs systematically among loci, within-genome conflict is expected (Cosmides and Tooby 1981; Dawkins 1982a; Grafen 1991; Hurst et al. 1992). So the cooperation of genes for altruism and other genes cannot always be taken for granted. On the other hand, since autosomes outnumber sex chromosomes, genes will usually be selected to cooperate in social behavior, conferring the individual's characteristic unity of purpose. Therefore gene selectionism does not deny the integrity of individuals. But it does show that it flows from gene selectionist principles themselves, and according to these same principles can break down. So the integrity of individuals cannot be regarded as a given fact of nature (Dawkins 1982a, 1990; Hamilton 1987a; Section 2.5).

A final point is that kinship is about the only phenomenon in nature likely to bring about uniform average relatednesses between individuals over all loci in the genome (Grafen 1985). For example, say individuals shared a genetic preference for a particular habitat (e.g. Hamilton 1975a). Then those meeting at the habitat would be related at the locus for habitat preference, but not at their other loci, since they would not be kin. So altruism between individuals within the habitat could not be achieved without provoking within-genome conflict, despite the interacting individuals having greater than average genetic similarity. This is why relatedness due to kinship is by far the most plausible means of achieving the genetic similarity represented by the relatedness term in Hamilton's rule (Dawkins 1982a; Grafen 1985).

1.9 Parental Manipulation Theory

Alexander (1974) proposed a parental manipulation hypothesis for the evolution of altruism in social insects. This suggests that workers help raise the queen's young because the queen manipulates them into doing so. For example, she may underfeed them (making them poor potential foundresses) or dominate them and so prevent their reproduction directly (Alexander 1974; Michener and Brothers 1974). This theory has strong appeal because social insect queens do appear to hinder worker reproduction (e.g. Brian 1980; Fletcher and Ross 1985; Bourke 1988a; Section 7.4).

Parental manipulation theory has frequently been presented as an alternative to kin selection as an explanation for the evolution of eusociality. Alexander (1974) himself maintained that parental manipulation was more powerful than kin selection alone because parents had an intrinsic, genetic advantage in conflicts of interest with offspring. His argument was that any gene in an offspring causing it to rebel against the parent would disadvantage the offspring when it became a parent itself, since the gene would be present in its own offspring. So only lineages with pliable offspring would persist, and offspring altruism could be best explained by parental manipulation.

However, this argument contained a flaw, pointed out by Dawkins (1976 pp. 145–148), which is that it works equally in reverse. The reversed argument is that a gene causing an adult to act against the interests of an offspring will be counter-selected when it is present in juveniles, through these juveniles having a parent bearing the gene. This suggests that offspring should always win in parent–offspring conflicts, the opposite of Alexander's conclusion. In reality, there is no inbuilt genetic advantage to either party, because there is no genetic asymmetry between them (parent–offspring relatedness equals offspring–parent relatedness) (Dawkins 1976). This point has since been conceded by Alexander (1979 pp. 38–39). Population genetics models have also shown that genes which make offspring behave selfishly at the parents' expense can spread (Blick 1977; Parker and Macnair 1978; Stamps et al. 1978). The basic reason is that, although a gene for offspring selfishness reduces litter size (parental fitness), it may be disproportionately frequent within the litter (Godfray and Parker 1992). So the status of parental manipulation as an independent evolutionary principle has been undermined. In general, models suggest that the outcome of parent–offspring conflicts (which party wins, or whether a compromise results) varies with the costs to offspring and parents of, respectively, behaving selfishly and opposing the offsprings' demands (e.g. Trivers 1974; Blick 1977; Macnair and Parker 1978, 1979; Parker and Macnair 1978, 1979; Stamps et al. 1978; Harpending 1979; Parker 1985; reviewed by Clutton-Brock 1991) (Section 7.2).

Parental manipulation in the form of queen hostility to worker reproduction, or queen underfeeding of female larvae, is certainly feasible in social Hymenoptera. But it should not be considered to lie outside kin selection theory. First, as pointed out by several authors (e.g. West-Eberhard 1975, 1981; Maynard Smith 1982a; Michod 1982; Vehrencamp 1983a; Andersson 1984; Sudd and Franks 1987 p. 7), parental manipulation by definition involves social actions performed among kin, and so falls within the theory. Second, if parental manipulation occurs, offspring will not necessarily accept it passively. They may be selected to resist. The strength of this selection will then depend on the terms of kin selection theory, namely benefit, cost, and relatedness (Trivers 1974; Charlesworth 1978; Charnov 1978a; Craig 1979; Crozier 1979, 1982; Metcalf 1980a). The possibility of offspring counter-manipulation is particularly important given that there is no general (genetic) reason to assume parental victory. In fact, parental manipulation is best seen as part of kin conflict theory (Trivers 1974; Trivers and Hare 1976; Yamamura and Higashi 1992), the branch of kin selection theory dealing with evolutionary conflicts of interest between relatives. In this context, parental manipulation may well have been important in insect social evolution (Sections 3.9, 7.4).

1.10 Conclusion

Kin selection theory is not just a fashion in evolutionary biology, as it is sometimes perceived to be. Kin selection is a logical consequence of natural selection (Dawkins 1979; Ridley and Dawkins 1981). This is because kin selection follows from considering how natural selection will act on genes for social actions directed at co-bearers of these genes. In fact kin selection arguably includes conventional natural selection, since it is hard to see how a system of natural selection could exist without the appearance of the feedback phenomena implicit in the idea, from kin selection, that genes can influence their own spread via effects on copies of themselves in other individuals. The main purpose of this chapter has been to argue for the fundamental importance of kin selection theory.

A consequence of viewing kin selection as a logical corollary of natural selection theory is the conclusion that it is a deep-level theory, rather than the "local" hypothesis it is sometimes treated as. Deep-level theories are both more powerful, and harder to prove or disprove, than more superficial ones. This does not mean that kin selection is unfalsifiable (universal, unreciprocated altruism towards nonrelatives would falsify it). The point is that many ideas which previously have been regarded as inimical to kin selection theory can instead be incorporated into its deep-level gene-selectionist logic. This chapter has argued that this is the case for parental manipulation theory. The next one extends the argument to group selection, colony-level selection, and levels-of-selection theory.

1.11 Summary

1. Darwin's theory of natural selection forms the basis of the evolutionary study of animal behavior and sociality. The fundamental element of natural selection theory, the "unit of self-interest," is Dawkins' (1976, 1982a) replicator. Replicators make copies of themselves, head potentially immortal lineages, faithfully transmit mutations, and store information. Adaptation arises from the cumulative selection of favorable traits in replicators or mutually stable sets of replicators. Adaptations are therefore for the replicators' benefit. Genes are replicators. Genes inhabit "vehicles" such as bodies and groups over which they have causal priority. Adaptations therefore primarily benefit genes, and only benefit bodies and groups incidentally.

2. The eusocial insects (those with a reproductive division of labor) exhibit biological altruism, which involves one individual reducing its survival or offspring production and increasing those of a neighbor. Altruism poses a problem for individual-centered natural selection theory, which predicts that individuals should maximize their offspring number. This problem was solved by Hamilton's theory of kin selection, which is a replicator (gene)-centered theory.

3. Kin selection is the natural selection of genes for social actions (genes affecting the offspring output of their bearers and their bearers' neighbors) as a result of the sharing of genes among relatives. Kin selection theory states that a gene for altruism spreads if the condition, (relatedness) × (benefit) − (cost) > 0, is met (Hamilton's rule). Here, benefit means the number of extra offspring gained by the beneficiary and cost means the number of offspring lost by the altruist. Relatedness is formally defined as the regression coefficient obtained when the gene frequency among potential recipients of a social action is regressed across groups of interactants against the gene frequency among the

potential actors. Informally, relatedness is the probability of gene sharing between individuals independently of the average probability, which
is set by the gene's average frequency in the population. Therefore,
according to kin selection theory, altruism can evolve because a gene
for kin altruism promotes care to individuals with an above-average
probability of sharing the same gene, through relatedness. That is, the
gene promotes care for copies of itself.

4. Hamilton's rule implies that both genetic and ecological factors are
important in explaining the evolution of altruism and eusociality.
Hamilton's rule can also be expressed as the statement that a gene for
altruism spreads if bearers have above-average inclusive fitness. A focal
individual's inclusive fitness is its personal fitness (offspring number) in
the absence of the average social effects of conspecifics, plus the individual's influence on neighbors' fitnesses devalued by its relatedness
with neighbors.

5. Kin selection can operate at high frequencies of a gene for altruism
because kinship always guarantees above-average levels of gene-
sharing, wherever the average lies. This principle follows from
Grafen's (1985) "geometric view of relatedness," which both demonstrates the validity of Hamilton's rule and emphasizes that relatedness
is measured relative to a gene's mean frequency in a population.
Geometric relatedness and the regression definition of relatedness are
formally equivalent. Kin selection also requires that bearers of genes
for altruism include both altruists and reproducers, and is therefore
consistent with nongenetic caste determination in eusocial insects.

6. Genes for altruism and other genes within individuals are selected
to cooperate if they lie on chromosomes with the same inheritance system, because this means that the average relatedness between all corresponding loci on these chromosomes in different individuals is equal.
Kinship is about the only agent producing uniform average genetic similarity across all loci. The theory that altruism among offspring arises
from parental manipulation is not an alternative to kin selection theory,
but part of it, namely kin conflict theory.

7. Kin selection is a fundamental theory in evolutionary biology, and
inseparable from natural selection theory. It is applicable whenever
conspecific individuals interact in ways that affect offspring output and
are nonrandom with respect to relatedness. In principle, it applies to all
living things. Many other selectionist principles can be assimilated by its
deep-seated logic.

2 Levels-of-selection Theory, Gene Selectionism, and Insect Societies

2.1 Introduction

The previous chapter discussed natural selection, kin selection, and the evolution of altruism in terms of gene selectionism. However, gene selectionist language is not universal in the scientific literature on social insects. For example, many authors invoke "colony-level selection" as a competing mode of evolution. This chapter examines whether colony-level selection is a legitimate concept, and whether it differs substantively from gene selection. To anticipate, our argument will be that there is no fundamental clash between gene and colony-level selection, with certain important qualifications.

Sections 2.2 and 2.3 reach this conclusion by considering the relation between colony-level, group, kin, and gene selection. Next, Section 2.4 puts the discussion in the wider context of "levels-of-selection" theory. Section 2.5 examines whether conflicts of interest in insect societies and other groupings can be analyzed with a single, shared set of principles. Lastly, Section 2.6 relates the chapter's findings to the idea, closely connected with colony-level selection, that insect societies are "superorganisms." The overall aim is to argue that recognizing ties between these concepts avoids unnecessary controversy over which is correct. Instead, they may represent correct, but different, ways of viewing gene selectionist natural selection theory. Similar conclusions, especially in the kin versus group selection controversy, have recently been reached by other authors (e.g. Wilson and Sober 1989; Ross and Carpenter 1991a; Queller 1992a,b; Ratnieks and Reeve 1992; Dugatkin and Reeve 1994).

2.2 Colony-level, Group, Kin, and Gene Selection

The concept of colony-level selection occurs in the social insect literature in both empirical and theoretical contexts (e.g. Darwin 1859; Sturtevant 1938; Wynne-Edwards 1962; Michener 1964; E.O. Wilson

1966, 1968, 1985a, 1990; Lin and Michener 1972; Crozier 1977; Oster and Wilson 1978; West-Eberhard 1981; Myles and Nutting 1988; Hillesheim et al. 1989; Page et al. 1989a,b; Robinson and Page 1989; Hölldobler and Wilson 1990 p. 336; Frumhoff and Ward 1992; Moritz and Southwick 1992). In addition, genetic (allele frequency) models of colony-level selection have been presented by, for example, Williams and Williams (1957), Crozier and Consul (1976), Owen (1986, 1989), and Moritz (1989). But colony-level selection appears to have several meanings. For example, it is sometimes equated with selection on queens (E.O. Wilson 1966; Oster and Wilson 1978 p. 98), or with the differential survival, multiplication, and extinction of whole colonies (West-Eberhard 1981). Alternatively, it is a force limiting the spread of selfishness (e.g. Sturtevant 1938; Hillesheim et al. 1989), or selection for features contributing to colony productivity (E.O. Wilson 1968), especially when these are properties expressed by whole colonies such as caste frequency distributions (Oster and Wilson 1978; Robinson and Page 1989; Frumhoff and Ward 1992). Owen (1989) explicitly defined colony-level selection as selection on the genotypes of the founding pair of the colony, acting via colony features determined by the genotypes of their worker offspring. Many authors who invoke colony-level selection also recognize the possibility of within-colony or "individual-level" selection. This is held to arise from the presence of multiple queens (Sturtevant 1938), reproductive workers (West-Eberhard 1981; Hillesheim et al. 1989; Frumhoff and Ward 1992; Tsuji 1994), nepotism in sexual rearing by workers (Page et al. 1989a), or more generally from genetic conflicts of interest within colonies (e.g. Leigh 1991; Ratnieks and Reeve 1992).

This and the following sections consider the validity of colony-level selection. The starting point is that colony-level selection is a special case of group selection, since a colony is a group of individuals. So much of this section examines the legitimacy of group selection. It does this in the context of the evolution of altruism, since this is the key theoretical issue. In addition, group selection's claim to explain altruism is the major source of controversy. This discussion concludes that a form of group selection (and hence colony-level selection) exists that is compatible with kin selection. Section 2.3 then shows how previous usages of colony-level selection, despite their superficial differences, fall within the framework that will be set out.

To begin, then, with group selection. This means entering the minefield of the group selection controversy, which has preoccupied evolutionary biologists for the past thirty years. For a comprehensive survey, see the reviews of Wynne-Edwards (1962, 1986, 1993), G.C. Williams (1966, 1971, 1992), Lewontin (1970), E.O. Wilson (1975), Dawkins

(1976), Maynard Smith (1976, 1982a), Alexander and Borgia (1978), Wade (1978b), Starr (1979), Boorman and Levitt (1980), Grafen (1980, 1984), Uyenoyama and Feldman (1980), D.S. Wilson (1980, 1983, 1990), B.J. Williams (1981), Sober (1984a), Nunney (1985a), Trivers (1985), Pollock (1989), Brandon (1990), and Dugatkin and Reeve (1994). This section concentrates on the area where the controversy overlaps with social insect biology. However, a useful way of doing this is with an account of the controversy.

The first requirement is to dispose of a type of "group selection" that few now endorse, or which is only supported in qualified form (Leigh 1977, 1991; D.S. Wilson 1976, 1980; Wilson and Sober 1989). This is the idea that animals might act for the good of their species, community, or ecosystem. Naive versions of this idea are pre-Darwinian in origin and in flavour, in that they do not recognize selectionism of any kind, at least below the species level. For this reason, species-advantage reasoning was heavily criticized from an individual selectionist perspective by Fisher (1958 p. 49) and from a gene selectionist one by G.C. Williams (1966). Williams called the idea he was attacking "biotic adaptation," but he and others (e.g. Dawkins 1976; Trivers 1985) also at times classified it as group selectionism. However, group selection is better understood as the theory that animals might act for the good of their group or local population. Lumping group selection with species-advantage theory unfairly tarnishes it by association.

The idea that launched the group selection controversy was Wynne-Edwards' (1962, 1963) group selection theory of population regulation. Wynne-Edwards suggested that animal populations evolved restrained breeding to prevent themselves from over-exploiting their food supply and so going extinct. The process worked because populations were structured into groups, and groups of altruistically restrained breeders outlived groups of selfish, unrestrained breeders. Wynne-Edwards' theory invoked no explicit role for relatedness. It therefore suggested that groups of altruists can evolve without relatedness, whereas nowadays this is usually considered highly unlikely, unless extra factors are involved such as reciprocity (Section 1.3).

Wynne-Edwards' theory was attacked on several grounds (e.g. Maynard Smith 1964, 1976; Lack 1966; G.C. Williams 1966; Dawkins 1976; Trivers 1985). One argument was that his examples of altruistic population regulation were illusory, and that animals which seemed to be curtailing their breeding were in fact reproducing maximally given the trade-offs imposed by their conditions (e.g. Perrins 1964; Lack 1966). Another was that group selection for altruism would be a slow and weak process compared to individual selection for selfishness, since the first process occurs on a timescale of group extinctions, whereas the second operates on the shorter timescale of organismic death (e.g. Lewontin 1970; Ridley and Dawkins 1981; Ridley 1985 p. 48). Lastly, genetic models, starting with Maynard Smith's (1964) "haystack" model, suggested that to achieve groups of altruists that were stable against invasion by selfish immigrants required small, isolated, poorly-mixed groups with high rates of group extinction (see also Wright 1945). For example, Wynne-Edwards' theory seems to require the fixation of the altruist gene in groups through random sampling effects (genetic drift), which in turn requires small groups. If so, group selection – although not impossible – could only occur under restricted conditions (e.g. Maynard Smith 1964, 1976). However, later authors have argued that these models either had inappropriate assumptions (Wade 1978b) or misinterpreted Wynne-Edwards' verbal description of his theory (Pollock 1989). In short, the controversy over Wynne-Edwards' brand of group selection concerns: (1) whether Wynne-Edwards correctly interpreted population regulation phenomena; (2) whether it would be strong enough to have important effects; (3) the assumptions underlying its mathematical formulation; and (4) the plausibility of the conditions it requires (e.g. Maynard Smith 1964; Wynne-Edwards 1964a,b; Wade 1978b; Ridley and Dawkins 1981).

The next development in the controversy was a series of models claiming not to require restrictive conditions to achieve workable group selection. An early example was D.S. Wilson's (1975) model of "trait-group" selection within "demes," which are populations within which mating occurs (see also Matessi and Jayakar 1976). These models were therefore termed "intrademic" (Wade 1978b) or "structured deme" (D.S. Wilson 1977) models, in contrast to the earlier "interdemic" (G.C. Williams 1966 p. 96) or "traditional" models (Wade 1978b). The major technical difference is that in interdemic models mating occurs mainly within groups (so selection between groups is interdemic), whereas in intrademic ones mating is population-wide, and is followed by reassortment into intrademic groups (among which selection then occurs). The important finding of the new models was that altruism could evolve by group selection, provided the interactants within groups showed a greater than average genetic similarity (e.g. D.S. Wilson 1975). (This ensures that between-group genetic variance, a prerequisite for any form of group selection, is achieved without genetic drift. A positive association of bearers of the altruist gene also means that this gene benefits preferentially from the altruism.) In this, they were unlike Wynne-Edwards' group selection, which made no mention of genetic similarity.

Therefore, critics have claimed that the intrademic models are correct, but are not genuine models of group selection, since the most plausible cause of genetic similarity within groups is kinship (Maynard Smith 1976, 1982a; Grafen 1984; Nunney 1985a). The crucial genetic similarity required by the models is at the locus determining selfish or altruistic behavior. And kinship, as well as being a universal factor generating genetic similarity, is about the only one that can form the basis for the preferential allocation of social behavior without provoking within-genome conflict between the locus for social behavior and other loci (Section 1.8). Therefore, Maynard Smith (1976), for example, regarded intrademic group selection models as valid models of kin selection. So the controversy over these models has been mainly semantic (D.S. Wilson 1983).

Additional semantic disputes have arisen over the meaning of "altruism" and of "group" (D.S. Wilson 1983). Some models (D.S. Wilson 1975, 1980; Matessi and Jayakar 1976) suggested that altruism did not always require genetic similarity or relatedness, but could evolve in genetically random groups of individuals. However, this result holds only if altruism is defined in terms of an individual's fitness relative to group-mates, which D.S. Wilson (1979) termed "weak altruism". The standard definition of altruism, D.S. Wilson's (1979) "strong altruism", is in terms of fitness relative to all members of the population. Put another way, "weak altruism" involves increasing one's own personal fitness but increasing one's group-mates' even more, whereas "strong altruism" entails an absolute negative effect on the actor's personal fitness (Section 1.3). Under the conventional definition of altruism as "strong," the previous conclusion – that altruism cannot evolve with random grouping – remains (D.S. Wilson 1979, 1983, 1990; Grafen 1980, 1984; Nunney 1985a). A possible example of "weak altruism" occurs in foundress associations of unrelated ant queens (D.S. Wilson 1990; Section 8.3). However, since weak altruism's net benefits eventually fall on the weak altruists themselves, other authors regard such cases as examples of "ordinary" individual selection (Grafen 1980; D.S. Wilson 1990).

The semantic argument over the meaning of "group" is this. Kin selection can operate both if the population is structured into physical groups (such as families) and if relatives are ungrouped but still interact preferentially. But, to authors such as Maynard Smith (1964, 1976, 1982a) and Dawkins (1979), the groups in "group selection" must be physical structures. So, to them, kin and group selection are distinct in that only kin selection can work without physical grouping. In contrast, supporters of intrademic group selection define "groups" either as physical structures or as networks of preferential interaction (e.g. Wade 1980). So they maintain that kin selection and group selection can both operate in the absence of physical grouping.

The conclusion so far, from both sides of the controversy, is that the evolution of altruism (under the standard definition) requires either genetic similarity (usually, relatedness) or restrictive conditions. At this point two features of social insects are relevant. First, nearly all social insect populations are (physically) grouped by relatedness, because colonies are extended families. Second, colonies of most social Hymenoptera are not demes, or groups within which mating occurs (Chapter 11). Therefore, group selection for altruism in social insects must involve relatedness and must be intrademic. This rules out Wynne-Edwards' group selection in the social insect case (at least the version of his theory as usually understood). The issue now becomes to what extent kin selection for altruism, which Chapter 1 presented as the key process in the social evolution of insects, is compatible with intrademic group selection involving relatedness.

This requires consideration of a further development in the group selection controversy. A set of modelers took the situation previously regarded as the province of kin selection theory, namely populations structured by relatedness. They showed that the kin-selected spread of a gene for altruism in such populations could be modeled using an alternative approach, which they labeled group selection (e.g. Wade 1980; Box 2.1). Briefly, a gene for altruism will always (by definition) decrease in frequency within groups. Therefore, the only way the gene can spread is if groups with altruists are more productive than those without them. More formally, there must be a positive between-group component of selection that exceeds the negative within-group component, with groups being either physical groups or patterns of nonrandom interaction (Box 2.1). However, this modeling method also shows that altruism is only favored if the condition in Hamilton's rule is met. In other words, the method is mathematically equivalent to kin selection theory and similarly demonstrates that altruism requires relatedness (e.g. Queller 1992a, b).

Both sides of the controversy therefore agree on the validity of Wade's (1980) "components of selection" method. But does it follow that kin selection implies a form of group selection, as is claimed? This depends on agreement over two points. The first, mentioned above, is that group selection can be said to occur even when there are no physical groups. This is asserted by, for example, Wade (1980), but denied by Maynard Smith (1982a) and Dawkins (1979, 1982b), who therefore do not accept that kin selection implies some kind of group selection (Maynard Smith 1976; Dawkins 1979, 1982b). (In any event, since social insect colonies are physical groups, this objection to invoking group selection does not apply in the present context.) Second, there needs to be agreement that between-group selection can be equated with group

BOX 2.1 KIN SELECTION AND GROUP SELECTION

Hamilton's rule (Box 1.2) describes the conditions under which a gene for altruism will be positively selected. This box demonstrates that this process can also be viewed as the outcome of a conflict of between-group selection and within-group selection. Therefore, provided certain qualifications hold, kin selection and group selection do not represent different evolutionary processes, but are alternative ways of viewing the same phenomena – in this case the spread of a gene for altruism (Section 2.2). The key qualifications are that: (1) between-group selection is equated with group selection; (2) "groups" need not be physical structures; and (3) it is recognized that kin selection and group selection for altruism both rely on genetic similarity (relatedness) at the locus for social behavior.

The equivalence of kin and group selection has been advocated by several authors, including Wade (1980, 1985a), Breden (1990), Goodnight et al. (1992), and Queller (1992a,b); see also Williams and Williams (1957), Hamilton (1975a), Wade (1979), Wade and Breden (1981), Michod (1982), Pollock (1983), D.S. Wilson (1983), Grafen (1984, 1985), Owen (1989), Wilson and Sober (1989), Ross and Carpenter (1991a), and Dugatkin and Reeve (1994). Similar arguments show that sex allocation in structured populations is the result of either individual selection on female parents (Section 4.9) or of "group selection," where this means selection between groups of differential productivities (e.g. Colwell 1981; Wilson and Colwell 1981; Charnov 1982; Maynard Smith 1983; D.S. Wilson 1983; Grafen 1984; Harvey 1985; Nunney 1985b; Bulmer 1986a; Frank 1986; Herre et al. 1987; Wilson and Sober 1989; Leigh 1991; Avilés 1993; Godfray 1994).

The starting point is a basic equation derived by Price (1970) describing the change in frequency of any gene under natural selection. This equation is also derived by Grafen (1985), whose symbolism will be followed:

$$w\Delta p = \text{Cov}(w_i, p_i) + E(w_i \Delta p_i) \qquad [2.1]$$

where

$w =$ the average number of offspring per individual in the parental generation.

$\Delta p =$ the change in frequency of a given ("focal") gene, i.e. its frequency in the offspring generation minus its frequency in the parental generation.

$w_i =$ the number of offspring of the ith parent.

$p_i =$ the focal gene's frequency in the ith parent.

$\Delta p_i =$ the difference in gene frequency between the ith parent and its gametes.

BOX 2.1 CONT.

$Cov(X,Y) =$ the covariance between N pairs of two variables X and Y, which is defined as $\Sigma(X - \bar{X})(Y - \bar{Y})/N$, with $\bar{X}$ and $\bar{Y}$ being the means of X and Y respectively.

$E(X) =$ the average ("expected") value of a variable X.

This equation can be interpreted as follows. If Δp is positive, the focal gene's frequency increases; if Δp is negative, it falls; and if Δp is zero, the gene's frequency remains unchanged. Now the change in gene frequency between an individual and its gametes is zero, unless a non-Mendelian process like meiotic drive is occurring. (Meiotic drive occurs when a gene gains greater than random [50%] representation in the gametes independently of any effects on offspring number [Section 2.5].) Therefore Δp_i is zero on average, so $E(w_i \Delta p_i)$ is also zero and Equation 2.1 simplifies to

$$w\Delta p = Cov(w_i, p_i)$$
$$\Delta p = [Cov(w_i, p_i)]/w. \qquad [2.2]$$

The term w is always positive. Therefore, Equation 2.2 states that the direction of selection on a gene is given by the sign of the covariance between two variables – number of offspring of an individual in the parental generation, and the individual's gene frequency.

Also, the covariance between X and Y equals the product (regression coefficient of Y on X) × (variance of X). Variance is always positive. So the sign of Δp depends on just the sign of the regression coefficient. This means that a gene is positively selected if the slope of a plot of individual offspring number and gene frequency is positive, negatively selected if this slope is negative, and unchanged in frequency if there is no systematic relation between offspring number and gene frequency. This makes intuitive sense, since a positive slope means that individuals with an above-average frequency of the gene have an above-average number of offspring, which in turn means that the gene's frequency will rise.

Now say the gene frequency of an individual makes no difference to its offspring number (the covariance term in Equation 2.1 is zero) but there is meiotic drive. The second term, $E(w_i \Delta p_i)$, in Equation 2.1 would then be positive and so the focal gene would increase in frequency despite the lack of the covariance term. This is why the second term needs to be included. In sum, the above arguments have shown that Price's equation correctly describes the direction of gene frequency change under natural selection.

Equation 2.1 may now be extended to cover the situation in which a population is structured into groups (Price 1972; Hamilton 1975a; Grafen 1985). First, an index "g" for groups is substituted for the index "i" for

BOX 2.1 CONT.

individuals in Equation 2.1. The next equation then follows from the same logic as Price (1970) used to reach Equation 2.1:

$$w\Delta p = \text{Cov}(w_g, p_g) + E(w_g\Delta p_g) \qquad [2.3]$$

where w_g = the average number of offspring of individuals in the gth group, and p_g is the average gene frequency of individuals in this group.

Each group represents a local population within which gene frequency changes can be represented by reapplying Equation 2.1, giving:

$$w_g\Delta p_g = \text{Cov}(w_{gi}, p_{gi}) + E_g(w_{gi}\Delta p_{gi}) \qquad [2.4]$$

where the index "gi" denotes individuals within groups, so, for example, w_{gi} is the number of offspring of the ith individual in the gth group.

Substituting Equation 2.4 into the last term of Equation 2.3 gives:

$$w\Delta p = \text{Cov}(w_g, p_g) + E[\text{Cov}(w_{gi}, p_{gi}) + E_g(w_{gi}\Delta p_{gi})]. \qquad [2.5]$$

This is Grafen's (1985 p. 37) equation (with a misprint in the last term corrected). In effect, the "individual" version of Price's equation, Equation 2.1, has been "nested" within the "group" version, Equation 2.3. Theoretically, this process could be continued indefinitely for any depth of population structuring.

By the previous argument assuming the absence of meiotic drive, Δp_{gi} equals zero on average. Therefore Equation 2.5 simplifies to:

$$w\Delta p = \text{Cov}(w_g, p_g) + E[\text{Cov}(w_{gi}, p_{gi})] \qquad [2.6]$$

(cf. Price's [1972] Equation A17). This states that in a population structured into groups, the direction of gene frequency change is given by two terms. The first is the covariance between mean offspring number and mean gene frequency across groups, and so refers to differences between groups. The second is the average value of the covariance for individuals within groups between offspring number and gene frequency, so it refers to differences between individuals. Therefore, selection in a structured population can be partitioned into a between-groups component and a within-groups component.

At this point an argument from Wade (1980) demonstrates how the foregoing applies to kin selection and altruism. Suppose that the focal gene is a gene for altruism (Section 1.3). Then the term, $E[\text{Cov}(w_{gi}, p_{gi})]$, in Equation 2.6 must be negative. This is because in both all-altruist or all-selfish groups, no genetic variance exists on average, so this term is

BOX 2.1 CONT.

zero. But in mixed groups, altruists (bearers of the gene for altruism) receive a benefit (from other altruists in the group) and may pay a cost (if they show the altruist phenotype), whereas nonaltruists (nonbearers) receive a benefit but never pay a cost, resulting in bearers having a lower average offspring number. So $E[\text{Cov}(w_{gi}, p_{gi})]$ is negative in mixed groups. Therefore, across all groups, if the focal gene is a gene for altruism, this term must be negative (Wade 1980).

For the gene for altruism to be positively selected, the right hand side of Equation 2.6 must be positive. Since $E[\text{Cov}(w_{gi}, p_{gi})]$ is negative, this condition must therefore be as follows (where $|X|$ means the positive value of X):

$$\text{Cov}(w_g, p_g) > |E[\text{Cov}(w_{gi}, p_{gi})]|. \qquad [2.7]$$

So the spread of a gene for altruism depends on between-group selection on the gene being greater than the opposing force of within-group selection (Wade 1980). Biologically, this can occur because although within mixed groups altruists have fewer offspring than nonaltruists, groups consisting of many or only altruists may be much more productive than those made up mostly or solely of nonaltruists, which may be very unproductive. Groups with many altruists would then make a larger contribution to the next generation.

In a population in which relatives interact preferentially, the condition for the spread of a gene for altruism is given by Hamilton's rule (Box 1.2). Now, from Equation 2.6 and similar formulations it is possible to derive Hamilton's rule (e.g. Wade 1980, 1985b; Seger 1981; Grafen 1985; Breden 1990; Queller 1992a, b). This means that both the present "components of selection" method of modeling the evolution of altruism, and the kin selection method, reach the same conclusion – namely that relatedness is necessary for a gene for altruism to spread.

Wade (1980) further equated between-group selection with group selection and within-group selection with individual selection (see also Wilson and Colwell 1981). However, this is controversial because, according to Grafen (1985; see also Hamilton 1975a and Grafen 1984), the between-groups covariance term in Equation 2.6 contains an implicit individual selection subcomponent. Imagine a gene of sole benefit to its bearer. The between-groups covariance term would still be positive because groups containing a high proportion of the gene bearers (groups with a high mean frequency of the gene) would have more offspring on average than groups with few bearers (low frequency of the gene), due to the high productivity of each individual bearer. So "ordinary" individual-level selection always contributes to between-groups covariance. Hence,

BOX 2.1 CONT.

Grafen (1985) argued, between-group selection cannot be equated purely with group selection, nor within-group selection with individual selection.

However, the important conclusion is that the spread of a gene for altruism can be analyzed either with the mathematical machinery of kin selection, or with the methods outlined above that invoke a conflict of between-group selection and within-group selection (which under one interpretation is a conflict of group with individual selection). So these methodologies do not describe different evolutionary processes, although disagreement exists over their relative usefulness (Breden 1990; D.S. Wilson 1990; Queller 1992a,b; Wilson et al. 1992; Dugatkin and Reeve 1994). To sum up, kin selection and group selection (in the above sense) are not fundamentally antithetical: they both describe the basic process of gene selection.

selection, and within-group selection with individual selection. This is supported by Wade (1980) and Wilson and Colwell (1981) among others, but opposed by Grafen (1984, 1985; see also Nunney 1985a). Grafen's argument, which Box 2.1 explains more fully, is that the between-group component of selection contains an inseparable individual selection subcomponent, and so cannot purely represent group selection.

Whatever position one adopts on these points, the important conclusion is that (strong) altruism can be regarded as evolving either via kin selection or via a conflict of between- and within-group selection, where the groups are physical groups of relatives or networks of relatives. Therefore, provided group selection is understood in this sense, it is misleading to regard it and kin selection as competing theories for the evolution of altruism. Similar points have been made by Wilson and Sober (1989) and Queller (1992a,b). This argument does not imply that group selection is *only* valid as an alternative way of viewing kin selection. For example, when groups are composed of nonrelatives (as in foundress associations of ant queens), kin selection clearly cannot occur, but between-group selection for traits boosting group productivity may be important (Rissing et al. 1989; D.S. Wilson 1990; Dugatkin et al. 1992; Section 8.3). However, as discussed earlier, group selection cannot produce (strong) altruism in the absence of genetic similarity or relatedness (D.S. Wilson 1990). Therefore the conclusion above remains. In a population of interacting relatives both kin selection and group selection represent alternative means of analyzing the evolution of altruism. The two concepts need not be opposed.

To return now to colony-level selection. The preceding argument shows that if colony-level selection is equated with between-group selection in the sense just described (Wade's [1980] "components of selection" framework [Box 2.1]), it represents a viable approach to the study of social insects. This is despite its occasional presentation as contrary to kin selection. So, in answer to the question raised at the start of this chapter, colony-level selection is not a substantively different process to the gene selectionist processes considered in Chapter 1. Colony-level and kin selection are alternative ways of looking at gene frequency change in a population structured by relatedness. Therefore pitting the two modes of selection against one another, as is sometimes done in the social insect literature, is illogical.

2.3 Two Examples of Colony-level Selection

The previous section concluded that colony-level selection is fully compatible with kin selection when colonies are groups of relatives. This conclusion was reached using Wade's "components of selection" view of kin and group selection (Box 2.1). To reinforce this point, this section considers two examples from the literature in which colony-level selection has been invoked to explain the evolution of traits in social insects. The choice of examples is directed by the generality of the cases they represent. The aim is to show that in these cases the authors' interpretation in terms of colony-level selection fits the "components of selection" interpretation already described.

But first it is important to realize that a colony-level selection perspective is not confined to the question of the spread of genes for altruism, as the arguments above might imply. Colony-level selection (or equally, kin selection) could be invoked to explain the spread of almost any trait in social insects. This is because both modes of selection deal with the spread of genes for social actions, that is genes affecting the survival or offspring numbers of others (Section 1.4). It is also because, in a population structured into groups, almost all genes affecting an individual bearer *are* genes for social actions. The reason why colony-level selection deals with the spread of genes for social actions is that the between-group and within-group components of selection in its methodology (Box 2.1) can each take positive or negative signs, representing actions that are good or bad to groups, to individuals, or to both. And the reason why nearly all genes in a structured population are genes for social actions is that, in a group, almost all traits of an individual affect other group members to some degree. Even traits that apparently benefit only an individual will benefit the group by prolonging its survival, since the individual forms a fraction of the group (Grafen 1985).

With this background, the first example of colony-level selection to be considered is as follows. E.O. Wilson (1968, 1985a) and Oster and Wilson (1978) proposed that, in species with polymorphic workers (physical worker castes), the size frequency distribution of workers within colonies has evolved by colony-level selection. This proposal can be rephrased in the following way, as E.O. Wilson (1968) indicated. (For simplicity it is assumed that colonies have a single, once-mated queen and sterile workers, so within-colony selection is absent.) The founding queen and male of some colonies carried genes that led them to produce workers conforming to a particular caste profile. Colonies with this profile then produced more sexuals than those without it (this is the "social action" element of the genes in question). So genes "for" that profile spread. In other words, there was a positive between-colony component of selection (positive covariance of the frequency of the caste profile genes with colony productivity [Box 2.1]). As a matter of fact, Beshers and Traniello (1994) recently showed that the size frequency distribution of workers did affect colony sexual production in populations of attine ants.

This explanation is very close to Darwin's (1859) original description of the evolution of worker polymorphism (Section 1.3), and to Owen's (1989) definition of colony-level selection from the previous section. In addition, although it is the genes "for" the favored caste profile that ultimately benefit from it (they spread in the population), the profile itself is the property not of any one individual but of the colony as a whole. So although the profile represents a "colony-level adaptation" (Ratnieks and Reeve 1992 p. 36), it is still an adaptation serving the interests of the genes "for" it. One could also argue that the colony itself receives the benefit. However, if benefit to the colony is to have any evolutionary significance, it must result in more sexuals being produced by the colony, with these sexuals being bearers of the causative genes. So benefit to the colony reduces to benefit to genes. Therefore, colony-level selection for the worker size frequency distribution is a way of talking about gene selection for this (colony-level) trait.

The second example illustrating a general case in which colony-level selection is invoked concerns its acting as a brake on selfishness. So this example differs from the preceding case in allowing a degree of within-colony selection. In fact it is very close to the situation outlined in Box 2.1 (in which negative within-group selection is opposed by positive between-group selection). Sturtevant (1938) discussed why colonies with multiple queens are not destabilized by queens trying to dominate the colony's reproduction, so leading to single-queen colonies (monogyny). He reasoned that too great a level of inter-queen competition would reduce colony efficiency. Therefore, under colony-level selection, colonies with selfish queens would be outcompeted by colonies with more cooperative ones. This argument can be rephrased as follows. Genes existed whose queen bearers tended to monopolize reproduction in the colony, leading to affected colonies producing a greater proportion of sexuals also bearing the genes for selfishness. (Here the involvement of genes for social actions is clear.) This promoted the spread of the selfishness genes (there was a negative within-colony component of selection for the cooperative alleles of the selfishness genes). But colonies with selfish queens produced fewer sexuals than colonies with noncompeting queens due to reduced colony efficiency. So, overall, colonies with cooperative queens made a greater genetic contribution per colony to the whole population (positive between-colony component of selection). Given suitable magnitudes of the components of selection, the result might be a population of mostly cooperative queens. Therefore, here too colony-level selection makes sense in the components of selection framework, and is another way of referring to gene selection.

In sum, cases invoking colony-level selection in the literature fit a components of selection interpretation. It needs stressing that these two examples represent the main situations for which colony-level selection is usually deployed – namely selection for group-level properties boosting colony productivity, and the situation when within-group selection is countered by selection between groups. So the arguments showing that these cases fall within the components of selection perspective represent a fairly general demonstration of this point. Furthermore, as the previous section argued, in populations structured into groups of relatives, a components of selection analysis is equivalent to a kin selection one. So from these examples one can again conclude that colony-level selection and kin selection are not antagonistic concepts in social insect biology. The language of one concept may sometimes be more useful than that of the other, but this does not imply that the basic processes are different. In addition, these examples show that any appeal to colony-level selection that claims explanatory power as a piece of evolutionary reasoning must involve gene selection. So like kin selection, individual selection, and group selection, colony-level selection is a framework for analyzing gene frequency change.

Finally, the frequent appearance of colony-level selection in the social insect literature suggests that it is a useful term. The traditional categories of individual-level and colony-level selection will produce different kinds of adaptation (e.g. Ratnieks and Reeve 1992; Frumhoff and Ward 1992), and it has clearly helped social insect biologists to have words with which to recognize this. But the current "individual-level versus colony-level selection" language implies that individual-level and colony-level selection are two fundamentally different evolutionary processes, and that both are incompatible with kin and gene selection. Neither of these things is true. Therefore, to avoid these implications but to retain a useful terminology, we suggest that social insect biologists substitute "within-colony" for "individual-level" and "between-colony" for "colony-level" selection. (This would also avoid the difficulty posed by the old language's failure to recognize that individual selection strictly speaking also contributes to between-colony selection [Grafen 1985; see also Section 2.2 and Box 2.1].) Furthermore, both "within-colony" and "between-colony" selection should be explicitly recognized as means by which genes can increase their frequencies in a structured population. Lastly, for all these reasons, contrasting comparisons of individual, colony, kin, and gene selection should be abandoned.

2.4 Levels-of-selection Theory

An influential approach to the study of natural selection is levels-of-selection or hierarchical selection theory (e.g. Lewontin 1970; Gould 1980, 1982; Hull 1980; D.S. Wilson 1980; Arnold and Fristrup 1982; Wade 1982a; Sober 1984a; Eldredge 1985; Buss 1987; Vrba 1989; Wilson and Sober 1989; Brandon 1990; G.C. Williams 1992). Living things are hierarchically organized. For example, genes occur in individuals and individuals may occur in groups. Similarly, groups are found in populations, and populations in species. Levels-of-selection theory has several aspects. First, it regards as its chief strength its explicit recognition of the biological hierarchy. Second, it claims that the legitimate unit of selection is the unit of organization at a specified level. Therefore genes can be a unit of selection, but so can groups, for example. Third, the theory sees a correspondence between the operation of selection at the different levels. So, in the theory, individual selection involves the differential survival and fecundity of individuals subject to heritable variation; and, analogously, the differential persistence and productivity of groups with heritable variation constitute group selection (e.g. Lewontin 1970; Arnold and Fristrup 1982; D.S. Wilson 1983, 1990; Sober 1984a).

Lastly, levels-of-selection theory maintains that selection can act simultaneously at different levels (e.g. Wade 1982a). When selection acts in opposite directions at adjoining levels, the result will be a compromise between the strengths of selection at each level. Consider the example of altruism in social groups. Individuals in a group can be selfish (promoting their personal fitness), or altruistic (promoting the productivity of the group at their own expense). According to levels-of-selection theory, there is a conflict between selection on individuals within groups to be selfish (selfish individuals increase in frequency within groups, or obtain a higher proportion of group reproduction), and selection on groups to contain a high frequency of altruists (because such groups are more productive than those with few altruists). The outcome will be a balance between these two forces that results in a stable mixture of selfish and altruistic individuals within the population.

Social insects occur in an especially obvious biological hierarchy of individual, colony (group), and population. So they seem ideal candidates for a levels-of-selection interpretation of their evolutionary biology (e.g. Wilson and Sober 1989). But some advocates of levels-of-selection theory regard it as hostile to gene selectionism (e.g. Gould 1984, 1992; Sober 1984a). Can these two outlooks be reconciled?

The answer is yes, so long as within-species processes are being considered. For one thing, it is apparent from the opening description of levels-of-selection theory that the "components of selection" methodology outlined in Section 2.2 and Box 2.1 represents a formalization of the theory. The signs and magnitudes of the components of selection give, respectively, the directions and strengths of selection at the different levels (within-group and between-group). The previous sections argued that kin selection and group (or colony-level) selection are compatible if group selection is understood as a components of selection approach to kin selection. They also argued that both are ways of modeling gene selection. In this sense they have already reconciled levels-of-selection theory with gene selectionism.

But this section also argues for a broader position. This is the idea that gene selectionism and levels-of-selection theory are compatible given the following conditions. These are that gene selectionists concede that gene frequency change occurs within the biological hierarchy, levels-of-selection theorists concede that genes are a special level within the hierarchy, and all parties agree that natural selection concerns gene frequency change. This last point should be uncontroversial, since even levels-of-selection models (below the species level) deal with changes in gene frequencies (Dawkins 1978, 1982a). The other concessions can be made, as Dawkins has shown, via the replicator-vehicle distinction (Dawkins 1982a).

Groups can be vehicles for genes just as bodies can (Section 1.2). For example, social groups can be regarded as vehicles formed to benefit genes for social actions. Therefore the vehicle concept is gene selectionism's way of recognizing the biological hierarchy (Dawkins 1982a; Dugatkin and Reeve 1994). To consider the relation of the replicator-vehicle concept with levels-of-selection theory requires touching on the debate over the identity of the "unit of selection" (e.g. Lewontin 1970, 1974; Alexander and Borgia 1978; Hull 1980; Wimsatt 1980; Dawkins 1982a; Brandon and Burian 1984; Sober 1984a; Buss 1987; Gliddon and Gouyon 1989; Vrba 1989). Chapter 1 avoided this controversial term. On one hand, a largely semantic disagreement exists (Wilson and Sober 1989) because Dawkins (1978, 1982a) calls the gene the unit of selection since it is the replicator, whereas levels-of-selection theory defines all levels of the biological hierarchy as potential units of selection. On the other, the substantial issue is whether the levels in the biological hierarchy are all of the same kind. Levels-of-selection theory regards the different levels as in some sense equivalent. It does this explicitly by considering individual and group selection to be analogous (see above), and implicitly by not overtly recognizing the replicator-vehicle distinction. In fact, some levels-of-selection theorists apparently deny any special quality of the gene level (e.g. Gould 1984, 1992; Sober 1984a). However, the levels of biological organization do not form a scalar ranking (Dawkins 1982a p. 82, b, 1989a pp. 273–274). Instead, a fundamental difference exists between the gene (replicator) level, and all others up to species level, which are levels of vehicles (Section 1.2). (Another qualitative discontinuity may occur at the species level. Whether species and higher taxonomic units are best regarded as replicators or vehicles, and can undergo "species selection" or the equivalent, are aspects of the controversy in this area covered, for example, by Stanley [1979], Arnold and Fristrup [1982], Dawkins [1982a, 1986, 1989b], Gould [1982], Ridley [1985], Vrba [1989], G.C. Williams [1992], and Lloyd and Gould [1993]. Species selection is not further considered here because it is unlikely to have contributed to the origin of complex adaptations [Dawkins 1982a p. 108, 1986; Maynard Smith 1983].) Therefore, gene selectionism and levels-of-selection theory are reconcilable provided that the replicator-vehicle distinction is added to levels-of-selection theory. (Eldredge [1985] and Brandon [1990] adopt a different approach to reconciling replicators, vehicles, and hierarchy, arguing for various kinds of parallel hierarchy. This is partly because they have a looser concept of replicators than Dawkins.) By doing this, biologists can legitimately maintain a gene-centered perspective, while still invoking a levels-of-selection perspective if it is convenient.

The differences between gene selectionism and levels-of-selection theory are therefore largely differences of language (with the reservation that adaptive complexity requires replicators). It is agreed that, within species, natural selection theory concerns the rise and fall in the frequencies of genes of interest. But when populations are structured into different levels, how genes change frequencies may be complex.

For example, genes for selfishness may prosper in groups at low within-group frequencies, but at high frequencies they may perform badly because groups with many selfish individuals go extinct. In this situation one could preserve strictly gene-centered language by saying that the genes for selfishness do poorly in the presence of themselves, or have a frequency-dependent success where the relevant frequency is the local one (cf. Dawkins 1982a pp. 117, 239–249). The result could therefore be a stable mix of genes for selfishness and genes for altruism in the population. This way of viewing gene frequency change is a population genetics equivalent of Maynard Smith's (1982b) concept of the evolutionarily stable strategy or ESS (Dawkins 1976 p. 91). Alternatively, one could describe the same situation and reach the same conclusion in terms of the strength of selection acting on the different levels of individual and group (strictly speaking, in terms of the signs and sizes of the within-group and between-group components of selection). This would be to use the language of levels-of-selection theory. Therefore, gene selectionism and levels-of-selection theory represent not opposing theories of natural selection, but alternative ways of talking about it, respectively replicator-centered and vehicle-centered (Dawkins 1982a p. 82). (Buss [1987 pp. 176–183] discusses this language difference from a levels-of-selection stance.) Their perspectives differ, but in essence they both deal with the same process – changes in the frequencies of genes (replicators).

To summarize, there is scope for pluralism in the analysis of natural selection (Maynard Smith 1983; Wilson and Sober 1989). Gene selectionism is reconcilable with levels-of-selection theory provided the levels higher than the gene are recognized as vehicles and the common object of modeling gene frequency change is acknowledged. How biologists choose to model gene frequency change, or talk about it, will probably depend on the kind of situation they are dealing with. No single method is necessarily always superior to another. Therefore, levels-of-selection theory, like its subset, colony-level selection, is not a rival of gene selectionism for understanding the evolution of social insects or indeed of any other group.

2.5 Gene Selectionism, Levels-of-selection Theory, the Evolution of Individuality, and Suppression of Within-unit Conflict

A fruitful synthesis of levels-of-selection theory and gene selectionism is under way regarding the issue of the "evolution of individuality" (Buss 1987), although theorists in this area have not always expressed themselves in terms of these concepts. The basic idea is that individuality, or the quality of behaving like an integrated organism, emerges at any given level of units via the suppression of within-unit conflicts of interest (e.g. Leigh 1977, 1991; Alexander and Borgia 1978; West-Eberhard 1979; Cosmides and Tooby 1981; Dawkins 1982a, 1990; Buss 1987; Wilson and Sober 1989; Ratnieks and Reeve 1992; G.C. Williams 1992). The interesting question that then arises is how this suppression is achieved. As in other facets of the debate over hierarchy in natural selection theory, the social insects feature prominently in this unfolding area, which in turn helps illuminate their social evolution.

To see what this development involves, consider the current understanding, from the gene selectionist perspective, of why individual organisms in sexual diploid species act in their characteristically unitary fashion (Dawkins 1982a, 1990). The previous chapter addressed this subject when discussing within-genome conflicts of interest (Section 1.8). To expand the argument, the explanation is that each (adult) individual consists of: (1) a huge aggregation of cells, every one of which is genetically identical (ignoring somatic mutations) because all derive by mitosis from the same single zygote; plus (2) a handful of genetically diverse cells (the gametes) derived by meiosis from the zygote via the "germ line." Take any one (nuclear, autosomal) gene in a somatic cell. It has an allele at the corresponding part of the chromosome homologous to the chromosome it occupies. The gametes represent both alleles' only means of leaving their body and entering the next generation, because somatic cells are by definition not germ-line. All loci in the cell will be represented in every gamete, so each locus is selected to cooperate in gamete production. Similarly, across somatic cells, all loci are selected to cooperate because all somatic cells are clones of each other. Lastly, although each locus will be represented in gametes by only one of its alleles, each allele has exactly the same chance (0.5) of entering the gamete, because of a fair meiosis.

The result is that, in any one individual, between-allele, between-locus and between-cell conflict are all absent. So an individual acts as a unit dedicated to gamete dissemination, even though it consists of a collection of separate replicators (genes).

From the replicator selectionist viewpoint, the integrity of an individual organism is therefore conditional on the lack of within-individual conflict between potentially opposed groups of replicators inside it. But, as already indicated (Sections 1.2, 1.8), within-organism conflict has not entirely disappeared. Possible examples include conflicts between the following groups of replicators: meiotic drive (segregation distorter) genes and genes at other loci in the nucleus; genes on sex chromosomes and genes on autosomes; "selfish" DNA and protein-coding DNA in the nuclear genome; and nuclear genes and genes in organelles in the cytoplasm (e.g. Hamilton 1967; Cosmides and Tooby 1981; Dawkins 1982a; Trivers 1985; Hurst 1992, 1993a). On the other hand, although within-organism conflicts occasionally break out, the net outcome is that they are held in check, conferring on individuals their defining quality of integrity.

The insight that individuality stems from a lack of within-unit conflict has been expressed in several ways. For example, Alexander and Borgia (1978 p. 455) wrote that "Long-continued potent selection at any level in the hierarchy of life is likely to enhance the integrity of entities at that level and reduce the likelihood of conflicts of interest with units at lower levels." Dawkins (1982a p. 134) wrote that "There is a sense in which a 'vehicle' [individual] is worthy of the name in inverse proportion to the number of outlaw replicators that it contains." (The "outlaw" concept was discussed in Section 1.8.) The same essential point has also been made within the levels-of-selection tradition. Buss (1987 p. 171) states that "To the extent that control over replication of the lower unit is required for effective interaction with the external environment, organizations must appear in the higher unit to limit the origin or expression of variation at the lower unit." Lastly, there is Wilson and Sober's (1989 p. 343) view that "When between-unit selection overwhelms within-unit selection, the unit itself becomes an organism in the formal sense of the word."

These quotations illustrate the opening point that both gene selectionism and levels-of-selection theory are contributing to the unfolding field of the evolution of individuality. So, to return to the question also mentioned at the start of this section, what factors contribute to the suppression of within-unit conflict? Regarding the evolution of individuals in the standard sense of organisms, the basic, underlying questions must be about why replicators aggregate in cells and organisms to begin with, why germ lines become separated from somatic cell lines, and why meiosis is "fair." The first of these issues has been addressed by Dawkins (1982a), the next by Buss (1987), while the last is discussed by, for example, Leigh (1971, 1977, 1991), Crow (1991), Haig and Grafen (1991), and Hurst (1993b). Hurst (1992) and Hurst et al. (1992) review within-genome conflicts as a general force in evolution. Setting these issues aside, the rest of this section concentrates on some emerging principles most relevant to social insect biology.

We propose that units acquire individuality through the suppression of within-unit conflict among selfish subunits according to the following principles. They may be classified as involving either the effects of rival subunits or effects on self. The latter are essentially aspects of the overriding of within-unit selection by between-unit selection invoked by

Wilson and Sober's (1989) quotation above and discussed in Sections 2.2 and 2.3. The principles as a whole derive from the references cited at the beginning of this section. They also resemble the "possible mechanisms limiting increase of selfish genes" described by Werren et al. (1988 p. 301):

(1) Suppression of selfish subunits through effects of rivals.
(a) Opposition from a rival set of subunits sharing among themselves a coincidence of interest.
(b) Opposition from rival subunit(s) holding greater power.
(2) Suppression of selfish subunits through effects on self.
(a) Self-limitation through too-high costs inflicted on the whole.
(b) Self-limitation due to frequency-dependent success.

For an illustration of these principles, consider the fate of meiotic drive genes within an individual (e.g. Alexander and Borgia 1978; Dawkins 1982a; for reviews on meiotic drive see Crow 1979; Hurst and Pomiankowski 1991; Lyttle 1991). A meiotic drive gene is one that, at the expense of its allele, somehow gains a higher than fifty percent chance of entering a gamete. Imagine such a gene that is also pleiotropically harmful to the individual body it is in, for example by reducing the gamete count. As long as its negative effects on the gamete count do not cancel out its extra chances of being in a gamete, it will be positively selected. However, all the other (unlinked) genes in the genome will suffer from the reduced gamete count. Therefore, they will have a common interest in suppressing the effects of the drive "outlaw." This means that any gene that happens to nullify the drive gene's effects (a "modifier"), so restoring the higher gamete count, will be positively selected, with the "agreement" of its fellow nondriving genes. Since they seem fairly rare, meiotic drive genes are presumably often suppressed. So the question is why this comes about.

The answer seems to lie in the principles given above. First, all the nondriving genes stand to benefit from the suppression of drive (principle 1a: coincidence of interest). Second, there are more nondriving genes than drive ones, so collectively they are more powerful (1b: unequal distribution of power). Third, the meiotic drive gene will spread despite inflicting some cost on the individual through its pleiotropic harmful effects, but only if the overall level of harm is not too great, in which case the reduced gamete count would adversely affect the drive gene itself (2a: cost to individual of subunit selfishness damaging subunit). Fourth, some meiotic drive genes work by destroying their allele. But if this kind of gene becomes too common, it will tend to occur homozygously, and so damage copies of itself (Dawkins 1982a p. 136), illustrating the principle (2b) of the frequency-dependent success of selfishness. Therefore, there is good reason to think that meiotic drive genes will usually be suppressed (a process that could itself help keep meiosis fair: Leigh 1971; Crow 1991).

The importance of sharing common interests and holding power have been expressed in various ways by different authors. For example, Leigh (1971 p. 249, 1977 p. 4543) talked of "a parliament of genes", Alexander and Borgia (1978 pp. 458 and 470) of the "power of the collective" and a "commonality of interest", and Cosmides and Tooby (1981 p. 87) of "communities of interest". However, these phrases tend to merge the two principles of coincidence of interest and unequal distribution of power, attributing power to the set of subunits sharing a common interest both because they share an interest and because they are more numerous. Yet, it is helpful to separate the two properties, because this leaves open the possibility that a single entity might "win" over the collective because, though one against many, for some reason it holds more power. Also, it is important to qualify the sense in which nonmeiotic drive genes form a "community of interest". The point is not that these genes should cooperate in all circumstances, but rather that they are "agreed" *with respect to their attitude* to a drive gene. In addition, any one gene acting as a modifier does so because it benefits itself, not because of some "solidarity" with its fellow genes, although these would benefit indirectly and so would "agree." In sum, "community of interest" can be a context-dependent concept, and need not imply a lack of fundamental selfishness in each community member (Leigh 1971 p. 249; Dawkins 1982a pp. 138–139).

These arguments may now be extended to levels other than the organism. Several authors, reasoning mainly in levels-of-selection mode, have asserted that individuality can emerge at any level of units if within-unit conflict is suppressed (e.g. Leigh 1977, 1991; Alexander and Borgia 1978; Buss 1987; Wilson and Sober 1989). This conclusion was present in some of the quotations given above. Buss (1987) and Leigh (1991) present the most ambitious schemes in this line of reasoning. Buss (1987) attributes the evolution of the organizational hierarchy itself to successive waves of consolidation of subunits into individuality via within-unit suppression, followed by the same process at the new, higher level. Leigh (1991 p. 257) explicitly sets out to identify common principles explaining the evolution of "harmony" at all levels of organization up to the ecosystem. However, this kind of thinking will now be illustrated as it applies to the social insects.

Specifically, the issue is whether the principles above can explain the phenomena surrounding conflicts of interest in insect societies (see also Chapter 7). Ratnieks and Reeve (1992), in their analysis of conflict evolution in social Hymenoptera, similarly drew parallels between processes acting in social insects and ones acting within other groups and within individuals. M.D. Breed (cited in Page et al. 1989a p. 578) also recognized a correspondence between conflict in insect societies and individuals, proposing kin-biased queen rearing (Section 7.3) as a "social analogue of meiotic drive." Three social insect examples will be considered, the first of which involves sex allocation.

Trivers and Hare (1976) showed that in social Hymenoptera with random mating, worker sterility, and colonies headed by a single, once-mated queen, the stable population sex ratio for the queen is $1:1$ females : males and for the workers is $3:1$ females : males (Section 4.3). So there is a within-colony conflict of interest over sex allocation. The data indicate that population sex ratios in ant species with the appropriate biology are female-biased, suggesting that workers win the sex ratio conflict (Section 5.2). Trivers and Hare (1976) argued that this happens because the workers outnumber the queen and so have the power to enforce their preferred sex ratio. These authors, along with Alexander and Borgia (1978) and Ratnieks and Reeve (1992), added that the workers all share the same sex ratio preference and so should act as a unit. So, using the earlier terminology, workers both hold the most power and have a coincidence of interest. Ratnieks and Reeve (1992) pointed out that victory to workers also implies low costs of sex ratio manipulation to colony productivity. So at least three principles (1a, 1b, 2a) that were applied above to conflict resolution among genes are useful for understanding conflict within ant colonies. An explicit linkage of between-gene conflict and Trivers–Hare sex ratio conflict was also made by Alexander and Borgia (1978) and Leigh (1991).

A second example concerns within-colony conflict over male production (Section 7.4). Under a single, once-mated queen, workers' relatedness with an average worker-produced male (c. 0.375) is higher than their relatedness with a queen-produced male (0.25). (The relatedness values used in this example are "life-for-life" relatednesses [Box 3.1].) So, if able, workers should lay male-producing eggs, or tolerate laying by other workers. Workers in such colonies, however, are often nonreproductive. The reasons for this are still controversial, but one possibility is that queens produce substances enforcing sterility on workers (Section 7.4). If true, this could be an example of power lying in the hands of one against many, as mentioned above. When queens mate multiply, workers' interests change. Their relatedness with the average worker-produced male (c. 0.125) now falls below that with the queen's sons (0.25 still). So workers should raise queen-derived males and be intolerant of other workers laying. This intolerance is termed "worker policing," and there is good evidence for it in honey bees (Ratnieks 1988; Ratnieks and Reeve 1992; Section 7.4).

One way of looking at this situation is to imagine that the queen mates multiply *in order to* force worker policing on the workers. In other words, by mating multiply the queen turns the previously hostile workers into a "community of interest" as regards raising the queen's males, in that the workers can now no longer agree on raising their own males, but can agree on rearing the queen's. Therefore, perhaps the queen exploits the principle of "coincidence of interest" to resolve the conflict over male production in her favor (in contrast to the previous cases where the community of interest was the winner). This suggestion was made with reference to gene-level conflicts by Leigh (1991). The idea that queens evolve multiple mating as a result of queen–worker conflict over worker reproduction and sex allocation is fully discussed by Starr (1984), Moritz (1985), Woyciechowski and Lomnicki (1987), Pamilo (1991c), Queller (1993b), and Ratnieks and Boomsma (1995).

However, in this case, extending the gene-level argument to social insects is more controversial. Multiple mating could evolve for many reasons, so worker policing could just be a consequence of it (e.g. Crozier and Page 1985; Ratnieks and Reeve 1992; Section 11.4). The current point is again to illustrate that a basic principle of within-unit suppression (the coincidence of interest principle) at one level (genes in individuals) may be applicable at another (individuals in colonies). Of course, in all kinds of colony, selfish worker egg laying would also be prevented by high costs to colony productivity (Cole 1986; Hillesheim et al. 1989; Ratnieks and Reeve 1992) or by laying workers damaging colony productivity if they become too frequent (Tsuji 1994), suggesting that principles 2a and 2b above may also be important.

Our third example of within-colony conflict in social insects examined in the light of within-organism processes is more speculative. One type of within-organism conflict that has not yet been mentioned involves sex ratio distorters (e.g. Werren et al. 1988). Typically, replicating cytoplasmic elements can only be transmitted through eggs, because sperm contains very little cytoplasm. Therefore, to maximize their transmission, these elements should bias an individual's offspring sex ratio in favor of the egg-producing sex, namely females. Such sex ratio distortion indeed occurs (Werren et al. 1988). Now, many ant species have symbionts in their colonies such as other invertebrates or fungi which in some cases, crucially, are transmitted to new colonies only on winged queens taking their mating flight. For example, virgin leaf-cutter *Atta* ant queens carry a pellet of their symbiotic fungus specially placed in a cavity in the head (E.O. Wilson 1971 p. 45). Similarly, Buschinger et al. (1987) and Klein et al. (1992) observed winged *Acropyga* and *Tetraponera* queens on a mating flight each carrying a symbiotic coccid bug in its jaws. Males represent a useless means of spreading from the ant colony for such symbionts, because males die after mating and do not enter new-founded colonies. Therefore, we hypothesize that symbionts borne by queens might act as colony-level sex ratio distorters. That is, if they could influence colony sex allocation, both fungus and coccid would favor extreme female bias, by analogy with cytoplasmic elements.

It is not known if such effects exist, or how symbionts might alter sex allocation. But they might exist, just as sex ratio distortion occurs in individuals. Earlier, it was assumed that either the queen or the workers control sex allocation. But perhaps neither does. By analogy with individuals, colony sex ratios may not always serve the interests of colony members (Chapter 4). On the other hand, they probably usually do, since workers should assert their sex ratio preferences over the symbionts', in a conflict that – as a community of interest with perhaps greater power – workers might usually win. Similarly, conflict over an individual's sex ratio exists between cytoplasmic elements and autosomal genes (Werren et al. 1988). So colony-level sex ratio distortion is simply a testable idea arising from the analogical reasoning of this section. In addition, Shykoff and Schmid-Hempel (1991a), without explicitly invoking the concept, describe a trypanosomal internal parasite of bumble bees that appears to be a sort of colony-level sex ratio distorter. It is apparently transmitted via queens only and reduces the ovary development of infected workers, so inhibiting male-production by workers that might otherwise detract from the colony's production of queens.

To return to more general issues concerning the evolution of individuality. An important concept is clearly that of the "community of interest." This poses the question of how such a community arises. Cosmides and Tooby (1981 p. 87) originally equated their "community of interest" with a "coreplicon," a group of genes replicating together. At the higher levels, Wilson and Sober (1989) suggested that individuality emerges when there is either total genetic uniformity (clonality) or suppression of within-unit conflict. But genetic uniformity might itself be a quality promoting the formation of a community of interest. So a pure clone should show no internal conflict, but even a partial clone may achieve within-unit suppression of conflict because the largest or most powerful clonal element acts as a community of interest. However, Ratnieks and Reeve (1992) point out that clonality is not necessary in a community of interest, because genetically diverse subunits might still form such a community if they have a coincidence of interest, as do sterile ant workers with respect to sex allocation. But a nonclonal community of interest is likely to be context-dependent, whereas a clonal one should always show internal agreement. Finally, genetic uniformity might also act as a force working *against* the evolution of individuality, because genetically uniform units might be over-susceptible to destruction by disease through lacking genetic variation for disease resistance (Hamilton 1987b). In the social insects, this suggests that the evolution of individuality in colonies through high within-colony relatedness may be opposed by selection for resistance to parasites (e.g. Shykoff and Schmid-Hempel 1991b,c; Sections 8.4, 11.4).

To conclude, searching across the different biological levels for common principles with which to analyze the evolution of individuality by within-unit suppression is, if nothing else, a useful comparative exercise. It is more controversial whether these principles operate just by analogy or as deep-level common rules. For example, Dawkins (1982a p. 134) described a genes-in-individuals and individuals-in-groups analogy as "superficial." These reservations arise because all such comparisons liken genes as subunits to individuals as subunits. But genes are replicators and individuals are not (Section 1.2). This makes a difference because, for example, in a community of interest made of clonal organisms, the coincidence of interest of the organismic subunits is total (see above). But in the genic community of interest, Cosmides and Tooby's (1981) "coreplicon" of separate replicators, this is not necessarily the case. Put another way, the genic community of interest is likely to be always context-dependent, unlike the individual one. Despite these reservations, the arguments of this section do help shed light on both conflict resolution in social insects and on the superorganism concept that is next discussed.

2.6 The Superorganism

The idea of the insect society as a superorganism appears frequently in the social insect literature, often in connection with the concept of colony-level selection, to which it is intimately related. The idea has a long history (reviewed by E.O. Wilson 1971; Ghiselin 1974), but in its modern form it began with Wheeler (1911). Wheeler viewed an insect colony as formally analogous to a single multicellular organism. Both share a division into "germ-plasm" (sex cells in organisms, sexual forms in insect colonies) and "soma" (somatic cells and workers respectively). Both also share the properties of individuality, self-nutrition, self-protection, and regeneration. The superorganism has never ceased to be a common popular conception of insect societies. However, among professional social insect researchers, the idea fell into disrepute. First, since organisms were held to be subject mainly to individual-level selection, the superorganism concept implied that the legitimate level of selection in social insects was the colony. But this was at the time when group selectionism was being most heavily criticized (e.g. West-

Eberhard 1975; Section 2.2). Second, the evident genetic variation inside insect colonies, and the within-colony conflict it engenders, appeared contrary to the idea of a colony as a single, integrated entity (e.g. West-Eberhard 1975; Starr 1979). Third, the superorganism concept failed to generate detailed hypotheses open to experimental test (E.O. Wilson 1971; Starr 1979).

Now, however, the superorganism concept is again being increasingly invoked (e.g. Oster and Wilson 1978 p. 21; D.S. Wilson 1980, 1990; West-Eberhard 1981; Lumsden 1982; Brian 1983 p. 321; Jaisson 1985; Page et al. 1989b; Seeley 1989; Wilson and Sober 1989; Hölldobler and Wilson 1990 p. 358; E.O. Wilson 1990; Leigh 1991; Southwick 1991; Moritz and Southwick 1992; Ratnieks and Reeve 1992). To begin with, the partial rehabilitation of group selection (Sections 2.2, 2.3) has meant that to treat the colony as a focus of selection now seems less unpalatable (West-Eberhard 1981; Moritz and Southwick 1992). Next, Wilson and Sober (1989) have explicitly set out to "revive the superorganism" (see also Lumsden 1982). According to Wilson and Sober (1989 p. 339), the defining characteristic of both organisms and superorganisms is "functional organization," which insect colonies clearly possess. In addition, Wilson and Sober espouse levels-of-selection theory. This means that within-colony conflict can be regarded as just one element in the hierarchy of levels of selection, and that colonies are more or less superorganismic according to the strength of within-colony selection. So the conflict between colony members that previously seemed to undermine the superorganism concept now becomes an aspect of it, via levels-of-selection theory. Wilson and Sober (1989) correctly point out that this approach merely achieves consistency with implicit practice in the use of organism-level (individual selection) phraseology, since organisms are also subject to within-unit conflict. In terms of the last section, Wilson and Sober's (1989) point is that if individuals arise from within-unit suppression of conflict, then the existence of largely suppressed conflict cannot be used to deny individuality to insect colonies. To consider the colony a superorganism subject both to colony-level selection (or now, superorganism-level selection) and to within-colony (within-superorganism) selection, is therefore evolutionarily valid. In light of the earlier reconciliation of colony-level selection, levels-of-selection theory, and kin selection (Sections 2.2–2.4), it is also compatible with gene selectionism.

Another element in the superorganism's revival is the recognition that it is sometimes useful to view the colony as a unit. For example, in ant ecology, it makes sense to regard whole colonies as competitors for space (e.g. Sudd and Franks 1987). In addition, consider the growth phase of a colony when only workers are being produced, or the performance of tasks like foraging. In both cases, little conflict of interest exists between colony members even in species with severe conflict in other facets of colony life, such as sex allocation. So treating the colony as a unit again becomes a useful stance (e.g. Seeley 1989; Schmid-Hempel 1990). Lumsden (1982) and E.O. Wilson (1985a) make a strong case for considering colonies to be like organisms (in particular, self-regulating) in caste allocation. Finally, several researchers on communication and the organization of work among workers have proposed analogies between colonies and units like brains or computers, since in all these the performance of the whole emerges from the intercommunication of subunits (e.g. Franks 1989a; Seeley 1989; Section 12.7). Therefore, as the superorganism concept has regained credibility, it has also begun to generate fruitful hypotheses.

However, Ratnieks and Reeve (1992) have sounded a note of caution. The degree of within-colony conflict may vary with time and across tasks. So it may be better to regard a colony pluralistically as a mosaic of more or less superorganism-like elements (colony-level adaptations), rather than typologically as either a superorganism or not. In the terminology of the previous section, the colony is a set of overlapping context-dependent "communities of interest," rather than a single, monolithic individual. Ratnieks and Reeve (1992) also stress that a community of interest in an insect society need not exhibit Wilson and Sober's (1989) "functional organization." In addition, the best multicellular analogue of an insect society may often be not a vertebrate body, but a fungus or plant (Rayner and Franks 1987; López et al. 1994). This is because of the widespread occurrence of multiple queens and reproductive workers in insect colonies (Chapters 7, 8), which means that many insect colonies do not have the equivalent of the vertebrate body's single "germ line," but multiple germ lines like a fungus or plant. Insect colonies also exhibit modular growth, as do plants (Section 9.6). To conclude, in parallel with colony-level selection and with similar qualifications, the superorganism concept cannot be dismissed from the study of social insects. It may at times be desirable. But it is also not antithetical to gene selectionism.

2.7 Conclusion

A lot of confusion has surrounded the interconnected ideas of kin selection, group selection, colony-level selection, and the superorganism, both in the social insect literature and in evolutionary biology as a whole. This chapter has tried to remove this confusion. In this, an attempt has been made to separate semantic issues from arguments over substance. A key theme is that several ideas that have sometimes been regarded as opposing one another (such as kin selection and group selection, kin selection and colony-level selection, kin-selected conflict and the superorganism concept, and gene selectionism and levels-of-selection theory), need not be so regarded. We reached these conclusions by adhering to the gene selectionist insight that natural selection is about the rise and fall of gene (replicator) frequencies. But the insight from the replicator-vehicle distinction and from levels-of-selection theory – that gene selection occurs in a complex biological hierarchy – was also acknowledged. This means that biologists need to be sophisticated and pluralistic about the modeling of gene frequency change. It also means that fruitful analogies can be drawn between processes concerning genes-in-individuals and those to do with individuals-in-groups. With this background, the following chapters examine from an adaptationist perspective detailed facets of ant sociality.

2.8 Summary

1. The form of group selection (interdemic) proposed by Wynne-Edwards (1962) has little relevance to the evolution of altruism in social insects. Most insect societies are not demes (populations within which mating occurs). In addition, Wynne-Edwards' group selection makes no reference to genetic similarity within groups, whereas social insect colonies are nearly always composed of relatives. D.S. Wilson's (1975) intrademic group selection does invoke within-group genetic similarity. However, critics consider it to be equivalent to kin selection, since the most plausible cause of genetic similarity is relatedness. The spread of a gene for altruism among physical groups of relatives or networks of relatives can be modeled either as kin selection or by Wade's (1980) "components of selection" method, which involves the overwhelming of negative within-group selection ("individual selection") by positive between-group selection ("group selection"). In this sense, kin and group selection are not distinct processes.

2. Consequently, since colony-level selection in social insects is a form of group selection where the groups are usually colonies of relatives, there is no fundamental disagreement between colony-level selection and kin selection. Reinforcing this point, previous usages of the colony-level selection concept implicitly deal with the spread of genes for social actions in a "components of selection" framework. Therefore, colony-level, group, individual, and kin selection are all aspects of gene selection. This means that the practice of attributing traits to, say, either colony-level selection or kin selection is illogical. It may be helpful to

replace the terminology that pits colony-level selection against individual selection, with one invoking processes of between-colony and within-colony selection of genes.

3. Levels-of-selection theory is a way of treating natural selection in terms of the hierarchical levels of biological organization. Below the species level, levels-of-selection theory can be reconciled with gene selectionism via the replicator-vehicle distinction. That is, the levels of levels-of-selection theory are vehicles, and describing selection in terms of these levels is a way of talking about changes in the frequencies of replicators in a structured population. Therefore, many of the differences between levels-of-selection theory and gene selectionism are differences of language. The underlying agreement between levels-of-selection theory and gene selectionism is the wider basis for the reconciling of colony-level, group, and individual selection on the one hand, with kin and gene selection on the other.

4. In units at any level of organization, individuality arises to the extent that within-unit conflict is suppressed. This can come about either via rivalry among selfish subunits, with victory going to the set of subunits having greater power or shared interests (the "community of interest"), or via self-limiting effects of selfish subunits. These principles arguably apply both to within-organism suppression of conflict, and to the maintenance of harmony within insect societies. For example, the settling of queen–worker and worker–worker conflicts over sex allocation and male parentage is in some cases explicable in these terms.

5. The concept of insect colonies as superorganisms also relies on an analogy between organismic and social processes. Given that within-organism conflict occurs, the presence of within-colony conflict cannot be a valid argument against the superorganism concept. Equally, since there is no basic disagreement between colony-level selection and gene selectionism, the view of insect colonies as superorganisms subject to a balance of between-colony and within-colony selection is not incompatible with gene selectionism. For some purposes, such as the study of colony ecology and the division of labor, the superorganism concept may be useful. However, it may be more accurate to regard colonies as sets of shifting "communities of interest" rather than as functionally organized individuals.

3 Kin Selection, Haplodiploidy, and the Evolution of Eusociality in Ants

3.1 Introduction

This chapter discusses the detailed application of kin selection theory (Chapters 1, 2) to the origin of eusociality in the Hymenoptera, particularly in the ants. Although the precise definition of eusociality is controversial (Gadagkar 1994; Crespi and Yanega 1995; Sherman et al. 1995), we take the key feature of a eusocial society to be an altruistic worker caste (Section 1.3). The existence of altruistic workers makes the evolution of eusociality a topic of basic importance, since they provide one of the main examples of the problem of altruism that kin selection theory was devised to address (Section 1.3). Many previous authors have reviewed eusocial evolution in insects. A partial list includes E.O. Wilson (1971), Hamilton (1972), Lin and Michener (1972), Alexander (1974), West-Eberhard (1975, 1978a), Trivers and Hare (1976), Evans (1977), Crozier (1979, 1982), Starr (1979), Vehrencamp (1979), Brian (1983), Andersson (1984), Brockmann (1984), Gadagkar (1985a,b, 1990a, b, 1991a), Jaisson (1985), Grafen (1986), Krebs and Davies (1987, 1993), Sudd and Franks (1987), Alexander et al. (1991), Pamilo (1991a), Seger (1991), and Crespi (1994).

The chapter starts by describing some relevant concepts and issues (Section 3.2). Section 3.3 then proposes a scenario for the origin of eusociality in ants. Next, Section 3.4 considers West-Eberhard's (1987a,b, 1988) "epigenetic" theory of eusocial evolution. The following section introduces Hamilton's (1964b) "haplodiploidy hypothesis" of Hymenopteran eusociality (Section 3.5). After this comes a critique of the haplodiploidy hypothesis (Sections 3.6, 3.7). Section 3.8 deals with the influence of sex ratio and related factors on the origin of worker behavior in haplodiploids. Section 3.9 considers factors promoting worker evolution in all sorts of populations. Finally, Section 3.10 assesses the role of kin selection and haplodiploidy in the origin of Hymenopteran eusociality.

3.2 Concepts in the Origin and Evolution of Eusociality

EVOLUTION OF SOCIALITY VERSUS EVOLUTION OF ALTRUISM

In any discussion of the evolution of eusociality, a distinction needs to be drawn between three pairs of related issues. The first is the difference between the evolution of sociality (group-living) and the evolution of reproductive altruism (Alexander 1974; West-Eberhard 1978a). Sociality can occur without reproductive altruism, as in a group of communal breeders, but altruism always occurs in a social context. This is true by definition, since helping others requires contact with them. Therefore, evolutionary reasons for living in groups exist independently of reasons for reproductive altruism. Factors promoting sociality include the need for defense against predators and parasites, resource patchiness, reduced variance in reproductive success, and the possibility of inheriting group resources (e.g. Hamilton 1971b; Lin and Michener 1972; Alexander 1974; Evans 1977; Myles 1988; Alexander et al. 1991; Reeve 1991; Wenzel and Pickering 1991). However, once group-living is established, these factors can also promote reproductive altruism by their effects on the benefit and cost terms in Hamilton's rule (e.g. West-Eberhard 1975).

Three theories have traditionally been regarded as possible explanations for eusocial evolution – kin selection, parental manipulation, and mutualism (e.g. Andersson 1984). Chapters 1 and 2 discussed the first two, and argued that parental manipulation is in fact a kin selectionist theory (Section 1.9). The mutualism theory envisages altruism evolving after the appearance of mutualistic group-living (Lin and Michener 1972; Itô 1993). It has also been supported as the "hopeful reproductive" theory by West-Eberhard (1978a,b, 1981) and as the "gambling" hypothesis by Gadagkar (1990b, 1991c).

Consider a group of individuals breeding together (communal breeding). Such a society will evolve when an individual's personal fitness (average offspring number) from joining the group exceeds its personal fitness from attempting to breed alone (e.g. Vehrencamp 1983a,b). This condition could be fulfilled even if some joiners failed to breed. For example, if individuals had a very low expectation of successfully breeding alone, they might join a group in which they had only some chance of replacing the breeders, as "hopeful reproductives." If the joiners' presence boosted the productivity of the residents, the society would be defined as mutualistic (cooperative), because on average both parties would gain in personal fitness from the association (Table 1.1). The nonbreeding joiners would not be altruists, except in the loose, short-term sense of "delayed benefits" altruism (Section 1.3). They would not be altruists strictly speaking, because they would not suffer a fall in personal fitness relative to their fitness from solitary breeding.

Eventually, the nonbreeders could be selected to be truly altruistic if the breeders they joined were relatives and the condition in Hamilton's rule were satisfied. The result would be a eusocial society that had evolved from a mutualistic group.

A key conclusion from this argument is that the mutualism theory is not in fact a competitor with kin selection theory as an explanation of eusocial evolution. As already indicated, the mutualism theory cannot, by definition, account for altruism, which involves a net fall in offspring to the actor (e.g. Crozier 1979; Andersson 1984; Section 1.3). The mutualism theory instead states that a likely first step towards eusociality is mutualistic group-living. Altruism must still eventually emerge through interactions between relatives under conditions satisfying Hamilton's rule.

Therefore the three traditional theories of eusocial evolution are not true alternatives at all. As just argued, parental manipulation is a kin selectionist theory and the mutualism theory cannot explain the evolution of altruism. At present, then, the only explanation for true reproductive altruism in eusocial societies is kin selection.

PATHWAYS TO EUSOCIALITY VERSUS THEORIES FOR EUSOCIALITY

A second distinction is between pathways to eusociality and theories of its evolution. Kin selection is a theory for the evolution of altruism in eusocial societies. However, kin-selected eusociality conceivably arose via two pathways, namely the subsocial and semisocial routes (Michener 1969; E.O. Wilson 1971 p. 99; Lin and Michener 1972; Starr 1979; Seger 1991). Many species of Hymenoptera – the so-called solitary or nonsocial species – show parental care. This involves females rearing their own young alone by building a nest for their larvae and provisioning it with food. In the subsocial route, some offspring remain in the natal nest to help care for sibs. The nondispersal of these helpers leads to the overlap of adult generations that occurs by definition in eusocial societies. In the semisocial route, members of the same generation in a species with parental care form aggregations. Some individuals then become helpers that rear the offspring of their nestmates. Eventually some of these helpers are present when members of the second generation are adult, some of which may also become helpers, again leading to an overlap of generations in the colony.

West-Eberhard's (1978a) "polygynous family hypothesis" proposes a more elaborate route to eusociality, suggesting that, in wasps, semi-sociality could lead to a subsocial society (West-Eberhard 1978a; see also Carpenter 1989). Michener (1985) pointed out that the transition from solitary to eusocial, whichever route it takes, need not involve a series of intermediate species. This point is supported by populations of bees containing both solitarily and jointly nesting females (Michener 1985). It is also consistent with recent ideas that propose between-colony sex ratio variation within populations of facultatively social Hymenoptera as an important facilitator of worker evolution (e.g. Seger 1991; Sections 3.6, 3.8).

Although routes to eusociality need to be kept conceptually separate from theories for social evolution, the two issues are nonetheless connected. For example, the idea that kin-selected altruism evolves within mutualistic groups usually envisages a semisocial route. More generally, in kin selection theory, altruism evolves at a lower benefit : cost ratio when directed at sibs than when directed at nieces and nephews. So, other things being equal, a subsocial route is more conducive to kin-selected altruism than a semisocial one, as Section 3.6 further discusses.

Both the subsocial and the semisocial routes presuppose a species showing parental care. This matters because, as explained in Section 1.4, it indicates that kin-selected genes for altruism are already present within populations, and that the altruism is behaviorally elaborate. This facilitates the later appearance of genes for equally elaborate altruism towards, say, sibs (Dawkins 1979). Therefore, altruism towards kin other than offspring need not arise spontaneously in eusocial evolution, as is sometimes argued. The occurrence of parental care also helps explain the taxonomic distribution of eusociality (Alexander et al. 1991; Section 3.7). But the current point is frequently neglected because parental care is often mistakenly viewed as not involving altruism at all (Section 1.4).

ORIGIN VERSUS MAINTENANCE OF EUSOCIALITY

A final distinction is between the origin of eusociality and its mainte-nance (Trivers and Hare 1976; Charnov 1978a; West-Eberhard 1978a). This is important because the factors affecting the origin of eusociality could differ from those maintaining it. Therefore, the study of advanced eusocial species like ants may be relatively unhelpful in clarifying euso-ciality's origin (e.g. Crozier 1982). For example, their kin structure may differ from the ancestral one (e.g. Ross and Carpenter 1991a). There are also no living solitary ants with which to compare the social species. So studies of kin selection in ants are implicitly about the maintenance of eusociality (Crozier 1982). To study kin selection and eusocial origins, it is better to look at primitively eusocial species (Crozier 1989; Packer and Owen 1994), and especially at bee and wasp taxa that contain both primitively eusocial and solitary representatives (e.g. Michener 1985; Yanega 1988; Packer 1990; Seger 1991). Any findings could then be useful in reconstructing the likely evolution of early ants.

3.3 The Origin of Eusociality in Ants

Ants probably evolved eusociality by the subsocial route (e.g. Jaisson 1985). The evidence for this is necessarily indirect, since ants originated from solitary ancestors more than 80 million years ago (Hölldobler and Wilson 1990 p. 23). However, a subsocial route is indicated because modern ant societies typically develop by this pathway (Brockmann 1984), with a foundress queen whose offspring include nondispersing, worker individuals. Some ant species, including species belonging to "primitive" subfamilies, show colony foundation by semisocial groups of queens (Hölldobler and Wilson 1990 p. 28). But such groups lack a reproductive division of labor, are usually temporary, and typically consist of unrelated individuals (Strassmann 1989; Section 8.3). They therefore represent poor models for the origin of altruism in early ants. The second reason for supposing that ants followed a subsocial route is that the tiphiid wasps and their relatives, the modern vespoid wasps most closely resembling the ancestors of ants (E.O. Wilson 1971 pp. 32, 72; Hölldobler and Wilson 1990 p. 24), show rudimentary parental care but no semisociality (E.O. Wilson 1971 p. 72).

Assuming a subsocial route, this section outlines how eusociality may have originated in the ants. Biologists cannot know the details of this process. E.O. Wilson (1971 pp. 72–74) and Hölldobler and Wilson (1990 pp. 27–29) make plausible suggestions. Instead, the aim is to present the logical structure of the evolutionary sequence that ants may have followed. Note that the proposed sequence is very general, since it could equally apply to birds and mammals with subsocial helpers (Dawkins 1979; Bourke 1992). In ants, the sequence could have occurred twice, since ants may form two independently evolved eusocial complexes (E.O. Wilson 1971 p. 32; but see R.W. Taylor 1978; Baroni Urbani 1989a; Hölldobler and Wilson 1990 p. 27).

Consider a wasp-like ancestor of the ants that exhibited maternal care of the young like today's solitary Hymenoptera. Genes for (parental) altruism therefore existed within its populations. Daughters of the solitary mothers originally dispersed from the nest to breed themselves. (Daughters alone are considered because all Hymenopteran workers are female, although reasons for this remain controversial [Sections 3.6 and 3.7].) Suppose that genetic variation then arose for the timing of dispersal. Some daughters therefore departed later than others, and so were present as adults when their mothers were still laying eggs. (Delayed departure might itself have been favored for ecological reasons: e.g. Myles 1988; Alexander et al. 1991.) It seems likely that solitary Hymenopteran females are behaviorally preprogramed to feed young in a nest, since these young would usually be their own. Therefore nondispersing daughters might have shown misplaced parental care for the foundress female's young, through a passive triggering of the neural machinery affecting care behavior (e.g. West-Eberhard 1987a).

Assuming care to be costly, the nondispersing females would then have been promoting their sibs' welfare at a cost to their own. So the genes for late departure would, by definition, be genes for altruism (Sections 1.3, 1.4). Further, this sib altruism would have been just as elaborate as the previously existing parental care, because, as regards its neural machinery, it would have been the same behavior. This is how parental care can facilitate the appearance of sophisticated altruism towards kin apart from offspring (Dawkins 1979; previous section).

Two genotypes would now have existed in the population. The first would have been the ancestral genotype, present in mothers of "dispersing-and-breeding" daughters and in these daughters themselves. The second would have been a genotype involving both "dispersing-and-breeding" and "nondispersing-and-helping". In other words, it would have been a genotype specifying facultative altruism, present in mothers of helpers, in their helper daughters, and in their dispersing daughters, but differentially expressed in these classes of individual. (Section 1.7 and the next section discuss the facultative nature of genes for altruism.) Provided Hamilton's rule was satisfied, the altruist genotype would have spread through the population. In this, it could have been aided by sex ratio variation in haplodiploid populations (Sections 3.6, 3.8), and by factors enhancing the benefit: cost ratio of helping (Section 3.9). The result would have been a population consisting entirely of subsocial colonies headed by reproductive mothers and containing worker daughters that reared dispersing, reproductive sibs. These were the first ants.

3.4 The Epigenetic Theory of Insect Sociality

West-Eberhard (1987a,b, 1988) proposed an "epigenetic" theory of the origin and initial spread of eusociality in insects. West-Eberhard characterized the theory as a scheme for eusocial evolution "without a gene for altruism" (1987b p. 42). It could therefore be construed as opposed to kin selection. This section argues that the epigenetic theory is in fact consistent with kin selection. Crozier (1992) made a similar point, to which West-Eberhard (1992) responded.

The epigenetic theory envisages that eusociality starts with casteless mutualistic groups (West-Eberhard 1987a,b, 1988), which are probably common in wasps (Section 3.2). Suppose that in such groups reproductive competition leads some egg-laying females to dominate others. Dominated wasps tend to resorb the eggs in their ovaries and, at the same time, to show (misplaced parental) care of the larvae. So dominated females would effectively become workers. If all groups eventually had them, the result would be a fully eusocial population.

Therefore, the epigenetic theory makes the following claims. (1) Group-living and within-group competition are sufficient to explain the origin of a nonreproductive worker caste. (2) This caste may appear as a side-effect of these phenomena, and so initially be a by-product of selection in contexts other than selection for worker behavior. (3) This process does not involve the competition among alternative alleles invoked by standard kin selection theory. Rather, since worker behavior is facultative (dominated females remain potential egg-layers), it involves the expression of alternative phenotypes of a single genotype. This is why the epigenetic theory argues for the absence of genes for altruism. (4) The role of kin selection is to maintain worker behavior and to fine-tune the regulatory mechanism that determines when worker behavior is expressed.

We do not dispute the natural history behind West-Eberhard's proposals, but its interpretation. First, groups of jointly-nesting females may form mutualistically to begin with, but the later emergence of altruism requires kin selection (Section 3.2). This argument applies even if group-nesting females remained mutualists after they had started to exhibit worker behavior (that is, acquired more offspring on average from working than from not working, perhaps because of the possibility of inheriting the nest), as West-Eberhard (1978b, 1992) suggested. However, it seems unlikely that in all cases early workers were mutualists, and available data indicate that workers are indeed altruists (suffer a net fall in offspring output) in some primitively social wasps (e.g. Grafen 1984). Second, the epigenetic theory does involve genes. At the start of any process of worker evolution, females must have a choice between actions that lead them to become workers or breeders (for example, staying in or leaving the nest). As with other forms of behavior, this choice behavior is likely to be subject to genetic variation. If working is altruistic, then the gene associated with the decision that leads to working must be the gene for altruism (Section 1.4; previous section). This is the case even for a gene that influences a female to stay in the nest, become reproductively inhibited, and show misdirected parental care. So, third, whether or not worker behavior originates as a "side-effect" is irrelevant to the fate of the gene for it. Perhaps a confusion over what logically constitutes a gene for altruism (Dawkins 1979) resulted in the apparent absence of genes from the epigenetic theory.

Fourth, the issue now is whether the gene for worker behavior spreads, that is whether there is selection of the gene for facultative staying-and-working behavior at the expense of the gene for dispersing-and-breeding. This will depend on the terms in Hamilton's rule. So what happens depends on orthodox kin selection, as Crozier (1992) also argued.

Lastly, a gene for altruism virtually has to be facultatively expressed to undergo positive selection according to kin selection theory (Section 1.7). (Again, Crozier [1992] argued as much.) This is because carriers of a gene for obligate altruism would help reproductive noncarriers, so such a gene could never spread. If a gene for facultative helping spreads to fixation, the result is a eusocial population with nongenetic caste determination. So, nongenetic caste determination is expected from kin selection theory, and is not a unique feature of the epigenetic approach (Section 1.7).

West-Eberhard (1992 p. 224) repeated her belief that "*all* models depicting genes for altruism, including those with conditional expression, may be irrelevant to the evolutionary establishment of conditional worker behavior in a population" (emphasis as in original). This is because she maintained that the evolution of sterile workers occurs as a side-effect of group-living and within-group aggression, and so does not involve genes for altruism (or, implicitly, genes for mutualism if working is mutualistic). However, in sum, none of the key elements of the epigenetic theory – mutualism as a factor promoting group-living, worker behavior arising as a side-effect, and the facultative nature of worker behavior – is inconsistent with kin selection. On the contrary, these features are either compatible with kin selection theory or demanded by it. Moreover, genes (such as the gene for facultative altruism required by kin selection theory) are implicitly present in the epigenetic theory. The epigenetic theory has been useful in emphasizing a possible natural history context of social evolution, and the facultative nature of worker behavior. But it should not be viewed as an alternative to standard kin selection as a theory of eusocial evolution.

3.5 The Haplodiploidy Hypothesis

A complete explanation of eusocial evolution in the Hymenoptera should address three features of eusociality in this group. These are that: (1) all eusocial species are in the aculeate division, the members of which either bear stings or have secondarily lost them; (2) there have been multiple, independent origins of eusociality; and (3) workers are always female, whereas termite workers are both male and female (E.O. Wilson 1971 pp. 104, 327).

Hamilton (1964b, 1972), as well as devising kin selection theory, proposed a hypothesis to account for the second and third of these features. This is now known as the "haplodiploidy" or "3/4 relatedness" hypothesis (West-Eberhard 1975 p. 23). All Hymenoptera are haplodiploid (have males developing from unfertilized haploid eggs and females from fertilized diploid eggs [Box 3.1]). This creates unusual coefficients of relatedness between family members compared to those found in diploid families (Table 3.1). In particular, haplodiploidy makes sisters in a simple family structure especially highly related (by three-quarters), because they always share genes contributed by their haploid father (Box 3.1). In haplodiploids, sisters are in fact more closely related with each other than with offspring. *The haplodiploidy hypothesis therefore proposes that high sister–sister relatedness facilitates the evolution by kin selection of reproductive altruism among Hymenopteran females, so explaining both the prevalence of eusociality and the restriction of worker behavior to females in the Hymenoptera* (Hamilton 1964b).

The haplodiploidy hypothesis has been interpreted in several ways, not all of which are equally valid. First, as discussed by Dawkins (1989a p. 316), it has sometimes been equated with kin selection as a whole. But kin selection theory applies under many regimes of relatedness (Section 1.4), so does not stand or fall with the truth or falsity of the haplodiploidy hypothesis. The hypothesis is instead a subset of kin selection theory (e.g. West-Eberhard 1975, 1978a; Gadagkar 1991a). Second, the haplodiploidy hypothesis has been taken to mean that a relatedness asymmetry in favor of sibs explains the multiple origins of eusociality in the Hymenoptera. (Here, relatedness asymmetry refers to females being more closely related with sisters than with offspring. The term can also mean a greater relatedness with sisters than with brothers [Section 4.2].) A lack of relatedness asymmetry would therefore falsify the hypothesis (e.g. Gadagkar 1991a). However, even with no relatedness asymmetry, haplodiploidy might still predispose species to eusociality relative to diploidy. For example, multiple mating by queens reduces the relatedness asymmetry in favor of sibs in haplodiploids, yet

BOX 3.1 RELATEDNESS AND REPRODUCTIVE VALUE
IN THE SOCIAL HYMENOPTERA

Relatedness (Boxes 1.1, 1.3) in the social Hymenoptera is greatly affected by haplodiploidy in this group. Haplodiploidy is the sex determination system in which males develop from unfertilized, haploid eggs, and females develop from fertilized, diploid eggs. (The production of males and the production of females from unfertilized eggs are termed *arrhenotoky* and *thelytoky* respectively. Exceptionally, Hymenopteran females develop by thelytoky [e.g. E.O. Wilson 1971 p. 325; Crozier 1975; Bourke 1988a; Cook 1993]. Also, some Hymenopteran males arise from fertilized, diploid eggs, but they are sterile [Cook 1993; Box 8.2].) Haplodiploid invertebrates aside from the Hymenoptera include mites, ticks, beetles, whiteflies, scale insects, thrips, nematodes, and rotifers (Crozier 1979; Bull 1983 p. 148; Andersson 1984).

Consider relatedness among sisters in a Hymenopteran colony with a single, once-mated, outbred queen and male-producing workers (Figure 3.1). Haploid males produce genetically uniform sperm by mitosis, so sisters inherit any one gene in their father with a probability of 1. Diploid females produce gametes by meiosis, so sisters inherit a gene in their mother with a probability of 0.5. A female's genome is in two equal portions, consisting of chromosomes inherited from the father's sperm and the mother's egg respectively. Averaging over these, sisters share any one gene with a probability of $(1 + 0.5)/2 = 0.75$. This contrasts with the coefficient of relatedness between sisters in diploids of 0.5. Similar considerations reveal that Hymenopteran families with the above structure have the relatedness levels shown in Table 3.1.

Table 3.1 shows two schemes for expressing relatedness. The first involves regression relatednesses, as defined in Box 1.1. The second involves "life-for-life" relatednesses (Hamilton 1972 p. 203), as used for example by Trivers and Hare (1976) and many other writers on social insects. The existence of two relatedness measures is also a consequence of haplodiploidy. The life-for-life relatedness values differ from the regression relatedness ones because they include an implicit term for *relative sex-specific reproductive value* (Hamilton 1972; Grafen 1986). To be precise, the two types of relatedness are connected by the following formula:

Life-for-life relatedness = Regression relatedness × (Sex-specific reproductive value of the recipient : Sex-specific reproductive value of the actor).

Reproductive value is a concept from life history theory and is defined as the long-term per capita fitness payoff from a class of individuals in the absence of factors such as selection, mutation, and genetic drift (e.g.

BOX 3.1 CONT.

Grafen 1986). Fisher (1930) derived an expression for the reproductive value of individuals of a given age class (e.g. Horn and Rubenstein 1984; Stearns 1992 p. 26). Age-specific reproductive value is relevant to kin selection studies in that, other things being equal, altruists should prefer to help when their reproductive value is low and that of the recipients is high, as, for example, in old and young individuals respectively (e.g. Fisher 1930 p. 27; Hamilton 1964b; West-Eberhard 1975; Milinski 1978; Horn and Rubenstein 1984; see also Charlesworth and Charnov 1981; West-Eberhard 1981; Mumme et al. 1989; Taylor 1990).

However, as already indicated, the relevant concept in the present context is sex-specific reproductive value. In haplodiploids, by definition, females have twice the ploidy (chromosome number) of males. As a result, females pass on a focal gene twice as efficiently as males, since (assuming worker sterility) female genes enter both sexes in the next generation, whereas male genes enter females (daughters) only (Figure 3.1). Therefore, females have twice the reproductive value of males, as Box 4.4 formally shows (Hamilton 1972; Pamilo and Crozier 1982; Grafen 1986; Taylor 1988, 1990; Pamilo 1991b; Crozier and Pamilo 1993). Given this, any expression for a sex-specific fitness payoff in haplodiploids must be weighted by reproductive value. Life-for-life relatedness, as has been seen, is a compound measure incorporating regression relatedness and an implicit reproductive value weighting.

The fact that the $2:1$ ploidy ratio of the sexes sets the ratio of their reproductive values explains why, in Table 3.1, the life-for-life coefficients are the same as the regression ones within sexes (there is no ploidy difference within sexes). It also explains why, between sexes, the conversion factor for changing between the two sets of coefficients is the ratio of ploidies of actor and recipient (Grafen 1986). In diploids, the sexes have equal ploidies and hence reproductive values, so regression relatedness and life-for-life relatedness are the same (Grafen 1991). If Table 3.1 referred to diploids, all values would be 0.5, except those in the last two rows, which would be 0.25.

If Hymenopteran workers reproduce, the relative reproductive value of females and males alters, because male genes are now passed on to males of the next generation (workers' sons) (Box 4.4). This means that it is often more useful to use regression relatedness than life-for-life relatedness when analyzing fitness strategies in social Hymenoptera. Only then can the relatedness and the reproductive value components of fitness be explicitly separated (Grafen 1986). The above calculations assumed discrete generations. Complex patterns of generation overlap also alter reproductive values and so again make regression relatednesses necessary for analytic purposes (Grafen 1986).

BOX 3.1 CONT.

Finally, another factor affecting relatedness is inbreeding (e.g. Hamilton 1972; Michod 1979; Seger 1981; Grafen 1985; Pamilo 1985, 1991a; Queller 1989a; Lessard 1992). Inbreeding means that mating pairs are related, whereas all previous calculations assumed that they are unrelated. Consequently, inbreeding may also influence social evolution (e.g. Starr 1979; Wade and Breden 1981; Michod 1982, 1993; Uyenoyama 1984). However, its effects are not always clear or obvious (Michod 1993). Furthermore, inbreeding does not seem to have had important effects on the origin of Hymenopteran eusociality. There is little evidence for inbreeding in primitively social Hymenoptera (Gadagkar 1991a; Ross and Carpenter 1991a). In addition, inbreeding does not affect relatedness with offspring and sibs differentially (Hamilton 1972; Trivers and Hare 1976; Craig 1982a; Pamilo 1991a). Blows and Schwarz (1991) also found that inbreeding did not contribute to between-group genetic variance (relatedness) in a facultatively social bee. Inbreeding would affect relatedness with offspring and sibs differentially if it alternated with outbreeding. Myles and Nutting (1988) review the significance of such cyclical inbreeding for social evolution in the termites.

---

sisters remain more closely related than diploid ones for any given mating frequency (Page and Metcalf 1982). Therefore, the interpretation of the haplodiploidy hypothesis as requiring a relatedness asymmetry is restrictive. This interpretation also lead to ingenious, but largely unsuc-

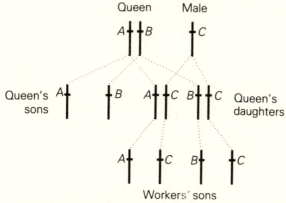

Figure 3.1 The pedigree of a Hymenopteran colony, in which a single queen mates with a single male and produces sons and daughters (workers and new queens), with the workers in turn producing sons. A, B, and C are alleles on representative chromosomes (vertical bars).

cessful attempts to find widespread genetic conditions creating relatedness asymmetries among the wholly diploid termites (reviewed by
Myles and Nutting 1988; Seger 1991).

Table 3.1

Relatedness Levels in a Colony of Social Hymenoptera

Actor, Recipient	Regression Relatedness	Life-for-Life Relatedness
Queen, daughter	0.5	0.5
Any female, sons	1.0	0.5
Father, daughter	0.5	1.0
Queen's mate, queen's sons	0.0	0.0
Sister, sister	0.75	0.75
Sister, brother	0.5	0.25
Brother, sister	0.25	0.5
Brother, brother	0.5	0.5
Queen, workers' sons	0.5	0.25
Nonlaying worker, workers' sons	0.75	0.375

NOTES: These relatedness values are for colonies with single, once-mated, outbred queens (Figure
3.1). Regression and life-for-life relatedness are as defined in Boxes 1.1 and 3.1 respectively. After
Grafen (1986, 1991).

A final interpretation is that haplodiploidy – through its effects on
relatedness – predisposes a species to eusociality under the given biological conditions, relative to diploidy (Page and Metcalf 1982;
Gadagkar 1985b). This is the one favored here. With these points stated, this and the next two sections explain the logic of the haplodiploidy
hypothesis in more detail and evaluate its status.

As already described, a key consequence of haplodiploidy is that, in
the simplest kind of family, females are more closely related with sisters
(by $r = 0.75$) than with any offspring they might produce ($r = 0.5$)
(Table 3.1). So Hamilton (1964b, 1972) suggested that if the benefit and
cost terms in Hamilton's rule (Box 1.2) were simply assumed to be
equal (meaning a female can raise one sister as efficiently as one daughter), the genetic gain to a female from raising a sister would exceed that
from raising a daughter. More formally, a gene for caring for females
would be more strongly selected if sisters rather than daughters
received care, because for the same effort more copies of the gene
would be added to the population. From this followed Hamilton's
(1964b) conclusion that high sister–sister relatedness promotes worker
behavior among female Hymenoptera.

Hamilton also recognized three other consequences of the peculiar relatedness values brought about by haplodiploidy. (The following account employs the "life-for-life" coefficients of relatedness used by Hamilton [1964b] [Box 3.1]. The essential arguments remain the same regardless of whether these or regression relatedness coefficients are used.) First, females are more closely related with sons ($r = 0.5$) than with brothers ($r = 0.25$), suggesting that workers should retain their male-producing ability and try to replace the queen's male eggs with their own (Hamilton 1964b, 1972; Section 7.4). Second, females are more closely related with sisters ($r = 0.75$) than with brothers ($r = 0.25$), suggesting that workers should concentrate on raising females and that sex ratios in haplodiploids will be female-biased (Hamilton 1964b, 1972). Hamilton did not fully state his reasoning on this issue, which was later fully explored by Trivers and Hare (1976) and many other authors (Section 3.8; Chapter 4).

Finally, males are not more closely related with sibs ($r = 0.5$) than with offspring ($r = 0$ with the sons of mates and $r = 1$ with daughters). (Hamilton [1964b] originally calculated male-sib and male-daughter life-for-life relatednesses as, respectively, 0.25 and 0.5, an error he [1971c] and Crozier [1970] later corrected. The basic point remains unaffected.) Therefore, unlike the case for females, there is no genetic predisposition to helper behavior in the male sex (Hamilton 1964b, 1972). This was how the haplodiploidy hypothesis explained female-only workers in the Hymenoptera.

3.6 A Critique of the Haplodiploidy Hypothesis (I)

This and the following section present a critique of Hamilton's (1964b) haplodiploidy hypothesis for frequent eusocial evolution among female Hymenoptera. Many previous publications have also reviewed the status of this idea (e.g. E.O. Wilson 1971; Alexander 1974; West-Eberhard 1975; Trivers and Hare 1976; Crozier 1979; Starr 1979; Vehrencamp 1979; Andersson 1984; Gadagkar 1985a,b, 1990a,b, 1991a; Grafen 1986; Stubblefield and Charnov 1986; Hölldobler and Wilson 1990; Itô 1993). This section describes various arguments against the hypothesis, then assesses their significance both for the hypothesis and for kin selection in general.

To begin with, several spurious criticisms of the haplodiploidy hypothesis need addressing. One is the argument that, since many haplodiploid taxa – including many Hymenoptera – are not social, factors aside from haplodiploidy must be important in eusocial evolution. A similar argument is that haplodiploidy cannot be necessary for eusociality, since the diploid termites are eusocial (e.g. E.O. Wilson 1971 p. 329; Alexander 1974). These points correctly show that haplodiploidy does not inevitably lead to eusociality, and that eusociality does not require haplodiploidy. But Hamilton (1964b) never claimed otherwise. Hamilton (1972) also warned that the influence of haplodiploidy in the Hymenoptera should not be overemphasized. Therefore, these are valid criticisms of an exclusive preoccupation with haplodiploidy, but not of the haplodiploidy hypothesis itself.

The haplodiploidy hypothesis has also been criticized for focusing on the genetics of eusociality, and so ignoring ecological promoters of social evolution (e.g. Lin and Michener 1972; Alexander 1974; Evans 1977; Andersson 1984). Undoubtedly, the hypothesis did lead to a preoccupation with genetic factors among researchers in insect social biology, as Andersson (1984) discussed. But this does not represent a logical criticism. This is because the haplodiploidy hypothesis is a subset of kin selection theory, which acknowledges in Hamilton's rule the importance of both genetic and ecological factors in eusocial evolution (Section 1.4). The rest of this section now considers three legitimate criticisms of the haplodiploidy hypothesis.

CRITICISM 1

The haplodiploidy hypothesis assumes a subsocial route to eusociality (social groups consist of parent and offspring), but some bees and wasps seem to have followed a semisocial one (groups are composed of individuals from one generation) (Lin and Michener 1972; Evans 1977; West-Eberhard 1978a).

This point is not a problem for kin selection theory, which can easily accommodate semisociality provided that aggregating females are related (West-Eberhard 1975). In primitively social bees and wasps, this is nearly always the case. Reports of groups of unrelated females (e.g. Lin and Michener 1972) do not appear to have been substantiated (West-Eberhard 1978a). Alternatively, unrelated pairs of females may occur in mutualistic associations (Stark 1992), or represent attempts at usurpation (Field and Foster 1995). But lifetime altruism inside semisocial groups of unrelated Hymenopteran females has never been demonstrated.

This objection is also not a problem for the haplodiploidy hypothesis, except in its restrictive sense requiring potential workers to be more closely related with sibs than with offspring (relatedness asymmetry). In the simplest kind of semisocial route, females initially help raise nieces and nephews, with which they are related by a life-for-life relatedness of 0.375 (Table 3.1). The relatedness asymmetry in favor of rearing relatives other than offspring that exists under subsociality is therefore abolished. However, in diploids, the corresponding relatedness value would be 0.25. So haplodiploids require a lower benefit : cost ratio than diploids for the evolution of semisocial helping (Trivers and Hare 1976; Andersson 1984; Sudd and Franks 1987 p. 5). This means that, relative to diploidy, haplodiploidy may still predispose a species to eusociality via a semisocial route.

In addition, the semisocial route is itself not problem-free. Under subsociality, above-zero relatedness is achieved within colonies by the failure to disperse of a proportion of sibs (prospective workers). In the simplest model, all the reproductive individuals in a new generation disperse as adults, either to mate (males) or to found new colonies (females). In the semisocial route, however, to achieve aggregations of related females of the same generation, many reproductive females must routinely fail to disperse, and instead must attempt to join a colony in the natal area. This leads to genetic "population viscosity" (Hamilton 1964b p. 36). A viscous population consists of pockets of individuals that have more recent common ancestry than random individuals, due to incomplete mixing. Hamilton (1964b) invoked this population structure to explain foundress associations in wasps, and it was explicitly proposed as a factor in semisocial routes to eusociality by West-Eberhard (1978a).

However, recent models suggest that, although population viscosity increases local relatedness, it may also act *against* the evolution of altruism (Pollock 1983; Grafen 1984; Kelly 1992a,b, 1994; Queller 1992c, 1994a; Taylor 1992a,b; D.S. Wilson et al. 1992). One way of seeing why is with Wade's (1980) "components of selection" method. This shows that, for altruism to evolve by kin selection, groups with altruists must be more productive than those without (positive between-groups component of selection [Box 2.1]). But say group size is constrained in a viscous population (there is what D.S. Wilson et al. [1992 p. 332] term "local population regulation"). Then groups with many altruists cannot benefit fully from their differential productivity, since their extra offspring – to the extent that they fail to disperse – will displace existing members of the group. This diminishes the between-group component of selection, and so hinders the evolution of altruism. Put another way, in a viscous population, like genotypes tend to remain in competition with each other, cancelling out the beneficial effects of the altruistic behavior. Population viscosity need not always hamper the evolution of altruism (e.g. Goodnight 1992). But the conclusion from models like that of D.S. Wilson et al. (1992) is that it often does.

In fact, kin selection for altruism seems to operate most easily when "viscous" and dispersal phases alternate within the lifecycle (West-

Eberhard 1988; Queller 1992c; Taylor 1992a,b; D.S. Wilson et al. 1992). This is because these two phases allow, respectively, the build-up of relatedness within groups, and the full expression of the differential productivity due to altruism between groups. The subsocial route to eusociality involves such an alternation, but the semisocial one does not, or not to the same extent. Therefore, the harmful effects of population viscosity on the evolution of altruism are a greater problem for the semisocial route than for the subsocial route (Pollock 1983; Queller 1992c).

CRITICISM 2

Multiple mating by queens (polyandry) and multiple queens within colonies (polygyny) are common in modern social Hymenoptera. But eusocial evolution is facilitated under monogamy, since – in both haplodiploids and diploids – monogamy makes daughters of a single queen genetically indifferent as to whether to raise sibs or offspring (Hamilton 1972; Charnov 1978a). (In the Hymenoptera, whether this is strictly true depends on sex ratio effects [Box 3.2].) In addition, both multiple mating and polygyny reduce within-colony relatedness, particularly among females, so diminishing the relatedness asymmetry in favor of rearing sisters and weakening the haplodiploidy hypothesis (e.g. E.O. Wilson 1971 pp. 329, 332–333; Wade 1982b). This point was also made by Hamilton himself (1964b).

The existence of low within-colony relatedness is not necessarily harmful to kin selection theory, since Hamilton's rule can operate at low relatedness (Section 1.4). For example, in advanced eusocial species with multiple mating, workers may help despite low relatedness because of their and their queens' morphological specialization for their roles. This effectively means that the benefits of helping are high and the costs low (West-Eberhard 1975; Gadagkar 1991b), and these are precisely the conditions in which altruism may be favored when relatedness is low. Effects on the benefit : cost ratio due to ecological factors (see below) or demographic factors (Nonacs 1991a; Section 3.9) may also promote worker behavior at low relatedness.

Some authors have suggested that low relatedness levels would not be a problem for the haplodiploidy hypothesis if there is kin discrimination. Perhaps female Hymenoptera in colonies with low relatedness recognize their full sibs and preferentially direct their altruism towards them. This would effectively restore the situation to one of high relatedness (Hamilton 1964b; Page and Metcalf 1982; Gadagkar 1985b, 1990b). However, current opinion holds that within-colony kin discrimination in social Hymenoptera is in fact unusual, involves only small effects, and may be nonadaptive (e.g. Carlin 1988, 1989; Venkataraman et al. 1988;

Grafen 1990a; Barnard 1991; Section 7.3). So kin discrimination cannot be invoked to overcome the criticism of the haplodiploidy hypothesis that relatedness is not high enough (Gadagkar 1991a).

However, multiple mating and polygyny may still not represent a problem for the haplodiploidy hypothesis. For one thing, under any given number of queen matings, relatedness between sisters is still higher in haplodiploids than in diploids (e.g. Page and Metcalf 1982). So haplodiploidy remains more conducive to helping than diploidy (Trivers and Hare 1976; Andersson 1984; Sudd and Franks 1987 p. 5).

Gadagkar (1990a, 1991a) tried to quantify the effect of low relatedness on the haplodiploidy hypothesis. He aimed to test the hypothesis in its sense that requires a relatedness asymmetry in favor of sibs, and argued that, for sib-rearing to be favored over offspring-rearing, females need to be related with sisters by at least 0.604. (This "*haplodiploidy threshold*" was calculated as the minimum relatedness among females that would yield an average relatedness [relatedness with the sexes weighted by abundance] of workers with sibs of 0.5, since this is the average relatedness of offspring-rearers with brood. It was assumed that female–female relatedness falls due to multiple mating and that workers skew the sex ratio to equal the ratio, relatedness with females : relatedness with males.) But a survey of both primitive and advanced social Hymenoptera suggested that, because of multiple mating and similar factors, relatedness rarely exceeds 0.604. Gadagkar (1990a, 1991a) concluded that the relatedness asymmetry created by haplodiploidy does not explain either the origin or the maintenance of Hymenopteran eusociality.

However, the calculation of the haplodiploidy threshold is flawed because it ignores the influence of mating success (per capita number of mates). Box 3.2 demonstrates that, when worker–sister relatedness is 0.75 (there is single mating) and the population sex ratio is 3 : 1, workers and offspring-rearers have equal fitness payoffs, because the effect on the workers' payoff of raising more females is cancelled out by the lower mating success of females (e.g. Grafen 1986). It also shows that workers have a lower fitness payoff than offspring-rearers at all mating frequencies greater than one, which includes the frequency corresponding to a worker–female relatedness of 0.604. (The fitness payoffs of workers and offspring-rearers [Expressions 3.1 and 3.2 in Box 3.2] are independent of sex ratio because of the effect of mating success, and the only relatedness value that is changed by multiple mating is the worker–female relatedness of 0.75 in Expression 3.1, which falls [Box 4.3].) Therefore, a haplodiploidy threshold of 0.604 is not meaningful; if anything, the threshold should equal 0.75.

This conclusion arguably challenges the haplodiploidy hypothesis even more than Gadagkar's, because few social Hymenoptera have a female–female relatedness of 0.75 (Gadagkar 1990a, 1991a). But the whole approach of testing the hypothesis with a haplodiploidy threshold is problematic. First, as already discussed, the haplodiploidy hypothesis does not necessarily require the existence of a relatedness asymmetry. Second, a high fraction of measured relatedness values are from advanced social species, whose colony kin structure may have altered over evolutionary time (e.g. Ross and Carpenter 1991a). For example, multiple mating could be a secondary trait (e.g. E.O. Wilson 1971 p. 330). Similarly, at least in ants, polygyny could be a derived condition (Hölldobler and Wilson 1977; Section 8.4). Therefore, genetic tests of the haplodiploidy hypothesis should concentrate on the kin structure of solitary and primitively eusocial Hymenoptera (Gadagkar 1985b; Jaisson 1985). In these groups, levels of relatedness among females approach 0.75 in some cases, for example the halictine bee *Lasioglossum*, the sphecid wasp *Microstigmus*, and the polistine wasp *Polistes jadwigae* (Gadagkar 1991a; Ross and Carpenter 1991a; Packer and Owen 1994; Tsuchida 1994).

Third, the significance of observed relatedness values for eusocial evolution cannot be judged without a knowledge of the benefit and cost terms in Hamilton's rule. For example, the survivorship of colonies founded by single females in primitively eusocial wasps is probably low (Gadagkar 1990b; Reeve 1991; Queller 1994b), reducing the cost of helper behavior and so making it more likely to evolve. Other authors have also stressed the importance for assessing the haplodiploidy hypothesis of ecological factors affecting the benefits and costs of help-ing (e.g. Hughes et al. 1993a). However, current knowledge of benefits and costs is very poor.

Of course, ecological promoters of eusociality should also operate among diploid species (Section 1.4). However, lastly, various effects may enhance the gain from sib-rearing under haplodiploidy, such as effects involving split sex ratios (see below). For all these reasons, it remains possible that haplodiploidy, despite low relatedness, promotes eusocial evolution on average.

CRITICISM 3

Genetic models of the origin of eusociality have concentrated on three basic questions (Crozier 1977, 1979; Andersson 1984; Grafen 1986; Pamilo 1991a; Seger 1991). (1) Is sib-rearing favored over offspring-rear-ing among females in haplodiploid populations? (2) Is haplodiploidy more conducive to eusocial evolution among females than diploidy? (3) Are females more likely to evolve worker behavior than males in hap-

lodiploids? The answer in all three cases turns out to be no, given a uniform, stable population sex ratio. These findings represent the third challenge to the haplodiploidy hypothesis.

Trivers and Hare (1976) first pointed out that if female workers in Hymenoptera raised equal numbers of sisters and brothers, their average (life-for-life) relatedness with brood would be 0.5, because the higher relatedness with sisters (0.75) is offset by lower relatedness with brothers (0.25) (Box 3.1). This equals the females' average relatedness with offspring. So Trivers and Hare (1976 p. 250) argued that haplodiploidy cannot promote eusociality unless workers "capitalize on the asymmetries" in relatedness with sibs by raising more sisters than brothers.

Trivers and Hare (1976) also showed that in a fully eusocial Hymenopteran population (with a simple kin structure), the stable population sex ratio for workers is female-biased (Section 4.3). In view of the preceding point, some understood this to mean that a female-biased population sex ratio was sufficient for the *origin* of eusociality. But they failed to distinguish between the origin and the maintenance of eusociality (Section 3.2). A female-biased population sex ratio alone is *not* sufficient to favor the origin of worker behavior, since the effect on the workers' fitness of raising a bias of sisters is cancelled out by the lower mating success of females (e.g. Craig 1979, 1980a; Grafen 1986). Box 3.2 explains this argument fully. It shows that, under any uniform, stable population sex ratio, the fitness payoff to haplodiploid females from rearing sibs equals that from rearing offspring. Furthermore, this equals the payoff that diploid females gain from both sib-rearing and offspring-rearing. Put another way, the critical benefit : cost ratio in Hamilton's rule (the value that b/c must exceed for the behavior to be favored) is the same (1.0) for female sib-rearers and offspring-rearers in both haplodiploids and diploids. Therefore, haplodiploidy seems no more likely to lead to worker behavior in females than diploidy (Box 3.2).

In response to this argument, several factors have now been proposed that could plausibly promote worker behavior in Hymenoptera. These factors effectively permit Hymenopteran workers to "capitalize" on their high relatedness with sisters. They are mostly inoperative among diploids, because in diploids differential relatedness among sibs does not exist. So these factors also make haplodiploidy more conducive to worker evolution than diploidy, thereby meeting the challenge to the haplodiploidy hypothesis from the models described above.

Section 3.8 describes the suggested factors in detail. They are also reviewed by Grafen (1986), Pamilo (1991a), and Seger (1991). One important way in which sib-rearing can be favored among haplodiploids

BOX 3.2 SEX RATIOS AND THE ORIGIN OF EUSOCIALITY IN HAPLODIPLOIDS

The Relative Gain from Sib-rearing and Offspring-rearing in Female Haplodiploids

Consider a population of Hymenoptera with singly mated, outbred females. Daughters choose between staying in the maternal nest and rearing sibs (subsocial eusociality), or leaving the nest and rearing offspring with equal efficiency (solitary breeding). The following calculations show that under any given, uniform sex ratio in this population, sib-rearing and offspring-rearing are equal in profitability (e.g. Craig 1979, 1980a; Grafen 1986, personal communication in Krebs and Davies 1987 p. 310). Therefore, a female-biased population sex ratio alone does not make worker behavior more profitable in female haplodiploids.

The full proof is as follows. First, the fitness payoff from the rearing of a certain number of sexuals is defined as the product, (number × regression relatedness × relative sex-specific reproductive value × mating success), summed over both sexes. (Regression relatedness and reproductive value are defined in Boxes 1.1 and 3.1. Mating success is the per capita number of mates.) This expression has been developed and used by several authors (e.g. Oster et al. 1977; Grafen 1986; Taylor 1988; Boomsma and Grafen 1991; Pamilo 1991a,b; Ratnieks and Reeve 1992). Its justification lies in its power and versatility, and in the fact that it yields results similar to those of traditional allele frequency models (e.g. Oster et al. 1977; Matessi and Eshel 1992). Formal similarities of the two modeling approaches have been proved by Taylor (1989).

Let each female rear N sibs or N offspring. Regression relatedness is denoted by r, reproductive value by V, and mating success by MS, with the subscript F standing for female and M for male. Let the population sex ratio be s, expressed as the proportion of females reared (or $s : [1 - s]$ as the ratio females : males). So if $MS_F = 1/s$, then $MS_M = 1/(1 - s)$. For example, say $s = 0.75$. Then the female : male sex ratio is 3 : 1, and each male gets 3 mates on average if each female gets one mate.

Therefore, for the sib-rearing females, number of females reared is Ns, and the number of males reared is $N(1 - s)$. In addition, $r_F = 0.75$ and $r_M = 0.5$ (Table 3.1), and, if $V_F = 1$, $V_M = 0.5$ (Box 3.1). So the fitness payoff to sib-rearers is

$$[Ns \times 0.75 \times 1 \times 1/s] + [N(1 - s) \times 0.5 \times 0.5 \times 1/(1 - s)] = N. \qquad [3.1]$$

For offspring-rearing females, $r_F = 0.5$ and $r_M = 1.0$ (Table 3.1). All the other quantities are as for the sib-rearers. So the fitness payoff to offspring-rearers is

BOX 3.2 CONT.

$$[Ns \times 0.5 \times 1 \times 1/s] + [N(1 - s) \times 1 \times 0.5 \times 1/(1 - s)] = N. \qquad [3.2]$$

Therefore, under any population sex ratio, sib-rearing and offspring-rearing are of equal value. This is true even if the relatedness among female sibs is 0.75, and the female : male sex ratio is 3 : 1. In fact, the population sex ratio evidently does not affect the fitness payoffs, because the lower mating success of the more common sex exactly cancels out the effect on the payoff of its higher numbers (e.g. Craig 1979; Iwasa 1981). This follows from Fisher's (1930) sex ratio theory, which shows that the mating success of a sex falls as it becomes more frequent (Section 4.2).

To look at the situation another way, one can reinterpret Hamilton's rule (Box 1.2) as the statement that sib altruism evolves if $(b/c) >$ (per capita payoff from offspring-rearing)/(per capita payoff from sib-rearing), where b and c equal, respectively, the numbers of sibs and offspring reared. (This expression substitutes the earlier fitness function for the simple relatedness coefficients in Hamilton's rule, which is necessary if sex ratio and reproductive value effects are to be included.) This inequality's right hand side represents the critical benefit : cost ratio that must be exceeded for worker behavior to be favored. So the above calculations show that in a haplodiploid population at sex ratio equilibrium, the critical benefit : cost ratio for females is 1.0. Therefore sib-rearing cannot be favored unless workers raise more sibs than offspring.

Conditions for Worker Evolution in Male Haplodiploids and Male and Female Diploids

By the same method, and noting from Box 3.1 that, for males as actors, $V_F = 2$ when $V_M = 1$, the payoff gained by male sib-rearers in haplodiploids (using regression relatedness values from Table 3.1) is

$$[Ns \times 0.25 \times 2 \times 1/s] + [N(1 - s) \times 0.5 \times 1 \times 1/(1 - s)] = N.$$

The payoff from producing offspring (recalling that males in haplodiploids have no sons) is

$$[Ns \times 0.5 \times 2 \times 1/s] = N.$$

Therefore, the critical benefit : cost ratio for males to become workers is 1.0, as for female haplodiploids.

Similarly, in diploids the payoff to females and males from both sib-rearing and offspring-rearing is

$$[Ns \times 0.5 \times 1 \times 1/s] + [N(1 - s) \times 0.5 \times 1 \times 1/(1 - s)] = N$$

BOX 3.2 CONT.

since in diploids the regression relatedness of females and males with all offspring and all sibs is 0.5, and the reproductive values of the sexes are in equal ratio (Boxes 1.1, 3.1). So the critical benefit : cost ratio for worker evolution among diploids is also 1.0.

In sum, assuming a uniform, equilibrial population sex ratio, females do not favor sib-rearing over offspring-rearing in haplodiploids, and diploids evolve sib altruism as easily as haplodiploids. Many previous models have shown this (e.g. Charlesworth 1978; Charnov 1978a; Craig 1979, 1980a, 1983a; Wade 1979; Aoki and Moody 1981; Iwasa 1981; Kasuya 1982; Grafen 1986). In addition, males in haplodiploids are no less likely than females to be workers. Again, this is the conclusion of several previous models (e.g. Trivers and Hare 1976; Charnov 1978a; Bartz 1982; Craig 1982b; Pamilo 1984, 1991a; Grafen 1986).

Effect of Greater than Average Female-biased Colony Sex Ratio on Worker Evolution

The above methods also show under what circumstances sib-rearing can be favored over offspring-rearing among females in haplodiploids. The right conditions arise when workers occur in colonies that have a greater fraction of females in their brood than the average colony in the population (Grafen 1986; Pamilo 1991a; Seger 1991; Sections 3.6, 3.8).

Consider a population of solitary breeders in which the population sex ratio (proportion of females) is $s = 0.5$. This is as expected under maternal control of sex allocation in solitary Hymenoptera (Box 4.2). Imagine mutant daughter females arise that are sib-rearers, and let the sex ratio in their colonies be $(s + t) = (0.5 + t)$. Because these colonies are initially rare, they do not affect the population sex ratio.

The payoff of the sib-rearers is

$$[N(0.5 + t) \times 0.75 \times 1 \times 1/0.5] + [N(1 - 0.5 - t) \times 0.5 \times 0.5 \times 1/(1 - 0.5)].$$

Similarly, the payoff of the offspring-rearers is

$$[N(0.5) \times 0.5 \times 1 \times 1/0.5] + [N(1 - 0.5) \times 1 \times 0.5 \times 1/(1 - 0.5)].$$

The sib-rearers' payoff minus the offspring-rearers' payoff therefore equals Nt. If this is positive, sib-rearing will be favored. N is always positive. So sib-rearing is favored if the sib-rearers' colony sex ratio is more female-biased than the population sex ratio (positive t).

When sib-rearing colonies produce one sex ratio, and offspring-rearers produce another, the population is said to exhibit a split sex ratio (Grafen 1986; Sections 3.8, 4.5). As sib-rearing colonies become more frequent,

BOX 3.2 CONT.

they contribute increasingly to the population sex ratio, and so alter this ratio itself. This would reduce the sib-rearers' fitness advantage due to their higher sex ratio, because the population would come to resemble the earlier case of an entirely eusocial population in which the population sex ratio is uniform (unsplit). So the present methodology refers to the initial spread of worker behavior. However, Trivers and Hare (1976), who closely anticipated split sex ratio theory (Boomsma 1991; Pamilo 1991a), suggested that sib-rearers might grow more efficient once their evolution had started (their N could increase). Sib-rearing could then spread to fixation, giving a population of entirely eusocial colonies.

occurs when workers raise a brood whose sex ratio is *more* female-biased than the population sex ratio (e.g. Grafen 1986). This could come about through split sex ratios, which are found when females contribute systematically different sex ratios to the same offspring generation (Grafen 1986; Section 3.8, Box 3.2). However, firm evidence for split sex ratios in facultatively social Hymenoptera is still required (Section 3.8). Therefore, a complete assessment of the importance for the haplodiploidy hypothesis of split sex ratios and similar phenomena awaits more data on sex allocation in bees and wasps.

The finding that an above-average female bias in the sex ratio could promote worker behavior also removes a potential problem for the haplodiploidy hypothesis (Seger 1991). This is the claim that – if evolving workers had to raise female-biased broods – females in solitary Hymenoptera should be able to tell the sexes apart as larvae or pupae (Crozier 1977). There has been a controversy over whether this ability is plausible (Evans 1977; Charnov 1978a; Crozier 1979). But, even if it is not, the problem is sidelined by the more recent sex-ratio work. For potential workers might now only have to tell whether they were in colonies with a higher fraction of females in the brood than the population average. As Seger (1991) indicates, this could require no more than noting the presence of many females within colonies.

Genetic models also failed to confirm the prediction from the haplodiploidy hypothesis that males are less likely to evolve worker behavior than females. Specifically, as with females, males are genetically indifferent between rearing sibs and having offspring, assuming a uniform population sex ratio. This point is proved in Box 3.2. Trivers and Hare (1976) again demonstrated it by observing that if the sex ratio is 1 : 1, the average relatedness of males with sibs equals their average relatedness with their mates' offspring. Hamilton (1964b) had argued that males should be reluctant to be workers by comparing males' relatedness with sibs to their relatedness with daughters. But this comparison is inappropriate, because males that became workers would exchange both sexes of their potential mates' offspring for sibs (Craig 1982b).

In males, in contrast to female haplodiploids, it is unlikely that special sex ratio effects might still create a preference for working. Assume that worker behavior first arises in a population of solitary Hymenoptera where the sex ratio is under maternal control (Boxes 3.2, 4.2). Then the population will also be in sex ratio equilibrium for any hypothetical male workers. This is because the equilibrial sex ratio for any party is set by the ratio, life-for-life relatedness with females : life-for-life relatedness with males (Box 4.1), and this ratio is the same for females with respect to offspring as for males with respect to sibs (but not for females with respect to sibs). Therefore, incipient male workers, unlike female ones (Box 3.2), could not profit by manipulating the sex ratio of their sibs (Grafen 1986).

In sum, both females and males in haplodiploids neither favor nor disfavor worker behavior under a uniform, stable population sex ratio. But, for females alone, extra conditions plausibly exist that would promote working. In this sense, haplodiploid relatedness levels may still encourage worker behavior in females but not in males.

To conclude, theoretical objections to the haplodiploidy hypothesis have not proved fatal to either kin selection theory or the hypothesis itself. However, as the next section describes, other traits of the Hymenoptera and of haplodiploidy could explain some of the special features of eusociality in the group.

3.7 A Critique of the Haplodiploidy Hypothesis (II)

As described in Section 3.5, Hamilton proposed the haplodiploidy hypothesis as an explanation for why eusociality has arisen many times in the Hymenoptera, and for why workers are always female. Therefore, some authors have asked whether other features of the Hymenoptera apart from haplodiploidy, or features of haplodiploidy aside from unusual relatedness levels, could account for these phenomena. This section considers these issues.

First, it needs noting that there has been a controversy over whether in fact eusociality is especially frequent in the Hymenoptera compared to other taxa. It used to be accepted that eusociality evolved around eleven times in the Hymenoptera and just once in the termites (e.g.

E.O. Wilson 1971 pp. 327, 329). However, it now seems that both these figures are too low (e.g. Evans 1977). In bees alone, there have been multiple, separate origins of eusociality in the halictine group (E.O. Wilson 1971 p. 327; Evans 1977), and molecular phylogenies suggest that advanced eusociality in honey bees and stingless bees has evolved independently (Cameron 1993). Termite biologists have pointed out that in termites the evolution of a nonreproductive, defensive soldier caste, and of a nonreproductive helper caste (workers), are separate processes. This contrasts with the case in the Hymenoptera, since "soldier" ants are modified workers. It appears that throughout the termites the soldier caste has arisen once, and a sterile worker caste at least three times (Noirot and Pasteels 1987, 1988; Myles 1988; Myles and Nutting 1988; Higashi et al. 1991). This gives a minimum of four origins of reproductive altruism in termites. Therefore, the concentration of origins of sociality and altruistic behavior in the Hymenoptera could be more apparent than real. Biologists may now need explanations for multiple origins of eusociality in both the Hymenoptera and the termites. In addition, evidence is growing that altruism and eusociality have evolved several times independently within other taxa, such as aphids and mole-rats (Benton and Foster 1992; Jarvis et al. 1994; Stern 1994). So neither the social Hymenoptera nor the termites are as unusual as they once seemed.

FEATURES OF THE HYMENOPTERA PROMOTING EUSOCIALITY INDEPENDENTLY OF HAPLODIPLOIDY

In the Hymenoptera several explanations (other than haplodiploidy) have been advanced for the frequent occurrence of eusociality.

1. Nest-building and nest-provisioning (parental care) are very common in the solitary Hymenoptera. A nest provides a valuable resource encouraging young not to disperse (Alexander 1974; Myles 1988; Seger 1991). In addition, both nest-building and brood care supply a behavioral "preadaptation" to worker behavior, and represent a practical means of exhibiting care by potential workers (E.O. Wilson 1971 p. 328; Eickwort 1981; Andersson 1984; Hansell 1987). Similarly, nesting could help explain multiple eusocial origins in termites (Alexander 1974; Andersson 1984; Myles 1988; Alexander et al. 1991).

2. The aculeate Hymenoptera are equipped with a sting, which might facilitate social evolution because colony life requires the colony to be defended (Starr 1985b, 1989; Kukuk et al. 1989; Seger 1991; Fisher 1993). This could also be partly why Hymenopteran eusociality is confined to the aculeates (Andersson 1984; Jaisson 1985; Starr 1985b; Stubblefield and Charnov 1986).

3. Assuming eusocial origins occur more or less randomly in insect orders, the high frequency of eusociality in Hymenoptera could just be a statistical artefact of the large number of (solitary) species in this group (Andersson 1984; Alexander et al. 1991). However, it is unknown how large a group the Hymenoptera formed in the distant past, when most eusocial lineages evolved.

4. The social Hymenoptera (and termites) apparently have above-average chromosome numbers. Sherman (1979) suggested that this facilitates eusocial evolution by parental manipulation (Sections 1.9, 3.9). He argued that high chromosome numbers reduce the variance in the fraction of genes shared among sibs, and so make them less able to recognize those individual sibs genetically closest to themselves. This would reduce costly discrimination in the sibship, so making offspring easier for their mother to manipulate. Various objections to this idea have been raised. For example, if a mutant queen arose with a high chromosome number, any selected response of workers would only occur after many generations (e.g. Dawkins 1982a pp. 151–153; Andersson 1984). This weakens the case for the spread of the mutation. It is also not helpful to the hypothesis that a species in the primitive ant genus *Myrmecia* has a haploid chromosome number of one (Crosland and Crozier 1986).

FEATURES OF THE HYMENOPTERA ACCOUNTING FOR FEMALE-ONLY WORKERS INDEPENDENTLY OF HAPLODIPLOIDY

Why only females become workers in Hymenoptera has also been explained in several ways.

1. Parental care by males is rare in the solitary Hymenoptera, so a transition to worker behavior would be harder for this sex (Hamilton 1964b; Lin and Michener 1972; Alexander 1974; Eickwort 1981; Craig 1982b).

2. Males lack strong mandibles in solitary Hymenoptera, whereas females are preadapted for helping by possessing these tools (E.O. Wilson 1971 p. 328; Crozier 1979; Eickwort 1981). However, some male Hymenoptera have been observed working with their mandibles, suggesting that males are not always constitutionally work-shy (Hamilton 1972; West-Eberhard 1975; Kukuk et al. 1989).

3. Only female Hymenoptera possess a sting (a modified ovipositor), which assists in the important worker task of colony defense (e.g. Eickwort 1981; Bartz 1982; Andersson 1984; Starr 1985a, b; Kukuk et al. 1989). Again, however, male wasps are capable of defense, at least against invertebrates (Brockmann and Grafen 1989).

FEATURES OF HAPLODIPLOIDY PROMOTING EUSOCIALITY
INDEPENDENTLY OF EFFECTS ON RELATEDNESS

Several authors have suggested that haplodiploidy involves features that promote eusocial evolution independently of effects on relatedness levels.

1. Seger (1991) proposed that male production from unfertilized eggs (arrhenotoky) is one such feature. Arrhenotoky means that a female losing the opportunity to mate through specializing on worker functions does not forfeit all chances of reproduction. This could facilitate her initial move to a worker role, especially if individuals first join groups as "hopeful reproductives" (Section 3.2). An important point about this reasoning, not emphasized by Seger (1991) but hinted at by Kukuk et al. (1989), is that it applies just to females. Males in haplodiploids must always mate to have offspring. So arrhenotoky could also be a reason for female-only workers in the Hymenoptera.

2. Reeve (1993) put forward the "protected invasion hypothesis" as another possible explanation for why haplodiploidy might facilitate eusociality. A model showed that dominant alleles for sib care by females are less likely to be lost from small populations by genetic drift in haplodiploids than in diploids. A similar model predicted easier fixation of dominant sib care alleles in females than in males among haplodiploids. However, as Reeve (1993) himself pointed out, it is not certain that the assumption of dominance is valid, since nothing is known about the genetics of sib care.

3. Saito (1994) argued that haplodiploid genetics promote eusociality by ensuring a low frequency of harmful recessive genes. In haploid males, such genes are immediately exposed to selection (Box 4.2). Therefore, haplodiploids should be more prone to inbreeding than diploids, in which inbreeding – since it increases homozygosity – is discouraged by a high frequency of harmful recessives. So, assuming inbreeding facilitates kin selection, haplodiploids should evolve eusociality more easily. Similarly, Saito (1994) argued, females should be especially likely to be sterile in haplodiploids, through homozygosity for recessive genes reducing fertility. However, a problem with this scheme is that inbreeding may not always facilitate kin-selected altruism, and there is little evidence that early social Hymenoptera were inbred (e.g. Michod 1993; Box 3.1). As Saito (1994) pointed out, female sterility due to homozygosity for recessive genes is also poorly supported. Lastly, it seems very unlikely that a gene for altruism would spread more easily if linked to a recessive sterility allele, as Saito's scheme requires (Keller 1995a).

ASSESSMENT OF CURRENT STATUS OF
HAPLODIPLOIDY HYPOTHESIS

The conclusion from this and the previous section is that the validity of the haplodiploidy hypothesis remains uncertain at present. But this is not because the hypothesis has any major theoretical defects (Section 3.6). Instead, to evaluate the hypothesis fully, biologists need more information on genetic structure, social structure, and sex allocation in solitary and facultatively social wasps and bees. This is required to confirm that factors (Sections 3.6, 3.8) that might permit female Hymenoptera to exploit their high relatedness with sisters actually occur (cf. Seger 1991). In addition, an experimental approach to measuring the benefits and costs of facultative helping in bees and wasps is required (e.g. Gadagkar 1985a, 1990b; Queller and Strassmann 1988, 1989; Reeve 1991).

The status of the haplodiploidy hypothesis is also uncertain because the number of eusocial origins may not in fact be uniquely high in the Hymenoptera. In addition, the key features of Hymenopteran eusociality could stem from traits aside from haplodiploidy, or from features of haplodiploidy other than those affecting relatedness. However, if haplodiploidy does promote eusocial evolution through its effects on relatedness, these other features would simply reinforce this effect. In other words, they might operate in concert with kin selection and haplodiploidy to make the evolution of female-centered eusocial societies especially likely in the Hymenoptera. As E.O. Wilson (1976b p. 208) put it, many factors may combine to push a species over the "eusociality threshold".

Lastly, despite the questionable status of the haplodiploidy hypothesis, it needs stressing that kin selection still remains essential to understanding eusocial evolution, because relatedness is necessary for altruism to evolve.

3.8 Factors Promoting Worker Evolution via Relatedness and Sex Ratio Effects in Haplodiploid Populations

Section 3.6 described how sib-rearing was no more likely to be favored than offspring-rearing in haplodiploid populations assuming a uniform population sex ratio. It then stated that a number of additional factors had been proposed that might nonetheless promote worker behavior. This section gives details of these factors.

ASSOCIATION OF WORKER BEHAVIOR WITH SEX
RATIOS OF ABOVE-AVERAGE FEMALE BIAS

Recent work has shown that incipient workers in haplodiploids can "capitalize" on their greater relatedness with females if they raise a brood whose sex ratio is more female-biased than the population sex ratio (e.g. Charnov 1978a; Craig 1980a; Seger 1983, 1991; Grafen 1986, personal communication in Krebs and Davies 1987 pp. 309–310; Stubblefield and Charnov 1986; Pamilo 1987, 1991a). Box 3.2 demonstrates this point in detail.

There are various ways in which prospective workers could find themselves in colonies raising a greater female bias than the population (Pamilo 1991a; Seger 1991). For example, Seger (1983) showed that this could occur in halictine bees with two, overlapping generations per year (partial bivoltinism). Male offspring from the first generation may survive to mate with female offspring in the second one. To the foundress females, this makes sons less valuable than daughters in the second generation, so second-generation brood sex ratios should be female-biased. Since first-generation males are still present, second-generation broods are therefore more female-biased than the population sex ratio (Andersson 1984; Brockmann 1984; Krebs and Davies 1987 pp. 310–313; Seger 1991). (A greater relative reproductive value [Box 3.1] among second-generation females created by the pattern of generational overlap would also enhance the profitability of sib-rearing in this system [Grafen 1986; Krebs and Davies 1987 p. 312].) In support of Seger's argument, halictine bees include both partially bivoltine, solitary species, and primitively eusocial species (e.g. Brockmann 1984). In addition, Seger's proposed between-generation sex ratio differences do occur (Brockmann and Grafen 1992; Boomsma and Eickwort 1993).

Workers might also raise an above-average fraction of females if there are split sex ratios (Grafen 1986). These occur when separate classes of colony in a population raise different sex ratios in the same offspring generation (Grafen 1986; Section 4.5). Examples would be sib-rearing and offspring-rearing colonies. (Seger's [1983] partial bivoltinism model does not involve split sex ratios, since here sex ratio differences occur across generations [Grafen 1986].) Split sex ratio theory proposes that females opting to be workers in colonies with a brood sex ratio more female-biased than the average gain a fitness advantage. This is the situation considered in Box 3.2.

Split sex ratios could arise in several ways in solitary and facultatively social Hymenoptera. One involves unmatedness (Godfray and Grafen 1988). Say unmated females become reproductive and produce males (arrhenotoky). Then the population will contain two classes of reproductive female that each raise different sex ratios. Mated females will produce both sexes, but unmated ones can only produce males. Therefore, if daughters of mated females express a sib-rearing gene, they will automatically raise an above-average female bias, and this will be due to a split sex ratio (Godfray and Grafen 1988). More generally, a population might exhibit split sex ratios if separate classes of female had different cost ratios (energetic cost of raising a female : energetic cost of raising a male [Box 4.1]). Females should concentrate on producing the sex that is relatively less costly for them. Variable cost ratios might themselves arise from the sexes requiring different levels of resources, with resource availability depending on microhabitat (Grafen 1986; Boomsma 1993).

Evidence for these types of split sex ratio is required to confirm these ideas. Specifically, it needs to be confirmed that colonies with workers in facultatively social Hymenoptera raise a more female-biased brood than the average. However, good evidence for very similar effects already exists in primitively eusocial bees and wasps (e.g. Yanega 1989; Boomsma 1991; Mueller 1991; Boomsma and Eickwort 1993; Queller et al. 1993; Mueller et al. 1994; Packer and Owen 1994). Split sex ratios may also arise in the advanced eusocial ants, as the next chapter discusses (e.g. Section 4.5).

Positive Feedback Between Above-average Female-biased Sex Ratios and Selection for Worker Behavior

In the models considered above (Seger 1983; Grafen 1986), the extra female bias required for worker evolution does not stem from worker behavior itself, but from selection on the foundress female. Frank and Crespi (1989) suggested that worker behavior and above-average female bias could become causally linked. Say colonies with workers arise in a population of solitary Hymenoptera, for example through delayed departure of daughters. If helping is effective, the larvae reared are likely to be larger than average. In solitary Hymenoptera, mothers producing large offspring tend to concentrate on daughter production, because the rate at which fitness increases with body size is greater for females than for males. So a causal association arises between the presence of workers and a greater than average female bias in the brood sex ratio. The extra female bias itself enhances the selective advantage of worker behavior, for the reasons given earlier and in Box 3.2. Hence there is a positive feedback between worker behavior and the conditions favoring it (a synergism, in Frank and Crespi's term), which further promotes the spread of worker behavior. Frank and Crespi (1989) and Boomsma and Eickwort (1993) provide evidence that these synergistic effects actually occur.

EFFECT OF MALE PRODUCTION BY WORKERS

A number of genetic models suggest that if evolving workers added sons to the colony brood, worker behavior would have been favored, even among nonlaying workers (Charnov 1978a; Aoki and Moody 1981; Iwasa 1981; Bartz 1982; Pamilo 1984). The basic reason is that nonlaying workers would then be exchanging brothers (life-for-life relatedness = 0.25) for nephews ($r = 0.375$), provided that worker-laid males were raised at the expense of queen-laid ones. In addition, in Iwasa's (1981) model, the presence of worker-laid males means that female mating success does not fall sufficiently to cancel out the effect on workers' fitness of raising a bias of sisters. This would occur if workers were exclusively sib-rearers and the population sex ratio were uniform (Box 3.2).

ABOVE-AVERAGE FEMALE BIAS DUE TO SELECTION ON PARASITES?

The previous chapter suggested that symbionts in ant colonies might induce a more female-biased sex ratio than workers favor (Section 2.5). Similarly, we speculate that selection on parasites might promote the origin of eusociality by causing an association between an above-average female bias and sib-rearing. This idea is similar to Hurst and Vollrath's (1992) suggestion that sex ratio distorters in spiders might allow the evolution of female-biased sex ratios that benefit autosomal genes under conditions of local mate competition (Hamilton 1967; Section 4.9). It is specific to haplodiploids in that a sex ratio more female-biased than average favors sib-rearing by females in haplodiploids and not diploids.

For parasites to induce eusociality, the following conditions must hold. (1) Some females in a population of solitary Hymenoptera possess parasites that are transmitted via daughters only, so that selection on the parasites produces a female-biased sex ratio among brood in parasitized nests. The bumble bee trypanosome studied by Shykoff and Schmid-Hempel (1991a) could resemble such a parasite. (2) An association arises between sib-rearing behavior and presence of the parasite. This sounds improbable, but could occur very simply if parasites slowed daughters' growth and so delayed their dispersal. These daughters might then show sib-rearing behavior through misdirected parental care (Section 3.3). (3) The parasite does not substantially reduce the brood-rearing efficiency of affected females.

These conditions would create an association between sib-rearing and a female-biased colony sex ratio which would favor the spread of sib-rearing (Box 3.2). Eventually a sex ratio conflict would arise between parasites and hosts. But by that stage, sib-rearing might be widespread in the population, and selection might then both promote worker efficiency and moderate the sex ratio biasing effects of the parasite. So a search among solitary and facultatively social bees and wasps for parasites with the proposed effects might yield interesting results.

3.9 Factors Promoting Worker Evolution in Diploid and Haplodiploid Populations

Previous sections have concentrated on factors that promote worker behavior in haplodiploid populations alone. But the origin of eusociality may also be strongly influenced by factors operating on both diploid and haplodiploid populations (reviewed by Gadagkar 1991b; Pamilo 1991a). These form the subject of this section.

PARENTAL MANIPULATION

Parental manipulation theory suggests that offspring become workers because they are compelled to by their parents (Alexander 1974; Michener and Brothers 1974). Although sometimes viewed as an alternative to kin selection theory, parental manipulation involves social actions among kin and so comes under the theory, as discussed in Sections 1.9 and 3.2.

Several genetic models have examined whether sib-rearing in female Hymenoptera evolves more easily if it is imposed on sibs by their mother, or if it is voluntary. All conclude that the critical benefit : cost ratio in Hamilton's rule (Box 3.2) is lower for parentally-imposed than for voluntary sib altruism (e.g. Charlesworth 1978; Charnov 1978a; Craig 1979; Pamilo 1987). An argument from Charnov (1978a) shows why. If haplodiploid parents force daughters to be workers, they trade the grandchildren their daughters would have raised (life-for-life relatedness $= 0.25$) for offspring ($r = 0.5$). So the parents' critical benefit : cost ratio (Box 3.2) is 0.5. But from the daughters' viewpoint, the trade is of offspring ($r = 0.5$) for sibs (average $r = 0.5$, assuming single mating and a $1 : 1$ sex ratio). Therefore, the daughters' critical ratio is 1.0. So, for a fixed cost, voluntary altruists must raise more sibs than manipulated ones. Charnov (1978a) also stressed that daughters are genetically indifferent between raising sibs or offspring (both $r = 0.5$), and so should be easily manipulated into being workers. Stubblefield and Charnov (1986) added that each daughter would want her sisters to be manipulated, since this represents a trade of nephews and nieces ($r = 0.375$) for sibs ($r = 0.5$). So daughters will not unite to oppose parental manipulation. In the terminology of Section 2.5, daughters are not a community of interest with respect to resisting manipulation.

Charnov's (1978a) argument also applies to diploid populations, because the corresponding average relatedness values are the same. Therefore, parental manipulation represents a plausible, general factor in the origin of worker behavior by kin selection (Seger 1991). However, if in haplodiploids the sex ratio biases discussed in the previous section occur, daughters may favor worker behavior anyhow, and manipulation would be unnecessary. Furthermore, if daughters are coerced into sib-rearing when the benefit : cost ratio is less than 1.0, they are expected to resist the manipulation, even if they do so only singly. Finally, in fully eusocial populations workers may raise a female-biased population sex ratio that is contrary to their mother's interests (Trivers and Hare 1976; Sections 4.3, 5.2). Therefore, the period of queen–daughter agreement may often be short-lived.

SUBFERTILITY

Craig (1983b) proposed a means by which reduced fecundity ("subfertility") of females might promote worker behavior through effects on the benefit : cost ratio. Suppose that an allele for helping arises that is facultatively expressed in females of below-average fecundity ("subfertile" females). Subfertile females would have few offspring if they attempted to nest alone. Therefore, by not nesting alone, they make a relatively small sacrifice. Put another way, the cost of their altruism is small. Hence, provided subfertility does not diminish the females' ability to feed larvae (the benefit of helping stays the same), their benefit : cost ratio increases, and the allele for helping can spread.

Craig suggested that this process could interact with parental manipulation. If subfertile females act as helpers, mothers would be selected to underfeed or maltreat some of their female offspring to make them subfertile, since the mothers themselves would benefit from worker daughters. This would lead to even lower fecundity among the daughters, so promoting their worker behavior and facilitating the transition to complete worker sterility (Craig 1983b; see also West-Eberhard 1981). However, although in the polistine wasp genus *Ropalidia* some females are inherently less likely to become foundresses than others (Gadagkar 1991c), Reeve (1991) suggested that there is little evidence that helper individuals are subfertile in natural populations of polistines.

LIFE HISTORY EFFECTS

Like Craig (1983b), Queller (1989b) investigated how effects on benefits and costs might promote helping behavior. Consider a female Hymenopteran with the choice of working in the natal nest or leaving to breed alone. If her productivity in both roles is the same ($b = c$), helping is not selected. However, the female may die between making her decision and, acting as either a worker or mother, raising a larva to independence. In the natal nest there are probably already larvae close to reaching adulthood. But if the female bred alone, she would have to pass through a long period of colony growth before acquiring larvae at an equivalent stage of development. Therefore, the length of time during which mortality will reduce the fitness of females that become workers is shorter than for females opting to be mothers. So Queller (1989b) argued that, for a given mortality schedule, workers would raise more larvae to independence than mothers (provided the females heading nests with potential workers laid extra eggs that could be reared into extra larvae). This would effectively raise the benefit of working. In Queller's (1989b) phrase, workers obtain a "reproductive head start." Suggested examples include the epiponine wasps (Hughes et al. 1993b).

Gadagkar (1990c) criticized Queller's model for implying that an almost completely grown larva reared to adulthood by a worker should count as the fitness equivalent of one offspring. This inflates the worker's fitness, because the worker has clearly not contributed to all the larva's development. So the head start advantage of workers cannot be as great as Queller supposed. In addition, Gadagkar (1990c, 1991b) proposed another life history benefit that workers would enjoy. When a worker dies, its previous brood-rearing effort is not lost, because fellow workers continue to rear the brood to adulthood. But if a solitary female dies, all her earlier investment in rearing brood comes to nothing. Therefore, workers have an "assured fitness return" absent in solitary breeders, which is proportional to the fraction of the brood developmental period for which workers survive (Gadagkar 1990c, 1991b). Similar advantages to incipient sociality were proposed by Queller (1994b).

Nonacs (1991a) argued that both Queller's (1989b) and Gadagkar's (1990c) models assumed that the amount of brood reared increases linearly with worker number. But efficiency per worker probably falls as the workforce grows (Michener 1964). So Nonacs concluded that both models could only account for the first few females staying to work in a colony, since subsequent females would not gain sufficient fitness.

The overall conclusion from these models is that life history effects may indeed promote worker behavior, although their precise quantitative benefits remain controversial. Gadagkar (1991b) added another effect, which arises due to variations in the expected time taken by young females to attain reproductive maturity (late-maturing females are more likely to become workers than early-maturing ones). The probability that a young female finds a mate must also affect her decision to be a worker or solitary foundress (Seger 1991). Finally, as Nonacs (1991a) and Queller (1994b) pointed out, all the life history factors represent additional reasons why helping may evolve at quite low relatedness in both diploid and haplodiploid populations.

3.10 Conclusion

Both haplodiploidy and other traits of the Hymenoptera (nest-building, maternal care, stings) are likely to have contributed to the frequency and nature of eusociality in the group. Furthermore, haplodiploidy has probably promoted eusociality through several genetic effects, not just its effect on relatedness (Sections 3.6–3.8). Other authors have reached similar conclusions (e.g. Oster and Wilson 1978 p. 24; Andersson 1984; Pamilo 1991a; Seger 1991).

However, kin selection is still essential for eusocial evolution. This is because none of the above traits would promote eusociality if helping were not directed at relatives. Kin selection requires both relatedness and appropriate benefits and costs (Section 1.4). So it makes no sense to ask whether relatedness or ecology is the most important influence on eusocial evolution. Both must be considered together (e.g. Hughes et al. 1993a). Another way of seeing this is by recognizing that eusocial evolution in the Hymenoptera is an extreme example of selection for high reproductive skew (unequal allocation of reproduction) among communal breeders (Keller and Reeve 1994a; Sherman et al. 1995). So, again, its study needs to address both ecological and genetic factors, just as in the wider study of skew evolution (Vehrencamp 1983a; Keller and Reeve 1994a; Section 8.6).

Finally, haplodiploid relatedness values, as well as having probably contributed to the origin of eusociality, have without doubt affected its further evolution. This becomes clear from considering their influence on sex allocation and kin-selected conflicts of interest in established colonies, as the next four chapters show.

3.11 Summary

1. The evolution of group-living is distinct from the evolution of reproductive altruism within groups (eusociality). The mutualism theory of

eusocial evolution can explain group-living, but not lifetime reproductive altruism. Therefore, neither mutualism nor parental manipulation (a facet of kin selection) represent alternatives to kin selection as an explanation of eusocial evolution.

2. In the subsocial route to eusociality in the Hymenoptera, workers arise from offspring in parent–offspring associations. In the semisocial route, they arise from individuals originally in single-generation associations.

3. The factors involved in the origin of eusociality may differ from those influencing its maintenance. The origin of eusociality in ants cannot be studied directly, because all ants are highly eusocial (or derived from such species). Ants probably became eusocial by the subsocial route. The initial gene for sib altruism could have been a gene for delayed departure among daughters, since nondispersing daughters would plausibly have fed their sibs through misdirected parental care.

4. West-Eberhard's (1987a,b, 1988) "epigenetic theory" of eusocial evolution neither contradicts, nor replaces, kin selection theory. The epigenetic theory stresses the facultative nature of worker behavior, and claims to dispense with genes for altruism. However, kin selection theory also predicts that worker behavior should be facultative. Furthermore, the epigenetic theory does involve genes for worker behavior, since evolving workers would have had a choice, which would almost certainly have been subject to genetic variation, between actions leading to breeding or working.

5. In the Hymenoptera, fertilized, diploid eggs develop into females, and unfertilized, haploid ones develop into males (haplodiploidy). This leads to unusual relatedness levels compared to diploidy. Hamilton's (1964b) "haplodiploidy hypothesis" proposes that high sister–sister relatedness in haplodiploids promotes the evolution by kin selection of reproductive altruism among females in the Hymenoptera. This arguably explains the multiple origins of eusociality and the presence of female-only workers in the group.

6. Criticisms of the haplodiploidy hypothesis are that it assumes a subsocial pathway, single mating, and single foundresses. However, even with semisociality, multiple mating, and multiple foundresses, haplodiploid relatedness levels may still promote helping relative to diploidy.

7. Genetic models show that the critical benefit : cost ratio for the evolution of sib altruism is the same for females and males in both diploids and haplodiploids given a uniform population sex ratio. However, selection favors sib-rearing if haplodiploid females raise broods of sibs with a more female-biased sex ratio than the population sex ratio. Thus, haplodiploidy may still promote eusociality via its

effects on relatedness. Female-biased colony sex ratios may arise through partial bivoltinism (two overlapping generations per year) or split sex ratios (females contribute systematically different sex ratios to the same offspring generation).

8. Other features of the Hymenoptera, including nest-building, maternal care, and possession of a sting in females, could account for the multiple origins of female-only workers. Features of haplodiploidy independent of unusual relatedness levels (e.g. the ability of females to reproduce without mating) could also promote eusocial evolution. However, despite uncertainty over haplodiploidy's precise contribution, kin selection remains essential for eusocial evolution.

9. Various factors could assist the evolution of altruism in both diploid and haplodiploid populations. These include parental manipulation and diminished fertility among potential helpers. They also include life history effects, such as differential influences of mortality on the payoffs to altruists and breeders owing to: (a) the presence of partly grown larvae in the nest available for immediate rearing by altruists (Queller's [1989b] "reproductive head start"); and (b) the guarantee to altruists that their brood-rearing effort will not be wasted if they die (Gadagkar's [1990c, 1991b] "assured fitness return").

4

Sex Ratio Theory for the
Social Hymenoptera

4.1 Introduction

This and the next two chapters deal with sex ratio evolution in the social Hymenoptera, and in ants in particular. The present chapter considers theory, and the following one examines the evidence for the theory. Chapter 6 concentrates on kin conflict over sex allocation. Chapter 3 discussed how sex ratio might influence the origin of eusociality. This one is concerned with sex ratio evolution in populations of established eusocial colonies.

The study of sex ratio evolution is a huge and active area of evolutionary biology. Charnov (1982, 1993), Trivers (1985), Karlin and Lessard (1986), Herre et al. (1987), Bull and Charnov (1988), Frank (1990), Clutton-Brock (1991), Antolin (1993), and Wrensch and Ebbert (1993) review both theory and data. This chapter first describes the basic theory for the field, which is Fisher's sex ratio theory (Section 4.2). It then considers how this applies to the haplodiploid, social Hymenoptera (Section 4.3) in the simplest case (monogyny, singly mated queens, sterile workers, and random mating). The following sections (4.4–4.10) deal with variants of this case (e.g. multiple mating, reproductive workers, polygyny, nonrandom mating). Chapter 5 examines whether the predictions of the various models are met in nature.

The study of sex ratios in the social Hymenoptera is not just relevant to social insect biology. In diploids with random mating, Fisher's theory predicts a 1 : 1 population sex ratio. But this would also occur if sex ratio were simply determined by the mechanics of chromosomal sex determination in diploids. So it is difficult to know whether 1 : 1 ratios among diploids are really confirmation of Fisher's theory (e.g. G.C. Williams 1979; Maynard Smith 1984). By contrast, sex ratios in haplodiploids cannot be constrained by chromosomal sex determination, since this is absent. In haplodiploids an egg's fertilization – and hence its sex – depends on the mother's control of sperm release from the sperm receptacle (e.g. Gerber and Klostermeyer 1970). In addition, solitary Hymenoptera frequently show patterns of nonrandom mating such as local mate competition (Section 4.9) in which Fisherian theory predicts sex ratios other than 1 : 1. Therefore, haplodiploids have long been recognized to be particularly suitable for testing Fisher's theory (e.g. Boomsma 1988). This is reflected by the vast amount of work on the sex ratios of the parasitoid wasps (e.g. Charnov 1982; Wrensch and Ebbert 1993; Godfray 1994).

On top of this, in the social Hymenoptera, applying Fisher's theory involves explicitly invoking a role for relatedness in influencing sex ratios. This role is implicit in the theory, but was not overtly stated by Fisher. The theory also predicts kin-selected conflicts of interest over sex allocation (Trivers and Hare 1976; Chapter 6). So tests of sex ratio theory in the social Hymenoptera are of wide interest because they are also tests of kin selection theory.

4.2 Fisher's Sex Ratio Theory

Fisher's (1930 p. 142) sex ratio theory is often presented in a form different from the one adopted here. We favor the following version because it is more general and is easily adapted to cases other than the diploid case that Fisher originally considered.

Let the population sex investment ratio be defined as the total amount of energy invested in females in a population, divided by the total amount of energy invested in males. Let a controlling party be any group of individuals (e.g. parents, offspring) who might control sex allocation. *Fisher's theory states that the population sex investment ratio is evolutionarily stable from the viewpoint of a controlling party when the fitness payoffs to this party from each sex, per unit of investment, are equal* (e.g. Benford 1978; Trivers 1985 p. 272). The theory deals with returns per unit cost because the sexes may differ in their average cost, for example if there is sexual size dimorphism. If costs are equal, Fisher's theory gives the condition for the stable numerical population sex ratio. Box 4.1 presents the theory algebraically.

Why does Fisher's condition of equal fitness returns give the stable sex ratio? The answer is in two parts. First, if the condition holds, there is no incentive for the controlling party systematically to overproduce either sex. Therefore, the population sex ratio will not be changed and so, by definition, is stable. Second, if the condition does not hold, the population sex ratio is unstable, and will change until the condition holds again (e.g. Benford 1978). This is because the fitness payoffs from producing a female or male are themselves functions of the sex ratio.

The fitness payoff to the controlling party from producing a member of either sex is defined as the product, regression relatedness × sex-

BOX 4.1 FISHER'S SEX RATIO THEORY

This box gives an algebraic version of Fisher's sex ratio theory (Section 4.2). First, a number of terms and variables need defining.

Sex ratio. This refers to the *primary sex ratio*. This is the sex ratio at conception or birth (e.g. among eggs), or – if young receive care – up to the end of the period of investment (Trivers 1985 p. 457). In social Hymenoptera, investment ends just before the release of sexuals for mating. Some authors use the primary sex ratio to mean only the sex ratio among eggs (e.g. Aron et al. 1994).

Numerical sex ratio. Total number of females divided by the total number of males.

Investment sex ratio. Total energy invested in females divided by the total energy invested in males. All sex ratios may also be expressed as proportions, e.g. investment in females divided by total investment in the sexes.

Cost ratio. Average energetic cost of raising a female divided by the average energetic cost of raising a male. So, if cost is measured by dry weight, the investment sex ratio equals the biomass of females divided by the biomass of males.

Controlling party. The party (e.g. parents, workers) that controls the allocation of sex investment.

Per capita fitness payoff. The product, regression relatedness $\times$ sex-specific reproductive value $\times$ mating success (Box 3.2), i.e. $r \times V \times MS$ in the notation below. The relevant relatedness values are those between the controlling party and each sex in the brood.

Mating success. Average per capita number of mates (Box 3.2).

Relatedness asymmetry. Life-for-life relatedness with females divided by that with males, i.e. $r_F.V_F/r_M.V_M$ in the notation below (Boomsma and Grafen 1991; Section 4.2).

Random mating. Within a sex, every adult has the same chance of mating with any other adult of the opposite sex.

Fisher's theory states that the population sex investment ratio is at equilibrium for the controlling party when the genetic return to the party per unit of energy expended is equal for both sexes (Benford 1978; Trivers 1985 p. 272; Section 4.2), that is when

$$(r_F.V_F.MS_F)/c = r_M.V_M.MS_M \qquad [4.1]$$

where

BOX 4.1 CONT.

r_F, r_M = regression relatedness of controlling party with females and males respectively (Boxes 1.1, 3.1).

V_F, V_M = sex-specific reproductive value of females and males respectively (Box 3.1).

MS_F, MS_M = mating success (see above) of females and males respectively.

c = female : male cost ratio (see above).

Say the stable numerical population sex ratio is $X : 1$ females : males. Then if the mating success of females equals one unit, that of males equals X units. Therefore:

$$(r_F . V_F . 1)/c = r_M . V_M . X$$
$$Xc = r_F . V_F / r_M . V_M. \qquad [4.2]$$

Xc is, by definition, the stable female : male population sex investment ratio. So Equation 4.2 states that this ratio equals the relatedness asymmetry (Section 4.2). Equations 4.1 and 4.2 will be the guiding principles for establishing the stable sex ratio throughout this chapter (cf. Grafen 1986; Boomsma and Grafen 1991; Pamilo 1991b). Note that if costs are equal ($c = 1$), then the stable numerical sex ratio equals the relatedness asymmetry. In derivations to follow in this chapter (Boxes 4.3–4.5), equal costs are assumed for simplicity. But the expressions obtained will generally be implicitly of the type "$Xc = \ldots$" with $c = 1$, and so will be expressions for investment sex ratios.

Fisher's theory applies to randomly-mating diploids with parental control of sex allocation as follows. Substituting appropriate values (Box 3.1) for relatedness (0.5) and reproductive value (1) in Equation 4.1 for the stable female : male sex ratio (X) gives:

$$(0.5 \times 1 \times 1)/c = (0.5 \times 1 \times X)$$
$$Xc = 1.$$

There are several points to note about this result.

1. The result states that the stable population sex investment ratio is 1 : 1, as is the numerical sex ratio if costs are equal. This is Fisher's familiar result for diploids.

2. If the result is rearranged as $X = 1/c$, it is evident that when $c > 1$ (females cost more than males), then $X < 1$ (females are rarer than males). And if $c < 1$ (females cost less than males), then $X > 1$ (females are more common than males). This is Fisher's prediction that, if costs are unequal, the cheaper sex is produced in greater numbers.

BOX 4.1 CONT.

3. In diploids, the relatedness and reproductive value terms in Equation 4.1 cancel out (Section 4.2). So, for diploids, the equation could be rewritten as $Xc = 1$ from the start. This means that (number of females × cost of a female) = (number of males × cost of a male), which is Fisher's statement that at equilibrium the total expenditure on each sex is the same.

The idea of equal total investment in the sexes is sometimes presented as the core of Fisher's theory. However, it does not apply to social haplodiploids with worker control of investment, where at equilibrium total expenditure on females exceeds that on males (Box 4.2). Nor does it apply, in diploids, to cases in which the fitness returns on investment in the sexes are differentially nonlinear (Frank 1990). Therefore, equal total investment is a *result* in a special case of Fisher's theory (the diploid, random mating case), not the theory itself. Instead, Fisher's theory is better regarded as the principle of equal fitness returns at equilibrium (Equation 4.1; Section 4.2).

---

specific reproductive value × mating success, where mating success means the per capita number of mates (Box 3.2). Therefore, another way of expressing Fisher's theory is to say that at equilibrium this product (per unit cost) is equal for both sexes (Box 4.1). The mating success of a female or male is inversely related to its relative frequency in the population. Put another way, the rarer sex has the higher mating success. For example, if the sex ratio is 2 : 1 females : males, and females have an average mating success of one unit, males must have an average mating success of two units (each male gets two mates on average). The frequency dependence of mating success shows why the fitness payoffs are functions of the sex ratio itself. It also shows why Fisher's theory represents an application of game theory, or ESS theory (evolutionarily stable strategy theory), to sex ratio evolution (Hamilton 1967; Charnov 1982; Maynard Smith 1982b).

In addition, it follows that the stable female : male sex ratio equals the male : female ratio of mating successes. If the equal returns principle holds, the male : female mating success ratio must itself equal the product, regression relatedness with females × female reproductive value, divided by the product, regression relatedness with males × male reproductive value. Regression relatedness multiplied by reproductive value equals life-for-life relatedness (Box 3.1). Thus, according to Fisher's theory, the stable female : male sex ratio equals life-for-life relatedness with females divided by that with males (Box 4.1).

Boomsma and Grafen (1991 p. 386) termed this relatedness ratio the *relatedness asymmetry*. (Note that this usage differs from that in Chapter 3, where relatedness asymmetry referred to the ratio, related-ness with sibs : relatedness with offspring [Section 3.5].) Therefore, *Fisher's theory states that the stable female : male sex ratio equals the relatedness asymmetry* (Box 4.1).

Now consider the application of Fisher's theory to the simple diploid case. Imagine a nonsocial, diploid species with random mating and no sexual dimorphism. The controlling party is the parent that raises the offspring. Neither parental relatedness with offspring (0.5), nor repro-ductive value, differ between the sexes (Box 3.1). Therefore the stable sex ratio occurs simply when the sexes have equal mating success. This will be when the sex ratio equals 1 : 1. For, if the mating success of a female equals one unit, male mating success also equals one unit, so sat-isfying Fisher's condition. Therefore, the stable sex ratio for diploids is 1 : 1 (Box 4.1).

This example shows why Fisher (1930) did not have to consider relat-edness and sex-specific reproductive value explicitly when presenting his theory. These are equal between the sexes in diploids, and therefore drop out of the equal returns equation. However, this is not the case in haplodiploids. So relatedness and reproductive value need including in the general theory.

This example also illustrates how any sex ratio at which the sexes had unequal returns would be unstable, and would change until they were equal again. Say the population sex ratio were 2 : 1 females : males (the argument applies in reverse if males are initially in excess). Then the mating success of males (two units) exceeds that of females (one unit). This will select for parents to overproduce the more valuable sons, making the sex ratio more male-biased. But as males become more numerous, their relative mating success will decline, since mating suc-cess is inversely related to frequency. When the sex ratio reaches 1 : 1 again, the advantage in overproducing males will have been abolished. Returns on each sex will be equal once more, and the sex ratio will have stabilized through frequency-dependent selection as Fisher's theory predicts. A final point highlighted by this example is that Fisher's theo-ry assumes the existence of genetic variation for sex allocation. Clearly, if individuals are to respond appropriately to an unstable population sex ratio, genes for producing different brood sex ratios must exist with-in populations.

Fisher's theory has been confirmed and extended by many authors using several modeling approaches (e.g. Shaw and Mohler 1953; Bodmer and Edwards 1960; Benford 1978; Uyenoyama and Bengtsson 1979; Taylor 1985; Frank 1987a; Maynard Smith 1989 p. 258; reviewed by Antolin 1993). It needs stressing that the predicted sex ratio in the theory is the *population* sex ratio (e.g. Kolman 1960; Benford 1978; G.C. Williams 1979). This is because mating success is set by relative abundance within the population-wide mating pool. For example, in diploids, the theory does not imply that all parents should produce a 1 : 1 ratio, but only that the population as a whole should have equal numbers of females and males (or equal investment in them). In fact, at equilibrium, it does not matter what sex ratio individual parents produce among their offspring, since by definition sons and daughters yield equal returns (e.g. G.C. Williams 1966 pp. 152–153). One might therefore expect simply random (binomial) variation in brood sex ratios at equilibrium (G.C. Williams 1979).

Strictly speaking, parents are only indifferent to their brood sex ratios at equilibrium in infinite or very large populations. In small populations, brood sex ratios should approach the population equilibrium value (Verner 1965; Taylor and Sauer 1980). Parents may also vary systematically in their offspring sex ratios if variations in parental quality affect the quality of offspring, and the sexes differ in the rate of fitness yield with rising quality (Trivers and Willard 1973; Frank 1990; Section 6.2).

Finally, these arguments show why the stable sex ratio is an ESS. An ESS is defined as a strategy resistant to displacement by alternatives when all or most of the population are following it (Maynard Smith 1982b p. 204). In a large population at sex ratio equilibrium, the strategy adopted by all individuals – the putative ESS – must be to produce (on average) the population mean sex ratio. Also, as stated above, all possible individual sex ratios yield equal fitness if Fisher's condition holds. Therefore, the trait of producing a sex ratio other than the population mean would not be fitter than, and so could not displace, the trait of producing the mean. Hence Fisher's predicted sex ratio must be an ESS.

4.3 Sex Ratios in Social Haplodiploids: Basic Theory

Trivers and Hare (1976) combined Fisher's (1930) sex ratio theory, Hamilton's (1964a,b) kin selection theory, and Trivers' (1974) parent–offspring conflict theory to produce a sex ratio theory for the social Hymenoptera. Trivers and Hare's insight was to realize that Fisher's original theory could be modified for when: (1) parties other than parents – namely workers – control sex allocation; and (2) the relatednesses between the controlling party and the sexes are not equal.

These modifications are implicit in the presentation of Fisher's theory so far (Section 4.2; Box 4.1).

Many factors aside from relatedness and the identity of the controlling party affect sex investment in social Hymenoptera (Herbers 1990; Table 4.5). This section presents the Trivers–Hare theory in the simplest case. The following ones then show how changing the theory's assumptions predicts the sex ratio when some of these additional factors apply. Theory and data on sex allocation in ants have also been reviewed by Trivers and Hare (1976), Alexander and Sherman (1977), Crozier (1979), Joshi and Gadagkar (1985), Nonacs (1986a), Frank (1987b), Boomsma (1988, 1989), Hölldobler and Wilson (1990 p. 192), Pamilo (1990b, 1991b), and Crozier and Pamilo (1993). Frank (1987b) and Pamilo (1991b) in particular present unified theoretical treatments of sex ratio evolution in the social Hymenoptera. This chapter's approach is essentially that of Pamilo (1991b).

The simplest situation with which the Trivers–Hare theory deals is when there is a randomly-mating population of colonies each headed by a single, once-mated queen, and in which workers are sterile. Trivers and Hare showed that in this case the equilibrium population sex investment ratio is 3 : 1 females : males if workers are the controlling party, and 1 : 1 if queens are. This result is derived in Box 4.2. It implies that a queen–worker conflict exists over sex allocation.

Note that investment in females refers to investment in new queens, not workers. This is because workers do not form part of the breeding population (in advanced eusocial Hymenoptera). In addition, investment in workers represents investment in future sexuals, which will be allocated according to the prevailing stable sex ratio. So, with respect to sex allocation, investment in workers by a colony is rather like investment in somatic cells as opposed to gametes in a single (hermaphrodite) organism (Nonacs 1993a). An exception occurs when workers accompany queens during colony fission or budding. As discussed later, these workers are usefully regarded as investment in females (Pamilo 1991b; Nonacs 1993a; Section 4.8).

The reason why 3 : 1 females : males is the equilibrial sex ratio for workers is that, at this ratio, females have one-third of the mating success of males. But the (life-for-life) relatedness of workers with females (0.75) is three times worker–male relatedness (0.25). This exactly makes up for the lower mating success of females, causing the per capita fitness gain to workers from females to equal that from males. Since this is the criterion for sex ratio equilibrium, 3 : 1 must be the stable ratio.

This reasoning shows that it is the workers' asymmetric relatedness with sisters and brothers (Table 3.1) which makes their stable sex ratio female-biased. By contrast, no relatedness asymmetry exists for the

BOX 4.2 BASIC SEX RATIO THEORY FOR SOCIAL HAPLODIPLOIDS

The 3 : 1 Prediction

This box presents an explanation of Trivers and Hare's (1976) sex ratio theory for the social Hymenoptera in its simplest form (Section 4.3). The method (from Box 4.1) shows how Trivers–Hare theory is based on Fisher's sex ratio theory.

Assume a population of social Hymenoptera with monogyny (one queen per colony), singly mated queens, sterile workers, and random mating. Both queens and workers are potential controlling parties. Let the stable numerical sex ratio be $X : 1$ females : males, and let the female : male cost ratio equal $c : 1$ (Box 4.1).

If queens are the controlling party, Equation 4.1 for the stable sex ratio is as follows (using values for relatedness and reproductive value from Table 3.1 and Box 3.1):

$$(r_F.V_F.MS_F)/c = r_M.V_M.MS_M$$
$$(0.5 \times 1 \times 1)/c = 1.0 \times 0.5 \times X$$
$$Xc = 1.$$

So the stable investment ratio (Xc) for queens is $1 : 1$ as for the diploid case (Box 4.1).

If workers are the controlling party, the relevant relatedness values are sister-sib ones (Table 3.1), giving:

$$(0.75 \times 1 \times 1)/c = 0.5 \times 0.5 \times X$$
$$Xc = 3.$$

This is Trivers and Hare's (1976) well-known result that the workers' stable population sex investment ratio is $3 : 1$ females : males.

Male ants are frequently limited in their number of matings (Nonacs 1986a; Section 11.2). Note, however, that the workers' ESS sex ratio is unaffected by male mating frequency. Thus, say that the stable numerical sex ratio is $3 : 1$ females : males (costs of the sexes are equal) and all males mate just once. Then only one in three females will be mated. Therefore, female mating success remains at one-third that of males, so the workers' condition for sex ratio equilibrium when the sexes are equal in cost is still met (cf. Trivers 1985 p. 273). Similarly, say males mate once but the stable numerical sex ratio is, for example, $1 : 2$ females : males (because females are 6 times as costly as males, yielding the stable invest-ment sex ratio of $3 : 1$ females : males). (Such numerical male bias approximates the usual situation in monogynous ants [Section 5.2].) Now

BOX 4.2 CONT.

half the males will be unmated when all females are mated, so if female mating success equals 1 unit, that of males will be 0.5 units. Applying Equation 4.1, the per capita fitness payoff from a female then equals $(0.75 \times 1 \times 1)/6 = 0.125$, and from a male equals $(0.5 \times 0.5 \times 0.5)/1 = 0.125$. Thus, there is again sex ratio equilibrium. The essential point is that the ratio of the mating successes of the sexes is always set by their comparative abundance, and not by male mating frequency.

Other Possible Influences on Social Hymenopteran Sex Ratios

The present chapter describes how sex ratios evolve when various assumptions of the Trivers–Hare theory are altered. The rest of this box briefly reviews other kinds of factor that might influence sex ratios in social Hymenoptera. No data exist on whether they are operative.

1. GENOMIC IMPRINTING

When calculating relatedness levels, one implicitly averages across the maternally-inherited and paternally-inherited portions of the genome. Thus, the 3/4 relatedness between sisters in social Hymenoptera is an average of a chance of gene-sharing of 0.5 for maternal alleles, and 1.0 for paternal ones (Box 3.1). (This averaging also occurs in the formal calculation of relatednesses as regression coefficients [Box 1.1].) But say alleles could have different effects (expression) according to the sex of the contributing parent. Then expectations based on averaged relatednesses would change. For example, at a locus for altruism towards sisters in the social Hymenoptera, a paternally-derived allele would be selected to deliver more aid than a maternally-derived one. There would then be intragenomic conflict over the level of aid.

Alleles whose effects depend on their "sex" are said to show genomic imprinting, or parent-specific gene expression (Moore and Haig 1991; Haig 1992a). There is exciting evidence that imprinted alleles influence parent–embryo conflicts in mammals and plants (Moore and Haig 1991). Therefore, similar effects on kin conflicts in the social Hymenoptera cannot be ruled out. Consider a locus for biasing sex allocation in an ant worker (Haig 1992b). Paternally-inherited alleles will be present in sisters but absent in brothers, and so will favor an all-female sex ratio. But maternally-inherited alleles, which are equally likely to be in both sexes of sib, favor an unbiased sex ratio. The sex ratio would then alter according to which alleles "win" (Haig 1992b). On the other hand, intragenomic conflict could result in a stalemate in which averaged relatednesses still determined the expected outcome (Haig 1992b). Clearly, the search for an influence of imprinted alleles on ant kin conflicts is a fascinating challenge for future investigation.

BOX 4.2 CONT.

2. INTRACELLULAR SEX RATIO DISTORTERS

In solitary Hymenoptera, there exist a variety of intracellular elements that bias sex ratios to their own advantage against the interests of the nuclear genes (e.g. Werren et al. 1988; Ebbert 1993; Godfray 1994; Section 2.5). As Crozier and Pamilo (1993) point out, such elements plausibly also influence sex ratios in social Hymenoptera.

3. SYMBIOTIC OR PARASITIC ORGANISMS

Many intracellular sex ratio distorters occur in the cytoplasm and are therefore inherited via females only. By analogy with these, it was earlier suggested that any colony symbiont or parasite that disperses via a single sex might act as a colony-level sex ratio distorter (Sections 2.5, 3.8).

4. MALE MORTALITY DUE TO RECESSIVE LETHAL GENES

In haplodiploid organisms, lethal recessive alleles are fully expressed in the haploid males, while remaining masked in heterozygote diploid females (e.g. Crozier 1977). Therefore, if mutations yielding lethal recessives are frequent, males suffer greater mortality than females. This process could affect sex ratios (Metcalf 1980b; Smith and Shaw 1980). For example, if mortality occurs during the egg stage, an increase in relative female bias could result. Differential mortality after the end of investment has no effect (e.g. Trivers 1985 pp. 273–274). Smith and Shaw (1980) estimated the level of male mortality due to lethal recessive mutations as 10% in parasitoid Hymenoptera. These mutations could partly contribute to female bias in social species (Smith and Shaw 1980; Hölldobler and Wilson 1990 p. 185; Godfray 1994 p. 202). However, their frequency does not seem high enough to account for the degree of bias observed in many cases (Table 5.11).

---

queen (Table 3.1), so her stable sex ratio is even. Hence the different patterns of comparative relatedness also explain why there is queen–worker sex ratio conflict.

Following Trivers and Hare, the predictions of 3 : 1 or 1 : 1 sex investment have been obtained by numerous authors using many modeling techniques (e.g. Oster et al. 1977; Benford 1978; Charnov 1978b; Macnair 1978; Oster and Wilson 1978 p. 86; Craig 1980b; Bulmer 1981, 1983a; Bulmer and Taylor 1981; Orlove 1981; Uyenoyama and Bengtsson 1981; Pamilo 1982a, 1991b; Joshi and Gadagkar 1985). The conclusions hold for both annual and perennial Hymenopteran societies meeting the stated conditions (e.g. Pamilo 1991b). Some disagreement occurs in that a number of allele frequency models find ratios other than $3:1$ to be favored by workers, depending on the models' detailed genetic assumptions. However, these models still predict some degree of female bias under worker control (Craig 1980b; Pamilo 1982a).

Another complication is that intermediate population sex ratios may result if queens and workers share control of allocation. In fact, mixed control possibly leads to a lack of an ESS sex ratio altogether (Matessi and Eshel 1992; Section 6.3). The models in this chapter assume that usually either one party or the other totally controls sex allocation. There are some grounds for thinking this correct, or at least a good approximation (Section 5.10).

Lastly, recall from the previous section that Fisher's theory presupposes genetic variation in the sex ratio. Does this occur in the social Hymenoptera? The answer is almost certainly yes. For one thing, although no direct evidence exists (Pamilo 1982a), it is easy to imagine that queens vary genetically in their tendency to fertilize eggs. Genetic sex ratio variation has been found in parasitoid Hymenoptera (e.g. Antolin 1992, 1993; Orzack and Gladstone 1994). It is similarly plausible that queens and workers vary genetically in the degree of sex-biased treatment of brood. In honey bees, Page et al. (1993) found evidence for genetic variation in the fraction of comb devoted to raising males. For these reasons, both queen and worker control of sex allocation should be capable of evolving.

The following sections describe how Trivers and Hare's (1976) theory can be modified for situations in which some basic assumptions do not hold. Detailed workings for many of these cases are presented in Boxes 4.3–4.5, and the results are shown graphically in Figures 4.1–4.7.

4.4 Sex Ratios in Slave-making Ants

The first modification to Trivers and Hare's (1976) theory concerns the identity of the party controlling sex investment. Trivers and Hare (1976) predicted a queen–worker conflict over the sex ratio. They further suggested that since the workers both rear the sexual brood and outnumber queens, they should usually win the conflict. Consequently, under conditions of monogyny, single queen mating, worker sterility, and random mating, sex ratios should be female-biased ($3:1$).

However, Trivers and Hare (1976) also suggested that slave-making ants represent an exception to this prediction. In these species, workers capture brood of other species. These mature in the slave-maker's nest to become "slave" workers performing all the brood-rearing tasks within the nest (e.g. Sudd and Franks 1987; Hölldobler and Wilson 1990). This is termed nonconspecific brood care. Trivers and Hare argued that the slave-maker workers – being uninvolved in brood care – lack the practical power to bias the brood to their preferred sex ratio. In addition, the slaves have no genetic stake in the slave-maker brood and so should be indifferent to the sex ratio. Therefore, queens should control sex allocation in slave-makers. With the assumptions above, this would predict 1 : 1 instead of 3 : 1 investment. Hence ants with nonconspecific brood care provide a special opportunity to test the idea of queen–worker sex ratio conflict.

4.5 Sex Ratios under Multiple Mating

Any factor that alters a controlling party's relatedness with sexual brood is liable to change the stable sex ratio. One such factor is multiple mating by queens. Sex ratio evolution with multiple mating has been examined by Benford (1978), Charnov (1978b), Boomsma and Grafen (1990, 1991), and Pamilo (1991c). Two cases arise. If every queen has the same mating frequency, there will be no sex ratio splitting. The same applies if mating frequency varies but workers cannot assess it. However, if workers can assess a variable mating frequency, sex ratios will be split.

Multiple Mating without Split Sex Ratios

Multiple mating does not alter queens' relatedness with sexual offspring. So multiple mating never affects the sex ratio under queen control, which remains at 1 : 1 as in the single mating case (Box 4.2). However, workers' average relatedness with the females they rear declines, because some are now half sisters rather than full sisters (Box 4.3). This reduces the workers' fitness payoff from females, and so decreases the female bias of their stable sex ratio. For example, if all queens mate twice, workers under monogyny favor a population sex ratio of 2 : 1 females : males. Box 4.3 derives the formula for the workers' sex ratio as a function of the mating frequency of queens (e.g. Pamilo 1991c). This formula applies when queen mating frequency does not vary between colonies, or if it varies but workers cannot assess it. In the second case, workers should raise the sex ratio appropriate to the average mating frequency.

Multiple Mating with Split Sex Ratios

Boomsma and Grafen (1990, 1991) showed that if queen mating frequency varies, and workers both control sex allocation and can assess

BOX 4.3 SEX RATIO IN SOCIAL HYMENOPTERA WITH MULTIPLE MATING

This box considers the situation when there is monogyny, multiple mating by queens, worker sterility, random mating, and (for simplicity) equal costs of the sexes. Under these conditions, sex ratios may be uniform or split (Section 4.5).

Multiple Mating without Split Sex Ratios

Let k = the number of males mating with an average queen. Each male has equal paternity among the queen's female offspring. Queen mating frequency is either constant or not assessable by workers, so sex ratios are not split.

Multiple mating alters workers' relatedness with sisters but not brothers (workers are only related with brothers through their mother). The relatedness of workers with sisters via the paternal half of their genome is $1/k$. Via the maternal half, it is 0.5 (unaffected by multiple mating). So, averaging across the two portions of the genome, worker–sister regression relatedness equals $0.5(0.5 + 1/k)$ (e.g. Hamilton 1964b), which falls as k rises.

The stable female : male sex ratio (X) for workers is found by substituting this relatedness value in Equation 4.1 with equal costs:

$$r_F.V_F.MS_F = r_M.V_M.MS_M$$
$$(0.5)(0.5 + 1/k) \times 1 \times 1 = 0.5 \times 0.5 \times X$$
$$X = (k + 2)/k.$$

Or, expressed as the proportion of females, the workers' stable sex ratio is $(k + 2)/(2k + 2)$. This result was derived by Benford (1978), Charnov (1978b), Joshi and Gadagkar (1985), and Pamilo (1991c) and is illustrated in Figure 4.1.

Multiple Mating with Split Sex Ratios

Boomsma and Grafen (1990, 1991) showed that when queens mate multiply and their mating frequency varies, the result can be split sex ratios (Section 4.5). This assumes that workers control allocation and can assess the queen's mating frequency.

The starting point of Boomsma and Grafen's argument is that variable mating frequencies make workers in different colonies have unequal relatedness asymmetries. For example, in colonies with a singly mated queen, worker–female regression relatedness (r_F) equals 0.75, and worker–male regression relatedness (r_M) equals 0.5 (Table 3.1). Relatedness asymmetry is defined as $r_F.V_F/r_M.V_M$ (Box 4.1), and so equals $(0.75 \times 1)/(0.5 \times 0.5) = 3$. By contrast, in colonies with a doubly mated

BOX 4.3 CONT.

queen, worker–female relatedness (from the formula in the first part of this box) equals 0.5. Worker–male relatedness is unchanged, so relatedness asymmetry equals $(0.5 \times 1)/(0.5 \times 0.5) = 2$.

Another major point in Boomsma and Grafen's argument is that the workers' optimal sex ratio is a function of the level of their relatedness asymmetry relative to the population sex ratio. This can be seen with the following algebraic argument. The notation (as above) is from Equation 4.1, with X being the female : male population sex ratio:

Say $r_F.V_F/r_M.V_M > X$
Now $X = MS_M/MS_F$ (Section 4.2)
So $r_F.V_F/r_M.V_M > MS_M/MS_F$
Or $r_F.V_F.MS_F > r_M.V_M.MS_M$

Therefore, if the relatedness asymmetry ($r_F.V_F/r_M.V_M$) exceeds the female : male population sex ratio, the per capita fitness payoff from females is greater than that from males. So workers should produce only females. By the same calculation, but with the inequality sign reversed, workers should produce only males if the relatedness asymmetry falls below the population sex ratio. Boomsma and Grafen (1991) derived these results by a more rigorous method.

Boomsma and Grafen (1991) also presented a graphical method for working out the exact population and class-specific sex ratios when workers assess a variable relatedness asymmetry. This is illustrated in Figure 4.2 for the case when colonies with singly and doubly mated queens make up 80% and 20% of the population respectively. Applying this method across the whole range of relative abundances of these colony classes yields Figure 4.3.

mating frequency, colonies under queens with different mating frequencies should produce different sex ratios. In other words, sex ratios will be split (Grafen 1986). The previous chapter introduced split sex ratios in the case where one sex-ratio class consisted of solitary breeders and the other of colonies with sib-rearing workers (Section 3.8; Box 3.2). Now the classes are colonies with different worker relatedness asymmetries. By contrast, all situations considered so far in this chapter have predicted a random, unimodal distribution of sex ratios (Section 4.2).

Boomsma and Grafen's argument begins with the point that a variable level of multiple mating causes variation in the workers' relatedness asymmetry (Section 4.2). For example, workers under singly mated

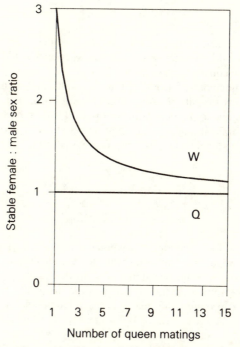

Figure 4.1 The stable female : male sex ratio for the queen (Q) and the workers (W) with multiple mating and no split sex ratios. The workers' sex ratio tends to 1 : 1 as queen mating number rises. Workers' curve calculated from the formula in Box 4.3.

queens have a relatedness asymmetry of 3 : 1, whereas those under doubly-mated queens have a relatedness asymmetry of 2 : 1 (Box 4.3).

The next key point is the very general one that the workers' optimal sex ratio depends on the level of their relatedness asymmetry relative to the population sex ratio. When the workers' relatedness asymmetry exceeds the female : male population sex ratio, the per capita fitness gain from females is higher than that from males. So these workers should produce all females. Conversely, if the relatedness asymmetry is less than the population sex ratio, gains from males are higher and workers should produce males only. Put another way, workers should raise only the sex that is rare in the population relative to the sex ratio dictated by their relatedness asymmetry (Boomsma and Grafen 1991; Pamilo 1991b). Box 4.3 shows this algebraically. The biological reason can be seen from the following example. Say workers' relatedness asymmetry is 3 : 1 but the population sex ratio is 1 : 1. Female mating success then equals that of males. But a relatedness asymmetry of 3 : 1 means that the workers are three times more closely related (using life-

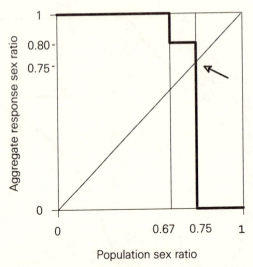

Figure 4.2 An example of Boomsma and Grafen's (1991) graphical method for calculating the stable population and class-specific sex ratios when there are split sex ratios due to variable mating frequency. (From Boomsma and Grafen 1991; by permission of Birkhäuser Verlag, Basel)

Sex ratio is expressed as proportion of females, and it is assumed that all colonies have the same average sexual productivity. Eighty percent of colonies have a singly mated queen (Class 1) and 20% have a doubly mated queen (Class 2). Class 1's relatedness asymmetry is therefore 3 : 1 (or 0.75 expressed as a fraction), and Class 2's is 2 : 1 (or 0.67 expressed as a fraction). The bold line shows the net ("aggregate") sex ratio response of each class as a function of the population sex ratio. Thus, when the population sex ratio is below 0.67, both classes favor all-female production (because the class-specific sex ratio depends on the level of its relatedness asymmetry relative to the population sex ratio). Similarly, when it exceeds 0.75, both favor all-male production. Between 0.67 and 0.75, Class 1 favors all-female production still, but Class 2 favors all males. Therefore the net sex ratio equals the average of these values weighted by the class-specific frequencies, i.e. $(1 \times 0.8) + (0 \times 0.2) = 0.8$.

The population sex ratio is at equilibrium when the aggregate response sex ratio equals the population sex ratio (because if the two classes respond to a given population sex ratio by collectively producing that same ratio, clearly it will not change). Geometrically, this point is given by the intersection of the bold line with the line at 45° to the X axis $(Y = X)$, as shown by the arrow. So, in this example, the population sex ratio equals 0.75. The Class 2 sex ratio must then equal 0 (all males), because 0.75 exceeds the relatedness asymmetry of this class. If the Class 1 sex ratio is s, then $(s \times 0.8) + (0 \times 0.2) = 0.75$, so $s = 0.94$. In sum, at equilibrium, Class 1 colonies produce a sex ratio of 0.94, Class 2 colonies produce all males, and the population sex ratio is 0.75.

for-life relatednesses) with females than with males (Box 4.1). Therefore, females yield three times the fitness payoff of males and so are favored. Similar reasoning shows why male production is favored when relatedness asymmetry is comparatively low.

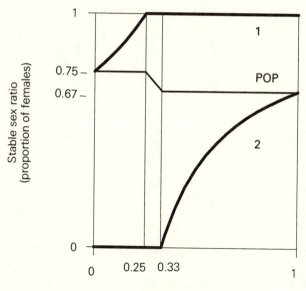

Figure 4.3 Split sex ratios when a population contains both colonies with singly mated queens (1) and colonies with doubly mated queens (2), as a function of the fraction of colonies with doubly mated queens. POP is the population sex ratio. (From Boomsma and Grafen 1991; by permission of Birkhäuser Verlag, Basel)

The final step in Boomsma and Grafen's (1990, 1991) argument is this. Suppose a population contains two classes of colony with different relatedness asymmetries. Then the relatedness asymmetry of at least one of the classes must be unequal to the population sex ratio. From the argument above, this class will be selected to produce single-sex broods. Thus, worker assessment of variable relatedness asymmetry should lead to single-sex broods and so to split sex ratios.

Consider now the example of a population containing colonies with either singly or doubly mated queens. What will be the mutually stable population and class-specific sex ratios in such a system? This needs asking because, as the colony classes respond to the population sex ratio, their responses must alter the population sex ratio itself, since this equals the average of the class-specific sex ratios weighted by the frequency of the classes.

Boomsma and Grafen (1991) showed that the sex ratios can reach equilibrium in two ways. First, they will stabilize when the population sex ratio equals the relatedness asymmetry of one of the classes (3 : 1 or 2 : 1 in the above example). For, at this point, this class has no incentive to respond further (by definition, females and males yield equal returns). And the other class cannot change its sex ratio further, since it must be producing its optimal single-sex response to the population sex ratio, as the population sex ratio is unequal to its relatedness asymmetry. Therefore, neither class changes its sex ratio and so the population sex ratio does not change.

Consequently, in this case, the population sex ratio equals one of the relatedness asymmetries and one class produces single-sex broods. So the other must produce mixed-sex broods biased in the opposite direction. Boomsma and Grafen (1991) termed this second class the *balancing class*, because it balances the contribution of the single-sex class to achieve a population sex ratio equal to its preferred relatedness asymmetry value. Another term for this is *sex ratio compensation* (Taylor 1981a). This outcome results when the frequency of one class exceeds a critical frequency that is a function of its relatedness asymmetry. Thus, in the current example, colonies with singly mated queens must exceed a frequency of 75% to form the balancing class, whereas those with doubly mated queens must exceed a frequency of 33% (Figure 4.3).

The second outcome is when neither class is frequent enough to be a balancing class. In this case, the population sex ratio equilibrates between the relatedness asymmetry values of each class (between 3 : 1 and 2 : 1). This situation can be stable because both classes then produce single-sex broods that collectively equal the same population sex ratio to which they are responding (Box 4.3).

To summarize, the Boomsma–Grafen theory predicts: (1) a population sex ratio lying at or between the class-specific relatedness asymmetries; (2) split sex ratios, with at least one class producing single-sex broods; and (3) population and class-specific sex ratios that are functions of the frequency and relatedness asymmetries of the classes (Figure 4.3). By contrast, if workers cannot assess a variable relatedness asymmetry, they respond to its average value only. Sex ratios will not then be split and the population sex ratio will always lie between the relatedness asymmetry values. Sex ratios are also unsplit if there is a variable mating frequency but queens control sex allocation, since the queens' relatedness asymmetry is unaffected by multiple mating.

Ratnieks (1991a) confirmed Boomsma and Grafen's (1990, 1991) inclusive fitness model with an allele frequency model. Ratnieks (1990a, 1991a) also considered by what means, and how well, workers might assess queen mating frequency. Recall that, if workers are totally unable to tell what type of colony they belong to, sex ratios will not be split. But equally, if workers recognize each class of relative within a colony with perfect accuracy, they should raise full sibs in a 3 : 1 ratio even in colonies with a multiply mated queen, so again sex ratios would

not be split. Split sex ratio theory requires workers to have an intermediate sense of their kin environment, and be able to detect what class of
colony they are in (with a singly or multiply mated queen) without
being able to recognize all classes of relative (Boomsma and Grafen
1990). This feature of the theory is arguably one of its strengths, as current data suggest that within-colony kin discrimination is poorly-devel-
oped in ants (e.g. Carlin et al. 1993; Section 7.3).

In addition, Ratnieks (1991a) confirmed that an allele for biasing
colony sex ratios could spread even if workers' assessment of their
colony class is quite inaccurate, and only more accurate than random
guessing (Ratnieks 1991a). This means that split sex ratios should
evolve despite the likelihood that assessment errors occur, given assessment based on genetic odor differences among females (Ratnieks
1990a). Since workers may sometimes judge their class wrongly, it also
means colonies may not always produce the appropriate class-specific
sex ratio (Ratnieks 1991a). In addition, one might observe deviations
from the expected class-specific sex ratios if the frequency of multiply
mated queens varies from year to year within populations (Boomsma
and Grafen 1990).

Boomsma and Grafen's (1990, 1991) split sex ratio theory provides a
powerful test of the idea of worker control (Boomsma 1993). This is
because it predicts a strong pattern of colony-level sex allocation, rather
than just a particular population sex ratio. In addition, the theory is
very general. As the following sections show, it predicts split sex ratios
due to many traits, including worker reproduction and polygyny.

4.6 Sex Ratios under Worker Reproduction

The production of males by unmated workers affects sex ratios in two
ways. The first is by increasing the relative sex-specific reproductive
value of males, and the second by altering the relatedness of the controlling party with males. Under worker control, the overall effect is to
make population sex ratios relatively more male-biased (Benford 1978;
Taylor 1981a; Boomsma and Grafen 1991; Pamilo 1991b). However, the
details vary according to whether workers reproduce in colonies with a
queen (*queenright colonies*) or in queenless colonies. In the second
case, sex ratios will be split.

When workers are sterile, females have twice the reproductive value
of males (Boxes 3.1, 4.4). The reason is that no genes from males that
are mated with queens (paternal males) are transmitted to males of the
next generation. However, if workers reproduce, paternal males do
transmit genes to males of the next generation, namely to the workers'

sons, their grandsons (Figure 3.1). Therefore, the relative sex-specific reproductive value of males rises (e.g. Boomsma and Grafen 1991; Box 4.4). The result is an increase in the per capita fitness payoff from producing males.

WORKER REPRODUCTION IN QUEENRIGHT COLONIES

If there is queenright worker reproduction, a queen's average relatedness with males decreases, because some males are now her grandsons. However, the queen's stable sex ratio turns out to be independent of the level of worker reproduction. This is because her reduced relatedness with males is exactly offset by their higher reproductive value.

Workers' relatedness with males increases if they are reproductive, because some males are now the workers' sons and nephews, rather than their less closely related brothers (Table 3.1). This raises the workers' per capita payoff from rearing males. Given that the increased reproductive value of males has the same effect, the result is that, under worker control, the sex ratio grows relatively more male-biased (e.g. Benford 1978; Pamilo 1991b). For example, if all males are produced by a single laying worker, its stable sex ratio is 3 : 4 females : males, and that of the nonlaying workers is 1 : 1 females : males. However, the actual number of laying workers turns out to make little difference to the workers' stable sex ratio. This is because, if there are multiple laying workers, the expected sex ratio lies between that for a single laying worker and the nonlaying workers (Box 4.4), and these are close anyhow (Figure 4.4). Box 4.4 derives formulae for the sex ratios of all parties under queenright worker reproduction (see also Table 4.1 and Figure 4.4).

WORKER REPRODUCTION IN QUEENLESS COLONIES

Consider a population where queenright colonies produce only queen-derived sexuals, and queenless colonies produce only males from reproductive workers. Many monogynous ant species probably occur in such populations (Bourke 1988a; Choe 1988; Section 7.4). In this situation, males still have a relatively greater reproductive value than under worker sterility. However, in queenright colonies, worker–male relatedness is unaffected, because workers in these colonies raise only brothers. In addition, workers in the queenless colonies – confined to raising all male broods – can only influence the population sex ratio via the frequency of their colony class in the population. So queenless workers cannot be a party controlling the sex ratio. Instead, the possible controlling parties are the queen and the workers in queenright colonies.

BOX 4.4 SEX RATIO IN SOCIAL HYMENOPTERA WITH WORKER REPRODUCTION

This box considers sex ratio evolution in the case of monogyny, single mating by queens, male-producing workers, random mating, and equal costs of the sexes. Workers may reproduce in queenright or queenless colonies (Section 4.6).

Reproductive Value with Worker Reproduction

Worker male-production increases the relative sex-specific reproductive value of males. This can be seen with a method from Boomsma and Grafen (1991). See also Oster et al. (1977), Benford (1978), Taylor (1988), Pamilo (1991b), and Ratnieks (1991a).

If workers are sterile, queens transmit genes through new queens (daughters, which are also genetically half derived from males) and males (sons, which are genetically only derived from females) (Figure 3.1). Males transmit genes through daughters alone, which are also half derived from queens. Let V_F and V_M equal the sex-specific reproductive values of females and males in one generation, and let V'_F and V'_M equal these values in the next generation. Therefore:

$$V_F = (1/2)V'_F + V'_M$$
$$V_M = (1/2)V'_F.$$

At equilibrium, $V_F = V'_F$ and $V_M = V'_M$. So, substituting into either equation above yields:

$$V_F/V_M = 2.$$

Therefore, when workers are sterile, females have twice the reproductive value of males (Box 3.1).

Now assume there is worker reproduction, and let p = the average proportion of males derived from the queen. Queens obtain genetic representation in both their own sons and the workers' sons, and paternal males gain representation in males for the first time (in workers' sons), half of which they share with queens (Figure 3.1). Therefore:

$$V_F = (1/2)V'_F + p.V'_M + (1/2)(1 - p).V'_M$$
$$V_M = (1/2)V'_F + (1/2)(1 - p).V'_M.$$

So at equilibrium:

$$V_F/V_M = 1 + p. \qquad [4.3]$$

BOX 4.4 CONT.

Therefore, as worker reproduction rises (p falls), the relative reproductive value of males increases. Note that if $p = 1$ (workers are sterile), $V_F/V_M = 2$, as expected.

Sex Ratio with Worker Reproduction in Queenright Colonies

There are three possible controlling parties in a colony – the queen, the nonlaying workers, and a laying worker. Let p = the average proportion of males produced by the queen. Let X = the stable female : male population sex ratio. Substituting Equation 4.3 into Equation 4.2 for the stable sex ratio (with equal costs) gives:

$$X = r_F.V_F / r_M.V_M$$
$$X = (r_F/r_M)(1 + p). \qquad [4.4]$$

For each party, relatedness with males is the average of relatedness with the queen's sons and relatedness with workers' sons, weighted by the frequency of the two types of male, which are p and $(1 - p)$ respectively. For example, for a laying worker (with regression relatedness values from Table 3.1):

$$r_F = 0.75$$
$$r_M = 0.5p + 1.0(1 - p) = 1 - 0.5p.$$

Substituting these values into Equation 4.4 gives the stable sex ratio for a laying worker:

$$X = (0.75 / [1 - 0.5p])(1 + p) = (3 + 3p)/(4 - 2p).$$

The equivalent calculations for the nonlaying workers and the queen yield the results shown in Table 4.1 and Figure 4.4.

It may be unrealistic to assume that a single laying worker can control sex investment (Benford 1978). So say there are N laying workers per colony, sharing worker male-production equally. The relatedness of a focal laying worker with the colony's males now equals a weighted average of its relatedness with the (p) queen-derived males (brothers) and the $(1 - p)$ worker-derived ones, $1/N$ of which are sons and $(1 - 1/N)$ of which are nephews. Therefore:

$$r_M = 0.5p + (1)(1 - p)(1/N) + 0.75(1 - p)(1 - 1/N)$$

$$= 0.5p + (1 - p)(3 + 1/N)/4$$

BOX 4.4 CONT.

Substituting into Equation 4.4 gives:

$$X = (3 + 3p)/[2p + (1 - p)(3 + 1/N)].$$

This equation was derived by Benford (1978). It shows that the stable sex ratio when there are several laying workers lies between the ratios for a single laying worker and for the nonlaying workers. Thus, when $N = 1$, the above formula is as for a single layer (Table 4.1), as it should be. And if N is large, it approximates that for nonlaying workers (Table 4.1). This is because, with many laying workers, most worker-derived males will be nephews of any given one (Benford 1978).

The sex ratio formulae in Table 4.1 were derived by Benford (1978), Charnov (1978b), and Pamilo (1991b). Some of their results were obtained by Oster et al. (1977), Oster and Wilson (1978 p. 86), and Crozier (1979). Trivers and Hare (1976) and Starr (1984) presented similar formulae, but they are slightly inaccurate because reproductive value was not correctly incorporated (Benford 1978; Pamilo 1991b).

Sex Ratio with Worker Reproduction in Queenless Colonies

Consider a population consisting of queenright colonies that potentially produce both sexes of queen-derived offspring, and queenless colonies producing only worker-derived males. Boomsma and Grafen (1991) applied split sex ratio theory (Section 4.5; Box 4.3) to analyze how the sex ratio should evolve, as is now explained.

Let the sex ratio at equilibrium equal F females : M_Q queen-produced males : M_W worker-produced males. Let p = the proportion of males produced by queens, as before. So p equals $M_Q/(M_Q + M_W)$. It is assumed that, on average, the per colony productivities of queenright and queenless colonies are equal.

The potential controlling parties are the queen and the workers in queenright colonies. From Equation 4.4, the queen's relatedness asymmetry and hence her stable female : male population sex ratio is $0.5(1 + p)$, and the workers' is $1.5(1 + p)$.

If they can, the queenright colonies should produce a female-biased class-specific sex ratio that makes the population sex ratio equal to their relatedness asymmetry. The queenright colonies' class-specific sex ratio (females : males) at equilibrium is F/M_Q, by definition.

Also $\quad F/M_Q \times M_Q/(M_Q + M_W) = F/(M_Q + M_W)$
So $\quad F/M_Q = [F/(M_Q + M_W)]/[M_Q/(M_Q + M_W)]$
Or $\quad F/M_Q =$ (female : male population sex ratio)$/p$.

BOX 4.4 CONT.

Hence, at equilibrium, $F/M_Q = [0.5(1 + p)]/p$ in the case of queen control, and $[1.5(1 + p)]/p$ under worker control. These findings are summarized in Table 4.2.

These results were also derived by Taylor (1981a), Nonacs (1986a), and Pamilo (1991b). Boomsma and Grafen (1991) proved the equivalence of Taylor's (1981a) and Nonacs' (1986a) results with theirs. The general idea that queenright colonies should overproduce females when queenless workers produce males was stated by Noonan (1978), Metcalf (1980b), and Owen et al. (1980). Godfray (1990) modeled sex allocation in a population of female parasitoid Hymenoptera in which some mothers were unmated and produced males only. Again, the mated females compensated with an excess of daughters, but sufficiently to restore the population sex ratio to 1 : 1, unlike the case under queen control in the present situation (Table 4.2). This is because, in the parasitoid case, the reproductive value of males is unaffected (there is no male-producing, worker generation), so the mated females' relatedness asymmetry remains at 1 : 1.

The formulae for the population sex ratios given in Table 4.2 do not apply if the frequency of the queenless colonies exceeds 67% (under queen control) or 40% (under worker control) (Nonacs 1986a; Boomsma and Grafen 1991; Pamilo 1991b). These critical frequencies mark the point when the queenless colonies are so abundant that the queenright colonies start to produce all-female broods.

The critical frequencies are calculated as follows. Take the case of queen control. Let:

s = the queenright colonies' sex ratio in terms of the proportion of females.

$\beta = M_W/(F + M_Q)$, i.e. the ratio of the queenless colonies' male production to the queenright colonies' total sexual output (Taylor 1981a; Nonacs 1986a).

w = the frequency of queenless colonies. Note that β and w are not independent. Since per colony productivity is assumed equal across both colony classes, β is w expressed as a ratio. In other words, $\beta = w/(1 - w)$, or $w = \beta/(1 + \beta)$.

From these definitions, the stable sex ratio equals

$$F : M_Q : M_W = s : (1 - s) : \beta. \tag{4.5}$$

Since p equals $M_Q/(M_Q + M_W)$, Equation 4.5 gives:

$$p = (1 - s)/(1 - s + \beta). \tag{4.6}$$

BOX 4.4 CONT.

Also, from Table 4.2, at equilibrium:

$$s = (1 + p)/(1 + 3p).$$ [4.7]

Substituting Equation 4.7 in Equation 4.6 gives:

$$p = (2 - \beta)/(2 + 3\beta).$$ [4.8]

From Equation 4.7, it is apparent that queenright colonies produce females alone $(s = 1)$ when $p = 0$. (Clearly, if workers in queenless colonies are producing all the males, then queenright colonies cannot be producing any.) Substituting $p = 0$ into Equation 4.8 gives $\beta = 2$ and hence $w = 2/3$, or 0.67.

In the case of worker control, the same argument gives:

$$p = (2 - 3\beta)/(2 + 5\beta).$$ [4.9]

So here the critical value of β is 2/3, which corresponds to $w = 0.4$.

The population sex ratio above these critical frequencies is simply determined by the relative frequency of the two colony classes. Therefore, as the proportion of females, it must equal:

$$[1 \times (1 - w)] + [0 \times w] = (1 - w) = 1/(1 + \beta) \text{ or,}$$
$$\text{as the female : male ratio, } (1 - w)/w \text{ or } 1/\beta.$$

The possible sex ratios over the whole range of values for the proportion of queenless colonies are illustrated in Figure 4.5.

Table 4.1

Relatedness and Sex Ratio When Workers Produce Males in Queenright Colonies

Quantity	Queen	Laying Worker	Nonlaying Workers
		Controlling Party	
r_F	0.5	0.75	0.75
r_M	$(0.5 + 0.5p)$	$(1 - 0.5p)$	$(0.75 - 0.25p)$
X	1	$(3 + 3p)/(4 - 2p)$	$(3 + 3p)/(3 - p)$

NOTES: r_F, r_M are the regression relatednesses of the controlling party with females and males respectively. p is the proportion of males produced by queens. X is the equilibrial population sex ratio (females : males), and equals $(r_F/r_M)(1 + p)$ (Equation 4.4). After Benford (1978).

Table 4.2

Sex Ratios When Workers Produce Males in Queenless Colonies

	Stable Female : Male Sex Ratio	
	Queen Control	Worker Control
Queenright colonies	$(1 + p)/2p$	$(3 + 3p)/2p$
Population	$(1 + p)/2$	$(3 + 3p)/2$

NOTES: p is the proportion of males produced by queens. The formulae for the population sex ratios only apply when queenless colonies make up less than 67% (queen control case) or 40% (worker control case) of the total population. From models of Taylor (1981a), Nonacs (1986a), and Boomsma and Grafen (1991).

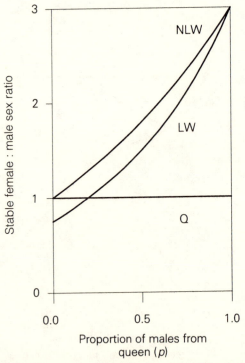

Figure 4.4 The stable population sex ratios for the queen (Q), a laying worker (LW), and the nonlaying workers (NLW), as a function of the level of worker reproduction when workers produce males in queenright colonies. Calculated from the formulae in Table 4.1. (From Benford 1978; by permission of Academic Press Ltd)

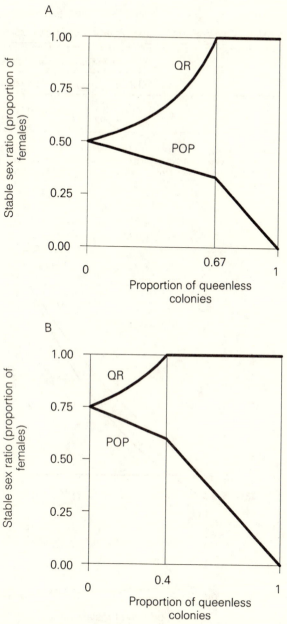

Figure 4.5 The stable sex ratios for queenright colonies (QR) and for the population (POP) as a function of the proportion of queenless colonies when workers produce males in queenless colonies. A. Queen control. B. Worker control. Calculated from the formulae in Table 4.2, with *p* (fraction of males from the queen) being expressed in terms of *w*, the fraction of queenless colonies (Box 4.4).

This type of population evidently consists of two classes of colony producing different sex ratios. Therefore, Boomsma and Grafen's (1991) split sex ratio theory applies. Furthermore, workers can almost certainly assess their colony class (as the theory requires) with a low error rate. Assessment would involve simply detecting the queen's presence or absence, and worker laying only in the queen's absence suggests this is easy.

Split sex ratio theory predicts that both queens and workers in the queenright colonies are selected to overproduce females in response to the all-male production of the queenless colonies. They should do this until the population sex ratio equals their respective relatedness asymmetries. In other words, there should be sex ratio compensation with the queenright colonies acting as the balancing class (Taylor 1981a; Boomsma and Grafen 1991). Note that the relevant relatedness asymmetries (Table 4.2) differ from those found when workers reproduce only in queenright colonies (Table 4.1), because in the present case relatedness with males in queenright colonies is not affected by the presence of worker-produced males. This is why the equilibrial population sex ratios differ according to whether workers reproduce in queenright or queenless colonies.

Because worker reproduction enhances the reproductive value of males, the relatedness asymmetries of both queens and workers are less than if workers are sterile (Equation 4.4 in Box 4.4). Therefore, for the current situation, the population sex ratio is always relatively more male-biased than, and never equal to, its level under worker sterility. This contrasts with the case of split sex ratios due to multiple mating, where the population sex ratio in some circumstances can stabilize at its level under single mating (e.g. Figure 4.3). Boomsma and Grafen (1991) refer to the present case as incomplete sex ratio compensation. In addition, at a certain critical frequency of queenless colonies, queenright colonies are selected to start producing all-female broods. Thereafter, if queenless colonies grow more frequent, the queenright colonies can no longer balance the queenless colonies' all-male broods. The population sex ratio then becomes a simple function of the relative frequency of the all-female producing queenright colonies and the all-male producing queenless ones (Pamilo 1991b).

Summarizing, the general outcome if there is queenless worker reproduction is that the queenright colonies' stable class-specific sex ratio is always more female-biased, and the population sex ratio is always relatively more male-biased, than under worker sterility (Taylor 1981a; Nonacs 1986a; Boomsma and Grafen 1991; Pamilo 1991b). These findings are derived algebraically in Box 4.4 (see also Table 4.2 and Figure 4.5).

4.7 Sex Ratios under Polygyny

Sex ratio evolution when there is polygyny (colonies contain multiple queens) has been considered by Trivers and Hare (1976), Benford (1978), Crozier (1979), Pamilo (1990b), Boomsma (1993), and Nonacs (1993a). Effects on the sex ratio differ according to whether coexisting queens are unrelated or related. Both these cases occur in ants (Chapter 8).

POLYGYNY WITH UNRELATED QUEENS

When queens are unrelated, sex ratio predictions are simple. For both queen and worker control, each party favors the ratio that it would prefer if there were only one queen per colony. This is 1 : 1 for queens and 3 : 1 females : males for workers (assuming single mating) (Box 4.2). The reason is that each party is only selected to adjust the sex ratio among those individuals in the sexual brood that are its relatives, namely its offspring (in the case of a queen) or its sibs (in the workers' case). Each party is indifferent to the sex ratio in unrelated sexuals, in which it has no genetic stake. So the net outcome will be as if only relatives were being reared, and hence as under monogyny (Benford 1978; Crozier 1979; Frank 1987b; Pamilo 1990b). The number of colony queens does not affect this argument, and so the sex ratio is 1 : 1 or 3 : 1 independent of the degree of polygyny.

POLYGYNY WITH RELATED QUEENS

Queens in polygynous colonies are often related. This arises because queens are readopted by their natal colonies (Section 8.2). Two effects on sex ratios follow. First, a controlling party now has a genetic stake in offspring other than its own (queens) or in sexuals that are not full sibs (workers). So more complex relatedness effects will operate. Second, the recruitment of queens by their natal colonies means that some queens disperse (nonrecruits) whereas others do not (recruits). Previous models have assumed that all sexuals disperse. A level of nondispersal means that the fitness payoff derived from female production needs to be partitioned between the fraction attributable to nondispersing queens, and the fraction derived from dispersing ones, which was termed q by Pamilo (1990b).

Pamilo (1990b) produced a comprehensive model for the sex ratio under polygyny, which was also the first to take into account both these effects. It therefore superseded earlier models of Trivers and Hare (1976), Benford (1978), and Crozier (1979), which ignored the effects of queen recruitment. A factor not included by Pamilo (1990b) was reproduction by colony budding, which appears common in polygynous ants (Section 8.2). Colony budding independently affects sex ratio, inducing male bias (Pamilo 1990b, 1991b; following section).

Pamilo's (1990b) model predicts the sex ratio as a function of the parameter q, the number of queens, the relatedness between them, and the identity of the controlling party (Box 4.5). Its conclusions are as follows. Note that the predicted sex ratio is that among dispersing sexuals only. However, this should differ little from the sex ratio among all progeny, because the number of recruited queens is likely to be small compared to the number of dispersing ones. A greater problem is how to measure the level of q (Pamilo 1990b).

When queens are the controlling party, Pamilo (1990b) showed that the female : male sex ratio equals q and is therefore male-biased (since, by definition, q is a fraction) (Box 4.5). In other words, the sex ratio is independent of both queen number and relatedness. This is because a focal queen is symmetrically related with both her own sexual offspring and with those produced by the other queens. So the only factor causing a deviation from 1 : 1 sex allocation will be the nondispersal of recruited queens. Since this reduces the fitness payoff from dispersing females in proportion to q, the stable sex ratio is $q : 1$.

Under worker control the situation is more complex and the sex ratio can be male- or female-biased depending on q, queen number, and queen relatedness (Box 4.5). First, falling q (rising recruitment) decreases relative female bias just as under queen control. Second, rising queen number also decreases female bias. This is because workers have the standard 3 : 1 relatedness asymmetry in favor of females among those sexuals that are their full sibs. But workers are symmetrically related with sexuals produced by queens other than their mother. As the number of such queens rises, their proportionate contribution to the sexual progeny also rises. Hence, with more and more queens, the female bias of the workers' stable sex ratio drops.

Third, falling relatedness among the colony queens causes the workers' sex ratio to grow more female-biased again. This occurs because the contribution to a worker's relatedness with sexuals that is made by the offspring of queens other than its own mother is reduced. So the sex ratio will be affected more by the worker's relatedness asymmetry with its full sibs, and hence will grow more female biased. Put another way, female bias rises because the situation increasingly resembles the unrelated queens case, in which 3 : 1 is the workers' stable ratio (see above).

In sum, Pamilo's (1990b) model finds that, under queen control, the sex ratio will range from male bias to equality, and depend on q alone. Under worker control, it will vary between a male-biased level and 3 : 1

BOX 4.5 SEX RATIO IN SOCIAL HYMENOPTERA WITH POLYGYNY

This box presents a simplified version of Pamilo's (1990b) model for sex ratio evolution under polygyny with queen recruitment to the natal colony (Section 4.7). The other assumptions are single mating by queens, worker sterility, random mating, and equal costs of the sexes. The original model allowed for multiple mating and worker reproduction, but here the focus is on how polygyny alone affects sex allocation. Let:

X = the stable female : male population sex ratio among the dispersing sexuals. Note that X in Pamilo's (1990b) terminology is different, namely average per colony absolute investment in females.

q = the contribution to the gene pool by dispersing queens as a fraction of the contribution to the gene pool of all new queens (dispersing and recruited).

N = the number of maternal queens per colony. These are assumed to share reproduction equally (in the terminology of Section 8.6, reproductive skew is zero).

G = the average regression relatedness of maternal queens.

As usual, the population sex ratio is at equilibrium when the per capita fitness payoffs from both sexes are equal. But because only the sex ratio among dispersing offspring is being considered, female per capita fitness must be weighted by the term q. Male per capita fitness is not weighted because all males disperse.

So Equation 4.1, with this amendment, gives:

$$q.r_F.V_F.MS_F = r_M.V_M.MS_M$$
$$q.r_F.V_F.1 = r_M.V_M.X$$
$$X = q.r_F.V_F/r_M.V_M. \qquad [4.10]$$

This is Pamilo's (1990b) Equation 6, allowing for Pamilo's expressing the sex ratio as proportionate investment in males. Pamilo derived this equation more rigorously. It makes intuitive sense in that if $q = 1$, it gives – as expected – the standard equation for the case when all sexuals disperse (Equation 4.2). And if $q = 0$, $X = 0$, which again is expected in that if dispersing queens yield no fitness, it does not pay to produce any. Instead, colonies should produce only dispersing males and queens that are all recruited (Pamilo 1990b).

The next step is to calculate the regression relatedness values for the present case. Assume queen control. A focal maternal queen's relatedness with new queens is the average, weighted by abundance, of her relatedness with daughters and her relatedness with the daughters of the other queens. The latter term is the product of the focal queen's related-

BOX 4.5 CONT.

ness with a fellow queen (G) and any queen's relatedness with a daughter (0.5). So:

$$r_F = 0.5(1/N) + G(0.5)(1 - 1/N) = [1 + (N - 1)G]/2N.$$

Similarly, relatedness with males is the weighted average of relatedness with own sons and other queens' sons:

$$r_M = (1)(1/N) + G(1)(1 - 1/N) = [1 + (N - 1)G]/N.$$

Substituting these values into Equation 4.10, and recalling that $V_F/V_M = 2$ (Box 4.4) yields:

$$X = q \qquad [4.11]$$

i.e. Pamilo's (1990b) Equation 7.

Under worker control, relatedness with females is the weighted average of relatedness with sisters and relatedness with daughters of queens other than one's mother. The latter is the product of relatedness with one's mother (0.5), her relatedness with fellow queens (G), and their relatedness with daughters (0.5). So:

$$r_F = 0.75(1/N) + (0.5)(G)(0.5)(1 - 1/N) = [3 + (N - 1)G]/4N.$$

Similarly:

$$r_M = 0.5(1/N) + (0.5)(G)(1)(1 - 1/N) = [1 + (N - 1)G]/2N.$$

Substituting into Equation 4.10 gives:

$$X = q.[3 + (N - 1)G]/[1 + (N - 1)G]. \qquad [4.12]$$

This is Pamilo's (1990b) Equation 8a with single mating and no worker reproduction. These findings are summarized in Table 4.3 and Figure 4.6. Note that if all queens disperse ($q = 1$), queen–queen relatedness will equal zero ($G = 0$), because there can be no recruitment of queens to their natal colonies. The sex ratio is then $1 : 1$ under queen control (from Equation 4.11) and $3 : 1$ under worker control (from Equation 4.12). In other words, as expected, the predictions are as for the case of unrelated queens (Section 4.7).

in favor of females, depending on q, the number of queens, and their relatedness (Table 4.3; Figure 4.6). Therefore, as with multiple mating and worker reproduction, the general effect of polygyny is to increase relative male bias.

Table 4.3

Sex Ratios Under Polygyny

	Stable Female : Male Sex Ratio in Dispersing Sexuals	
	Queen Control	Worker Control
Unrelated queens	1 : 1	3 : 1
Related queens	q	$\dfrac{q.[3 + (N - 1)G]}{[1 + (N - 1)G]}$

Notes: q = fitness contribution of dispersing queens to the gene pool relative to that of all queens, N = number of colony queens, G = queen–queen regression relatedness. From the model of Pamilo (1990b). The formula for the related queens, worker control case is from Pamilo (1990b Equation 8a), which in full states that the stable female : male sex investment ratio equals

$$q.[1 + 2/k + (N - 1)G]/[1 + 2(1 - p)/k(1 + p) + (N - 1)G]$$

where k = the number of queen matings, and p = the fraction of males produced by queens.

Split Sex Ratios due to Polygyny

In polygynous ants, queen numbers frequently differ between colonies, causing the workers' relatedness asymmetry to vary. Therefore, provided workers can assess queen number, polygyny may induce split sex ratios in the same way as multiple mating (Boomsma 1993; Section 4.5). This section has so far assumed either a constant level of polygyny, or a variable level that workers cannot assess. Boomsma (1993) suggested that workers can assess queen number more easily than mating frequency. For example, in a polygynous colony, workers that are not full sibs will differ in both their parents, and not only in their father. In addition, the presence of several queens itself provides information about the colony's status, although this cannot be fully independent of the information provided by odor diversity among the workers.

The qualitative predictions of split sex ratio theory under polygyny are as for the multiple mating case (Section 4.5). Consider a population consisting of some colonies with a single queen and others with two queens. Then workers in the first class of colony have a higher female : male relatedness asymmetry than in the second. So, under split sex ratio theory, one would expect the monogynous colonies to concentrate on female production, and the doubly-queened ones on male production (Boomsma and Grafen 1990; Boomsma 1993).

However, there are a number of complications. First, relatedness asymmetry may not vary with queen number in a simple fashion, since it depends on the detailed history and generational structure of the colony (Boomsma 1993). Second, other factors affecting the sex ratio may covary with queen number, such as local mate competition and local resource competition (Boomsma 1993; Sections 4.9, 4.10). Third, there is the problem of knowing the level of the parameter q (fraction of fitness from dispersing queens) in Pamilo's (1990b) population-level model (above). However, this should not greatly matter, since split sex ratio theory would still predict qualitative differences between the colony classes. Fourth, as queens are lost or recruited, colonies will change in their queen number, so the relative frequency of the different classes may vary from year to year. This might weaken selection for class-specific sex ratios in polygynous species (Herbers 1990).

In conclusion, a population's relatedness structure, breeding biology, and dispersal system must all be well known before split sex ratio theory allows quantitative predictions under polygyny (Boomsma 1993). But tests that examine these factors, or qualitative tests, remain a possibility.

4.8 Sex Ratios when there is Colony Fission, Colony Budding, or Polydomy

So far the discussion has dealt with cases where reproduction occurs via the emission of sexual forms (new queens and males). Colony fission and colony budding are different modes of reproduction. In colony fission, a monogynous colony first produces a batch of males and a number of queens. It then divides into two. Either each daughter colony is headed by a new queen produced by the parent colony, or one has a new queen while the other retains the old queen. Colony fission occurs in army ants (Franks and Hölldobler 1987), honey bees (Seeley 1985 p. 37; Page et al. 1993), and stingless bees (E.O. Wilson 1971 p. 92; Michener 1974 pp. 131–132). It has almost certainly evolved because the ecology of these groups dictates that colonies below a certain size, including incipient colonies founded by solitary queens, are not viable (e.g. Nonacs 1993a). For example, army ant colonies must be large for economically profitable group predation (Franks and Hölldobler 1987).

Colony fission is distinct from colony budding, which involves queens and workers leaving polygynous colonies to form new societies (Franks and Hölldobler 1987). First, fission coincides with the production of a

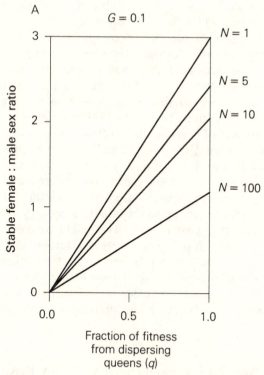

A

G = 0.1

N = 1
N = 5
N = 10
N = 100

Stable female : male sex ratio

Fraction of fitness
from dispersing
queens (q)

Figure 4.6   The stable sex ratio among the dispersing sexuals when there is polygyny with related queens and worker control. The sex ratio is a function of q (the fitness payoff from dispersing queens relative to that from all queens), N (number of colony queens), and G (regression relatedness between colony queens). A. G = 0.1. B (*facing page*). G = 0.5. Calculated from Equation 4.12. In all cases the stable sex ratio under queen control is given by the line (sex ratio = q), to which the worker control lines tend for high N.

new generation of sexuals, whereas budding may involve the departure of already mated queens. Budding species generally have reproduction by sexual emission as well, whereas colony fission is the exclusive mode of reproduction of some species. Second, fission is typical of some monogynous species, and budding of polygynous ones. Third, fission is a stereotyped process that typically results in two monogynous daughter colonies. Budding is more haphazard, and the number and composition of the resultant new colonies seem variable. Finally, in fission the daughter colonies are independent and often disperse (for example, army ant colonies are nomadic). Therefore, they probably compete with each other no more than with other colonies in the population. In budding, the colony buds usually remain close to each other and the parent colony (Franks and Hölldobler 1987). This often leads to polydomous

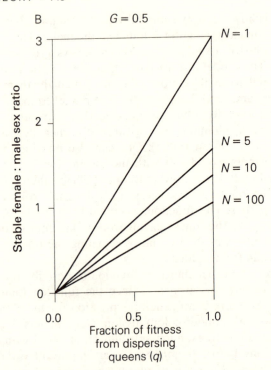

colonies (having multiple nest sites) rather than to fully independent daughter colonies (Pamilo 1991b).

COLONY FISSION

Theoretical treatments of sex ratio evolution under colony fission have been provided by Macevicz (1979), Bulmer (1983b), Pamilo (1991b), and Nonacs (1993a). The present assumptions include monogyny, single queen mating, and worker sterility. The first conclusion is that the investment in new queens by a parent colony can be very small. This is because at most only one new queen will head a daughter colony, and she will not compete with new queens from other colonies (since workers do not accept nonnestmate queens). But males from all colonies still engage in population-wide competition for matings. Therefore, a parent colony should produce only enough new queens to head their daughter colonies and as many males as possible (Oster et al. 1977; Oster and Wilson 1978 p. 92; Craig 1980b; Bulmer 1983b). So the numerical and investment sex ratio among the reproductive forms will be very male-biased.

Given a negligible investment in new queens, the central allocation problem is how a colony should divide its resources between the workers that form each daughter colony and the males. This was modeled by Pamilo (1991b), assuming that fission always produces two daughter colonies. The model also assumed that the genetic return on males is proportional to their number (the investment in them), weighted by relatedness and reproductive value. The payoff from a daughter colony is harder to specify. One can imagine that it depends on the colony being above a minimum size (as argued earlier, fission evolves when small colonies are not viable). One can also imagine that the colony's competitiveness would rise with increasing size, but at a diminishing rate. Pamilo (1991b), following Bulmer (1983b), therefore chose a function for the reproductive success of a daughter colony incorporating these properties. This function, weighted by relatedness and reproductive value, gave the fitness payoff from a daughter colony. The stable allocation ratio was then found by equating the rates of fitness payoff for males and for daughter colonies.

However, the exact choice of function to describe the payoff from a daughter colony was arbitrary, since the function found in nature is unknown. So there is no general expression for the allocation ratio of workers to males under colony fission. Nevertheless, Pamilo's (1991b) model allows one to examine how allocation varies with the size of the parent colony before fission. Size is an important variable because it specifies the total amount of resources available for making males or daughter colonies.

The main results were these. First, as parental colony size rose, relative investment in males increased. This is because there are diminishing returns from making larger and larger daughter colonies, but not from making more and more males (Bulmer 1983b). Second, for any given parental colony size, relative investment in males was greater under queen control than under worker control (Pamilo 1991b). This makes sense, since workers are comparatively more closely related with the new queens heading the daughter colonies than with males. Third, when a daughter colony retained the old queen, it was larger than its sister colony containing a new queen. Under queen control, this is because the mother queen prefers most workers to end up rearing her offspring, since she is more closely related with these than with her daughter's offspring. Under worker control, it is because workers are more closely related with their sibs than with nephews and nieces (Bulmer 1983b).

As mentioned earlier, the investment ratio among sexuals under fission is very male-biased. Hamilton (1975b) suggested that the workers that form each daughter colony should be regarded as a form of investment in females (the queens heading each colony). This would make the investment ratio far less male-biased (e.g. Page et al. 1993). In fact, in Pamilo's (1991b) numerical examples, it was usually female-biased when measured in this way.

Colony Budding

The key difference between budding and fission from the viewpoint of sex allocation is that, in budding, daughter colonies remain near each other and the parent colony. Therefore, queens and workers in these colonies will compete for the resources in the immediate neighborhood. Such competition between relatives is termed local resource competition (Clark 1978). When, as in this case, competing relatives are female, local resource competition promotes male-biased investment (Section 4.10). This should apply even if workers are counted as investment in females (Pamilo 1991b). Similarly, polydomy – to the extent that it involves colony budding – should also lead to male-biased investment (Pamilo and Rosengren 1983; Pamilo 1990b).

In the fission case, one could regard the low investment in new queens as the result of local resource competition between sister queens for the leadership of the daughter colonies (Bulmer 1983b; Ward 1983b; Franks and Hölldobler 1987). However, this is different from the local resource competition due to budding, which involves competition between related colonies (Pamilo 1991b). Thus, in fission, parent colonies should divide as often as possible, since their daughter colonies will not be competing preferentially with each other. However, returns on producing extra colony buds tail off, so limiting the expected frequency of budding.

In sum, colony fission leads to male-biased investment among the sexuals because competition among queens for heading daughter colonies is localized, whereas competition among males for mates is population-wide. But if workers are counted as female investment, the investment ratio may work out as female-biased, because daughter colonies do not compete. By contrast, colony budding leads to male-biased investment even if workers count as female investment, because related offspring colonies experience local resource competition.

4.9 Sex Ratio with Local Mate Competition

All the sex ratio models considered so far have assumed random mating. Local mate competition is a form of nonrandom mating in which relatives compete for mates (Hamilton 1967). It typically involves mating competition between related males. A huge literature exists on sex ratio evolution under local mate competition. Charnov (1982), Wrensch and Ebbert (1993), Antolin (1993), and Godfray (1994) provide reviews. General theoretical treatments are in Hamilton (1967), Taylor and Bulmer (1980), Bulmer (1986a, b), Taylor (1988), and Frank (1990).

FACTORS AFFECTING SEX RATIO UNDER LOCAL MATE COMPETITION

The parasitic ant *Epimyrma kraussei* provides a good example of local mate competition. Mated queens singly enter host nests of another species, kill the resident queen, and lay eggs yielding almost exclusively sexual forms. These are then raised by the host workers just as in slave-making ants (Section 4.4). Next the young adult parasite sexuals, instead of leaving on a nuptial flight, mate in the host nest. Finally, the newly-mated queens disperse to find other nests to parasitize (Winter and Buschinger 1983) (Section 10.4).

Several factors in this system influence the sex ratio. First, there is local mate competition itself. Related *Epimyrma kraussei* males (brothers) compete locally for mates (within the host nest) rather than competing with all males in the population. Second, there is sib-mating. Not only do brothers compete for mates, but their mates are provided for them by their mother, as sisters. Third, there is inbreeding. Since they are sibs, pairs of *E. kraussei* mates are genetically more similar than random pairs in the population. Local mate competition is often accompanied by sib-mating, inbreeding, or both, but in theory it need not be. For example, colonies might simultaneously release all their virgin queens into a population-wide swarm, then release their males one colony at a time. There would then be competition between related males for mates (local mate competition), but the mates would not be the males' relatives (no sib-mating or inbreeding) (Alexander and Sherman 1977; Crozier 1980).

Fourth, in *Epimyrma kraussei*, it is conceivable, though unlikely (Winter and Buschinger 1983), that some sexuals disperse to other nests before mating. This would produce a mixture of local mate competition and inbreeding, and population-wide mate competition and outbreeding. Lastly, since *E. kraussei* queens shed their wings before leaving the natal colony (Winter and Buschinger 1983), there is possibly limited dispersal of queens to fresh host nests after mating. This could lead to competition between related females for host nests (local resource competition).

All these factors could be combined in *Epimyrma kraussei*. But it is important to keep them conceptually distinct, because they affect the sex ratio in different ways. Since, as in *E. kraussei*, local mate competition, sib-mating, and inbreeding often go together, their effects have previously been confounded. That they influence the sex ratio separately has been shown by Bulmer and Taylor (1980), Taylor (1981b), Uyenoyama and Bengtsson (1982), and Herre (1985) (see also Antolin's [1993] review).

Qualitatively, local mate competition among males promotes female-biased sex ratios (Hamilton 1967). Sib-mating and inbreeding also frequently promote female bias. Partial premating dispersal mitigates these effects, and local resource competition leads to male bias (e.g. Bulmer and Taylor 1980; Frank 1990). From now on, it is assumed that there is no premating dispersal, and that postmating dispersal is complete and random. So the focus will be on the effects of just local mate competition, sib-mating, and inbreeding. Actual investment levels in *Epimyrma kraussei* are discussed in Section 5.8.

A Model of Local Mate Competition

Formulae for the sex ratio under local mate competition have been derived in the following model, originating with Hamilton (1967). Consider a habitat divided into patches. Let N unrelated females (foundresses) settle on a patch and let all mating occur within the patch among their progeny. (So both sib-mating and inbreeding are present.) Mated offspring females then disperse fully to occupy new patches at random. Then, assuming many patches, the expected sex ratios within patches for diploids and social haplodiploids under maternal and worker control are as shown in Table 4.4. The haplodiploid sex ratios are illustrated in Figure 4.7. The influence of local mate competition on the sex ratio may also be modeled in a "group selection" framework, with similar or identical results (e.g. Frank 1986; reviewed by Antolin 1993, Godfray 1994; Box 2.1).

Table 4.4

Sex Ratios Under Local Mate Competition

	Stable Female : Male Sex Ratio
Diploids	
Maternal control	$(N + 1)/(N - 1)$
Haplodiploids	
Queen control	$(2N^2 + 2N - 1)/(N - 1)(2N - 1)$
Worker control	$3N/(N - 1)$

NOTES: N equals the number of foundresses per patch within which mating occurs. These formulae were derived by Hamilton (1967, 1979), Taylor and Bulmer (1980), Uyenoyama and Bengtsson (1982), Joshi and Gadagkar (1985), Frank (1987b), and Taylor (1988, 1993).

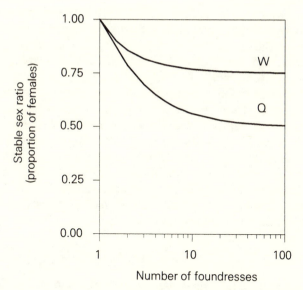

Figure 4.7   The stable sex ratio in haplodiploids with local mate competition under queen (Q) and worker (W) control. The horizontal scale is logarithmic. From the formulae in Table 4.4.

When there is one foundress per patch ($N = 1$, as in *Epimyrma kraussei*), the female : male patch sex ratio equals infinity in every case (Table 4.4). Biologically, this means that each foundress should produce a sex ratio of almost all females and hardly any males – only enough, in fact, to inseminate all the females (assuming males mate multiply) (Hamilton 1967). When there are several foundresses ($N > 1$), the sex ratio in each case is female-biased, and – for haplodiploids with worker control – greater than 3 : 1 (Table 4.4). Lastly, with very many foundresses (large N), the two maternal control sex ratios tend to 1 : 1, and the haplodiploid worker control one tends to 3 : 1 (Table 4.4; Figure 4.7). This is because the situation now approximates to the random mating case, where these sex ratios are expected (Sections 4.2, 4.3).

Why does local mate competition induce female bias? The reason is that the fitness payoff from males tails off with increasing investment in them, since the more numerous males are, the more severely they engage in unproductive competition with relatives bearing the same genes. By contrast, the payoff from females rises linearly with increasing investment, since the more females that are produced, the more new patches can be colonized at the expense of nonrelatives (e.g. Frank 1990). Therefore, within patches, Fisher's equal returns principle

(Section 4.2) will be satisfied when females outnumber males, because the males' high mating success then makes up for their otherwise poor returns.

Sib-mating also contributes to female bias. This is because it makes daughters (or sisters, in the case of worker control) more highly valued, since a daughter's (sister's) presence raises the chance of finding a mate, and hence the mating success, of a son (or brother) (Taylor 1981b). Lastly, in haplodiploids, inbreeding adds to female bias by increasing relatedness with females (Herre et al. 1987). In diploids, inbreeding increases mother–daughter and mother–son relatednesses equally, and so has no net effect. But in haplodiploids, it increases the relatedness between mothers and daughters disproportionately. This is because inbred haplodiploid mothers gain relatedness with daughters via their mates, whereas their relatedness with sons coming from unfertilized, fatherless eggs is unaffected (Herre 1985; Herre et al. 1987).

In diploids, Frank (1990) provided a neat demonstration of the separate effects of local mate competition and sib-mating. The expected female : male sex ratio, when rearranged, equals $(1 + 1/N) : (1 - 1/N)$ (Table 4.4). If mating were random, the sex ratio would be $1:1$. Assuming the N foundresses reproduce equally, a male's chance of experiencing mate competition with a brother is $1/N$. So a male's worth must be devalued by $-1/N$. Similarly, a female's chance of providing a mating for a brother is $1/N$. So her worth is increased by $+1/N$. Therefore, the stable sex ratio is $(1 + 1/N) : (1 - 1/N)$. Note that these two alterations to a $1:1$ sex ratio represent the effects of local mate competition and sib-mating respectively, and both contribute to female bias. Stable sex ratios for haplodiploids under local mate competition are more complex to derive (see references in Table 4.4 notes).

The Constant Male Hypothesis

Frank (1987b) suggested that local mate competition also influences between-colony sex ratio variance. His argument is best made with his own graphical model (Figure 4.8).

Local mate competition involves decreasing fitness returns with increasing investment in males, and linear returns on females. But average returns on both sexes must, by Fisher's theory, be the same at equilibrium. Therefore, if the slope of the returns/investment curve is lower for males than for females at high levels of investment, it must be higher for males than for females at low investment levels. Let the level of investment above which the rate of returns on males dips below that on females be termed K (Figure 4.8).

Colonies should produce the sex yielding the greater return. Frank

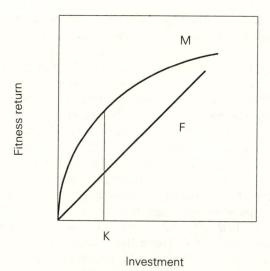

Figure 4.8 Curves describing the rate of fitness return with rising investment in females (F) and males (M) under local mate competition. K is the level of investment above which the rate of return on males falls below that on females. (From Frank 1987b, Figure 1; by permission of Springer-Verlag and the author. Copyright © 1987 Springer-Verlag)

(1987b) therefore argued that colonies with low resources (less than the amount corresponding to an investment level of K) should produce only males. But colonies with high resources (more than the amount corresponding to K) should produce K-worth of males, then females alone. Hence local mate competition should lead to sex ratios that are split by resource level (Section 6.2). Since colonies should never produce more than a K-worth of males, Frank (1987b) termed this the *constant male hypothesis* (see also Yamaguchi 1985).

Frank (1987b) pointed out that these effects should occur even when local mate competition is otherwise barely detectable. This is because local mate competition must be quite intense to affect the (local) population sex ratio appreciably. For example, say ten monogynous ant colonies contribute to a local mating swarm. Then the female : male sex ratio among these colonies would be 1.3 : 1 under queen control, and 3.3 : 1 under worker control (from the formulae in Table 4.4 with $N = 10$). These values are close to the expected sex ratios under random mating. Therefore, evidence for the constant male hypothesis should be sought at the colony level. It should include evidence for local swarming, and the observation that small, poorly-resourced colonies produce only males, while large ones produce mainly females, with a constant level of male production. However, male production may vary slightly if the degree of local mate competition is weak, or variable, or hard for colonies to assess accurately (Frank 1987b).

4.10 Sex Ratio with Local Resource Competition and Local Resource Enhancement

Local Resource Competition

Local resource competition occurs when relatives compete for resources (Clark 1978). In ants, this could happen if limited dispersal of new queens from the nest lead to competition between sisters for nest-sites. Local resource competition also occurs if there is colony budding, with buds remaining near the parent colony (Pamilo 1991b; Section 4.8). In both cases, the result would be diminishing fitness returns on investment in females, which should increase male bias in the sex ratio.

Local mate competition between related males causes female bias because of diminishing returns on investment in males. So local resource competition between related females is a mirror image of this situation. The basis of both processes is that whenever relatives of one sex compete in some way, returns on that sex will fall with rising investment, leading to sex ratio bias in favor of the opposite sex (e.g. Taylor 1981b; Bulmer and Taylor 1980; Charnov 1982 p. 73; Frank 1990).

Frank (1987b) proposed that local resource competition might affect colony variance in the sex ratio for the same reason, but with reversed effect, as local mate competition does according to the constant male hypothesis (see previous section). Thus, if there is local resource competition among females, small colonies should produce females only, and large colonies should produce mainly males beyond a certain level of female production (the *constant female hypothesis*). If both local resource competition and local mate competition occur, then how colony sex ratio variance is affected depends on their relative strengths (Frank 1987b).

Local Resource Enhancement

Local resource competition among related females favors male bias in the sex ratio. But what if related females cooperate with each other? There will then be increasing returns on investment in females, so the sex ratio should be female-biased. This is the idea of local resource enhancement (Schwarz 1988; Seger and Charnov 1988). Schwarz (1988) found evidence for female bias due to this factor in a semisocial bee whose female brood members form cooperative groups to rear their own brood. However, few ants have a natural history involving local resource enhancement. Conceivably, polygynous ants that released groups of sisters which founded colonies in associations would exhibit it. However, most foundress associations are of nonrelatives (e.g. Strassmann 1989; Section 8.3). Therefore, local resource enhancement is probably not an important influence on sex allocation in ants (Herbers 1990).

4.11 Conclusion

Clearly, many factors potentially affect the population sex investment ratio in the social Hymenoptera. Table 4.5 summarizes them, including some to be fully considered in the following two chapters. Herbers (1990) also tabulates and reviews these factors.

The second general conclusion from this chapter is that, despite their diversity, many of the factors influencing sex allocation can be incorporated into a unified body of theory. This is Fisher's (1930) sex ratio theory, with modifications by Trivers and Hare (1976), and permits sex ratio predictions for a wide variety of social and genetic structures.

Finally, given this wealth of contributing factors, testing sex allocation theory in social Hymenoptera is evidently not simple. However, many factors can be measured, so that critical tests of the theory remain possible. The next chapter reviews the existing data.

4.12 Summary

1. The fundamental theory in sex allocation is Fisher's (1930) sex ratio theory. This states that the population sex investment ratio is at equilibrium for a party controlling sex allocation when the fitness return from females per unit of investment equals that from males. This condition gives the stable sex ratio because, if it holds, there is no selection for overproduction of either sex, so the sex ratio is then stable. And if it does not hold, the sex ratio alters until the condition is met again.

2. More formally, the equal returns principle holds when the product, regression relatedness × sex-specific reproductive value × mating success, is the same for both sexes (per unit cost of the sexes). Relative mating success (per capita number of mates) is the inverse of relative abundance (the sex ratio). It follows that the stable female : male sex ratio equals (regression relatedness with females × reproductive value of females)/(regression relatedness with males × reproductive value of males). This is the female : male ratio of life-for-life relatednesses, or relatedness asymmetry. In diploids and for mothers in haplodiploids, the relatedness asymmetry – and hence the predicted sex ratio – is 1 : 1.

Table 4.5

Summary of Factors Possibly Affecting the Stable Population and Colony Sex
Investment Ratios in Social Hymenoptera

1. *Relatedness structure*
   Relatedness asymmetry (Boxes 4.1, 4.2)
   Number of queens (Section 4.7; Box 4.5)
   Number of queen matings (Section 4.5; Box 4.3)
   Worker male-production (Section 4.6; Box 4.4)
   Inbreeding (Section 4.9)

2. *Sex-specific reproductive value*
   Worker male-production (Section 4.6; Box 4.4)

3. *Identity of controlling party*
   Queen control (Section 4.3; Box 4.2)
   Nonlaying worker control (Section 4.3; Box 4.2)
   Laying worker control (Section 4.6; Box 4.4)
   Mixed control (Section 6.3)
   Nonconspecific brood care (Section 4.4)

4. *Cost ratio*
   Relative energetic cost of average female and male (Boxes 4.1, 5.1)

5. *Social structure and mode of reproduction*
   Colony fission (Section 4.8)
   Colony budding (Section 4.8)
   Polydomy (Section 4.8)

6. *Mating and dispersal systems*
   Local mate competition (Section 4.9)
   Local resource competition (Sections 4.8, 4.10)
   Local resource enhancement (Section 4.10)
   Sib-mating (Section 4.9)
   Inbreeding (Section 4.9)
   Fraction of fitness from dispersing females when there is queen readoption (Section
   4.7; Box 4.5)

7. *Split sex ratios*
   Variable relatedness asymmetry (Sections 4.5–4.7; Boxes 4.3, 4.4)
   Queenless worker male-production (Section 4.6; Box 4.4)
   Colony budding (Section 5.7)
   Variable cost ratio (Section 3.8)
   Mixed control (Section 6.3)
   Worker assessment (Section 4.5; Box 4.3)
   Resource levels and quality (Section 6.2)   *Continued overleaf*

Table 4.5 (continued)

8. *Resource levels*
   Proximate effect via queen–worker caste determination (Section 6.2)
   Relative rate of fitness gain of sexes with increasing body size? (Section 6.2)

9. *Other*
   Genomic imprinting (Box 4.2)
   Intracellular sex ratio distorters (Box 4.2)
   Symbiotic or parasitic multicellular sex ratio distorters (Box 4.2)
   Differential mortality of males due to lethal mutations (Box 4.2)

NOTES: Each major heading represents a category of effect. Listed beneath are some of the main factors that operate in these categories. See also Herbers (1990 Table 1).

3. Trivers and Hare (1976) modified Fisher's theory to apply to the social Hymenoptera. Here, workers can be a controlling party as well as mothers (queens). Assuming monogyny, single queen mating, sterile workers, and random mating, Trivers–Hare theory predicts 1 : 1 sex allocation under queen control, and 3 : 1 female : male allocation under worker control, because these are the respective relatedness asymmetries of the two parties. So there is a queen–worker conflict over sex ratio.

4. Factors altering either relatedness, reproductive value, the mating system, or the identity of the controlling party all influence the stable sex ratio. Trivers–Hare theory predicts their effects, as follows:

(a) In slave-making ants with nonconspecific brood care, workers are unlikely to be the controlling party. Hence the queen's preferred 1 : 1 sex allocation is expected.

(b) Multiple mating by queens reduces worker–female relatedness, leading to less female-biased sex investment (under worker control).

(c) Worker reproduction raises both the relative reproductive value of males and average worker–male relatedness. So, under worker control, it also decreases relative female bias.

(d) Polygyny with unrelated queens does not affect sex ratios, because both queens and workers are indifferent to the sex ratio among unrelated broods. Polygyny with related queens makes the sex ratio male-biased under queen control. This is because the nondispersal of adopted queens devalues the fitness return from females. Under worker control, the sex ratio is either male-biased or less female-biased. This is because of queen nondispersal and because the workers' relatedness asymmetry with brood other than full sibs is 1 : 1. Rising queen number increases male bias, but falling queen relatedness increases female bias.

5. Competition between relatives for mates (local mate competition) or resources (local resource competition) induces bias towards the non-competing sex, since fitness returns decline with increased investment in the competing sex. Therefore, local mate competition among males promotes female bias. Sib-mating and inbreeding may independently add to this effect. If colonies undergo fission (and daughter colonies do not remain in competition), there is local resource competition among new queens for the leadership of new colonies. The predicted sex ratio among reproductives is then male-biased. If colonies bud, local resource competition between nondispersing daughter colonies induces male-biased investment even if workers in buds count towards female investment. Polydomy (single colonies occupy multiple nests) may have similar effects. Local resource enhancement (sex-ratio bias among cooperating relatives of one sex) is probably unimportant in ant sex allocation.

6. Split sex ratios occur when discrete classes within a population contribute systematically different sex ratios to the same offspring generation. Boomsma and Grafen (1990, 1991) showed that split sex ratios should arise when workers' relatedness asymmetry varies between colonies. Workers are selected to raise only females if their relatedness asymmetry exceeds the female : male population sex ratio, and only males if it is lower. Therefore, under partial multiple mating or polygyny, workers should concentrate on female production in colonies with a below-average number of mates or queens, and on male production in colonies with an above-average number.

Split sex ratio theory predicts that at least one colony class produces single-sex broods. It also predicts that the population sex ratio stabilizes at or between the class-specific relatedness asymmetries. Above a critical frequency, one class (the balancing class) is selected to overproduce the sex opposite to that produced by the single-sex class, so making the population sex ratio equal to the relatedness asymmetry of the balancing class. Therefore, the sex ratio of a class depends on both its workers' relatedness asymmetry and its frequency.

For this reason, when workers reproduce in queenless colonies, colonies with a queen should produce a more female-biased sex ratio than under worker sterility, so restoring the population sex ratio to their relatedness asymmetry level.

7. In sum, sex ratio theory for the social Hymenoptera is a unified body of theory stemming from Fisher's (1930) equal returns principle, as modified by Trivers and Hare (1976) and adapted to the variable biology of the social Hymenoptera.

5

Tests of Sex Ratio
Theory in Ants

5.1 Introduction

The previous chapter set out the basic predictions of sex ratio theory for the social Hymenoptera. This one examines the data in the ants. Chapter 6 then focuses on the conflict between queens and workers over sex allocation. The data on ant sex ratios have also been reviewed by Trivers and Hare (1976), Alexander and Sherman (1977), Nonacs (1986a), Boomsma (1988, 1989), Hölldobler and Wilson (1990), and Pamilo (1990b).

The main aim of sex ratio studies in the social Hymenoptera has been to seek evidence for worker control of investment (Pamilo 1991b). This is because the idea that sex allocation could be influenced by a party other than the parent, and by asymmetries in relatedness, represents the most novel and controversial element that Trivers and Hare (1976) added to Fisher's (1930) sex ratio theory (Alexander and Sherman 1977; Section 4.3). It is also why sex ratio studies in ants are important for kin selection and sex ratio theory in general.

This lends significance to the conclusion of this chapter, which is that observed sex ratios in ants broadly conform to predictions from Trivers–Hare sex ratio theory. However, the data still contain various uncertainties and inconsistencies. In particular, some studies have not tested sex ratio hypotheses conclusively because of inadequate sampling or failure to measure relevant variables. So this chapter also aims to highlight what information is still needed for testing sex ratio theory in ants.

5.2 Tests of the Trivers–Hare Model for Monogynous Species

The basic model of Trivers and Hare (1976) predicted $3:1$ female : male sex investment under worker control, and $1:1$ investment under queen control. The model assumed monogyny, singly mated queens, sterile workers, and random mating (Section 4.3). An ideal test of these predictions would involve measuring sex investment ratios and genetically verifying the assumptions in several populations each of a number of species. Unfortunately, the labor and expense involved have largely prevented such tests. Instead, investigators have calculated sex investment ratios from data in the literature for species assumed to meet the conditions of the model. A few studies have attempted precise measurements of relevant genetic parameters within single species. This section deals with across- and within-species studies in turn. Pamilo (1991b) reviews the strengths and weaknesses of the different types of test of sex ratio hypotheses.

ACROSS-SPECIES TESTS IN MONOGYNOUS SPECIES

A succession of authors have analyzed sex investment ratios across a range of ant species (Trivers and Hare 1976; Nonacs 1986a; Boomsma 1989; Pamilo 1990b). Boomsma (1989) also introduced a new cost ratio (ratio of female : male per capita costs) for calculating investment ratios. Pamilo's (1990b) analysis is the most recent, adopts Boomsma's (1989) cost ratio, and involves the largest number of species. Pamilo found that 40 monogynous ant species have an average sex investment ratio of 0.631 (expressed as the proportion of investment in females), or 1.7 : 1 (expressed as the female : male investment ratio). This value is significantly female-biased (greater than 0.5), but is also significantly less than 0.75 (3 : 1 females : males) (Table 5.1).

Ideally, any comparisons of across-species averages should allow for the possibility that species with the same social organization exhibit similar sex ratios just through common descent. This exemplifies a general problem in comparative biology (e.g. Harvey and Pagel 1991). However, a comparative method that corrects for phylogeny is difficult to apply in this case, because ant phylogeny is too poorly known. On the other hand, as will be seen, sex ratio varies widely between species within genera, and even within species, suggesting that it is not usually constrained by phylogeny.

Numerical sex ratios in monogynous ants are significantly male-biased (Table 5.1). Nevertheless, investment ratios are female-biased because, without exception, females are heavier than males (Table 5.1). Fisher's theory in diploids predicts that the cheaper sex should be produced in greater numbers (Box 4.1). The same prediction applies in social haplodiploids under queen control, whereas under worker control the corresponding prediction is that, if males are the cheaper sex, the numerical sex ratio (proportion of females) should be less than 0.75 (Box 4.2). This is the case for monogynous ants (Table 5.1). Therefore, as Boomsma (1988) pointed out, these ants fulfil the Fisherian prediction of numerical overproduction (relative to the expected investment ratio) of the cheaper sex.

Table 5.1

Observed Sex Ratios in Ants

Category	Sex Ratio (proportion of females) (Mean [95% C.L.])	N Species
Numerical		
Monogynous	0.430 (0.377–0.484)	40
Polygynous	0.352 (0.254–0.450)	24
Slave-making	0.382 (0.219–0.545)	3
Investment		
Monogynous	0.631 (0.589–0.673)	40
Polygynous	0.444 (0.349–0.539)	25
Slave-making	0.483 (0.331–0.634)	3

NOTES: (a) The data for the monogynous and polygynous ants are from Pamilo (1990b Table 1), reexpressed as either the numerical proportion of females or the proportion of investment in females. The data for the slave-making ants are from the present study (Table 5.4). All investment ratios are calculated using Boomsma's (1989) energetic cost ratio (Box 5.1). 95% C.L. = 95% confidence limits.

(b) Pamilo (1990b) did not list the cost ratios for each species. However, Boomsma (1989) gives cost ratios for many species in Pamilo's (1990b) dataset. The geometric mean (range) (N) female : male dry weight cost ratios for monogynous and polygynous ants are these (the geometric mean being the antilog of the mean log): monogynous, 4.40 (1.02–37.65) (N = 32 species); polygynous, 2.43 (0.98–10.41) (N = 12 species). The corresponding figures for slave-making ants are 1.82 (1.67–2.00) (for the species in Table 5.4; N = 3 species).

What is the biological meaning of an observed sex investment ratio of 0.631 in monogynous ants? To begin with, note that all such ratios depend partly on the cost ratio used for converting numerical sex ratios to investment ones. Trivers and Hare (1976) used a cost ratio equal to the ratio of the dry weights of the sexes, whereas Pamilo (1990b) used Boomsma's (1989) estimated energetic cost ratio (Box 5.1). Boomsma's ratio raises the relative per capita cost of males. This makes sex investment ratios relatively less female-biased than those calculated using the dry weight cost ratio (Box 5.1). Thus, Trivers and Hare (1976) estimated the sex investment ratio in monogynous ants to be 0.764 (calculated from the data in Trivers and Hare [1976 Table 2]). This is both more female-biased, and closer to the 0.75 expectation, than Pamilo's (1990b) figure (0.631). Trivers and Hare (1976) therefore concluded that the data firmly supported worker-controlled allocation.

BOX 5.1 METHODS OF ESTIMATING POPULATION SEX INVESTMENT RATIOS

This box considers methodological aspects of calculating sex ratios. The first part deals with measuring the cost ratio, the second with how to estimate population mean investment ratios from individual colony data. These issues have previously been reviewed by Boomsma and Isaaks (1985), Boomsma (1988, 1989), and Helms (1994). The final part discusses the meaning of across-year averages of the sex ratio within single populations.

Cost Ratio

The female : male cost ratio is defined as the ratio, average energetic cost of raising a female : average energetic cost of raising a male (Box 4.1). The cost ratio is required to convert numerical sex ratios into investment ones. But how to measure it represents a general problem for sex ratio studies (Bull and Charnov 1988; Frank 1990). Trivers and Hare (1976) estimated it as the ratio of the average dry weights of the sexes, while noting that this might underestimate the relative energetic cost of males. Laboratory studies of ant energetics have confirmed this (Boomsma and Isaaks 1985; Passera and Keller 1987). The basic reasons are that: (1) males, being smaller, have higher metabolic rates than queens; and (2) males lack the metabolically inactive fat reserves of queens. Therefore, energy consumption per unit weight is higher in males than in queens. This means that female : male dry weight cost ratios are almost certainly higher than the real cost ratio, and that this discrepancy rises with increasing sexual size dimorphism (Boomsma and Isaaks 1985). As a result, and based on measurements of energy use by *Lasius niger* sexuals and on the relation between numerical sex ratios and dry weight cost ratios in a range of species, Boomsma (1989) suggested that dry weight cost ratios should be converted to energetic ones with the following power function:

estimated energetic female : male cost ratio = (female : male dry weight cost ratio)$^{0.7}$

This will be referred to as *Boomsma's (1989) energetic cost ratio*. Appropriately, it grows increasingly close to the untransformed dry weight cost ratio as sexual size dimorphism falls. Where possible, Boomsma's ratio is the one used in this chapter (e.g. Tables 5.1–5.10; cf. Pamilo 1990b). Therefore, any conclusions based on degrees of female bias in the sex investment ratio are conservative relative to those based on dry weight investment ratios. Of course, further work may reveal other transformations to be more accurate (Crozier and Pamilo 1993).

BOX 5.1 CONT.

Calculating Population Mean Sex Ratios

Sex ratio data are usually gathered as the numbers of females and males produced by each colony within a sample of colonies from a population. How should the population sex ratio be calculated from these data? Several previous authors calculated the mean per colony sex ratio (e.g. Boomsma et al. 1982; Bourke et al. 1988; Van der Have et al. 1988; Bourke 1989). This method gives equal weight to unproductive and productive colonies. However, provided the colony sample is sufficiently large and representative (Boomsma 1988), equal weighting is undesirable. Sex ratio theory deals with the sex ratio of the entire mating population, since this determines mating success (Section 4.2). The greater contribution of productive colonies therefore needs to be reflected in the estimate of the population sex ratio.

This can be achieved by calculating the population mean as simply the ratio of the total number of females to the total number of males produced in the population. Thus, if F = total number of females produced, M = total number of males, and c = female : male cost ratio, the population numerical sex ratio is simply F/M, or, as the proportion of females, $F/(F + M)$. And the population investment sex ratio (as the proportion of investment in females) is

$$Fc/(Fc + M). \qquad [5.1]$$

This will be referred to as the *lumped population mean*. Note that it can be calculated when F, M, and c are the only data available, i.e. when no colony-by-colony data are present. Since productive colonies commonly produce proportionately more females (e.g. Boomsma et al. 1982; Section 6.2), the mean per colony sex ratio is usually less female-biased than the lumped mean (Boomsma and Isaaks 1985; Boomsma 1988).

When individual colony data are available, it is also possible to estimate confidence limits for the lumped mean sex ratio (J.J. Boomsma, personal communication). The method is to weight each colony sex ratio by the colony's relative sexual productivity. These weights are then treated as frequencies in frequency-structured data. Let

x = colony's numerical sex ratio as the proportion of females;

f = (number of sexuals produced by colony)/(average number of sexuals produced per colony in population).

Then the population mean numerical sex ratio is estimated as:

$$\Sigma fx/\Sigma f.$$

BOX 5.1 CONT.

Note that this equals the value calculated by simply summing the total production of females and males (the numerical lumped population mean). Also, Σf sums to the number of colonies in the sample, so that the degrees of freedom reflect this number. This is the reason for choosing relative rather than absolute productivity as the weighting.

The standard deviation is then obtained from the formula for frequency-structured sample data, as:

$$\sqrt{[(\Sigma fx^2)/(\Sigma f - 1) - (\Sigma fx)^2/(\Sigma f)(\Sigma f - 1)]}.$$

The mean and confidence limits for the numerical sex ratio calculated in this way can then, using the cost ratio, be converted to the investment ratio and its confidence limits. Whenever possible, this chapter presents population mean sex investment ratios as the lumped means with weighted confidence limits calculated using the methods above.

A possible problem arises when estimated average sex ratios, and their associated confidence limits, are used for statistical hypothesis-testing. Standard parametric techniques assume a normal distribution. However, in many if not most species, colony sex ratios within populations tend to a bimodal distribution, with colonies producing mainly females or mainly males (Nonacs 1986a; Section 6.2). Therefore, the confidence limits of the lumped population mean calculated from colony data are likely to be wider than if colony sex ratios were unimodally distributed. This raises the chance, when comparing observed and expected ratios, of finding the expected sex ratio to be met when it is in fact not met (Type 2 statistical error). The best way to test for a given population sex investment ratio would be to measure it in several populations in similar habitats in the same year (verifying the genetic structure of each population). But since this is difficult, the above method of using colony data from a single population remains the usual option. However, its potential limitations need bearing in mind.

Lastly, since large colonies tend to produce relatively more females (e.g. Boomsma et al. 1982), population female bias will be overestimated if only large colonies are collected. Small colony samples are more likely to contain large, more easily found colonies. Consequently, Boomsma (1989) recommended a minimum sample size of twenty colonies for the estimation of population sex ratios.

Annual Sex Ratio Data

Several studies have measured the sex ratio of single populations over more than one year (e.g. Brian 1979b; Herbers 1979, 1990; Pamilo and Rosengren 1983; Elmes 1987b; Sundström 1994a). In several of these

BOX 5.1 CONT.

cases, like some of the investigators, we have calculated across-year sex ratio averages for individual populations (Tables 5.2, 5.7, 5.8). But is averaging annual sex ratio data across years in fact justified?

As mentioned above, the key quantity in sex ratio theory is the population sex ratio, because this is what determines the relative mating success of members of each sex (Section 4.2). In temperate social insects with discrete breeding seasons, the sex ratio that determines how sex-ratio selection acts will therefore be the population (mating swarm) sex ratio in any one year. In tropical species that breed continually, it will be the population sex ratio at any given moment. Hölldobler and Bartz (1985) argued that the important quantity to be measured was a colony's lifetime sex investment ratio. But the above argument shows that this is not the case. For example, say a group of colonies produced nearly all females each season for the first three-quarters of their lives, then nearly all males for the remaining quarter. Their "lifetime" sex ratios would be roughly 3 : 1, but one could not conclude that sex allocation had reached a Fisherian equilibrium. In short, averaging sex ratios across years is potentially misleading, because it could obscure important annual variation.

Limited nonsystematic variation in annual sex ratios is expected from sampling error, since a single year's population sex ratio is always an estimate of a real but unknown population ratio. Across-year averages could then be regarded as attempts to obtain a better estimate of the population sex ratio for a typical year. Some variation might also result from time-lags in the process of frequency-dependent selection for the equilibrial sex ratio. For example, if unusual weather in one year affected the sex ratio in one direction, this could set up an oscillation in subsequent years as the sex ratio stabilized again.

In actual populations, the population sex investment ratio is often quite level over successive years (e.g. Brian 1979b; Table 5.2), with rare, exceptional years (e.g. Herbers 1979, 1990; Elmes 1987b; Tables 5.5, 5.7, 5.8). But the importance of year-to-year variation cannot always be judged, since confidence limits for single-year ratios are not available in all cases. This chapter assumes that, aside from exceptional years for which no explanation can be found, annual sex ratio variation arises either through sampling error or from processes inherent in Fisherian sex ratio evolution.

Although average sex investment in monogynous ants is less female-biased than Trivers and Hare concluded (Boomsma 1989; Pamilo 1990b), female bias is still significant (Table 5.1). Three explanations exist to account for this. First, Alexander and Sherman (1977) attributed it to local mate competition (Section 4.9). But most monogynous ants probably exhibit population-wide competition for mates (Section 5.8), ruling out local mate competition as a general explanation for female bias.

Second, workers may control sex investment and so cause female bias, but a bias of fully 3 : 1 may not be achieved because most species do not strictly meet all the assumptions of the basic Trivers–Hare model (e.g. Boomsma 1989). As others have pointed out (e.g. Nonacs 1986a; Boomsma 1988, 1989), the supposedly uniform species used for sex ratio analyses often have rather varied kin structures. For example, some of the "monogynous" species listed by Trivers and Hare (1976) and Pamilo (1990b) are in fact facultatively polygynous (e.g. *Leptothorax longispinosus*: Headley 1943; Herbers 1984). Several have a proportion of doubly mated queens (e.g. *Lasius niger*: Van der Have et al. 1988), and in most queen mating frequency is simply unknown. Many also have workers capable of producing males (Bourke 1988a; Choe 1988). Polygyny, multiple mating, and worker reproduction all tend to reduce female bias under worker control (Sections 4.5–4.7). Therefore, even if worker control exists, a sex investment ratio less than 3 : 1 is expected in the "monogynous" ant category.

The third reason for why monogynous ants show only moderate female bias is that both queens and workers might share control of investment, leading to a sex ratio lying between their respective equilibrial levels (e.g. Matessi and Eshel 1992; Section 6.3). The relative power held by the parties conceivably varies between species, populations, or colonies. For example, in the monogynous species *Apterostigma dentigerum*, the proportion of investment in females was 0.49 (Pamilo 1990b). This suggests queen control, even allowing for male production by queenless workers in this species (Forsyth 1981; Taylor 1981a; Section 5.5). Nonacs (1986a) proposed that queen control was more likely in *A. dentigerum* because of the small size of the workforce (Forsyth 1981). On the other hand, queen mating frequency was unknown in the study population. There was also a degree of functional monogyny (Box 8.1) and polydomy (Forsyth 1981), suggesting colony budding. Since multiple mating and colony budding would both reduce female bias under worker control, queen control is by no means certain in this case.

It is hard to know whether the observed level of female bias in monogynous ants results from exclusive worker control with a degree of polygyny, polyandry, or worker reproduction, or from mixed queen and worker control. However, Trivers–Hare theory also predicts greater relative male bias if there is polygyny with worker control (Section 4.7), and unbiased sex ratios in slave-making species with queen control (Section 4.4). Both these predictions appear to be met (Table 5.1; Sections 5.3, 5.6).

The contrast between the average sex investment ratios in these classes of species is evidence for a strong degree of worker control.

WITHIN-SPECIES TESTS IN MONOGYNOUS SPECIES

Several authors have recommended testing sex ratio hypotheses by examining within-species variation in populations whose genetic structure is known (e.g. Boomsma 1989; Pamilo 1990b). Others have suggested examining sex ratio variation over successive years (Herbers 1990). However, both these types of study are rare. In addition, surprisingly, there are no published population sex ratio data for monogynous species known from genetic evidence to have singly mated queens, totally sterile workers, and random mating. This could be because few such species exist. Nevertheless, they would clearly be valuable subjects for a sex ratio study. The best approximations to this case so far examined come from two very recent studies. One was on the monogynous species *Colobopsis nipponicus* in Japan (Hasegawa 1994), and the other on a largely monogynous population of *Leptothorax tuberum* in England (Pearson et al. 1995). In each case the investigators used allozyme analysis (Box 1.1) and field observations to establish that queens were singly mated (or mostly so), worker reproduction was negligible, and mating was random. The predicted population sex investment ratio under worker control, calculated from the observed workers' relatedness asymmetry (Box 4.1) as the expected proportion of investment in females, was 0.748 for *C. nipponicus* and 0.702 for *L. tuberum*. The observed values were 0.750 ± 0.027 S.D. for *C. nipponicus* and 0.747 (95% confidence limits, 0.625–0.870) for *L. tuberum*. Since the expected value under queen control was 0.5 in both cases, these results provide strong backing for worker-controlled investment (Hasegawa 1994; Pearson et al. 1995).

The rest of this section considers two more major studies of sex allocation in single monogynous species. The first is notable for covering a series of years, and the second considers between-population variation and is supported by extensive genetic data.

1. TETRAMORIUM CAESPITUM

Brian (1979b) studied sexual production over seven consecutive years in a population of this heathland ant in southern England. *Tetramorium caespitum* is known to have obligately sterile workers (e.g. Fletcher and Ross 1985). Although Brian (1979b) did not investigate his population genetically, he assumed monogyny (presumably from nest excavations). However, monogyny may not have been absolute, since recent allozyme data from a Danish population suggest a low degree of polygyny (J.J. Boomsma, personal communication). *T. caespitum* also has winged sexuals that fly from the nest (Sudd 1967 pp. 137, 139; Brian 1983 p. 229), suggesting that its mating system approximates to random mating. Queen mating frequency is unknown.

The study population consisted of two subpopulations (Brian 1979b). Although only 150 m apart, and therefore almost certainly in breeding contact, these differed in their ecology. The "Knoll" area was dry, bare, food-poor, and rich in a competing species, whereas the "Ditch" area was the opposite. Brian (1979b) therefore considered these areas to be suboptimal and optimal habitat respectively.

Sex ratio variation over successive years was slight. In the population as a whole, the mean per year proportion of investment in females was significantly higher than 0.5, but was also significantly less than 0.75. However, sex investment in the Ditch area was always more female-biased than in the Knoll one (Table 5.2). The consistency of this difference implies a genuine biological cause. Overall mean female investment in the Ditch area was not significantly different from 0.75, suggesting worker control, whereas in the Knoll one it was not significantly different from 0.5 (Brian 1979b; Table 5.2).

These differences were conceivably due to variable levels of mating frequency or queen number, assuming such variation was correlated with habitat type. However, the differences in habitat ecology suggest that the main factor inducing a persistent relative male bias in the Knoll population was a low level of resources (Brian 1979b). Several authors have proposed that resource stress induces male bias (e.g. Brian 1979b; Boomsma et al. 1982; Nonacs 1986a; Section 6.2).

2. *LASIUS NIGER*

Boomsma et al. (1982) collected sex ratio data from three Dutch populations of this monogynous species, Europe's common black garden ant. Van der Have et al. (1988) investigated the genetics of the same populations with allozyme analysis. Both the gene frequency data, and the extensive, synchronized nuptial flights of *Lasius niger*, suggested random mating (Van der Have et al. 1988). These data also indicated some double mating in all three populations, and worker male-production in two of them. Van der Have et al. (1988) implicitly assumed that sex ratios were not split through either multiple mating (Section 4.5) or worker reproduction in queenless colonies (Section 4.6). All worker reproduction was instead assumed to be in queenright conditions. However, Boomsma and Grafen (1990) later suggested that sex ratio splitting could have occurred in these populations.

Table 5.2

Sex Investment Ratios in *Tetramorium caespitum*

Year	Ditch Subpopulation		Knoll Subpopulation		Whole Population	
	N Colonies	Proportion of Investment in Females	N Colonies	Proportion of Investment in Females	N Colonies	Proportion of Investment in Females
1963	10	O.685	11	0.541	21	0.603
1964	10	0.555	13	0.535	23	0.543
1965	10	0.784	14	0.588	24	0.682
1966	10	0.755	14	0.492	24	0.641
1967	7	0.574	12	0.496	19	0.514
1968	10	0.634	13	0.473	23	0.557
1969	10	0.757	13	0.512	23	0.583
Mean		0.678		0.520		0.589
(95% C.L.)		(0.592–0.763)		(0.484–0.555)		(0.535–0.643)

NOTES: (a) Sex investment ratios were recalculated from the original data in Brian (1979b Table 2) and are lumped population means (Box 5.1). The cost ratio used was Boomsma's (1989) energetic cost ratio (Box 5.1) with a female : male dry weight cost ratio of 2.67 from Boomsma (1989 Appendix), since Brian (1979b) did not present dry weights.

(b) Within the subpopulations the colony samples over successive years are not independent, because the same colonies were repeatedly sampled (Brian 1979b). Box 5.1 discusses the analysis of across-year means such as those at the bottom of the table.

(c) Note that calculating an overall population sex ratio across the two subpopulations (far right column) assumes that the two types of habitat represented by each subpopulation occurred in the same relative frequency as the sampling frequency (i.e. here, 50 : 50).

In one population ("SV"), observed relative investment was close to the value expected under worker control, in another ("KID") it lay between the queen and worker control predictions, and in the third ("KBD") it was more male-biased than either (Table 5.3; Figure 6.2). Therefore, whereas two populations give qualified support for worker control in *Lasius niger*, results from the third are hard to explain on any hypothesis. However, food limitation in the KBD population arguably had a proximate effect by reducing queen production (Boomsma et al. 1982; Van der Have et al. 1988). In other words, as in *Tetramorium caespitum* (Brian 1979b), limited resources could have lead to greater relative male bias.

In sum, two extensive within-species investigations of sex allocation in monogynous ants have suggested worker control, but they have also implicated other factors such as resource levels. There remains a need for sex ratio studies of single monogynous species involving large sample sizes of colonies, several populations, explicit measures of genetic variables, and several years of data.

Table 5.3

Sex Investment Ratios and Relatedness in *Lasius niger*

| | Population | | |
	SV	KID	KBD
Workers' regression relatedness (S.E.) with:			
Females (r_F)	0.71	0.67	0.55
	(0.15)	(0.20)	(0.06)
Males (r_M)	0.56	0.32	0.56
	(–)	(–)	(–)
Fraction of males from queen (p)	1.00	0.78	0.45
Expected proportion of investment in females:			
Worker control	0.72	0.79	0.59
Queen control	0.5	0.5	0.5
Observed proportion of investment in females	0.683	0.651	0.357
N colonies	125	26	50

NOTES: (*a*) The data are from Boomsma et al. (1982), Boomsma and Isaaks (1985), and Van der
Have et al. (1988). These authors obtained sexual production and genetic data over the same, sin-
gle year in each population. The population names were Strandvlakte (SV), Kooiduinen (KID), and
Kobbeduinen (KBD).

(*b*) Relatedness and the fraction of queen-produced males were estimated from the genetic data.
Van der Have et al. (1988) presented worker–male life-for-life relatednesses. Their estimates
(±S.E.) were 0.28 ± 0.08 (for the SV population), 0.16 ± 0.15 (for KID), and 0.28 ± 0.18 (for KBD).
In the table, these values have been doubled to give regression relatednesses (Box 3.1).
Relatedness varied between populations because of variable levels of multiple mating. The
expected female : male investment ratio for workers equals $(r_F/r_M)(1 + p)$ (Equation 4.4). Van der
Have et al. (1988) also calculated the expected sex ratios with this equation. Note that the large
standard errors of the relatedness estimates reduce the precision of these expectations. Deriving
the sex ratios from the measured value of the relatedness asymmetry automatically allowed for
the expected effects of multiple mating and queenright worker reproduction. The expected ratio
under queen control is 0.5 since this is unaffected by multiple mating and queenright worker
reproduction (Sections 4.5, 4.6).

(*c*) The observed sex investment ratios have been recalculated from the data in Boomsma et al.
(1982 Table 3) as lumped population means, with Boomsma's (1989) energetic cost ratio (Box
5.1). The female : male dry weight cost ratio was 14.21 (Boomsma 1989 Appendix).

5.3 Sex Ratio Data in Slave-making Ants

Trivers and Hare (1976) predicted that in slave-making ants sex allocation should be queen-controlled (Section 4.4). Given monogyny, singly
mating queens, sterile workers, and random mating, the predicted
proportion of investment in females is therefore 0.5, not 0.75. This

assumes that slave-maker workers cannot influence sex allocation. Evidence for this comes from observations showing that, although slave-maker workers interact with brood, they do not feed it. So effective manipulation of the brood by slave-maker workers seems unlikely (Bourke 1989).

Sexual production has been measured in a number of obligately slave-making species (reviewed by Trivers and Hare [1976], Alexander and Sherman [1977], Nonacs [1986a], and Bourke [1989]). However, only three datasets are suitable for estimating population sex ratios. The others either consist of unrepresentative samples, or derive from data inappropriately pooled across populations (Bourke 1989). In the three suitable datasets, sex investment ratios conform with the Trivers–Hare prediction. All are close to 0.5, and the two with an error estimate are significantly less than 0.75 (Table 5.4).

Table 5.4
Sex Investment Ratios in Slave-making Ants

Species	Colony Category	Sex Investment Ratio (proportion of investment in females) (Mean [95% C.L.])	N Nests/ Colonies
Epimyrma ravouxi	Queenright	0.518 (0.381–0.642)	18
	Queenless	0.363 (0.065–0.610)	12
	Population	0.484 (0.366–0.592)	30
Harpagoxenus sublaevis	Queenright	0.540 (0.451–0.623)	25
	Queenless	0.553 (0.325–0.745)	12
	Population	0.543 (0.461–0.619)	37
Leptothorax duloticus	Population	0.421	105

NOTES: (a) The source references are Winter and Buschinger (1983) for E. ravouxi, Bourke et al. (1988) for H. sublaevis, and Talbot (1957) for L. duloticus. These studies each dealt with one population (in Germany, Sweden, and the U.S.A. respectively) over a single year. The sample sizes are for colonies in E. ravouxi and H. sublaevis, which are both monodomous, and for nests in L. duloticus, which is polydomous (Bourke 1989). Since Talbot (1957) did not present individual colony data, only a single population mean can be given. The mean proportion of investment in females in a sample of queenless nests can be calculated from Talbot (1957) as 0.241.

(b) All sex investment ratios were recalculated as lumped population means using Boomsma's (1989) energetic cost ratio (Box 5.1). Dry weights came from Bourke (1989). Weighted confidence limits were calculated with the method in Box 5.1.

(c) All figures differ from those given in Bourke (1989) for the same datasets. This is because Bourke (1989) (1) calculated the per colony (unweighted) means and confidence limits only (Box 5.1); and (2) used a different energetic cost ratio. The only substantial conclusion affected by the present recalculations is that in E. ravouxi the case for sex ratio compensation by the queenright colonies is weakened (Section 5.5).

The number of slave-making species providing these data is small. Moreover, all three are leptothoracine ants. Although the three species probably evolved independently within this group (Buschinger 1981; Bourke 1989), sex ratio data from additional slave-makers, including formicine ones, are clearly required. Pamilo and Seppä (1994) measured sex ratio in the formicine slave-maker *Formica sanguinea*, but this only keeps slaves facultatively and so is unsuitable for testing the queen control hypothesis (Section 5.4). The three species in Table 5.4 are monogynous and as far as is known have random mating. However, single queen mating has only been confirmed in *Harpagoxenus sublaevis* (Bourke et al. 1988). If the other species have multiple mating, unbiased sex investment would not be firm evidence of queen control (Bourke 1989), because multiple mating reduces relative female bias under worker control (Section 4.5).

The three species also all exhibit worker reproduction in queenless colonies. This is almost certainly why queenless colonies in *Epimyrma ravouxi* and *Leptothorax duloticus* show male-biased investment (Table 5.4). Male production by queenless colonies increases relative male bias in the population sex ratio (e.g. Boomsma and Grafen 1991; Section 4.6). Therefore, the lack of population female bias conceivably stemmed from worker reproduction, not from queen control. However, applying Taylor's (1981a) model (which makes the same predictions as Boomsma and Grafen's [1991] model) to the *Harpagoxenus sublaevis* and *E. ravouxi* data showed that the population sex ratio was more male-biased, and the queenright colonies' sex ratio less female-biased, than would be expected under worker control (Bourke 1989). In other words, the observed sex investment levels fitted the queen control prediction even allowing for male production by queenless colonies.

Another possible complication is that in *Harpagoxenus sublaevis* all queens were permanently wingless (Bourke et al. 1988). Therefore, local resource competition between poorly-dispersing related females (Section 4.10) arguably accounts for the absence of female bias. However, the lack of clumping of slave-maker nests in the study population suggested that females in fact disperse quite far (Bourke et al. 1988).

It is sometimes argued that the sex ratios of slave-makers should partly reflect the preferences of their slave species (e.g. Trivers and Hare 1976; Nonacs 1986a). If so, it is important to realize that slave-makers and their hosts often have very different social organizations, mating systems, and kin structures. For example, the principal host species of both *Harpagoxenus sublaevis* and *Leptothorax duloticus* are polygynous (Talbot 1957; Douwes et al. 1987; Bourke et al. 1988). So suppose that the slave workers within *H. sublaevis* colonies simply applied their own sex ratio preferences to the slave-maker brood. Then they would automatically exhibit a sex ratio close to the slave-maker queen's 1 : 1 preference, since polygyny makes sex ratios more even (Section 4.7). Trivers and Hare (1976) suggested that, in their early evolution, slave-maker queens would manipulate the slave workers to achieve her favored ratio. But the above argument shows that this manipulation might be minimal.

Nonetheless, evidence exists that the slave workers are not simply exercising their innate sex ratio preferences, and that active manipulation by the slave-maker queen has occurred. First, all the host species of *Epimyrma ravouxi* are monogynous (Winter and Buschinger 1983; Buschinger 1989a), yet sex allocation in *E. ravouxi* was not female-biased either in queenright colonies or in the population as a whole (Table 5.4). Second, slave-maker sex ratios tend to have a unimodal distribution across colonies (e.g. Bourke et al. 1988), whereas the colony sex ratios of free-living ants tend to be bimodally distributed (Boomsma and Grafen 1990; Table 6.1). It would be interesting to confirm that this difference exists between a monogynous slave-maker and free-living colonies of its polygynous host in the same habitat. This would further suggest that slave workers raise a sex ratio chosen by the slave-maker queen.

In sum, the population sex investment ratios of slave-making ants appear to be at the stable level for the queen, as Trivers and Hare (1976) predicted. By contrast, monogynous free-living ants exhibit female bias (Table 5.1). This difference forms striking evidence both for worker control and for queen–worker sex ratio conflict.

5.4 Sex Ratio Data in Multiply Mating Species

Multiple mating is widespread in ants and in the social Hymenoptera in general (Page 1986; Hölldobler and Wilson 1990 p. 156; Keller and Reeve 1994b). Furthermore, genetic data indicate that multiple mating appreciably reduces relatedness between workers and female brood, though not always by as much as the number of queen's mates alone suggests (e.g. Pamilo 1993; Section 11.3). So if many queens within a population are polyandrous, effects on sex allocation are expected (Section 4.5).

POPULATION SEX RATIO UNDER MULTIPLE MATING

Multiple mating is predicted to reduce the female bias of the population sex ratio under worker control (Section 4.5). Although few studies have assessed queen mating number, available evidence from monogynous species with partial multiple mating is consistent with this expectation. For example, in 15 nests from a monogynous population of *Formica exsecta*, the observed proportion of investment in females (95% confidence limits) was 0.61 (0.39–0.83) (Pamilo and Rosengren 1983; Table 5.6). However, the confidence limits of this estimate were too wide for unequivocal confirmation of the prediction. In *Lasius niger*, relative female investment was lowest in the population with the greatest level of multiple mating, but this could also have been due to ecological factors (Boomsma et al. 1982; Section 5.2). In Sundström's (1994a) study of monogynous *F. truncorum*, the population sex investment ratio (0.67) again suggested an effect of multiple mating on population sex allocation (see below).

Split Sex Ratios in Ants with Multiple Mating

Unless queen mating frequency is high, the expected change in the population sex investment ratio is small (Figure 4.1). Therefore, effects of multiple mating on sex allocation may be hard to detect at the population level. However, Boomsma and Grafen (1990, 1991) predicted that in populations with partial multiple mating, colonies with singly mated queens should concentrate on female production, whereas colonies with multiply mated ones should concentrate on producing males. This occurs if workers respond to variations in their relatedness asymmetry. Under queen control, sex ratios should not be split because queens lack variation in their relatedness asymmetry (Bourke and Chan 1994; Sundström 1994a; Section 4.5). So split sex ratio theory allows a particularly powerful test of the influence of multiple mating on sex allocation (Section 4.5).

1. *FORMICA TRUNCORUM*

Sundström (1994a) carried out the first investigation of this issue. In a population of *Formica truncorum* in Finland, allozyme analysis demonstrated monogyny, a lack of worker reproduction, random mating, and partial multiple mating. Thirteen of 23 colonies had a singly mated queen, and ten had a doubly or triply mated one. The same colonies showed the predicted pattern of split sex ratios (Figure 5.1). Moreover, this pattern was maintained over four consecutive years, with individual colonies producing a characteristic sex ratio year after year. This is expected if a colony's sex ratio is largely set by the workers' response to the mating frequency of their queen. This study therefore strongly supported Boomsma and Grafen's (1990, 1991) split sex ratio theory for the multiple mating case, and the idea of worker control.

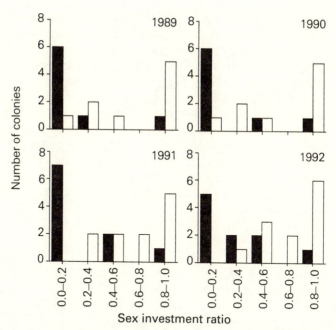

Figure 5.1   Sex ratios split by partial multiple mating, illustrated by the frequency distribution of colony sex investment ratios (proportion of dry weight investment in females) from one *Formica truncorum* population over four years. Colonies were headed by either a singly mated queen (white bars) or a multiply mated one (black bars). Colonies in these two classes had significantly different sex ratios each year, with relative investment in females falling with rising mating frequency (Mann–Whitney tests: all $p < 0.02$, $n = 17-22$). (Reprinted by permission of the author and publisher from *Nature* [Sundström 1994a]. Copyright © 1994 Macmillan Magazines Limited)

Sherman and Shellman-Reeve (1994) criticized this interpretation of Sundström's (1994a) results. They suggested that *Formica truncorum* queens both mate multiply, and produce largely male broods, in areas where colonies are likely to encounter disease. Multiple mating is favored because it increases genetic diversity within colonies, and this could protect colonies from disease (e.g. Sherman et al. 1988). Producing males is favored (by both queens and workers) because the colony's sexuals are selected to disperse from the diseased area, whereas in non-diseased patches colonies should produce females that found new nests in the same, healthy neighborhood. However, Sherman and Shellman-Reeve's (1994) idea suggests that colonies with queens having the same mating frequencies should occur near each other, which was not found (Sundström 1994a). In addition, as Sundström (1994b) pointed out, if females disperse less than males, there should be population male bias in the sex ratio due to local resource competition (Section

4.10), which did not occur (Sundström 1994a). Therefore, Sherman and Shellman-Reeve's (1994) hypothesis is unlikely to explain split sex ratios in *F. truncorum*.

Sundström's (1994a) results also suggested an effect of multiple mating on the population sex investment ratio. If workers controlled sex allocation but did not respond to queen mating frequency, the predicted proportion of investment in females in the population was simply 0.75. If they responded to the average mating frequency alone, it was 0.71 (calculated by Sundström [1994a] from the observed average relatedness asymmetry over all colonies). Lastly, under split sex ratio theory, it was 0.63 (the relatedness asymmetry of the multiply mating class). The observed fraction of dry weight investment in queens was 0.67 (Sundström 1994a). Alternatively, using Boomsma's (1989) energetic cost ratio, it was 0.65. This second estimate was close to that predicted by split sex ratio theory. In addition, the fact that both estimates fell below 0.75 suggests that multiple mating, as predicted, reduces relative female bias at the population level.

2. *FORMICA SANGUINEA*

Pamilo and Seppä (1994) investigated the genetics and sex allocation of another Finnish ant population. The study species, *Formica sanguinea*, is a slave-maker (Section 5.3). However, it only keeps slaves facultatively, since it also occurs as a free-living ant (e.g. Pamilo and Varvio-Aho 1979). So workers, since they often rear the brood themselves, probably retain the power to manipulate the sex ratio. In addition, in the population studied, most colonies lacked slaves. Therefore, in contrast to obligate slave-makers, there is no special reason for expecting *F. sanguinea* to exhibit queen-controlled sex allocation.

The relatedness of workers with females across all colonies was 0.40 ± 0.03 S.E., suggesting some degree of polyandry (or polygyny, or both). Allozyme data also indicated random mating and a lack of worker reproduction. In a study of a different *F. sanguinea* population, Pamilo and Rosengren (1983) detected sex ratio splitting. This occurred in the present population as well. Forty-seven percent of 91 nests produced nearly all females, whereas 27% produced nearly all males, with the population sex investment ratio being 0.60 (as the proportion of dry weight investment in females, weighting all nests equally). However, relatedness among workers was significantly *lower* in colonies specializing on females (0.41 ± 0.04 S.E.) than in those specializing on males (0.52 ± 0.06 S.E.). This is the opposite of the trend predicted by Boomsma and Grafen (1990, 1991), and suggested that split sex ratios were not due to variation in the workers' relatedness asymmetry (although relatedness asymmetry itself was not measured). Pamilo and Seppä (1994) proposed that intraspecific slave-raids by one nest on another (leading to nest-sharing by unrelated workers), and unpredictability in the level of polygyny, could have made worker assessment of relatedness asymmetry difficult in *F. sanguinea*. In addition, colony budding has been suspected in other Finnish *F. sanguinea* populations (Pamilo and Varvio-Aho 1979; Pamilo 1981), further indicating that this species lives in a complex social and genetic environment. Pamilo and Seppä (1994) also proposed that sex ratio specialization in their population may have stemmed from variation in resource levels (e.g. Frank 1987b; Sections 4.9, 4.10), but this is uncertain because the total sexual productivity of colonies could not be measured. In sum, although split sex ratios in this *F. sanguinea* population did not seem due to variation in workers' relatedness asymmetry, more information is required before firm conclusions can be drawn.

5.5 Sex Ratio Data in Species with Worker Reproduction

There are very few data on whether male production by workers affects sex investment as theory predicts (Section 4.6). Trivers and Hare (1976) and Nonacs (1986a) assumed worker reproduction to be quite unusual in ants, mainly from lack of information. In fact, it may be fairly common in queenright conditions, and even more common in queenless ones (Bourke 1988a; Choe 1988; Hölldobler and Wilson 1990; Section 7.4). However, accurately measuring worker reproduction in queenright colonies requires genetic analyses of high resolution. As a result, no studies have explicitly investigated the predicted effects of queenright worker reproduction on sex allocation.

A few authors have considered systems in which the effects of queenless worker reproduction could be studied, namely monogynous populations in which a fraction of colonies were queenless and produced males. These include Elmes (1974a) on the red ant *Myrmica sulcinodis*, Forsyth (1981) on the fungus-growing ant *Apterostigma dentigerum*, and Buschinger and Winter (1983) and Winter and Buschinger (1983) on the slave-maker *Epimyrma ravouxi*. In these cases, queenright colonies should have compensated by producing extra females, and the population sex ratio should have equilibrated at a level more male-biased than under worker sterility (Taylor 1981a; Nonacs 1986a; Boomsma and Grafen 1991; Section 4.6).

However, in none of these studies can sex ratio compensation in response to queenless worker reproduction be proved. The problem is that the proportion of worker-produced males has to be large for compensation to be substantial (Nonacs 1986a; Bourke et al. 1988; Bourke 1989). Therefore, to detect any effects requires large colony sample sizes. For example, in *Epimyrma ravouxi* (Winter and Buschinger 1983), neither the sex investment ratio of the queenright colonies, nor that of the population, can be shown to differ from 0.5 (the value expected if there were no sex ratio compensation at all), due to their wide confidence limits (Table 5.4). Therefore, showing that queenless worker reproduction affects sex allocation awaits more extensive studies.

5.6 Sex Ratio Data in Polygynous Species

ACROSS-SPECIES TESTS IN POLYGYNOUS SPECIES

As in monogynous species, average sex investment ratio across polygynous ants has been calculated by several authors (Trivers and Hare 1976; Nonacs 1986a; Boomsma 1989; Pamilo 1990b). Pamilo (1990b) found that polygynous ants have, on average, male-biased sex investment (0.444 proportion of investment in females). This value is significantly less than that for monogynous ants (0.631) (Table 5.1).

Pamilo's (1990b) model for sex ratio evolution when there is polygyny with related queens predicted that, under queen control, the population sex investment should always be male-biased (below 0.5). Under worker control, it could vary between male-biased values and 0.75 (Section 4.7). Polygyny often occurs alongside polydomy and colony budding (Section 8.2), which also increase relative male bias (Pamilo 1990b; Section 4.8). Therefore, the observed sex investment ratio across polygynous species is in qualitative agreement with Trivers–Hare sex ratio theory (Trivers and Hare 1976; Nonacs 1986a).

On the other hand, Pamilo (1990b) suggested that, since polygyny, polydomy, and colony budding all promote male bias, observed sex investment in polygynous ants might in fact be relatively more female-biased than expected. In other words, average sex investment should arguably be even lower than 0.444. Pamilo (1990b) therefore implied that an unknown factor might increase female bias in ants. If so, observed female bias in monogynous species (Section 5.2) need not be evidence of worker control (Pamilo 1990b).

However, this argument is not strong. First, the confidence limits for average sex investment ratio in polygynous ants are wide (Table 5.1), so conclusions about the precise degree of male bias cannot be firm. Second, other factors might raise relative female bias under polygyny with worker control. These include the occurrence of unrelated colony queens (Section 4.7), as Pamilo (1990b) himself acknowledged. They also include, as pointed out by Boomsma (1993), a degree of local mate competition. This might occur alongside polygyny because polygyny is often associated with mating near the nest (Section 8.2). As Boomsma (1993) also suggested, the female bias among polygynous ants could have been overestimated because several studies in the dataset involved a low colony sample size (Box 5.1). Finally, interpreting female bias in monogynous ants as evidence for worker control does not rely only on the comparison with polygynous ants. It also depends on the contrast between monogynous and slave-making species (Trivers and Hare 1976; Nonacs 1986a; Bourke 1989; Section 5.3).

Pamilo (1990b) further suggested that testing sex ratio hypotheses in polygynous ants, as in other cases, is best done with single-species studies. These should aim to separate out the contributions of polygyny, relatedness, polydomy, colony budding, and other factors. It is almost certainly because the across-species dataset for polygynous ants (Table 5.1) is heterogeneous in these respects, and in sample size, that the average sex investment ratio has such broad confidence limits.

WITHIN-SPECIES TESTS IN POLYGYNOUS SPECIES

As with the monogynous species, this survey focuses on studies in which both the sex ratios and the genetic structure of populations were investigated, or in which data were gathered in successive years.

1. *FORMICA OBSCURIPES*

Herbers (1979) measured up to three years' sexual output of two populations of this North American species, the western thatching ant, which is a polygynous, mound-building wood ant. *Formica obscuripes* is also polydomous with colony budding, and may have laying workers (Herbers 1979). The relatedness among queens was unknown. The two populations (Areas I and III) were 800 m apart.

In Area I, the population sex investment ratio was very male-biased in one year, and significantly more female-biased in the following year (Table 5.5). In Area III, it varied far less, and was slightly female-biased in two of three years (Table 5.5). However, overall, four out of five sex ratios were not significantly different from 0.5 (and three were not significantly different from 0.75). This was because of wide confidence limits due to the small number of nests sampled (Table 5.5).

Polygyny, polydomy with colony budding, and worker reproduction all promote male bias (Sections 4.6–4.8). To the extent that this was genuinely absent, the results from *Formica obscuripes* are unexpected. But the wide confidence limits of the observed sex ratios make firm conclusions impossible. Furthermore, if queens were unrelated, female bias would have been the predicted outcome under worker control (Section 4.7).

Table 5.5

Sex Investment Ratios in *Formica obscuripes*

| Year | Area I population | | Area III population | |
	N Nests	Sex Investment Ratio (proportion of investment in females) (Mean [95% C.L.])	N Nests	Sex Investment Ratio (proportion of investment in females) (Mean [95% C.L.])
1975	7	0.092 (−0.043–0.219)	10	0.609 (0.333–0.850)
1976	6	0.605 (0.312–0.860)	11	0.412 (0.134–0.656)
1977	—		9	0.612 (0.327–0.860)

NOTES: (*a*) The data are from Herbers (1979 Table 1). The same nests were repeatedly sampled in successive years. Note that sampling bias (Box 5.1) due to the low number of nests per sample may have not have been large, because even small *Formica* mound nests are conspicuous objects.
(*b*) Sex investment ratios were recalculated from Herbers' (1979) data as lumped population means, using Boomsma's energetic cost ratio (Box 5.1). The female : male dry weight cost ratio was 1.439 (Herbers 1979). Weighted confidence limits were calculated by the method in Box 5.1.

A more puzzling feature of the *Formica obscuripes* data is the switch in the population sex ratio over one year in Area I (Table 5.5). Herbers (1979) used this to support a theoretical argument that no stable sex ratio should be expected. But, again, small sample sizes cast doubt on the reality of the switch at the population level. In addition, sex ratio was reasonably stable over time in Area III, given the colony sample sizes. In other species, population ratios also remain fairly level in successive years (Box 5.1). Fisherian theory, of course, does predict population equilibria, while permitting wide between-colony variation in sex allocation (Sections 4.2, 6.2).

2. *FORMICA AQUILONIA* AND *F. EXSECTA*

Pamilo and colleagues examined the sexual production and (with allozyme analysis) the genetics of populations of *Formica* species from Finland (e.g. Pamilo and Rosengren 1983). Data for two of the best-studied species are summarized in Table 5.6.

Local mate competition, promoting female bias (Section 4.9), is

Table 5.6

Sex Investment Ratios and Relatedness in European *Formica* Ants

Species	Population (Year)	Social and Genetic Structure	Sex Investment Ratio (proportion of investment in females) Mean (95% C.L.) [N Nests]	Workers' Life-for-life Relatedness (S.E.) with: Females	Males
F. aquilonia	Espoo (1979)	Polygynous, polydomous	0.80 (0.67–0.91) [20]	0.01 (—)	0.08 (—)
F. exsecta	Joskär (1979)	Monogynous, monodomous, double mating, queenless worker reproduction, slight inbreeding	0.61 (0.39–0.83) [15]	0.54 (—)	0.23 (—)
	Pusula (1980)	Monogynous, monodomous, queenless worker reproduction	0.29 (0.06–0.49) [15]	—	—
	Espoo (1979)	Polygynous, polydomous	0.10 (−0.05–0.24) [18]	0.04 (0.07), 0.10 (0.07)	—
	Tuusula (1979)	Polygynous, polydomous	0.00 (0.00–0.00) [10]	—	—
	Tuusala (1980)	Polygynous, polydomous	0.24 (0.05–0.41) [20]	0.09 (0.08)	—

NOTES: (a) The source references are Pamilo and Rosengren (1983, 1984), Rosengren and Pamilo (1986), and Fortelius et al. (1987). These papers also presented sex ratio data for other species and populations of *Formica*. This table deals with the two species with both large sample sizes and accompanying genetic data.

(b) Pamilo and Rosengren (1983) estimated sex ratio from samples of pupae, introducing a possible source of sampling error. Pamilo and Rosengren (1983) and Rosengren and Pamilo (1986) review difficulties in the collection and analysis of sex ratio data from the mound-building *Formica*.

(c) Sex investment ratios and confidence intervals were recalculated from Pamilo and Rosengren (1983 Tables 1 and 4). These data only permit the calculation of mean per colony sex investment ratios that have unweighted confidence limits (Box 5.1), leading to a possible underestimation of female bias and wide confidence intervals. Boomsma's energetic cost ratio (Box 5.1) was calculated from the dry weight cost ratios in Pamilo and Rosengren (1983 Table 2) and Fortelius et al. (1987 Table 4).

(d) In the Espoo 1979 sample of *Formica exsecta*, the two worker–female relatedness values are from two different allozyme loci (Pamilo and Rosengren 1984).

(e) Two nests in the Pusula 1980 *F. exsecta* sample were artificially made queenless, and two were probably naturally queenless. These four nests all produced males only (Pamilo and Rosengren 1983).

arguably a significant influence on sex allocation in polygynous *Formica* species because their sexuals often mate on the mound. However, this mating system would not lead to significant local mate competition if the degree of polygyny were high (Pamilo and Rosengren 1983; Rosengren and Pamilo 1986). This is because, when the number of foundresses is large, the expected female bias under local mate competition falls (Section 4.9). So the potential role of local mate competition in *Formica* ants remains uncertain.

In *Formica aquilonia*, sex investment was significantly female-biased (Table 5.6). This species is polydomous and highly polygynous, and the relatedness of nestmates is effectively zero (Table 5.6). Polygyny with unrelated queens and worker control promotes female bias (Section 4.7), and could explain the observed sex ratio in *F. aquilonia* (local mate competition was probably negligible because of the high degree of polygyny). On the other hand, polydomy and colony budding should lead to male bias (Section 4.8). However, if a highly polygynous colony with unrelated queens undergoes budding, expected male bias could be less than in the case of budding in a weakly polygynous colony with related queens. This is because many new queens are presumably required to head each bud, and because low relatedness between the buds (due to low relatedness within the parent colony) reduces the severity of local resource competition between them. This makes female bias in a species like *F. aquilonia* less surprising. Significantly, other studies have found female bias in polygynous and polydomous *Formica* (*F. lugubris*: Cherix et al. 1991; *F. polyctena*: Yamauchi et al. 1994).

Formica exsecta occurs in two types of population, one monogynous and monodomous, the other polygynous and polydomous with within-nest relatednesses indistinguishable from zero (Table 5.6). A single monogynous population (Joskär) had female-biased investment, whereas a second one and three polygynous populations had very male-biased investment, with one polygynous population producing nothing but males (Table 5.6). Female bias in the Joskär population agreed with the predictions of Trivers–Hare theory even allowing for an effect of partial multiple mating (Section 5.4). Male bias in the other monogynous population needs interpreting with caution, because of the wide confidence limits and an apparently high fraction of queenless colonies (Table 5.6).

Therefore, the *Formica exsecta* data provide evidence of male-biased investment under polygyny and polydomy, as also pointed out by Pamilo and Rosengren (1983), and weak evidence of a sex ratio difference in the predicted direction between monogynous and polygynous, polydomous populations. However, why *F. aquilonia* and the polygynous *F. exsecta*, which share polygyny, polydomy, and unrelated queens, show female bias and male bias respectively is unclear. Unless these species somehow differ more in their biology, one of them must represent an anomaly.

3. *FORMICA TRUNCORUM*

Sundström (1993a,b, 1994a, 1995a) studied the genetic structure and sex allocation of two Finnish populations of the wood ant *Formica truncorum*, one monogynous and one polygynous. In the monogynous population, overall sex allocation was female-biased and colony sex ratios were split by queen mating frequency (Sundström 1994a, 1995a; Section 5.4). In the polygynous population, relatedness among cohabiting queens was very low (0.07 ± 0.03 S.E.), queen numbers per individual nest were high, and reproduction by colony budding almost certainly occurred (Sundström 1993a,b, 1995a). The population sex ratio was male-biased (0.16 as the fraction of dry weight investment in females, measured from 20 colonies: Sundström 1993b, 1995a). This contrast with the monogynous population is further evidence that, under polygyny, colony budding, or both, sex allocation grows relatively more male-biased.

4. *LEPTOTHORAX LONGISPINOSUS*

Herbers (1984, 1990) conducted a long-term sex ratio study of this small acorn-dwelling leptothoracine at two sites, one in New York State and one in Vermont (Section 10.4). *Leptothorax longispinosus* is facultatively polygynous, but with far fewer queens per colony than in the polygynous *Formica* species dealt with above. Allozyme analysis showed queens within nests to be related. Queens probably mate singly, and mating is believed to occur in a population-wide swarm. Worker egg-laying appears rare (Herbers 1984, 1990, 1993; Herbers and Stuart 1990). *L. longispinosus* is also polydomous and has colony budding. Colonies are spread over several acorn nests, and fuse or divide over time. However, each nest seems to be a fairly autonomous unit (Herbers 1984, 1990; Herbers and Stuart 1990). Therefore, although not permitting the analysis of colony-level sex allocation (Herbers 1990), this study allows examination of investment ratios for nests and populations.

Within each site, the population sex investment ratio stayed reasonably constant over time, except that the New York population showed a jump in female bias in 1983 (Table 5.7). The reasons for this are unclear. In each year, the investment ratio in the New York site was more female-biased than in the Vermont one, with the across-years average being slightly female-biased in New York and male-biased in Vermont (Herbers 1990; Table 5.7). The stability of this pattern implies a genuine biological cause (Herbers 1990).

What accounts for the observed population investment ratios, and

Table 5.7

Sex Investment Ratios in *Leptothorax longispinosus*

Year	New York Population			Vermont Population		
	N Queens per Nest	N Nests	Sex Investment Ratio (proportion of investment in females)	N Queens per Nest	N Nests	Sex Investment Ratio (proportion of investment in females)
1981	1.69	99	0.520	—	—	—
1982	1.43	37	0.488	2.43	10	0.172
1983	1.87	75	0.868	2.23	15	0.415
1984	—	—	—	1.71	70	0.137
1985	—	—	—	1.49	64	0.191
1986	—	—	—	1.23	143	0.310
1987	1.13	128	0.577	1.00	38	0.389
Mean	1.53		0.613	1.68		0.269
(95% C.L.)	(1.02–2.05)		(0.336–0.890)	(1.10–2.27)		(0.145–0.393)

NOTES: (*a*) The sex ratio data are from Herbers (1990 Table 12). Lumped population means were recalculated from the data, using Boomsma's (1989) energetic cost ratio (Box 5.1). Dry weight cost ratios were inferred from Herbers (1990 Table 12). The sample sizes (*N* nests) are from Herbers (1990 Table 8). The population sex ratios in each year are independent samples, because sampling involved removing nests from the population (Herbers 1990).

(*b*) The average number of queens per nest (arithmetic mean) in each sample was calculated from Herbers (1990 Table 5). Ideally, these should be harmonic means, since relatedness is more responsive to this variable than to the arithmetic mean (Wade 1985b; Queller 1993a). (The harmonic mean is the reciprocal of the arithmetic mean of the reciprocals.) However, the published data only permit calculation of the arithmetic means.

why did they differ between the two sites? Neither the average degree of polygyny nor relatedness among queens differed substantially between populations (Herbers 1990, 1993; Table 5.7), yet their sex ratios were clearly unequal. This could suggest that the parameter *q* (fraction of fitness from dispersing queens) in Pamilo's (1990b) polygyny model (Section 4.7) was different for each population. However, since *q* is unknown, the expected sex ratios under this model cannot be calculated. Furthermore, female bias in the New York population cannot be taken as evidence of worker control without an explanation for why this was apparently absent in Vermont.

Another possible explanation of the sex ratio difference was that the Vermont site was poorer in resources than the New York one because of its higher density of nests and its greater number of competing ant species (Herbers 1990). Resource stress has been suggested to promote male bias (Nonacs 1986a; Section 5.2). However, Backus and Herbers (1992) could not raise the relative investment in females in laboratory and field *Leptothorax longispinosus* colonies by experimentally adding extra food. This indicated that resource limitation was not responsible for the male bias in Vermont. A further possibility is that this population had a greater degree of polydomy and colony budding (Backus and Herbers 1992), which would again promote male bias. But there was no direct evidence for this. Therefore, Backus and Herbers (1992) concluded that the different population sex ratios in *L. longispinosus* remain unexplained.

5. *MYRMICA SULCINODIS*

Two populations of this red ant species in southern English heaths were studied over seven years by Elmes (1987a,b). In these populations, *Myrmica sulcinodis* was weakly polygynous. In a pattern first described by this work, queen number appeared to cycle from year to year, for reasons that are not understood (Elmes 1987a; Figure 8.1). The relatedness of queens within colonies was unknown, but in similar *Myrmica* species queens are probably related on average (Seppä 1992). Queen mating frequency was also unknown, although single mating appears to be characteristic of the *Myrmica* genus (Elmes 1987b; Snyder and Herbers 1991; Seppä 1992, 1994). *M. sulcinodis* also exhibits colony budding, along with an unknown degree of worker reproduction (Elmes 1987b). Lastly, *Myrmica* species tend to mate in swarms on the ground (Brian 1983 p. 228).

Within each population, annual variation in sex investment ratio was reasonably limited. In one population (Stoborough) relative sex investment was usually around 0.5, whereas in the other (Winfrith) it was always lower and male-biased (Elmes 1987b; Table 5.8). (Exceptions to these observations can be attributed to years in which few colonies were sampled. This leaves one true exception, the highly female-biased production in Stoborough in 1979 [Table 5.8].) Elmes (1991) suggested that sex ratios might track the cyclical changes in queen number, although this is not obvious from existing data (Table 5.8). Since the queen number changes were quite short-term (Figure 8.1), workers might instead respond just to the average level of polygyny.

Elmes (1974a) earlier studied sexual production in a monogynous *Myrmica sulcinodis* population in Scotland (Section 5.5). Here, the overall proportion of investment in females was 0.622. (This lumped population mean was calculated from the data from 22 colonies in Elmes [1974a], using Boomsma's [1989] energetic cost ratio [Box 5.1]. A dry weight cost ratio of 2.487 was first calculated from the fresh weights presented by Elmes [1974a] by assuming the same fresh : dry weight ratio as in the polygynous *M. sulcinodis* populations [Elmes 1987b].)

Table 5.8

Sex Investment Ratios in Polygynous *Myrmica sulcinodis*

	Stoborough Population			Winfrith Population		
Year	N Queens per Nest	N Nests	Sex Investment Ratio (proportion of investment in females)	N Queens per Nest	N Nests	Sex Investment Ratio (proportion of investment in females)
1979	1.56	11	0.867	0.52	14	0.235
1980	2.97	10	0.488	2.63	1	0.000
1981	3.57	5	0.309	3.05	9	0.178
1982	2.07	8	0.534	1.83	3	0.987
1983	1.39	11	0.605	1.07	11	0.321
1984	1.41	11	0.595	2.46	14	0.364
1985	3.01	17	0.411	2.86	17	0.004
Mean	2.28		0.544	2.06		0.298
(95% C.L.)	(1.46–3.11)		(0.381–0.708)	(1.18–2.95)		(−0.013–0.609)

NOTES: (a) The sex ratio data and the sample sizes (N nests) are from Elmes (1987b Table 1). Lumped population means were recalculated from the data, using Boomsma's (1989) energetic cost ratio (Box 5.1). The female : male dry weight cost ratio was 3.00 (Elmes 1987b). Yearly sex ratio samples were independent, because sampling involved the excavation of colonies (Elmes 1987a).

(b) The average numbers of queens per nest (arithmetic means) were back-transformed from the data in Elmes (1987a Table 2). Harmonic means (Table 5.7) could not be calculated from the published data.

(c) If years in which few (N < 10) sexual-producing nests were sampled are omitted, the across-years sex ratio averages become 0.593 for Stoborough, and 0.231 for Winfrith, leaving the basic pattern of the data unaffected.

This female-biased value was roughly as expected under worker control given monogyny, singly-mated queens, and worker reproduction by queenless colonies (Section 5.5). Therefore, the contrast between this value, and the more male-biased investment ratios of the polygynous populations (Table 5.8), supports worker control. A polygynous, budding population of *M. sulcinodis* in Denmark was also found to exhibit male-biased sex investment, though this bias was much more extreme than in the U.K. Winfrith population (Pedersen and Boomsma 1994, and personal communication).

Elmes' (1987b) study of polygynous *Myrmica* populations raises the same questions as Herbers' (1990) work on *Leptothorax longispinosus*. Why were the population investment ratios at their observed levels, and why did they differ? Again, without values for queen relatedness or the fraction of fitness from dispersing queens (q), it is impossible to know if the population investment ratios were at their expected levels or not.

However, to the extent that polygyny and colony budding reduce relative female bias under both queen and worker control, Elmes' (1987b) results agree with expectation. The sex ratio difference between the two populations did not stem from a difference in average queen number, which was small (Elmes 1987a; Table 5.8). But, as in *Leptothorax longispinosus*, the site with greater male bias (Winfrith) appeared poorer in resources, housing fewer arthropod prey (Elmes 1987a). Therefore Elmes (1987b) attributed male bias there to resource stress.

Overall, the studies of Elmes (1987b) and Herbers (1990) on weakly polygynous populations lead to fairly similar conclusions. (1) Population investment ratios are reasonably consistent over successive years, with some unexplained, exceptional seasons. (2) Populations within species vary between strongly male-biased and moderately female-biased, but this is not the result of differences in queen number or relatedness. (3) The lack of strong female bias provides some support for Trivers–Hare theory, but there is no unequivocal evidence of worker control, partly because incomplete knowledge of genetic variables prevents the calculation of expected sex ratios. (4) Male bias could stem from resource stress, although this could not be experimentally confirmed in *Leptothorax longispinosus*.

The general conclusion from this survey must be that polygynous species can vary widely in their sex ratios, even between populations. However, no single-species study has yet achieved the difficult task of combining large samples with measurements of all the factors that are likely to be important in determining population sex investment ratios. These include queen number, queen relatedness, the parameter q, the degree of colony budding, and resource levels. Until the relative influence of these is clarified, some of the current results will remain hard to interpret.

Split Sex Ratios in Polygynous Species

Split sex ratio theory predicts that, if queen number varies between colonies within populations and there is worker control, colonies with few queens should concentrate on female production, and colonies with many queens should produce mostly males (Boomsma 1993; Section 4.7). Consistent with this, queen number and male bias in the colony sex ratio were positively associated in some ant populations (e.g. Herbers 1984, 1990; Elmes 1987b; reviewed by Boomsma 1993). Chan and Bourke (1994) explicitly looked for split sex ratios in a population of the facultatively polygynous ant *Leptothorax acervorum*. Relatedness among workers was significantly lower in polygynous colonies than in monogynous ones (Heinze et al. 1995), suggesting that workers' relatedness asymmetry was reduced under polygyny. As expected, 21 monogynous colonies produced female-biased sex ratios (62% females), whereas 24 polygynous colonies produced male-biased ones (28% females). However, reproduction by colony budding almost certainly occurs in polygynous *L. acervorum* colonies (Stille and Stille 1993; Bourke and Heinze 1994). This may have contributed to male-biased sex allocation (Section 4.8) in concert with the workers' response to their reduced relatedness asymmetry. In addition, queen number had no discernible effect on sex allocation within the polygynous colony class (Section 6.2), suggesting that the difference in sex ratio between the monogynous and polygynous classes was largely due to worker control (Chan and Bourke 1994).

5.7 Sex Ratio Data in Species with Colony Fission, Colony Budding, or Polydomy

COLONY FISSION

Colony fission occurs when monogynous colonies reproduce by splitting into two or more new colonies, which then disperse and do not compete with each other. The sex ratio among the sexual forms should be very male-biased, because sister queens undergo local resource competition to head the new colonies (Pamilo 1991b; Section 4.8). Such a sex ratio occurs in the army ants, where colony fission is obligate (Section 10.8). Colonies typically produce about six new queens and thousands of males (e.g. *Aenictus*: Macevicz 1979; *Eciton*: Franks and Hölldobler 1987). Some ponerine ants are – like army ants – monogynous with wingless queens (ergatoids) and so reproduce by colony fission. These also show male-biased sex allocation (Peeters 1993). Similar sex ratios are found in the fissioning honey bees (Page and Metcalf 1984; Seeley 1985 p. 49). Whether other predictions of Pamilo's (1991b) fission model (Section 4.8) are ever met is unknown. However, contrary to expectation, colonies always seem to divide equally in *Eciton* army ants (Section 10.8).

COLONY BUDDING

In colony budding, polygynous colonies release groups of workers and queens which fail to disperse and so compete with one another. Here the expectation is very male-biased investment even if workers in the buds count towards female investment, because local resource competition now occurs between whole daughter colonies (Pamilo 1991b;

Section 4.8). This prediction is borne out by the data. For example, in colonies of *Rhytidoponera* ants that undergo colony budding (Ward 1983a,b), the sexual brood consists almost entirely of males (see below).

POLYDOMY

If a colony produces buds that not only remain close together, but also interchange workers and brood, the whole assemblage is regarded as one polydomous colony. Thus colony budding and polydomy are overlapping phenomena, and male-biased investment is also expected under polydomy (Pamilo and Rosengren 1983; Section 4.8). Male bias does occur in polydomous populations of *Formica exsecta* (Pamilo and Rosengren 1983). However, apparently similar polydomous populations of other *Formica* species exhibited female bias (e.g. *F. aquilonia*: Pamilo and Rosengren 1983; previous section).

SEASONAL POLYDOMY

Snyder and Herbers (1991) studied a monogynous North American red ant, *Myrmica punctiventris*, which is seasonally polydomous. Colonies occupy several nests each summer, but regroup again in the autumn. So, during the polydomous phase, colonies with a queen consist of one queenright and several queenless nests. However, workers and brood continue to be exchanged between the nests. Polydomy of this type may not greatly affect sex allocation at the colony or population level, because it is not a reproductive process and colonies retain their integrity from year to year. On the other hand, seasonal polydomy might lead to genuine colony budding, and this could affect sex allocation.

The *Myrmica punctiventris* study population had singly mating queens, sexuals that dispersed during nuptial flights, and possibly some worker male-production. The proportion of dry weight investment in females in 38 nests from 10 colonies was 0.569 overall, with 4 queenright nests producing males only, and 34 queenless nests exhibiting a sex investment ratio of 0.583 (Snyder and Herbers 1991). Snyder and Herbers (1991) suggested that queenright nests produced exclusively males because of queen control of sex allocation in these nests (cf. Sections 6.2, 6.3), with workers in queenless nests evading queen control to specialize on female production. However, the small sample of queenright nests makes it unclear if these always produced just males. In addition, as Snyder and Herbers (1991) pointed out, the presence of queens in *Myrmica* inhibits the sexualization of female brood (e.g. Brian 1983 p. 192). So male production by queenright nests could have been the proximate result of this phenomenon, with colony-level allocation remaining unaffected.

The expected population investment ratio under worker control, ignoring worker reproduction and any colony budding, was the standard 0.75. The lower actual investment ratio implies that both these traits had some male-biasing effect. On the other hand, observed investment was sufficiently close to 0.5 to prevent queen control being ruled out. Although this uncertainty exists, the lack of absolute male bias confirms that seasonal polydomy is not a major influence on the population sex ratio. In the monogynous population of *Leptothorax tuberum* studied by Pearson et al. (1995), the occurrence of a substantial fraction of queenless nests also suggested some polydomy. If so, it was probably seasonal polydomy, since this is known in other *Leptothorax* species of the same *Myrafant* subgenus (e.g. Herbers and Stuart 1990). Thus, significant female bias in the population sex ratio of *L. tuberum* (Section 5.2) again suggested that seasonal polydomy has little effect. Polydomy in *Colobopsis nipponicus* similarly failed to influence the population sex ratio strongly (Section 5.2). This is understandable, since an observed lack of relatedness between neighboring colonies indicated that polydomy did not lead to appreciable colony budding in this species (Hasegawa 1994).

SPLIT SEX RATIOS IN SPECIES WITH COLONY BUDDING

Split sex ratio theory should apply to systems in which some colonies reproduce by budding and others do not (Boomsma 1993; Nonacs 1993a). Ward (1983a,b) studied sex allocation in just such a system, although of course he did not interpret his data explicitly in terms of present-day split sex ratio theory. (Chan and Bourke [1994] also suggested that some sex ratio splitting in *L. acervorum* could be due to the effects of colony budding [Section 5.6].) Ward's system involved populations of the ponerine ant genus *Rhytidoponera* in Australia, and has several notable features. First, the workers include mated workers or "gamergates" (Peeters 1991a) capable of producing both diploid and haploid offspring like orthodox ("morphological") queens. Second, using allozyme analysis, colony censuses, and ovarian dissections, Ward (1983a) showed that populations of his study species, *R. confusa* and *R. chalybaea*, are mixtures of two types of colony. Type A colonies are headed by one, singly mated morphological queen and produce dispersing sexuals. Type B colonies are headed by several, related gamergates and reproduce by colony budding. The colony buds probably remain close to each other (Ward 1983a). Third, Ward's (1983a) genetic data demonstrated that, within each species, the two colony types interbreed. These data also showed that, within populations, mating was random. In sum, *R. confusa* and *R. chalybaea* each occur in single populations consisting of a monogynous colony class reproducing by sexual emission, and a polygynous class reproducing by colony budding.

What are the predicted investment ratios for this case? If populations consisted of just Type B colonies, very male-biased investment would be expected because of colony budding (Ward 1983b; Section 4.8). This is the case even if the unmated workers that make up most of a bud are counted as female investment (Pamilo 1991b; Section 4.8). In addition, females heading a bud presumably need not be morphological queens, as they do not require stored body resources for solitary colony foundation. In short, Type B colonies should produce just unmated workers, a few of which become gamergates heading the buds, and many males. By contrast, if populations were entirely Type A, the expected investment ratio under worker control would be the standard 0.75 proportion of investment in females. Therefore, under split sex ratio theory (Boomsma and Grafen 1990, 1991), Type B colonies in mixed populations should still concentrate nearly entirely on males, and the Type A colonies should compensate by overproducing females to restore the population investment ratio to their preferred level. So, with worker control, the Type A class-specific ratio should be between 0.75 and 1.00 (all females), and should rise as the frequency of Type B colonies increases. The population ratio should lie at or between the class-specific values, and so range from 0 (all males) to 0.75. Under queen control, the Type B investment ratios would still be male-biased, but the Type A ones should be between 0.5 and 1.00, and the population values from 0 to 0.5.

Ward's (1983b) data show that both *Rhytidoponera confusa* and *R. chalybaea* exhibited the expected pattern. However, the data do not unequivocally allow identification of the controlling party. Thus, Type B colonies produced almost entirely males, whereas Type A colonies had female-biased investment ratios. Population investment ratios ranged from roughly 0.5 to female-biased (Table 5.9). (If workers in the buds of Type B colonies were included as female investment, as Ward [1983b] proposed, the average investment ratios of the Type B colonies would be larger but should still be male-biased, and the population ratios would be more female-biased.) Across all the populations studied, female bias in the Type A colonies rose with the frequency of Type B colonies (Ward 1983a). This is as expected if there were sex ratio compensation by the Type A colonies, as pointed out by Ward (1983b) himself.

In *Rhytidoponera confusa*, the female bias of the Type A colonies and of the populations (Table 5.9) suggest worker control. However,

Table 5.9

Sex Investment Ratios and Relatedness in *Rhytidoponera* Ants

Species and Population	Type A Colonies				Type B Colonies				Population	
	Workers' Life-for-life Relatedness with:		N	Proportion of Investment in Females	Workers' Life-for-life Relatedness with:		N	Proportion of Investment in Females	N	Proportion of Investment in Females
	Females	Males			Females	Males				
R. confusa										
Royal National Park	0.80	0.01	21	0.839	0.17	0.06	11	0.000	32	0.688
Pearl Beach	0.59	0.14	20	0.753	0.29	0.17	5	0.057	25	0.664
R. chalybaea										
Whian Whian State Forest	0.75	0.33	9	0.748	0.29	0.18	18	0.000	27	0.474
Mary Cairncross Park	0.74	0.15	10	0.727	0.38	0.20	15	0.003	25	0.509

NOTES: (a) Type A colonies are headed by a single morphological queen, Type B ones by several mated workers. N is the number of colonies contributing to the sex ratio data. Investment in females refers to investment in morphological queens only. Workers in the buds of Type B colonies are not included as female investment because of uncertainties over how to quantify this component (Ward 1983b).

(b) The sex ratio data are from Ward (1983a Tables 6 and 7), and the relatedness data from Ward (1983b Table 1). Lumped mean sex ratios were recalculated from the data using Boomsma's (1989) energetic cost ratio (Box 5.1), with the dry weight cost ratios coming from Ward (1983b Table 1). Only datasets with total N greater than 20 are included. Relatednesses are averages across several loci except for the *R. confusa* Royal National Park population. In Type A colonies the expected life-for-life relatednesses of workers with females and males are 0.75 and 0.25 respectively. The observed values match these figures allowing for their associated standard errors (Ward 1983a Tables 6 and 7).

(c) It is assumed that each sex ratio represents the level found in one year, although it is unclear whether Ward (1983a,b) pooled data across years. The numbers of Type A and B colonies sampled represent their relative frequencies in the population (Ward 1983a Table 1). However, the population sex ratios are not simply averages of the class-specific sex ratios weighted by abundance, because the colony types differed in their average productivity. They also differed slightly in their cost ratios.

the Type A colonies' bias should arguably have been even larger, to restore the population level to fully 0.75. On the other hand, both an unpredictable frequency of Type B colonies, and a small amount of male production by unmated workers (Ward 1983a), could have made the Type A colonies' stable ratio slightly under 0.75. By contrast, in *R. chalybaea*, an unbiased population investment ratio (Table 5.9) suggests queen control. The female bias of the Type A colonies would then be simply the result of sex ratio compensation in response to the Type B colonies' male production. However, the relatively low frequency of Type A colonies in *R. chalybaea* populations (Table 5.9) perhaps meant that they were unable to restore the population sex ratio to their preferred level. In addition, in both species, a lack of data on within-class colony sex ratio variation prevents firmly inferring the controlling party from the observed mean sex ratios.

Ward (1983b) argued that all the investment ratios were sufficiently female-biased to indicate worker control, even allowing for the effects of sex ratio compensation. However, his use of a dry weight cost ratio increased the observed level of relative female bias (Boomsma 1989; Box 5.1). In addition, Ward (1983b) did not consider whether the overall population investment ratios were unbiased or not. Summing up, while leaving the issue of which party controls sex allocation uncertain, Ward's (1983a,b) data provide evidence of male-biased investment under colony budding, and of sex ratio compensation by nonbudding colonies in the direction predicted by split sex ratio theory.

5.8 Sex Ratio Data in Species with Local Mate Competition

Local mate competition on a scale intense enough to affect population investment ratios appreciably is almost certainly unusual in ants. Alexander and Sherman (1977) argued otherwise, suggesting that local mate competition rather than worker control accounted for the female bias in monogynous species detected by Trivers and Hare (1976) (Section 5.2). However, the mating system of most monogynous ants is not conducive to local mate competition. With their synchronous and extensive nuptial flights, most species approximate far more closely to random breeding (Oster and Wilson 1978 p. 89; Crozier 1980; Hölldobler and Bartz 1985; Boomsma 1988). Nonrandom mating was found in the swarming species, *Messor aciculatus* (Hasegawa and Yamaguchi 1994), but this seems atypical in this respect (see below). In addition, inbreeding is rare in ants (Gadagkar 1991a), yet in other species local mate competition and inbreeding frequently occur together (Section 4.9). Lastly, Nonacs (1986a) pointed out that the production by colonies of a single sex, which commonly occurs (Section 6.2), would not be expected if there were widespread local mate competition. If mating takes place within one or a few colonies, producing single-sex broods could mean that reproductives failed to find mates.

Alexander and Sherman (1977) were also mistaken in arguing that, if local mate competition could be demonstrated, Trivers and Hare's (1976) conclusion that workers control sex allocation in nonparasitic ants would automatically be undermined. For, if there is local mate competition (and foundress number is low), strong female bias is expected under both queen and worker control (Figure 4.7). A similar point was made by Joshi and Gadagkar (1985).

Among ants, local mate competition seems likeliest in species with mating in or near the nest (Section 11.5). But it does not necessarily occur in these species. For one thing, if nests are highly polygynous, local mate competition would be minimal (Crozier 1980). This is arguably the case in some *Formica* wood ants (Pamilo and Rosengren 1983; Section 5.6). In addition, army ants, ponerine ants, and the Argentine ant *Linepithema humile* (= *Iridomyrmex humilis*) often mate in the nest, but males enter foreign nests and so probably compete with nonrelatives for matings (Franks and Hölldobler 1987; Peeters 1991a; Keller and Passera 1993; Passera and Keller 1994). Similarly, in species where stationary females attract males with sexual pheromones (Crozier 1980; Hölldobler and Bartz 1985), it is likely that males disperse to find mates.

SEX ALLOCATION IN THE WORKERLESS PARASITE
EPIMYRMA KRAUSSEI

Although local mate competition may be rare in ants, there is accumulating evidence that, when it does occur, sex allocation grows more female-biased as predicted. A good example is found in one class of ants that clearly exhibits a social and mating system promoting local mate competition. These are monogynous social parasites with sib-mating in the nest. The likelihood of local mate competition in these species was pointed out by Hamilton (1972) and Trivers and Hare (1976).

The example is *Epimyrma kraussei* (Winter and Buschinger 1983; Bourke 1989; Sections 4.9, 10.4). Colonies are founded by single queens entering host species nests and killing the existing queen. The *E. kraussei* queens then produce almost exclusively sexual forms, which are raised by the host workers. Some also produce a few parasite workers. However, unlike workers in the related *E. ravouxi* (Sections 5.3, 5.5), these apparently neither slave raid in the field nor produce males

(Buschinger and Winter 1983; Winter and Buschinger 1983). In addition, laboratory observations suggest that mating in *E. kraussei* is entirely between sibs in the nest (Winter and Buschinger 1983). Therefore brothers must compete for mates, which should promote female bias. (Sib-mating and inbreeding would independently contribute to this effect [Section 4.9].)

This prediction is borne out. Winter and Buschinger (1983) measured sex ratios in *Epimyrma kraussei* colonies from North Italy and found significant female bias (Table 5.10). Given local mate competition, one would also expect all colonies to produce both sexes and, since *E. kraussei* males mate multiply (Winter and Buschinger 1983), to have a female-biased numerical sex ratio. These predictions also hold (Bourke 1989). Female-biased numerical and investment ratios apparently occur in other *E. kraussei* populations (Buschinger et al. 1986; Buschinger 1989b), and in additional parasitic *Epimyrma* species with similar mating systems (*E. corsica*: Buschinger and Winter 1985; *E. adlerzi*: Douwes et al. 1988; reviewed by Buschinger 1989a).

Table 5.10

Sex Investment Ratios in *Epimyrma kraussei*

Category	Sex Investment Ratio (proportion of investment in females) (Mean [95% C.L.])	N Colonies
With workers	0.718 (0.553–0.884)	12
Without workers	0.834 (0.665–1.004)	6
All colonies	0.757 (0.640–0.873)	18

NOTES: (a) Data from Winter and Buschinger (1983 Table 2). The means are per colony means (Box 5.1); calculating the lumped population mean is inappropriate in species with sib-mating in the nest, because the sex ratio of the population of colonies is not the quantity under selection (Section 4.9). The cost ratio used was Boomsma's (1989) energetic cost ratio (Box 5.1), with a female : male dry weight cost ratio of 1.655 (Winter and Buschinger 1983).

(b) Winter and Buschinger's (1983) data were pooled across several populations in North Italy (Buschinger and Winter 1983). This was justified because there was no significant difference between the mean colony ratios from the two main populations (Bourke 1989), and again because the population sex ratio is not the quantity under investigation.

(c) Colonies with workers and colonies without workers did not differ significantly in their average investment ratio (*t*-test: $t = 0.99$, $df = 16$, $P > 0.3$).

(d) The figures above differ slightly from the corresponding calculations made by Bourke (1989), mainly because a different cost ratio was used.

The female bias found in *Epimyrma kraussei* is in striking contrast with the unbiased investment of its slave-making congener, *E. ravouxi* (Table 5.4). Since both species have nonconspecific brood care (Section

4.4), their different sex ratios suggest that unbiased investment in *E. ravouxi* genuinely results from queen control of allocation under random mating, as Trivers–Hare theory predicts (Bourke 1989). However, it is hard to demonstrate the queen control expected in *E. kraussei*, since when the foundress number is one, extreme female bias should occur under local mate competition with both queen and worker control (Figure 4.7). On the other hand, *E. kraussei* colonies with workers failed to raise a more female-biased ratio than those without (Bourke 1989; Table 5.10). This suggests that worker control was truly absent.

SEX ALLOCATION IN *TECHNOMYRMEX ALBIPES*

Another, totally different case of local mate competition occurs in the Japanese ant *Technomyrmex alpibes* studied by Tsuji and Yamauchi (1994). This also has an unusual social and mating system, more like that of termites than of other ants. Colonies are founded by solitary mated queens in patches of bamboo. The queens then produce wingless sexuals that inbreed in the colony for several generations. During this stage, the now highly polygynous and populous colony expands to occupy many bamboo stems (polydomy). Finally, the colony produces winged sexuals that disperse and mate (Yamauchi et al. 1991; Tsuji and Yamauchi 1994).

Clearly, during the inbreeding phase, there is local mate competition among the wingless sexuals. However, the case does not conform to Hamilton's (1967) classical local mate competition model (Section 4.9), because neither sex of wingless reproductive disperses. On the other hand, the observed sex investment ratio among the wingless sexuals was very female-biased (0.91 proportion of dry weight investment in females), which Tsuji and Yamauchi (1994) attributed to local mate competition.

This sex ratio could be a function of the ecology of *Technomyrmex albipes* as well. The species presumably evolved its peculiar social and mating system because of the need to monopolize available bamboo patches (Yamauchi et al. 1991). If so, after foundation, each colony must be under selection to expand and fill its patch as rapidly as possible. So all parties should favor maximizing productivity by producing numerous wingless queens. Females would arguably then favor male production because rare males would have high mating success in the nest. However, this would not occur because males within nests are in local mate competition. Hence both ecology and the mating system probably contribute to female bias.

What is the expected sex ratio among the winged sexuals? Tsuji and Yamauchi (1994) presented a model showing that the relatedness asymmetry declines to roughly unity after several generations of inbreeding, predicting a 1 : 1 sex ratio. The observed value matched this expectation, being 0.52 (proportion of dry weight investment in females). Tsuji and Yamauchi (1994) also argued that, according to split sex ratio theory (Boomsma and Grafen 1991), small, young colonies should have produced winged females because of their relatively high relatedness asymmetry. However, as they themselves pointed out, the biology of *T. albipes* makes it unsuitable for testing split sex ratio theory. The lack of production of winged females by small colonies probably occurs for the ecological reasons proposed above, and is not evidence against split sex ratios.

Sex Allocation in *Messor aciculatus*

A third case of local mate competition is found in another Japanese ant, the monogynous harvesting ant *Messor aciculatus*. Hasegawa and Yamaguchi (1994) gathered genetic data suggesting that, although this species exhibits nuptial flights, each swarm was drawn from very few colonies. Therefore, males from any one colony were likely to be in local mate competition. Consistent with this, the sex investment ratio (0.83 as the proportion of investment in females) was significantly more female-biased than the 0.75 value expected under random mating and worker control. Hasegawa and Yamaguchi's (1994) study is important in supplying the first strong evidence for local mate competition in a nonparasitic, monogynous ant with nuptial flights. Therefore, interpreting female-biased sex allocation in such species as evidence for worker control in the absence of genetic information about the mating system is risky. However, *M. aciculatus* seems atypical in its mating system, and the more commonly noted pattern for monogynous species is for many colonies in any single area to release their sexuals into a small number of large swarms (Hölldobler and Wilson 1990 p. 146). Genetic evidence in species with this type of swarming suggests random mating (e.g. *Lasius niger*: Van der Have et al. 1988). Furthermore, as earlier mentioned, the common occurrence of colonies producing single-sex broods in monogynous species (Nonacs 1986a) is further evidence against widespread local mate competition.

Testing the Constant Male Hypothesis

Frank (1987b) suggested that, even with only a slight degree of local mate competition, colonies should produce all males up to a certain level as their resources rose, and should then concentrate on female production (the constant male hypothesis: Section 4.9). The first test of this was by Herbers (1990). Using path analysis, she concluded that the hypothesis could be ruled out in *Leptothorax longispinosus* (Section 6.2). The same conclusion was reached by Chan and Bourke (1994) for *L. acervorum* and Tsuji and Yamauchi (1994) for *Technomyrmex albipes* (although the unusual social system of this species is arguably unsuited to testing the hypothesis). However, in *Messor aciculatus*, a species known to exhibit local mate competition (see above), predictions of the hypothesis were met (Hasegawa and Yamaguchi 1994). Finally, Crozier and Pamilo (1993) pointed out that, across many species, the frequent occurrence of colonies producing only females (Section 6.2) contradicts the prediction of the hypothesis that every colony should produce some males (up to the threshold level). Overall, it seems that the constant male hypothesis is fulfilled when conditions are appropriate (there is local mate competition), but that, as earlier argued, these conditions are uncommon.

5.9 Sex Ratio Data in Species with Local Resource Competition

Local resource competition between competing sister colonies should promote male-biased sex investment (Sections 4.8, 4.10). The existence of male bias in some species with polydomy or colony budding, e.g. *Rhytidoponera* species (Ward 1983b), *Formica exsecta* (Pamilo and Rosengren 1983), and *F. truncorum* (Sundström 1993b, 1995a) supports this expectation (Sections 5.6, 5.7). Local resource competition could also involve poorly dispersing solitary females competing locally for new nest sites (Section 4.10). However, an influence on sex allocation from this cause has not been demonstrated in any species. Lastly, Frank (1987b) proposed a constant female hypothesis for the effects of local resource competition on colony sex ratios (Section 4.10). Herbers' (1990) path analysis of nest allocation ratios in *Leptothorax longispinosus* found no evidence for this idea. However, in polygynous colonies of *L. acervorum* (Chan and Bourke 1994) and *F. truncorum* (Sundström 1993b, 1995a), male bias rose with increased sexual production, as predicted by the constant female hypothesis.

5.10 Conclusion

The main aim of sex ratio studies in ants and other social Hymenoptera has been to test whether workers control sex allocation as Trivers and Hare (1976) predicted. The general conclusion from this chapter is that the case for worker control is strong, but it could be stronger. In particular, female-biased sex investment is not always found where expected, and sometimes occurs where it is unexpected (Table 5.11). The reason why the evidence is not more definitive is that few studies have both verified the critical genetic and biological parameters of populations, and measured their sex investment ratios using suitable sample sizes. Sex ratio data that are unaccompanied by a good knowledge of the genetic structure of colonies and populations cannot be properly evaluated. Studies that measure the critical parameters, along with more tests for worker control by examining split sex ratios (Boomsma and Grafen 1990, 1991; Boomsma 1993), are therefore essential. Evidence that workers actively manipulate sex allocation is also highly desirable (e.g. Aron et al. 1994; Section 6.4). Nevertheless, the balance of evidence currently available suggests that worker control is a biological reality in ants (Table 5.11).

The occurrence of worker control does not rule out the possibility that queens retain some control of sex allocation as well. This issue, along with mechanisms of control, is discussed in the next chapter. However, even evidence for a degree of worker control is significant, because control by a party other than the parent was one of the most novel and controversial parts of Trivers and Hare's (1976) sex ratio theory (Section 5.1).

This chapter has also shown that Trivers–Hare theory explains a large amount of sex ratio data across several types of social structure. There is evidence for the predicted population effects of nonconspecific brood care, multiple mating, polygyny, colony fission, colony budding, polydomy, local mate competition, and local resource competition (Sections 5.3–5.9). There is also evidence for split sex ratios due to multiple mating, polygyny, and colony budding (Sections 5.4, 5.6, and 5.7). However, theory has always run ahead of the data in the field of sex allocation as a whole, and this is also true of the study of sex ratios in ants. For example, the predicted effects of worker reproduction remain to be clearly demonstrated (Section 5.5). Moreover, there are still gaps and anomalies in the data, which only future work can address.

5.11 Summary

1. Across 40 monogynous, nonparasitic ant species, the average per species proportion of investment in females was 0.631, which is both significantly female-biased and significantly less than 0.75 (3 : 1 females : males). This result is not due to local mate competition, because most species lack an appropriate mating system. Instead, it is best explained by worker control of sex investment as predicted by

Table 5.11

Summary of Evidence For and Against Worker Control of Sex Allocation in Ants

Evidence for worker control

1. Female-biased average sex investment ratio across monogynous, nonparasitic species. Contrasts with male-biased average investment ratio in polygynous species (Pamilo 1990b) (Sections 5.2, 5.6; Table 5.1).

2. Population female bias as predicted by the measured workers' relatedness asymmetry in monogynous populations of *Colobopsis nipponicus* (Hasegawa 1994) and *Leptothorax tuberum* (Pearson et al. 1995) (Section 5.2).

3. Subpopulation female bias in monogynous *Tetramorium caespitum* (Brian 1979b) (Section 5.2; Table 5.2).

4. Female bias in some populations of monogynous *Lasius niger* (Boomsma et al. 1982; Van der Have et al. 1988) (Section 5.2; Table 5.3).

5. Unbiased average investment ratio in slave-making ants (Bourke 1989) (Section 5.3; Tables 5.1, 5.4).

6. Sex ratios split by mating frequency (and hence workers' relatedness asymmetry) in monogynous *Formica truncorum* with partial multiple mating, and a female-biased population sex ratio (Sundström 1994a) (Section 5.4).

7. In species with intraspecific variation in queen number, monogynous populations exhibit female bias (or relative female bias), and polygynous populations exhibit male bias. Examples are *Formica exsecta* (Pamilo and Rosengren 1983), *F. truncorum* (Sundström 1993b, 1995a), and *Myrmica sulcinodis* (Elmes 1974a, 1987b; Pedersen and Boomsma 1994) (Section 5.6).

8. Sex ratios split by queen number in *Leptothorax longispinosus* (Herbers 1984, 1990) and *L. acervorum* (Chan and Bourke 1994) (Sections 5.6, 6.2).

9. Female bias in monogynous Type A colonies and in whole populations of *Rhytidoponera confusa* (Ward 1983b) (Section 5.7).

10. Degree of sex ratio variance among colonies indicates sex ratios are split by variable workers' relatedness asymmetry in monogynous and polygynous species but not in slave-makers (Boomsma and Grafen 1990) (Section 6.2; Table 6.1).

11. Nonreproductive workers selectively destroy male eggs and so increase relative female bias in *Linepithema humile* (= *Iridomyrmex humilis*) and *Solenopsis invicta* (Aron et al. 1994, 1995; Section 6.4).

Evidence against worker control

1. Unbiased population sex investment ratio in monogynous *Apterostigma dentigerum* (Forsyth 1981) (Section 5.2).

2. Male bias in one subpopulation and one population of monogynous *Tetramorium caespitum* and *Lasius niger* respectively, unless due to poor resources (Brian 1979b; Boomsma et al. 1982) (Section 5.2).

3. Possible lack of sufficient male bias in average sex investment ratio across polygynous species (Pamilo 1990b) (Section 5.6).

4. Female bias in polygynous, polydomous *Formica obscuripes* and *F. aquilonia*, unless due to worker control with unrelated queens, or to effects of budding being

Continued overleaf

Table 5.11 (continued)

mitigated by high degree of polygyny and low within-nest relatedness (Herbers 1979; Pamilo and Rosengren 1983) (Section 5.6; Tables 5.5, 5.6).

5.   Unbiased population sex investment in *Rhytidoponera chalybaea* (Ward 1983b) (Section 5.7; Table 5.9)

NOTES: The interpretation of each piece of evidence is subject to the qualifications discussed in the text. Several major sex ratio studies are omitted because they give no indication either way of the controlling party (e.g. Elmes 1987b; Snyder and Herbers 1991).

Trivers and Hare (1976). Female bias could be less than expected because of either: (a) low levels of multiple mating, polygyny, or worker reproduction; or (b) a slight degree of queen control of sex allocation; or (c) both of these.

2. In single monogynous *Colobopsis nipponicus* and *Leptothorax tuberum* populations, female-biased population sex investment supported worker control. In the monogynous species *Tetramorium caespitum*, sex investment was consistently female-biased in one subpopulation, and unbiased in another. In the monogynous *Lasius niger*, investment was female-biased in two populations, and male-biased in a third. Female bias was attributed to worker control, and absence of female bias to poor resource levels.

3. Long-term data from *Tetramorium caespitum*, along with long-term studies of three polygynous species (*Formica obscuripes, Leptothorax longispinosus, Myrmica sulcinodis*), suggest that population sex investment ratios are reasonably stable from year to year, with some unexplained, exceptional years.

4. Across three species of monogynous, obligate slave-making ants, the average proportion of investment in females was 0.483. This matched the Trivers–Hare prediction of queen-controlled allocation under nonconspecific brood care, even allowing for the effects of worker reproduction in queenless colonies.

5. Under worker control, multiple mating should reduce female bias in the population sex ratio, and variable mating frequency should promote split sex ratios. Both these effects were detected in Sundström's (1994a) study of *Formica truncorum*.

6. No study has firmly demonstrated that either queenright or queenless worker reproduction affect sex allocation as predicted. This is because the predicted effects are small, whereas the confidence limits of observed investment ratios are large.

7. Sex investment in 25 polygynous species was male-biased (0.444 average proportion of investment in females), as expected according to Pamilo's (1990b) model for the sex ratio under polygyny, and from associated colony budding or polydomy.

8. In species with intraspecific variation in queen number, monogynous populations are characteristically female-biased in their sex allocation, and polygynous ones male-biased (e.g. *Formica truncorum*, *Myrmica sulcinodis*), as Trivers–Hare theory predicts. However, across species, extreme polygyny and polydomy can be associated with both strong male bias (*F. exsecta*) and strong female bias (*F. aquilonia*), which is hard to explain. In mildly polygynous species, between-population sex ratio variation has been attributed to resource level differences (*M. sulcinodis*), or remains unexplained (*Leptothorax longispinosus*).

9. Species with colony fission (e.g. army ants) and colony budding produce very male-biased sex ratios among their reproductive broods, as expected. Seasonal polydomy, in which colonies are only temporarily polydomous, need not affect population investment ratios greatly.

10. A mating system involving appreciable local mate competition is probably rare in ants. However, where it occurs, sex allocation is female-biased as theory predicts. Examples are socially parasitic ants with sib-mating in the nest (*Epimyrma* spp.), the highly inbred, polygynous *Technomyrmex albipes*, and the monogynous, nonparasitic swarming species, *Messor aciculatus*. In *M. aciculatus*, there is evidence for Frank's (1987b) constant male hypothesis under local mate competition.

11. Split sex ratios occur in species with variable levels of multiple mating (e.g. *Formica truncorum*), polygyny (e.g. *Leptothorax longispinosus*, *L. acervorum*, *Myrmica sulcinodis*), or colony budding (e.g. *Rhytidoponera* spp., *L. acervorum*). However, in *F. sanguinea*, split sex ratios appeared not be correlated with workers' relatedness asymmetry as theory predicts.

12. Overall, the ant sex ratio data support the concept of worker control in Trivers–Hare sex ratio theory, and the application of the theory to different social and genetic structures.

6 Kin Conflict over Sex Allocation

6.1 Introduction

Trivers and Hare (1976) proposed the existence of a queen–worker conflict over sex allocation in the social Hymenoptera (Section 4.3). This represents a key example of an important, general phenomenon in social evolution – kin conflict. Kin conflict occurs when classes of relatives have divergent evolutionary interests (Trivers 1974; Ratnieks and Reeve 1992). The previous chapter concluded that in nonparasitic ants the population sex ratio is often at the level expected under worker control (Section 5.10). This suggests that sex ratio conflict is a genuine phenomenon. The present chapter considers further aspects of the conflict. For example, is worker victory inevitable, or do queens and workers ever share control (Section 6.3)? What are the mechanisms for biasing investment, and how can they be deployed (Sections 6.4, 6.5)? And what factors affect the outcome of conflict (Section 6.6)? The first section considers why colonies vary in their sex ratios (Section 6.2), since one idea is that this is because of variations in the relative power of queens and workers (e.g. Herbers 1984, 1990). Further types of kin conflict are dealt with in Chapter 7. Sex ratio conflict is treated separately here because it forms such a large topic.

6.2 Why Do Colonies Vary in their Sex Ratios?

Sex ratios vary at many levels in ants – between species, between populations within species, and from year to year within populations (e.g. Tables 5.1–5.10). As the previous chapter showed, investigators have sought to explain this variation by invoking differences in genetic structure or ecological conditions. Within populations, there is also a striking amount of variation in sex ratio between colonies. In fact, the distribution of colony sex ratios tends to be bimodal, with colonies producing either mainly females or mainly males. This pattern is present within many species (e.g. *Formica*: Pamilo and Rosengren 1983; *Lasius*: Van der Have et al. 1988; *Leptothorax*: Herbers 1990), and is strong enough to be maintained when they are examined together (Nonacs 1986a; Figure 6.1). This section reviews ideas for why it occurs. Many previous authors have either pointed out or discussed the wide variation in colony sex ratios (e.g. Alexander and Sherman 1977; Herbers 1979; Pamilo 1982a, 1991b; Pamilo and Rosengren 1983; Nonacs 1986a,b; Frank 1987b; Boomsma 1988; Boomsma and Grafen 1990, 1991; Crozier and Pamilo 1993). A general review of sex ratio variation at the brood level is in Frank (1990), and Hardy (1992) considers the topic in the Hymenopteran parasitoids.

Fisher's (1930) theory in its simplest form makes no prediction about sex ratios among individual broods, apart from stipulating that collectively they should yield the stable population sex ratio (Section 4.2). So

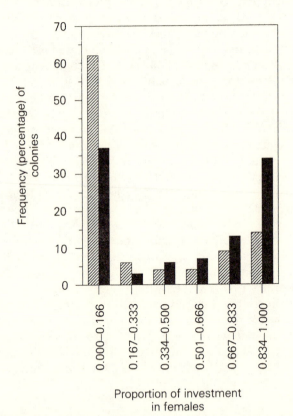

Figure 6.1 The frequency distribution of observed colony sex investment ratios in 21 monogynous ant species (black bars) and in 6 polygynous ant species (hatched bars). (From *The Quarterly Review of Biology* [Nonacs 1986a], by permission of The University of Chicago Press. Copyright 1986 © by the Stony Brook Foundation, Inc.)

one might expect that, in large populations, individual sex ratios should exhibit a random (binomial), unimodal distribution about the population equilibrium (G.C. Williams 1979). Therefore, in social Hymenoptera, colony sex ratio variation is not contrary to Trivers–Hare sex ratio theory (Nonacs 1986a). For example, say the population sex ratio is expected to be at the workers' stable level of 3 : 1 females : males. This does not mean that every colony should produce this sex ratio, but only that the population of colonies should collectively produce it. So a colony in this population that produced nothing but queen-laid males throughout its life could still be under worker control of sex allocation (Nonacs 1986b). What needs explaining is not simply that colony sex ratios vary, but that they vary more than predicted by a random, unimodal distribution (Figure 6.1).

A possible reason for the existence of male-only producing colonies is that they are all queenless colonies with reproductive workers. However, colonies often produce males alone even when diploid brood is present, indicating that females could be reared if the colonies chose (Nonacs 1986a). In addition, if only colonies containing queens are considered, colony sex ratios remain bimodally distributed (Boomsma and Grafen 1990; Table 6.1). Another idea is that single-sex broods evolved to prevent harmful inbreeding (Wheeler 1911; Marikovsky 1961; Sherman and Shellman-Reeve 1994). Yet in many species inbreeding seems adequately prevented by the release of sexuals into large population-wide swarms (Hölldobler and Bartz 1985). This is especially the case if the sexes leave the nest at different times, a common pattern being that the males leave before the females, although even then departure times may overlap (e.g. Hölldobler and Maschwitz 1965; Markin et al. 1971; Hölldobler 1976; Boomsma and Leusink 1981; Higashi 1983; Hölldobler and Bartz 1985; Mori et al. 1994). Furthermore, there is at least one species known to have unsplit sex ratios and random mating (*Harpagoxenus sublaevis*: Bourke et al. 1988). Herbers (1979) argued with a game theory model that colonies vary in their sex ratios because they have no stable global optimum. But Fisher's theory shows that a stable population sex ratio can exist in principle, and that when it does all colony sex ratios are equally optimal (Section 4.2). Remaining hypotheses for colony sex ratio variation, which are not mutually exclusive, are all linked to Fisher's theory and Trivers–Hare kin conflict theory.

MIXED CONTROL OF SEX ALLOCATION

Suppose that neither queens nor workers control sex allocation completely. Then colony sex ratios might vary with the relative amount of power held by each party in the sex ratio conflict (e.g. Pamilo 1982a; see also next section). A possible example comes from the work of Herbers (1984, 1990) on the facultatively polygynous *Leptothorax longispinosus* (Section 10.4).

In the case of polygyny with related queens, the population sex ratio is predicted to become relatively more male-biased as queen number rises across populations (e.g. Pamilo 1990b; Section 4.7). Herbers (1984) applied this argument to within-population sex ratio variation in *Leptothorax longispinosus*. Nests were classified as monogynous, oligo-gynous (2–4 queens), or polygynous (more than 5 queens). Across these classes, relative investment in males rose with increasing queen number (Herbers 1984). So Herbers (1984) proposed that, as their number rose, queens gained increasingly more power in the sex ratio conflict. Moreover, within each class, relative investment in females tended to rise as the number of workers per nest increased. So workers also apparently gained power over sex allocation in proportion to their number.

Herbers (1990) used path analysis to investigate between-nest sex ratio variation in an expanded set of data on *Leptothorax longispinosus* (see also Backus 1993) (Sections 5.6, 5.8). Path analysis is a multivariate statistical technique that examines the effect of factors on each other while controlling for the possibly confounding effects of other factors. Herbers (1990) again found evidence for conflict over investment ratio. The path coefficients (multiple regression coefficients) for the direct effects of queen and worker number on relative investment in males tended to be positive and negative respectively. They also tended to be of opposite sign for any given year and population.

However, these findings are open to other interpretations. One is Boomsma and Grafen's (1990, 1991) theory that workers control sex allocation overall and respond to variations in their relatedness asymmetry. This would also predict that colonies invest relatively less in females as queen number rises (Boomsma 1993; Section 4.7). Chan and Bourke (1994) found evidence for such sex ratio splitting in the facultatively polygynous *Leptothorax acervorum*. Furthermore, in this species, queen number had no influence on sex allocation within the polygynous colony class. This is as expected if workers, not queens, controlled allocation. Another explanation for Herbers' findings comes from Nonacs (1986a,b). He argued that workers might be primarily in control, with larger colonies investing more in females because of increased resources (see below). Nonacs (1986a) also pointed out that queen and worker sex ratio equilibria converge as queen number rises (e.g. Pamilo 1990b; Section 4.7). Therefore, under the conflict hypothesis, there should not have been greater female bias with rising worker number in the oligogy-nous and polygynous queen classes of *L. longispinosus*.

On the other hand, Herbers' (1990) path analysis showed that worker number still had a positive effect on relative female investment even when all other factors were statistically held constant. But the analysis also found evidence for the influence of resources, since as total investment rose, relative investment in males fell (Herbers 1990). Moreover, the effect of total investment seemed to exceed that of queen or worker number, since the relevant path coefficients were negative in every case, and total investment nearly always explained a greater percentage of the observed variance (Herbers 1990).

Using another multivariate statistical technique (partial correlation), Nonacs (1986b) attempted to separate the effects of productivity (resource level) and worker number on sex allocation across a range of species. In 16 of 18 monogynous species, the partial correlation of total sexual productivity and relative male investment (with worker number held constant) was negative. However, only 8 of 18 partial correlations between worker number and relative male investment (with total investment held constant) were negative. These findings supported the idea that larger colonies invest relatively more in females not because of queen–worker conflict, but because of resource abundance (Nonacs 1986b).

A final problem with attributing sex ratio variation between *Leptothorax longispinosus* nests to mixed control is that there is no obvious mechanism allowing queens to gain power over investment in proportion to their numbers. It is relatively easy to imagine that workers influence sex allocation more strongly as they grow more numerous, because they raise the brood. But it is less clear why, say, three queens in a colony of a hundred workers should be able to influence allocation more powerfully than a single queen.

To sum up, power-sharing by queens and workers within populations possibly contributes to colony sex ratio variation. However, the evidence that other factors such as relatedness asymmetry and resource level also contribute is strong.

RESOURCE ABUNDANCE HYPOTHESIS

Nonacs (1986a) suggested that resource abundance has a proximate effect on investment ratios by influencing queen–worker caste determination. When resources are poor, colonies raise female larvae as workers rather than new queens, so causing the sex ratio to be male-biased. If resources are plentiful, the opposite occurs. This predicts male specialization by small, unproductive colonies, and mainly female production by large, well-resourced ones (Nonacs 1986a).

There is some evidence that habitats with low resource levels do promote male bias in population sex ratios (Brian 1979b; Boomsma et al.

1982; Elmes 1987b; Herbers 1990; but see Backus and Herbers 1992; Sections 5.2, 5.6). Furthermore, within populations, it is commonly observed that total productivity and relative female investment rise together (e.g. Boomsma et al. 1982; Herbers 1990; Figure 6.2). This association was demonstrated across a sample of many species by Nonacs (1986a), and Nonacs (1986b) showed that it remained when the influence of worker number was controlled for by partial correlation (see above).

What is the ultimate (evolutionary) reason for an effect of resources on investment ratios (Frank 1987b)? In the case of whole populations in poor habitats, one might ask why colonies have not readjusted their thresholds for queen–worker caste determination. Any colony that did so,

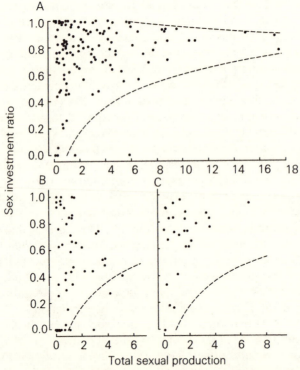

Figure 6.2   Relationship of colony sex investment ratio (proportion of investment in females) with total per colony sexual productivity (g) in three *Lasius niger* populations: A, Strandvlakte; B, Kobbeduinen; C, Kooiduinen (Section 5.2), illustrating the rise in female bias with increasing productivity. The dashed curves describe the limits of the investment frequency distribution in Strandvlakte. An energetic cost ratio (Boomsma and Isaaks 1985; Boomsma 1989) was not used for these figures, and consequently overall female bias is overestimated (Box 5.1). (From Boomsma et al. 1982; by permission of Blackwell Scientific Publications Ltd)

and so produced more females (new queens), would then gain a strong Fisherian advantage in the presence of a male-biased population sex ratio. Eventually, the sex ratio should stabilize at its Fisherian equilibrium.

On the other hand, male bias could persist for several reasons. First, colonies might be present in poor habitats only by "mistake", of no more significance than a seed germinating in unsuitable soil. In other words, colonies should have evolved to act in a manner appropriate to their average environment, of which poor habitats may not be a part. However, the persistent occurrence of colonies in "poor" habitats argues against this explanation. Second, colonies might be forced into poor areas by saturation of the best habitats. If the alternative is to die, occupying poor areas would be worthwhile even if physiological constraints on queen–worker caste determination prevented the production of the optimal sex ratio (Elmes 1987b). Third, male bias might be the stable sex ratio in poor habitats, if these favored reproduction by colony budding. However, this would explain male bias in poor areas in polygynous species only, since monogynous ones do not bud (Sections 5.2, 5.6).

Similarly, within populations, an ultimate explanation is required for why large colonies with rich resources produce mainly females. A possible reason is that poorly-resourced colonies opt for colony growth (worker production) as an investment for survival to better times. Male bias could follow as a by-product, since female larvae would be channeled into worker development. Large, well-resourced colonies might then compensate by overproducing females. As an alternative, Crozier and Pamilo (1993) suggested that uncertainty in the supply of resources might make rearing females relatively more expensive for small colonies.

Another explanation invokes Trivers and Willard's (1973) hypothesis for brood sex ratio variation in vertebrates. This suggests that when one sex gains greater fitness from an increase in body size than the other, parents in good condition should concentrate on producing that sex. By analogy, if ant queens gain more fitness from increased body size than males (cf. Frank and Crespi 1989), colonies with plentiful resources should produce mainly queens. However, the predictions of Trivers and Willard's (1973) hypothesis tend to break down when brood sizes are large (Frank 1987a, 1990). Therefore, the hypothesis may have limited applicability to ants. Nonacs (1993a) also discusses trade-offs between sex ratio, productivity, and the size of sexuals. Yet another explanation for why large colonies concentrate on female production is Frank's (1987b) constant male hypothesis, which assumes a degree of local mate competition within populations (see below).

Finally, resources levels could bring about different types of sex ratio variation to that described by Nonacs (1986a). For example, there could be variable cost ratios (Grafen 1986; Section 3.8). Boomsma (1993) proposed that, alternatively, one sex might require a particular resource for development (for example queen larvae might require a plentiful carbohydrate supply), and colonies could vary in their access to the resource. However, there are at present no data to test this idea.

Constant Male and Constant Female Hypotheses

According to Frank's (1987b) constant male and constant female hypotheses, poorly-resourced colonies are selected to produce only one sex if there is any amount of local mate competition or local resource competition in the population (Sections 4.9, 4.10). In the case of local mate competition, the general prediction – that small colonies with few resources should focus on male production – is the same as that from Nonacs' (1986a) resource abundance hypothesis (see above). Evidence supporting the constant male hypothesis was found by Hasegawa and Yamaguchi (1994) in *Messor aciculatus*, but in two *Leptothorax* species with no demonstrable local mate competition, sex ratio splitting seemed due to other causes (*L. longispinosus*: Herbers 1990; *L. acervorum*: Chan and Bourke 1994) (Section 5.8). In species with local resource competition due to colony budding, predictions of the constant female hypothesis were also met (e.g. *Formica truncorum*: Sundström 1993b, 1995a; *L. acervorum*: Chan and Bourke 1994) (Section 5.9).

Split Sex Ratios due to Variable Relatedness Asymmetry

Boomsma and Grafen (1990, 1991) attributed variable sex ratios among colonies to worker control of allocation with worker assessment of variable relatedness asymmetry, leading to split sex ratios (Section 4.5). Evidence for split sex ratios arising for this reason exists in the cases of partial multiple mating (Sundström 1994a) and polygyny (e.g. Elmes 1987b; Chan and Bourke 1994) (Sections 5.4, 5.6). On the other hand, Pamilo and Seppä (1994) found split sex ratios in *Formica sanguinea* that were apparently not correlated with relatedness asymmetry in the predicted direction (Section 5.4). In addition, Aron et al. (1995) reported colonies producing mainly single-sex broods in the monogynous form of the fire ant, *Solenopsis invicta*. Yet this form arguably lacks variation in the workers' relatedness asymmetry, because queens are obligately singly mated and the workers are sterile (e.g. Ross and Fletcher 1985a; Aron et al. 1995). However, since relatedness asymmetry was not directly measured in either the *F. sanguinea* or the *S. invicta* study, it may be premature to regard these cases as contradicting Boomsma and Grafen's (1990, 1991) theory. For example, in *S. invicta*, orphaned colonies are suspected of requeening themselves (Tschinkel and Howard 1978; Vargo 1990). Therefore, populations may contain some colonies headed by the workers' mother, and some headed by (say) the workers' sister, a situation that would cause relatedness asymmetry to vary (Boomsma 1991; Mueller 1991). In *Colobopsis nipponicus*, Hasegawa (1994) also reported an invariant workers' relatedness asymmetry due to obligate monogyny, single queen mating, and nonreproductive workers. In this case, consistent with Boomsma and Grafen's (1990, 1991) theory, colony sex ratios had a unimodal distribution.

Additional evidence for split sex ratio theory comes from a survey by Boomsma and Grafen (1990). They measured sex ratio variance (as the fractions of all males and females produced in single-sex broods) across species with different social organizations. Split sex ratio theory predicts that nonparasitic ants are more likely to show bimodal sex ratios than slave-makers. This is because in slave-makers the controlling party is assumed to be the queen (Sections 4.4, 5.3), whose relatedness asymmetry is unaffected by multiple mating or polygyny. (Note, however, that the slave-maker *Harpagoxenus sublaevis* has an invariant workers' relatedness asymmetry due to obligate monogyny and monandry [Bourke et al. 1988].) Polygynous ants should also show greater sex ratio variance within populations than monogynous ones, since queen number, and hence the workers' relatedness asymmetry, probably varies far more widely than mating frequency. Boomsma and Grafen (1990) found that sex ratio variance was indeed greatest in polygynous species, intermediate in monogynous ones, and smallest in slave-makers (Table 6.1). This represents further evidence for worker control in nonparasitic ants, queen control in slave-makers, and split sex ratio theory.

6.3 Do Queens and Workers Share Control of Sex Allocation?

In the previous section, one idea for variable sex ratios among ant colonies was that queens and workers share control over allocation within populations (e.g. Herbers 1984, 1990). This section explores the idea of mixed control in more detail.

Trivers and Hare (1976) proposed that, in ants, queens were usually in control of the parentage of males. The suggested reason was that workers would lose far more fitness from defeat in fights over male parentage than queens (Section 7.3). But these authors also suggested that, in nonparasitic ants, workers were in complete control of sex allocation, and female-biased investment in monogynous species supported this view (Section 5.2). However, Trivers and Hare (1976) also argued that the queens would control allocation in slave-making ants (Sections 4.4, 5.3).

Table 6.1

Sex Ratio Variation among Colonies in Different Classes of Ants

Category	N Species	Average Fraction of Sexuals Produced in Single-sex Broods within Populations	
		Females	Males
Monogynous	16	0.07	0.26
Polygynous	5	0.20	0.61
Slave-making	4	0.03	0.01

NOTES: (a) From Boomsma and Grafen (1990 Table 1). In monogynous ants and slave-makers, colonies known to have reproductive workers or to be (rare) polygynous colonies were excluded. In the polygynous ants, queenless colonies were excluded. Therefore, any effects are not due to worker reproduction in queenless colonies or (in monogynous ants) to variable relatedness asymmetry arising through polygyny.

(b) The slave-maker data include two datasets omitted from the consideration of population-level investment ratios in slave-makers in Section 5.3. These datasets are nevertheless suitable for examining between-colony sex ratio variation. However, here *Epimyrma kraussei* is excluded from the analysis, because a low variance in the colony sex ratios of this species is probably due not to an invariant relatedness asymmetry but to within-nest mating (Section 5.8), which means that members of single-sex broods would find no mates.

Subsequently, other authors speculated that the distribution of queen and worker control might vary both within and across nonparasitic species. This was in relation to both male parentage (e.g. Oster et al. 1977; Bulmer 1981; Ratnieks and Reeve 1992; Nonacs 1993b) and sex allocation (e.g. Macnair 1978; Craig 1980b; Bulmer and Taylor 1981; Dawkins 1982a p. 77; Pamilo 1982a, 1991b; Pamilo and Rosengren 1983; Ratnieks 1991a; Ratnieks and Reeve 1992; Matessi and Eshel 1992). This section concentrates on mixed control of sex allocation in nonparasitic species. Genetic models of this situation are provided by Bulmer and Taylor (1981), Pamilo (1982a), and Matessi and Eshel (1992). In general, these show that mixed control can lead to population ratios intermediate between those of queens and workers. Moreover, mixed control may make an ESS sex ratio unattainable, perhaps accounting for some of the temporal fluctuations observed (Box 5.1) in population sex ratios (Matessi and Eshel 1992; cf. Herbers 1979).

DOES THE CONTROLLING PARTY VARY ACROSS SPECIES?

In Trivers and Hare's (1976) and later authors' (e.g. Pamilo 1990b) datasets of sex investment ratios for monogynous, nonparasitic ants, there is considerable scatter among species. Several authors proposed that this stemmed from between-species variation in the degree of queen and worker control (e.g. Macnair 1978; Craig 1980b; Matessi and

Eshel 1992). Queen control in some species would explain why the average across-species proportion of investment in females does not reach 0.75 (Pamilo 1990b; Section 5.2). Mixed control among species was held to arise from different outcomes of an evolutionary arms race between queens and workers over sex allocation (e.g. Macnair 1978; Craig 1980b). Both early production of sexuals in the seasonal cycle (Pamilo and Rosengren 1983; Section 6.5), and small colony size (Nonacs 1986a; Section 5.2), have also been proposed to favor queen control among ants.

Alternatively, workers might be in overall control of sex allocation, and both the scatter in the species sex ratios, and the reduced average, be due to variations in the genetic and social structure of the species making up the "monogynous" dataset (Boomsma 1988, 1989; Section 5.2). This issue could be settled by careful verification of queen number, mate number, and level of worker reproduction in sex ratio studies (Section 5.2).

It has also been suggested that queen control is more likely in monogynous social Hymenoptera other than ants, where sex investment is not always female-biased. Examples are polistine wasps (Noonan 1978; Metcalf 1980b; but see Strassmann 1984; Suzuki 1986), and bumble bees (Owen et al. 1980; Fisher 1992). The typically smaller work-forces of these groups might make queen control more feasible (Noonan 1978; see above). In addition, in all the social Hymenoptera, queens potentially control allocation by influencing the ratio of diploid to haploid eggs that they lay. But ant workers could counter this by influencing queen–worker caste determination (Sections 6.4, 6.5), whereas workers in the primitively social bees and wasps probably cannot (Metcalf 1980b; Suzuki 1986), for example because caste is largely determined in the adult stage in these groups. On the other hand, queen control is not certain in these bees and wasps. Unmeasured factors such as partial polygyny, multiple mating, and worker reproduction could mean that the level of female bias expected from worker control is small. Furthermore, in other species of primitively social bees and wasps, both female-biased population sex allocation (Ross and Matthews 1989) and split sex ratios (e.g. Mueller 1991; Section 3.8) suggest the existence of worker control.

Does the Controlling Party Vary within Species?

One reason to expect mixed control within species is that when the population sex ratio is at the stable value for one party, selection for control by the other is very strong (Ratnieks 1991a). For example, consider a population with a simple kin structure (monogyny, singly mated queens, and random mating) in which the population sex ratio is at the workers' equilibrium of 3 : 1 females : males. Queens would then be very strongly selected to gain control and produce more males than the workers favored, because they would value each male three times more highly than each female. This is because, although queens' life-for-life relatedness with sons and daughters is equal, males would have three times the mating success of females (Section 4.3). Similarly, if the population sex ratio were initially under queen control, workers would value females at three times their valuation of males, because of their higher life-for-life relatedness with sisters. So workers would be strongly selected to regain control by the overproduction of females. For these reasons, over evolutionary time, one might expect control of sex allocation to swing from one party to the other.

For the queen to gain control could simply involve her laying a brood of unfertilized, all-male eggs (e.g. Bulmer 1981). On the other hand, producing too few diploid eggs might dangerously reduce the colony's future supply of workers (Bulmer and Taylor 1981; Pamilo 1982a). In addition, workers under a singly mated queen would be selected to replace the queen's male eggs with their own more closely related ones (e.g. Ratnieks 1988; Section 7.4). In extreme cases, workers might even be selected to kill a male-only producing queen (Trivers and Hare 1976; Bourke 1994a). Therefore, producing only males is not a risk-free means for queens of winning control (see also Section 6.5). Hardly any evidence exists on the composition of queen egg batches. However, Aron et al. (1995) found that *Solenopsis invicta* queens laid more haploid eggs in male-producing colonies than in female-producing ones, and that workers then reduced the fraction of haploid eggs developing in both types of colony (see next section). This suggests a degree of mixed control.

Another idea is that queens control allocation in only some colonies, for example small ones. An interesting consequence would be a split sex ratio in the absence of any variation in relatedness asymmetry for either party (Boomsma and Grafen 1990). Thus, if the population had a simple kin structure, the population sex ratio would lie between the relatedness asymmetry of queens (1 : 1) and workers (3 : 1). So, by the logic of split sex ratio theory (Section 4.5), queen-controlled colonies should tend to produce only males, and worker-controlled ones only females. In fact, looking for such effects might be a good way to test for mixed control.

On the other hand, if mixed control exists, it is possible that within each colony a compromise between the interests of the parties is reached (Oster et al. 1977; Macnair 1978; Bulmer 1981; Bulmer and Taylor 1981; Pamilo 1982a; Pamilo and Rosengren 1983; Matessi and

Eshel 1992). Herbers (1984, 1990) also suggested that nest sex ratios reflected a compromise between queen and worker control in *Leptothorax longispinosus*. However, as the previous section described, it proved difficult to separate the influence of this effect from those of the workers' relatedness asymmetry, resource abundance, or both. In sum, it has yet to be conclusively demonstrated that mixed control of sex allocation within populations actually occurs. Furthermore, many sex ratio data are at least consistent with complete worker control (Table 5.11).

6.4 What are the Mechanisms for Controlling Sex Allocation?

The investigation of the mechanisms queens or workers might use to control sex allocation is a neglected topic. Evolutionary biologists are traditionally more concerned with outcomes than with mechanisms. However, it must be legitimate to ask how workers or queens achieve control, if they do so. This section considers possible mechanisms in general terms. The following one then concentrates on the specific tactics either party might use to achieve victory. Frank (1990) discusses analogous mechanisms in vertebrates.

Pamilo (1991b) has reviewed the ways in which workers could influence sex allocation. These include (1) treating the sexes differently at any stage between egg and adult, for example by killing males or increasing the food supply to queen-destined brood. They also include (2) influencing queen–worker caste determination among female larvae, and (3) laying male-yielding eggs. Queens could in principle use all these methods, and they could also (4) vary the ratio of haploid to diploid eggs that they lay (e.g. Bulmer 1981; Pamilo and Rosengren 1983).

A particular expectation is that workers should destroy or underfeed queen-laid male eggs or larvae, since this would serve to concentrate investment in females (e.g. Trivers and Hare 1976; Craig 1980b). However, this assumes that workers can efficiently single out male brood for maltreatment. If they cannot, and mistakenly harm females as well, this manipulation would be costly (Nonacs 1993b; see next section and Section 7.4).

No study has yet determined how workers or queens achieve a predicted sex investment ratio. One reason is that there are severe technical difficulties in establishing the caste or sex of eggs or larvae, and of tracking or sampling brood items to ascertain their fate. However, in two pioneering studies, Aron et al. (1994, 1995) recently used chromosome squashes to measure the fraction of haploid eggs in colonies of the Argentine ant *Linepithema humile* and the fire ant *Solenopsis invicta*. (*L. humile*, formerly *Iridomyrmex humilis*, was renamed in a taxonomic revision by Shattuck [1992].). In both these species, all viable eggs are laid by queens (Ross and Fletcher 1985a; Passera et al. 1988a). In *L. humile*, Aron et al. (1994) found that males reached adulthood in lower frequency than queens laid haploid eggs. This suggested that workers, which are known to eat eggs, selectively eliminated male eggs to raise a greater bias of females. In *S. invicta* (monogynous form), colonies tended to produce single-sex broods, although the reasons for this are unclear (Aron et al. 1995; Section 6.2). Queens in male-producing colonies laid proportionately almost twice as many haploid eggs (19% ± 0.1% S.D.) as those in female-producing ones (11% ± 0.1% S.D.). In addition, as in *L. humile*, workers reared a smaller fraction of males to adulthood than was present at the egg stage in both colony types. These findings constitute strong evidence for worker control and queen–worker sex ratio conflict.

Aron et al. (1994, 1995) assumed (safely) that, in both *Linepithema humile* and *Solenopsis invicta*, workers should always favor greater female bias than queens within colonies. However, studies like theirs deserve to be extended to cases where there are clear *a priori* predictions of the colony and population sex investment ratios for queens and workers. In *L. humile*, such predictions are hard to make because this ant has a rather complex social and mating system (Aron et al. 1994; Section 10.6). *S. invicta* is a relatively recent introduction to North America (where the study population of Aron et al. [1995] was located) (e.g. Ross and Fletcher 1985a; Sections 8.4, 10.2). This complicates sex ratio predictions because it means that sex allocation may not have attained an equilibrium.

On the other hand, problems remain for investigating the mechanism of sex ratio control even if an exact population investment ratio can be predicted. Consider a population with a simple kin structure. If workers always control allocation, should one expect the queen to have abandoned the sex ratio conflict, and so lay queen-destined and male eggs in a 3 : 1 ratio? Or should the conflict take place over each batch of eggs? The answer might depend on the costs of manipulation (Ratnieks and Reeve 1992; Nonacs 1993b), and the frequency with which mutations that permit sex ratio biasing in either party arise (Craig 1980b), all of which are unknown. Finding that queens laid eggs in an unbiased ratio that the workers then altered to 3 : 1 females : males would be strong evidence for conflict (cf. Aron et al. 1994, 1995). But finding that queens just laid the workers' preferred ratio to begin with would not mean sex ratio conflict had never occurred. These difficulties are made more acute by the fact that the controlling party could often be indifferent to the colony sex ratio, as long as the population sex ratio is at that party's equilibrium (Sections 4.2, 6.2). Therefore, evidence for manipulation might be best sought in species with split sex ratios, in those colony classes where workers are predicted to favor producing nothing but one sex, and queens nothing but the opposite sex (Section 6.2).

The work of Aron et al. (1994, 1995) represents the first explicit examination of mechanisms of sex allocation control. However, in addition to their findings, there is already a lot of evidence that workers and queens can influence caste determination, recognize brood by caste and sex, and treat the castes and sexes differentially (Brian 1979a, 1980; Carlin 1988; Hölldobler and Wilson 1990 p. 201; Section 7.6). Therefore, each party almost certainly has some ability to manipulate sex allocation if selection acts on it to do so.

6.5 What Tactics could Queens and Workers use in the Sex Ratio Conflict?

Despite the rarity of direct observations of the manipulation of sex allocation, several authors have suggested various tactics that each party might use (reviewed by Ratnieks and Reeve 1992). These are now discussed.

Ratio of Diploid to Haploid Eggs Laid by Queen

An obvious way in which queens can control allocation is by influencing the frequency of haploid eggs. For example, in annual species, Bulmer (1981) argued that the queen could wrest some control of both male parentage and sex allocation from the workers by laying only male-yielding eggs in the final generation. Workers would then be forced, against their interests, to rear these males. However, in perennial species like ants, the queen does not have as much power of this kind. Too small a supply of diploid eggs would reduce the size of workforce, and producing only males might provoke damaging conflict with workers over male parentage (e.g. Pamilo 1982a; Bourke 1994a; Section 6.3). Nevertheless, queens might still lay all-male broods in occasional years (Dawkins 1982a p. 77; Pamilo 1982a).

If queens limit the supply of diploid eggs, another potentially influential factor is the per capita cost of queens relative to workers. A model by Bulmer and Taylor (1981) found that worker control grew more likely as this cost rose. The reason was that, if queens are much larger than workers, workers faced with a limited supply of diploid eggs can still generate heavily female-biased investment by converting diploid larvae from workers to queens. (This assumes workers can control caste determination.) But if queens are close in size to workers, this results in far less female bias. The relative size of queens and workers is presumably set by factors such as mode of colony founding, colony size, food type, and overall ecology. But no-one has examined whether relative size also influences sex ratio control.

Lastly, the power of queens to determine the ratio of diploid to haploid eggs among their broods could be particularly important if there is worker control and sex ratios are split (Section 6.2). In these conditions, workers in some colonies might favor all-female broods, and in others all-male broods (Boomsma and Grafen 1991; Section 4.5). Queens would favor all males in both types of colonies, because the population sex ratio would be more female-biased than queens prefer. Under these circumstances, it could be that only the requirement for queens to produce diploid eggs for developing into workers would supply workers with the females they need.

CASTE DETERMINATION IN THE EGG

Queens might also exercise control of allocation by laying diploid eggs which were already predestined to develop into workers (e.g. Pamilo 1982a, 1991b; Ratnieks 1991a). This would frustrate workers trying to make sex investment more female-biased. Some ant species have such "blastogenic" (in the egg) caste determination, but in most it appears that caste is determined by treatment of the larvae (reviewed by Hölldobler and Wilson 1990 p. 348; Section 7.6). However, the distribution of blastogenic caste determination in relation to sex ratio conflict has not been investigated.

RELATIVE TIMING OF SEXUAL AND WORKER PRODUCTION

Pamilo and Rosengren (1983) argued for an influence of this factor in *Formica* wood ants. In the *F. rufa* species group, sexuals are produced in spring, before the workers. In other species, sexuals and workers develop together in summer. Therefore, queens in the *F. rufa* group could lay an all-male batch of eggs to produce the sexual generation, without putting worker production at risk. By contrast, queens of other species must produce a mixture of diploid and haploid eggs, because diploid eggs are needed to yield workers. Thus, *F. rufa* group queens seem more likely to control sex allocation than queens in other *Formica*. Of course, this leaves unanswered the question of which party controls the timing of sexual production, and why this differs between species.

SEXUAL DECEPTION

One way in which queens could obstruct workers trying to destroy or underfeed queen-laid males would be to disguise the sex of their brood. The workers would then risk damaging queen-destined brood, which would be contrary to their interests. This idea has been most fully developed by Nonacs and Carlin (1990) and Nonacs (1993b) as the sexual deception hypothesis (Section 7.4). Factors affecting the success of sexual deception include the accuracy with which workers can discriminate the sex of eggs, and the stage in larval development at which sex becomes apparent (Nonacs 1993b; Box 7.2). Nonacs and Carlin (1990) found that *Camponotus floridanus* workers cannot tell the sex of brood except at a late stage, as the sexual deception hypothesis predicts (Section 7.4). But sexual deception could involve costs to queens and the colony as a whole. This is because it could hamper efficient queen–worker caste determination, if this depended on workers recognizing female larvae and treating them appropriately. On the other hand, the topic of brood discrimination by sex in relation to kin conflict is another almost unexplored field (Nonacs 1993b).

6.6 What Factors Affect the Outcome of Queen–Worker Sex Ratio Conflict?

The previous section discussed the tactics that queens and workers might use in their sex ratio conflict. But factors also exist that shape the overall nature and extent of the conflict. Such factors are discussed for kin conflicts in general in Section 7.3, and for sex ratio conflict here.

The most obvious of these factors is the relatedness structure of colonies. Multiple mating, worker reproduction, and multiple queening all make the workers' stable sex ratio closer to that of queens (Sections 4.5–4.7), and so reduce sex ratio conflict. Split sex ratios could also alter the conflict's intensity. As the previous section mentioned, sex ratio splitting can result in colonies where workers favor producing all females and queens all males, and other colonies where both castes favor all-male production. Thus, sex ratio conflict would be intensified in the first type of colony, and reduced in the second (Boomsma 1990; Ratnieks and Reeve 1992).

Asymmetry in the power held by each party also affects sex ratio conflict. Thus, Trivers and Hare (1976) attributed female bias in monogynous ants to the greater power of the more numerous workers (Section 5.2).

Another factor affecting conflict is the cost to colony productivity of manipulating brood composition (Craig 1980b; Pamilo 1991b; Ratnieks and Reeve 1992). Pamilo (1991b) suggested that for any party to evolve the habit of destroying brood to manipulate sex allocation, there must be a surplus of larvae. Such surpluses could indeed occur, since at least in some species queens appear to lay more eggs than are raised to adult-hood (e.g. Elmes 1973a; Bourke 1993). However, Ratnieks (1991a) and Ratnieks and Reeve (1992) showed that manipulation could evolve even if it involves a net cost to productivity, up to a specified limit. Assume a population with a simple kin structure. As shown in Section 6.3, queens value males three times more highly than females if the population sex ratio is worker-controlled, whereas if it is queen-controlled workers value females three times more highly than males. Therefore, under worker control, a mutant queen strategy that involved destroying female larvae to increase male bias could spread provided that no more than three females were killed for every male aided. Thus, manipulative behavior could evolve despite being costly to colony productivity. Similarly, under queen control, workers would be prepared to destroy up to three males per aided female.

6.7 Conclusion

Trivers and Hare (1976) proposed that there is no inherent tendency for any one party to be favored in kin conflicts. Instead, biologists have only a set of principles and practical considerations that can help determine the outcome of conflict. The primary influences on conflict are the fitness interests of each party, and the extent to which they overlap with those of different parties (set mainly by kin structure). Other influential factors include the physical power held by each party, the tactics available to them, the efficiency of these tactics, their costs to productivity, and the frequency with which genetic variation for them arises.

Whether queen–worker conflict over sex allocation is a continuing process in ant colonies, and leads to mixed control within species, is uncertain. Although mixed control is plausible, the evidence for it is patchy, and some has other interpretations. This evidence includes patterns of between-nest sex ratio variation in *Leptothorax longispinosus* (Herbers 1984, 1990; Section 6.2), less than 3 : 1 female : bias in the average sex ratio of monogynous species (Pamilo 1990b; Sections 5.2, 6.3), and the evidence for brood sex ratio manipulation (Aron et al.

1994, 1995; Section 6.4) and sexual deception (Nonacs and Carlin 1990; Section 6.5). In view of their importance, further studies of the mechanisms of control of sex allocation, and the extent to which the castes share control, are badly needed.

6.8 Summary

1. Within ant populations, colonies often tend to produce single-sex broods. Although between-colony sex ratio variation is not in itself contrary to Fisherian theory, which specifies the stable population sex ratio, explanations are required for why it is greater than random.

2. Such explanations include the following. (a) Colony sex ratios vary because of mixed control (queens and workers share sex ratio control within populations). (b) Small, poorly-resourced colonies convert female larvae into workers not queens (Nonacs' [1986a,b] resource abundance hypothesis). (c) Local mate competition or local resource competition lead to single-sex broods in colonies with few resources (Frank's [1987b] constant male and constant female hypotheses). (d) Workers adjust colony sex ratios in response to their relatedness asymmetry (Boomsma and Grafen's [1990, 1991] split sex ratio theory). Evidence (of variable quality) exists for all these ideas.

3. Mixed control could also account for variable population sex ratios across species. Mixed control within populations is arguably expected because each party is most strongly selected to regain control when the other controls allocation completely. Genetic models of mixed control suggest that the population sex ratio will fall between each party's stable values. However, an ESS sex ratio may be unattainable.

4. Queens and workers could manipulate sex allocation by differential treatment of brood by sex, influencing queen–worker caste determination, laying haploid eggs (workers), or changing the ratio of diploid to haploid eggs laid (queens). Two recent studies show that workers increase relative female bias by selectively destroying male eggs (Aron et al. 1994, 1995).

5. Possible tactics of queen control include: (a) laying broods of all-haploid eggs; (b) "blastogenic" (in the egg) queen–worker caste determination; (c) laying broods of sexuals only; and (d) disguising the sex of eggs and larvae (Nonacs' [1993b] sexual deception hypothesis). The following factors favor worker control. (a) The risk to queens of underproducing workers and provoking harmful conflict over male parentage if queens produce males alone. (b) Large queen size relative to worker size, which allows female-biased allocation even if diploid eggs are in short supply. (c) Caste determination in the larval stage. (d) Mixed

worker- and sexual-yielding broods. (e) Early apparency of the sex of larvae.

6. The main determinant of the strength and nature of queen–worker sex ratio conflict is the genetic structure of colonies or the population. Other important factors include the practical power held by each party, the tactics available to each for manipulation, the efficiency of these tactics, the genetic variation for them, and their costs to productivity. Both queens and workers can evolve manipulation of sex allocation even if it is costly to colony productivity, but only up to a certain limit.

7

Kin Conflict over Reproduction

7.1 Introduction

Kin conflict theory is the branch of kin selection theory dealing with evolutionary disagreements between individual kin or classes of kin within groups. Conflict occurs because these parties may differ in their relatedness with sexual offspring, or in their productivity, or both. Previous chapters have dealt with kin conflict over sex allocation in ants (Chapters 4–6) and over parental manipulation of daughters to become workers (Sections 1.9, 3.9). This chapter reviews other kin conflicts in ant societies, especially over the production of males. The theory and phenomena of kin conflict in social Hymenoptera have been discussed by Trivers and Hare (1976), West-Eberhard (1981), Brockmann (1984), Starr (1984), Pamilo (1991b,c), Seger (1991), Ratnieks and Reeve (1992), Visscher (1993), and Heinze et al. (1994).

The study of kin conflict in social insects is important because it supplies further ways to test kin selection theory, in addition to those provided by investigating eusocial evolution (Chapter 3) and sex allocation (Chapters 4–6). On top of this, fighting, egg-eating, and other forms of overt conflict are typical of many Hymenopteran societies, and require explanation. As will be seen, kin conflict theory helps explain many of these otherwise puzzling instances of internecine squabbling (Sections 7.4–7.6). Some social insect behavior could also represent the legacy of past conflict in lineages where it is now absent (West-Eberhard 1981). Furthermore, kin conflict may have influenced seemingly unconnected facets of colony life such as the division of labor (West-Eberhard 1981; Schmid-Hempel 1990) (Section 7.7). Validating kin conflict theory in social insects is also important because the application of the theory to vertebrates has proved controversial (Mock and Forbes 1992; Bateson 1994). Finally, the study of kin conflict illuminates the fundamental theme of conflict versus cooperation in all kinds of groups (Ratnieks and Reeve 1992; Section 2.5).

7.2 Basic Theory of Kin Conflict

The theory of kin conflict was implicit in Hamilton's (1964a,b, 1972) kin selection theory, but was first clearly presented by Trivers (1974) in the form of parent–offspring conflict theory (reviewed by Clutton-Brock [1991], Clutton-Brock and Godfray [1991], Godfray and Parker [1991], and Mock and Forbes [1992]). Consider a monogamous diploid species with parental care. Each offspring values itself (at one unit) twice as highly as its litter-mates (at half a unit), because relatedness between sibs is 0.5. Therefore offspring are selected to demand extra investment in themselves at the expense of their sibs. But parents favor treating each offspring equally, since they are equally related with all of them (by 0.5). So there is a conflict between parents and offspring over the allocation of parental investment.

To be more precise, imagine a gene for selfish behavior at a locus influencing resource acquisition in offspring. Trivers' (1974) argument is that such a gene can spread even though sibs incur costs (provided that for every unit of fitness gained by the selfish offspring, no more than two are lost by a sib). Parent–offspring conflict arises because the spread of this gene would promote selection for any gene expressed in parents causing them to suppress the selfish behavior.

Parent–offspring conflict theory has been supported by both gene frequency and game theory models (e.g. Blick 1977; Macnair and Parker 1978, 1979; Parker and Macnair 1978, 1979; Stamps et al. 1978; Feldman and Eshel 1982; Parker 1985; Eshel and Feldman 1991; Godfray and Parker 1991, 1992). There are two important findings. First, there is no general answer to the question of whether parents or offspring normally win the conflict (e.g. Stamps et al. 1978; Parker and Macnair 1979; Parker 1985; Clutton-Brock 1991 p. 200; Clutton-Brock and Godfray 1991; Godfray and Parker 1991). Instead, who wins depends on the conditions. For example, large costs to offspring of acquiring extra resources, and low ones to parents of counteracting selfish offspring, allow parental victory. By contrast, low costs to offspring and large ones to parents let offspring win (Stamps et al. 1978; Parker and Macnair 1979; Parker 1985). Intermediate levels of relative costs may result in compromise, with parents providing less equitable care than they favor, but individual offspring receiving less investment each than they prefer (e.g. Parker and Macnair 1979; Parker 1985). These findings are contrary to Alexander's (1974) argument that parents would normally win the parent–offspring conflict (Section 1.9), but support Trivers and Hare's (1976) intuition that conflict evolution favors no single party consistently. On the other hand, the meaning of "winning" may sometimes require qualification. Thus, in a model by Godfray (1991), parents "win" in that offspring are selected to signal their requirements honestly, allowing parents to allocate investment using accurate information about offspring condition. But parents "lose" in that their fitness is reduced, because the existence of parent–offspring conflict forces the honest signals of the offspring to be costly (Godfray 1991).

The second general finding of parent–offspring conflict models is that, costs aside, relatedness values alone do not always precisely determine the outcome (e.g. Metcalf et al. 1979; Parker and Macnair 1979; Feldman and Eshel 1982; Parker 1985). One reason is that, in gene frequency models, the outcome of conflict may depend on the exact genetic assumptions (e.g. Parker 1985). More importantly, relatedness is an imperfect guide to the outcome of conflict in models that deal with the fate of rare mutant genes of high penetrance (tending to obligate expression) or large effect. The expectation that parent–offspring conflict should evolve according to relatedness levels invokes Hamilton's rule. Hamilton's rule assumes additivity of gene effects, weak selection, and facultative gene expression (Section 1.7; Box 1.2). But genes with large effects are unlikely to act additively or be weakly selected. Therefore it is not surprising that relatedness is not always the sole predictor of conflict evolution (Queller 1984, 1989a; Grafen 1985; Godfray and Parker 1991).

These arguments create a potential problem for social insect studies, where most kin conflicts aside from sex ratio conflict (Section 4.3) have been analyzed by simple relatedness arguments. However, the problem is not serious. For one thing, the precise genetic basis of kin conflict behavior is not known in any real-life case, so exact genetic models cannot be applied (Grafen 1985; Godfray and Parker 1991, 1992). In addition, Grafen (1985) has argued that once a social character has been partly selected for, Hamilton's rule is probably a good guide to its further evolution. This is because further changes to an established behavior are likely to be of small effect and hence weakly selected and additive. Similarly, Godfray and Parker (1991, 1992) constructed parent–offspring conflict models based on applying Hamilton's rule to the fate of genes of small effect and low penetrance in the vicinity of the ESS. Godfray and Parker (1992) showed that this approach gave the same results as Macnair and Parker's (1979) gene frequency models. Therefore relatedness arguments still serve as good guides to kin conflict evolution.

7.3 Factors Affecting Kin Conflict in Social Hymenoptera

Ratnieks and Reeve (1992) introduced the important distinction between *potential conflict* and *actual conflict*. Potential conflict is when the kin structure of a group dictates that kin conflict should be present. Actual conflict is when conflict takes place. Potential need not lead to actual conflict because additional factors might intervene to prevent a potential conflict from being realized. These factors include costly manipulation or resistance, counter-manipulation by other parties, and constraints on perception (Ratnieks and Reeve 1992; Reeve and Ratnieks 1993). In light of this argument, this section examines the factors affecting the nature of kin conflict in social Hymenoptera. This theme was earlier considered in the specific context of the conflict over sex allocation (Section 6.6).

KIN STRUCTURE OF COLONIES

Kin structure is the prime determinant of the nature of the potential kin conflict within colonies (cf. Section 6.6). For example, multiple mating by queens in monogynous societies reduces the potential queen–worker conflict over sex allocation and male production (Sections 4.5, 7.4). However, as just argued, actual conflict may not be predictable from kin structure alone.

What kin structures occur among ants? Kin structure is affected by the number and relatedness of queens, the number of queen matings, the level of worker male-production, and the degree of inbreeding (e.g. Ross 1990). Queen number varies widely between and within ant species (Chapter 8). For example, it has been estimated that half of all European ant species are polygynous (Buschinger 1974a), although some authors consider monogyny more frequent (Hölldobler and Wilson 1977). Multiple mating is also frequent, with more than half of species for which data exist showing some multiple mating (Hölldobler and Wilson 1990 p. 156; Keller and Reeve 1994b). Worker reproduction is common as well (Bourke 1988a; Choe 1988; Section 7.4). On the other hand, inbreeding among ants seems rare (e.g. Gadagkar 1991a). Therefore, the kin structure of ant colonies varies mainly with the number of reproducing individuals (mothers and fathers) inside them (Queller 1993a; Ross 1993). Since this number is so variable, so is kin structure.

Tabulations of relatedness values measured with allozyme analysis (Ross 1988a; Box 1.1) confirm that kin structure varies widely in ants (Hölldobler and Wilson 1990 p. 187; Gadagkar 1991a; Herbers 1993; Rosengren et al. 1993). As expected, they also show that kin structure is largely determined by social structure. For example, monogynous ants with singly mated queens have high female–female relatedness, approximating the theoretical maximum under outbreeding of 0.75 (e.g. Ward 1983a; Bourke et al. 1988; Table 5.9). Multiply mated species (e.g. Van der Have et al. 1988; Table 5.3) and polygynous species (e.g. Stille et al. 1991; Seppä 1992) have reduced female–female relatedness, particularly in polygynous species with distantly related colony queens (e.g. Pamilo and Rosengren 1983, 1984; Table 5.6). This variety of relatedness regimes creates potential kin conflict of several kinds. Why the social structure might vary in the first place is considered in later chapters (Chapters 8, 11).

ASYMMETRIES IN KIN CONFLICTS

Various types of asymmetry between the parties may also affect the outcome of kin conflict. For example, there may be an asymmetry in the power held by each party. Thus, Trivers and Hare (1976) argued that workers in nonparasitic, monogynous ants win the queen–worker conflict over sex allocation because they are collectively more powerful (Sections 4.4, 5.2).

In colonies with single, once-mated queens, workers favor raising worker-produced males in place of the less closely related queen-derived ones. So there is potential queen–worker conflict over male production (Trivers and Hare 1976; Section 7.4). One reason why workers may nevertheless not always lay eggs is that queens physically inhibit their reproduction. This would involve another power asymmetry, because queen inhibition is, in Ratnieks and Reeve's (1992 p. 42) phrase, a "one against many" interaction. However, all workers agree on producing worker-laid males if the queen is singly mated (Box 7.1). So, for workers, it is a "many against one" interaction. In the terminology of Section 2.5, workers form a "community of interest" with respect to rearing nephews rather than brothers. By contrast, classical parent–offspring conflict (Section 7.2) is a "one against many" conflict from the viewpoint of individual progeny.

Trivers and Hare (1976) proposed another type of asymmetry. They argued that in any fights arising over worker reproduction, a queen that kills a worker suffers barely any loss of future productivity. But a worker killing the queen loses all the sibs it could have raised in the future. The workers' production of sons is only likely to offset this loss in special circumstances, for example if the queen has very low productivity (Bourke 1994a). Therefore, workers might be discouraged from laying eggs in monogynous societies by an asymmetry in the fitness consequences for each party of escalated fighting.

COUNTER-MANIPULATION AND COSTS OF MANIPULATION

If there is monogyny with multiple mating, workers are selected to prevent each other reproducing and instead to rear queen-derived males (Ratnieks 1988; Section 7.4). This is an example of actual conflict becoming attenuated because of counter-manipulation by other parties (Ratnieks and Reeve 1992).

Worker reproduction could also be costly, for example if laying workers imposed an energetic burden that reduced colony productivity (Cole 1986). Alternatively, workers trying to substitute their male eggs for the queen's may be poor at recognizing the sex of eggs, and so mistakenly damage valuable sisters (Nonacs 1993b). Both these types of cost would discourage the actual expression of worker reproductive behavior (Ratnieks and Reeve 1992). More general models have also identified costs of manipulation as an important influence on the outcome of kin conflict (e.g. Parker and Macnair 1979; Yamamura and Higashi 1992; previous section).

ECOLOGICAL CONSTRAINTS

Vehrencamp (1983a,b), Reeve and Ratnieks (1993), and Keller and Reeve (1994a) considered the evolution of reproductive skew (the allocation of reproduction within societies). They found that one or a few individuals should tend to dominate reproduction (high skew) as ecological constraints on breeding alone rise. High skew also induces aggression within the society, because it means that winning a dominant's position yields a high reward. By contrast, mild ecological constraints were predicted to promote low skew societies with little aggression. These findings are discussed more fully in Section 8.6. The present point is that factors external to societies, such as ecological constraints, also affect the level of actual kin conflict found within them.

WITHIN-COLONY KIN DISCRIMINATION

In a colony with a multiply mated queen, workers are faced with a mixture of full and half sisters among the brood of new queens. Workers should prefer to rear their more highly related, full sisters. Put more formally, a gene for rearing full sisters alone could spread through a population dominated by indiscriminate rearers because, for each unit of effort, more gene carriers would be raised. Such kin discrimination creates potential kin conflict, since workers would consist of different patrilines (sets of siblings deriving from the same father). Each patriline would favor concentrating investment in its own queens, and not in those of other patrilines. However, if kin discrimination does not occur, actual conflict could be absent (Ratnieks and Reeve 1992).

Social insects are well known for their ability to distinguish nestmates from nonnestmates, since they typically attack all conspecific individuals from colonies other than their own (e.g. Breed and Bennett 1987; Jaisson 1991). When within-colony relatedness is greater than zero, nestmate recognition is a form of kin discrimination, because it ensures that benefits are preferentially directed at kin. However, the key issue is whether social insects discriminate among classes of kin within colonies and whether such discrimination is adaptive. This is controversial (Carlin 1989; Carlin and Frumhoff 1990; Grafen 1990a; Barnard 1991). In ants, the meagre evidence suggests that within-colony kin discrimination is weak or absent. For example, in polygynous species, workers do not appear to favor their mothers or the closest relatives among their worker nestmates (Carlin et al. 1993; Snyder 1993), and egg-eating queens seem unable to discriminate between their eggs and those of nestmate queens (Bourke 1994b).

Previously, data from honey bees seemed to provide the best evidence for kin-selected nepotism within colonies (reviewed by Breed and Bennett 1987; Getz 1991; Moritz and Southwick 1992), although not all studies detected nepotism (Woyciechowski 1990). However, the degree of preference that workers show for close kin is small (Carlin 1988, 1989; Getz 1991). In addition, patrilines within colonies exhibit a large amount of genetic variation in their discrimination abilities. This is unexpected in adaptive characters, which (under directional selection) should spread to fixation (Moritz and Hillesheim 1990; Moritz 1991). Consequently, several new interpretations of the evidence for nepotism have been proposed.

The first is that within-colony kin discrimination is a nonadaptive byproduct of other forms of discrimination, such as nestmate or species recognition (Frumhoff and Schneider 1987; Carlin 1989; Grafen 1990a). The second is that it is costly and therefore kept at a low level by between-colony selection (Page et al. 1989a). Such costs plausibly arise from recognition errors (Ratnieks and Reeve 1991). Another idea is that within-colony kin discrimination fails to evolve because it depends on genetic variation in recognition cues, yet nepotism would itself tend to reduce this variation over time (Ratnieks 1991b). Finally, limitations on workers' sensory and information-processing capabilities (Getz 1991) might make discrimination unlikely or of low intensity (Ratnieks 1991b).

Which of these explanations best fits the honey bee data is undecided. However, these arguments, together with the absence of evidence for within-colony kin discrimination among ants and other groups (e.g. Queller et al. 1990), suggest that the level of actual kin conflict due to biased kin rearing is usually very low (Ratnieks and Reeve 1992). On the other hand, because of the technical difficulties, few studies have looked for within-colony kin discrimination in ants.

7.4 Kin Conflict over Male Production

THEORETICAL BASIS

A major type of kin conflict in social Hymenoptera arises over who should lay haploid, male-producing eggs. This is distinct from, but overlaps with, queen–worker conflict over sex allocation (Chapters 4, 6). Consider a colony with a single, once-mated queen. Workers' relatedness with their brothers is 0.25, but with sons it is 0.5 and with nephews it is 0.375. (For simplicity, these are life-for-life relatednesses [Table 3.1].) Therefore, workers should favor their own over the queen's and other workers' sons, and both queen–worker and worker–worker conflict over male production should occur. However, since even nonlaying workers favor nephews over brothers, the principal conflict is between workers and the queen. The existence of this conflict was proposed by Hamilton (1964b, 1972) and Trivers and Hare (1976). The theory and phenomena of conflict over male production have been further discussed by E.O. Wilson (1971 p. 333), Oster et al. (1977), Bulmer (1981), Brockmann (1984), Bourke (1988a, 1994a), Choe (1988), Ratnieks (1988), Baroni Urbani (1989b), Seger (1991), Ratnieks and Reeve (1992), and Heinze et al. (1994).

The argument above involves comparisons of relatedness coefficients, and ignores any sex ratio effects and the influence of worker-produced males on male reproductive value (Boxes 3.2, 4.4). However, imagine a rare mutant egg-laying worker arising in a population at sex ratio equilibrium. Provided it replaced queen-laid male eggs with its own at little cost to overall productivity, this worker's fitness gain would be accurately represented by its greater relatedness with sons compared to brothers. Therefore, arguments from relatedness alone are appropriate in this case. In addition, formal models confirm that worker male-production should evolve when queens mate singly (Cole 1986; Ratnieks 1988; Pamilo 1991c; Nonacs 1993b).

In ants, reproductive workers are usually unable to mate and produce female offspring because they lack a sperm receptacle (Bourke 1988a). Ponerine workers are notable exceptions (Peeters 1991a). If workers cannot mate, there is no queen–worker conflict over the parentage of females (although there may be sex ratio conflict). This is because there is no class of more closely-related kin that workers could substitute for females produced by the queen. In addition, even if workers can mate, they should still prefer females to be produced by their mother, since their relatedness (assuming single queen mating) with sisters (0.75) exceeds their relatedness with daughters (0.5) and nieces (0.375) (Ratnieks 1988). This could be why workers appear to have lost their female-producing ability quite readily (through loss of the sperm receptacle), but have in many cases retained their ability to produce males (by keeping their ovaries) (Ratnieks 1988).

Worker reproduction in the sense of the ability to lay unfertilized male-yielding eggs (arrhenotoky) is common in ants (e.g. Brian 1979a, 1980; Passera 1984; Fletcher and Ross 1985; Bourke 1988a; Choe 1988; Hölldobler and Wilson 1990). In fact, total sterility in workers due to the absence of functioning ovaries is known from relatively few genera, including *Solenopsis, Monomorium, Pheidole, Tetramorium*, and *Eciton* (Bourke 1988a). However, recent additions to this list, namely members of the genera *Anochetus, Brachyponera, Cardiocondyla*, and *Hypoponera* (Peeters 1991a; Heinze et al. 1993a; Ito and Ohkawara 1994), suggest that total sterility may be less rare than previously thought. Workers may have retained their male-producing competence because of selection for worker reproduction under monogyny and single queen mating, as just described. But under other social and genetic regimes, and even under monogyny with single mating in some circumstances, worker reproduction is counter-selected (see below). So workers may also have kept their ability to produce sons because it is advantageous to them if they are exposed to irreversibly queenless conditions (Bourke 1988a; Ratnieks 1988).

Workers in queenless colonies can promote their fitness either by producing males or by adopting a new, related queen (*queen replacement*). Monogynous ants tend to have a mating system involving outbreeding away from the nest (Sections 8.2, 11.5). This makes it unlikely that orphaned workers could adopt a mated, related queen, since she would have to disperse from her natal nest, mate, then return there. So, if a monogynous colony becomes queenless, there should be worker reproduction. Queen replacement is in fact rare in monogynous ants (e.g. Pamilo 1991d). On the other hand, totally sterile workers should attempt queen replacement, which is their only option. There is some evidence for this in the monogynous form of the fire ant *Solenopsis invicta* (Vargo 1990). Here, relatedness with replacement queens is possibly ensured either by workers permitting virgin queens reared in the colony to lay male eggs (Vargo and Porter 1993), or by some within-nest mating.

Under monogyny, queenless colonies are likely to occur (e.g. Elmes 1974a; Cole 1984; Bourke et al. 1988). But, in polygynous species, few colonies will lose all their queens, so reproduction by queenless workers is less likely. In addition, the mating biology of polygynous species (Section 8.2) usually permits queen replacement. These reasons could explain why ant species with reproductive workers are monogynous with an apparently disproportionate frequency (Bourke 1988a).

Worker Policing

When queens mate multiply, or if colonies are polygynous, worker reproduction may be counter-selected (Starr 1984; Woyciechowski and Lomnicki 1987; Ratnieks 1988; Pamilo 1991c). The reason for this is that multiple mating and polygyny dilute a focal worker's relatedness with an average worker-produced male more than its relatedness with an average queen-produced male (Box 7.1). For example, at mating frequencies greater than two, workers in monogynous colonies are less closely related with worker-derived males (nephews) than with queen-derived ones (brothers) (Figure 7.1). Hence they should rear brothers alone. Any one worker's sons are excluded from consideration because they are assumed to form only a small fraction of all worker-produced males in the colony. This could come about through workers actively preventing their fellow workers from laying eggs successfully. Ratnieks (1988) showed that such *worker policing* may be positively selected at high mating frequencies (Section 2.5; Box 7.1).

A factor affecting the evolution of worker reproduction and worker policing, independent of kin structure, is their cost. Not surprisingly, both traits evolve with greater difficulty if they decrease total colony productivity (Cole 1986; Ratnieks 1988; Pamilo 1991c). Costs also influence what happens if worker policing becomes prevalent. When this is the case, workers do not gain from even trying to reproduce. If this resulted in efficiency gains for the colony, it would be positively selected. Thus, the end product of worker policing may be *self-policing* (Ratnieks 1988). Both phenomena lead to the same outcome – a lack of worker male-production – but the behavior expected of workers might differ (Ratnieks 1988).

The worker policing hypothesis assumes that there is no single dominant worker. In ants, this is probably largely correct. Although workers may act as dominants, this tends to be in leptothoracine or ponerine species with small colonies (Heinze et al. 1994). In larger colonies, it is unlikely that one worker could police all the others. Large ant colonies seem more likely to have multiply mated queens (Cole 1983a), an association that would reinforce the evolution of worker policing.

BOX 7.1 THE EVOLUTION OF WORKER
REPRODUCTION AND WORKER POLICING

In a monogynous colony with single queen mating, workers are related to males in this order (using life-for-life relatednesses): own sons (0.5), sons of another worker (0.375), queen's sons (0.25). Therefore, workers should always favor worker- over queen-produced males (Section 7.4). This box examines the evolution of worker male-production when there is multiple mating, polygyny, or both.

Monogyny with Multiple Mating

In general, workers will favor worker-produced males when

$$r_{WM} > r_{QM} \qquad\qquad [7.1]$$

where r_{WM} = relatedness of workers with an average worker-produced male, r_{QM} = relatedness of workers with an average queen-produced male.

Under monogyny with multiple mating, worker relatedness with brothers is unaffected, so $r_{QM} = 0.25$ (life-for-life). But worker–sister relatedness falls, equalling $0.5(0.5 + 1/k)$ where k is queen mating frequency (Box 4.3). So worker–nephew relatedness, which is the product of worker–sister and female–son relatedness (0.5), equals $0.25(0.5 + 1/k)$. Assuming that no single worker can dominate worker male-production, this approximates the relatedness of a worker with a randomly-chosen worker-produced male (since most such males will be nephews). Hence $r_{WM} = 0.25(0.5 + 1/k)$.

Substituting in Expression 7.1 gives:

$$0.25(0.5 + 1/k) > 0.25$$
$$k < 2.$$

Therefore, workers favor rearing their own males at (effective) mating frequencies of less than two. But if the mating frequency exceeds two, workers should prefer rearing the queen's male offspring. The reason is that worker–nephew relatedness now dips below worker–brother relatedness (Figure 7.1). So, contrary to Bourke (1988a), multiple mating should be associated with a lack of worker reproduction. Since the queen prefers workers to rear her males, it should also reduce the level of queen–worker conflict over male production.

This idea was developed by Starr (1984), Woyciechowski and Lomnicki (1987), Ratnieks (1988), and Pamilo (1991c). It suggests that if queens mate many times, workers should actively inhibit each other's reproduction, because this guarantees that no single worker dominates

BOX 7.1 CONT.

male-production. This was termed worker policing by Ratnieks (1988), who confirmed that a rare police allele should spread when queens mate more than twice (see also Ratnieks and Reeve 1992).

Polygyny with Unrelated Queens

By a similar logic, worker reproduction may be counter-selected under polygyny (again, Bourke [1988a] mistakenly said otherwise). If queens of a polygynous colony are unrelated, workers have no interest in the off-spring of queens other than their mother. So worker male-production will evolve under the same conditions as in monogynous colonies, and will therefore depend only on queen mating frequency. Ratnieks (1988) showed formally that when there is polygyny with unrelated queens, a worker policing allele spreads under the same conditions as in his single queen model.

Polygyny with Related Queens and Single Mating

This situation was modeled by Pamilo (1991c) and Nonacs (1993b). Following Pamilo (1991c), let there be N singly mated queens per colony, reproducing equally, and let G equal the regression relatedness between them. Then, also expressed as regression relatednesses, and again assuming that a worker's own sons form a negligible fraction of all worker-produced males,

r_{QM} = (relatedness with full brothers $\times$ frequency) + (relatedness with sons of queens other than mother $\times$ frequency)

$= 0.5(1/N) + (0.5)(G)(1)(1 - 1/N)$

$= [1 + (N - 1)G]/2N$ (Box 4.5);

r_{WM} = (relatedness with nephews $\times$ frequency) + (relatedness with sons of workers other than full sisters $\times$ frequency)

$= 0.75(1/N) + (0.5)(G)(0.5)(1)(1 - 1/N)$

$= [3 + (N - 1)G]/4N.$

Substituting into Expression 7.1 shows that worker reproduction is favored when

$$N < 1 + 1/G \text{ or } G < 1/(N - 1).$$

This is Pamilo's (1991c) expression 7. Rising values of N or G would soon exceed the critical values given by the right hand side of each inequality. So as either queen number or relatedness increases, worker reproduction becomes less likely. First, as queen number rises, there is an increase in both the fraction of queen-produced males that are not a worker's full brothers, and the fraction of worker-produced males that are not sons of

BOX 7.1 CONT.

a worker's full sisters. Since the first class of male is more closely related with a focal worker than the second, worker-produced males become less attractive to rear. Second, as queen–queen relatedness rises, the relative contribution to overall relatedness made by both types of male increases, so worker reproduction again grows less profitable.

Pamilo's (1991c) result contradicts an argument of Trivers and Hare (1976) for a positive association between polygyny (with related queens) and worker reproduction, but supports the negative association predicted by Hamilton (1972; reviewed by Bourke 1988a). Trivers and Hare (1976) argued that, under polygyny with related queens, workers should be more inclined to initiate egg-laying than under monogyny, because other queens could substitute for any ones injured in fights over male parentage (Section 7.3). However, a counter-argument is that, as under monogyny, the loss of fitness to a worker killing a queen would still outstrip the fitness loss to a queen killing a worker. Therefore, contests over male production must still be weighted in favor of queens. The slender evidence also appears to support Pamilo's (1991c) prediction that workers are less reproductive when there is polygyny with related queens (Section 7.4).

Polygyny with Related Queens and Multiple Mating

Pamilo's (1991c) method can also be applied to this case, in which:

$r_{QM} = [1 + (N - 1)G]/2N$ as under polygyny with related, singly mated queens;

$r_{WM} = 0.5(0.5 + 1/k)(1/N) + (0.5)(G)(0.5)(1)(1 - 1/N)$
$= [1 + 2/k + (N - 1)G]/4N$.

So, from Expression 7.1, workers favor laying worker-produced males if

$$k < 2/[1 + (N - 1)G].$$

Note that if $G = 0$ (unrelated queens), the condition is as for monogyny, as stated above. In general, the result suggests that as mating frequency (k) rises, the right hand side of the inequality is exceeded and workers favor raising queen-laid males. The reason is this. Relatedness of workers with full brothers, and with all offspring of queens other than their mother, is unchanged by multiple mating, since these are only relatives on the maternal side. So relatedness with all queen-produced males, and with the grandsons of queens other than a worker's mother, is the same as under polygyny with single mating (see above). However, as before, multiple mating reduces the relatedness of workers with grandsons of their

BOX 7.1 CONT.

own mother (nephews). Therefore, an increased mating frequency makes workers favor rearing worker-produced males less.

Nonacs (1993b) considered a worker that attempted to "cheat" other workers by replacing an average queen-laid male egg with one of its own. He found that, for any level of polygyny, the "cheater's gain" rose with increasing mating frequency. Hence Nonacs (1993b) concluded that polygynous colonies with multiple mating might encourage worker egg-laying via strong selection for "cheating". This result differs from the one above, because Nonacs (1993b) assumed polygynous colonies consisting of an original foundress queen and her readopted daughters. Therefore, as mating frequency rises, more and more queen-produced males would be the offspring of half-sisters, and hence the payoff from these males, relative to the payoff from a worker's son, would decline. By contrast, Pamilo (1991c) assumed that queens, though related, are all members of the same generation (cf. Pamilo 1990b; Queller 1993a).

The various circumstances under which worker male-production is expected or not expected to occur are summarized in Table 7.1.

Table 7.1

Predicted Social and Genetic Correlates of Worker Reproduction (male production)

Queen Number Class	Queen Mating Frequency	
	Single (1–2)	Multiple (>2)
Monogyny	+	−
Polygyny		
Unrelated queens	+	−
Related queens	+ (low G, N)	+/−
	− (high G, N)	

NOTES: (a) Based on Starr (1984), Woyciechowski and Lomnicki (1987), Ratnieks (1988), Pamilo (1991c), and Nonacs (1993b). Under polygyny with related queens and multiple mating, the expected amount of worker reproduction depends on the exact pedigree structure of the colony (Box 7.1).
(b) +/− = workers favor or do not favor worker reproduction, respectively; G = queen–queen relatedness, N = number of queens; worker reproduction is assumed to have few or no costs.

Actual kin conflict over male production declines under multiple mating if there is worker policing or self-policing. This suggests that queens evolve multiple mating as a device to reduce conflict (Starr 1984; Moritz 1985; Woyciechowski and Lomnicki 1987). However, mul-

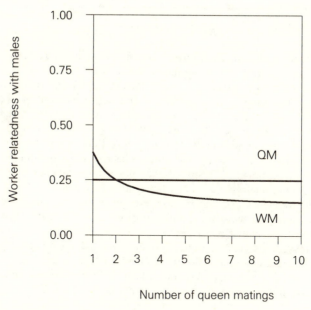

Figure 7.1 The life-for-life relatedness of workers in a monogynous colony with queen-produced males (QM) and average worker-produced males (nephews) (WM), as a function of queen mating frequency. Relatedness with nephews equals $0.25(0.5 + 1/k)$, where k is mating frequency (Box 7.1).

tiple mating is probably costly to queens, for example because it involves an extra risk of predation. So, to evolve, it needs to give queens immediate benefits (Ratnieks 1988; Ratnieks and Reeve 1992; Nonacs 1993b). But worker policing would evolve only gradually after multiple mating, as any genetic variation for policing underwent positive selection. Therefore, worker policing is probably a consequence, not a cause, of multiple mating, which may arise for other selective reasons (Sections 2.5, 11.4).

SOCIAL AND GENETIC CORRELATES OF QUEENRIGHT WORKER REPRODUCTION

As has been seen, kin selection theory predicts an association between monogyny with single mating and worker reproduction in queenright conditions (i.e. in the presence of queens). It also predicts that queenright worker reproduction should not occur under multiple mating, or polygyny with closely related queens (Box 7.1; Table 7.1). The evidence for these associations now needs examining. The correlates of worker reproduction have also been reviewed by Hamilton (1972), Trivers and

Hare (1976), Bourke (1988a), Ratnieks (1988), Pamilo (1991c), Keller and Vargo (1993), and Nonacs (1993b).

First, monogynous ants in fact display little worker reproduction in queenright conditions. Reasons for this are discussed below. On the other hand, there is good evidence for worker policing in honey bees. Honey bees are monogynous and queens mate many times. Consistent with worker policing, workers with developed ovaries are attacked by other workers (Van der Blom 1991; Visscher and Dukas 1995). In addition, honey bee workers preferentially eat worker-derived male eggs (Ratnieks and Visscher 1989; Ratnieks 1993). Consequently, the fraction of adult males produced by workers in queenright colonies is minuscule (Visscher 1989).

Monogynous ant species need to be examined for similar phenomena. In the monogynous *Leptothorax nylanderi*, Plateaux (1981) reported multiple mating (1–10 times) and a high level of worker male-production. This would seem to contradict the worker policing hypothesis (cf. Nonacs 1993b). However, confirming this requires investigators to know the average number of matings of *L. nylanderi* queens in natural populations, since worker policing should occur only when this average exceeds two (the relevant average being the harmonic mean: Queller 1993a).

Better evidence for worker policing in ants comes from the monogynous *Aphaenogaster cockerelli*. Hölldobler and Carlin (1989) divided a colony into a queenright and a queenless part. When these were reunited, workers from the queenright portion attacked egg-layers that had arisen in the queenless one. Whether worker policing prevents reproduction by queenright workers in this species is unknown (there is no information on queen mating frequency). However, these results show that egg-laying workers are sometimes recognized and "policed".

In principle, further evidence for worker policing among ants could be obtained by comparing monogynous with polygynous species. But information on the social correlates of worker reproduction is scarce. Worker reproduction in queenright conditions appears to occur disproportionately often in polygynous species (Bourke 1988a), which contradicts the worker policing hypothesis. However, this could be mainly because worker reproduction is especially likely in queenless conditions in monogynous species (see above), where it is also easiest to detect (Bourke 1988a; Pamilo 1991c; Nonacs 1993b). Some individual polygynous species apparently support worker policing. For example, the red ant *Myrmica rubra* has multiple queens of low relatedness (Pearson 1982; Pearson and Raybould 1993), which contrary to previous reports are probably singly mated, to judge from preliminary data in this species (Woyciechowski and Lomnicki 1987) and from genetic studies of other *Myrmica* (Snyder and Herbers 1991; Seppä 1992, 1994). These circumstances should lead to queenright worker reproduction (Ratnieks 1988; Pamilo 1991c; Box 7.1), and this occurs in *M. rubra* at high frequency (Smeeton 1981). (Bourke [1988a] and Nonacs [1993b] presented different interpretations of this case, based on the assumption of polygyny and multiple mating.) Conversely, the two leptothoracine species *Leptothorax acervorum* and *L. longispinosus* have multiple queens of comparatively high relatedness, which are probably also singly mated (e.g. Herbers 1986a, 1993; Herbers and Stuart 1990; Douwes et al. 1987; Stille et al. 1991; Heinze et al. 1995). These species have only low levels of worker egg-laying in queenright colonies (Herbers and Cunningham 1983; Bourke 1991), again as expected (Box 7.1). However, detailed data on genetic and social structure from more species are needed to test the worker policing hypothesis adequately.

Why are Workers usually Sterile in Colonies with a Single, Once-mated Queen?

As already mentioned, worker reproduction appears rare in monogynous ants under queenright conditions, although commonly occurring in queenless colonies (Bourke 1988a; Choe 1988; Ratnieks 1988; Nonacs 1993b). It seems unlikely that all monogynous species lacking queenright worker reproduction have multiply mated queens, and some almost certainly do not (e.g. *Aphaenogaster rudis*: Crozier 1974). Therefore, this finding contradicts the prediction of strong selection for queenright worker reproduction under monogyny with single mating (Ratnieks 1988; Pamilo 1991c). Three overlapping factors could explain it – productivity effects, queen policing, and sexual deception.

1. PRODUCTIVITY EFFECTS

One reason for workers failing to reproduce in the queen's presence could be that queens are usually much more fecund than workers. The productivity gains to workers of raising sisters and brothers would then outweigh their relatedness gains from raising sisters and nephews. This effect could be reinforced in ants, because queens in perennial societies usually have far greater future productivity than workers (Trivers and Hare 1976; Ratnieks 1988; Bourke 1994a). Alternatively, as mentioned earlier, workers might not reproduce if replacing the queen's male eggs with their own were costly to the colony's current productivity (Cole 1986; Keller and Nonacs 1993; Nonacs 1993b). Consistent with this argument, Cole (1986) estimated the cost to productivity caused by reproductive *Leptothorax allardycei* workers to be insufficient to prevent queenright worker reproduction evolving in this case.

2. QUEEN POLICING, QUEEN CONTROL, AND QUEEN SIGNALING

A second explanation for worker sterility under single, once-mated queens is that queens actively prevent worker reproduction. This would profit the queens because they favor the production of their more closely related sons. Such behavior has been referred to as *queen policing* (Ratnieks 1988 p. 228). It is also a type of *queen control* (e.g. E.O. Wilson 1971 p. 299) and of parental manipulation (Alexander 1974; Sections 1.9, 3.9). However, like parental manipulation, queen control is an aspect of kin conflict theory. Furthermore, as will be seen, workers almost certainly attempt to achieve their interests in the face of queen control.

The main question posed by queen control concerns how it is exercised. One way is for queens to dominate workers with physical aggression, and cannibalize any eggs that they lay. This system could be stable because of power asymmetries operating in the queens' favor. For example, queens are nearly always larger than workers. In addition, as Trivers and Hare (1976) argued, workers might respect the queen's safety more than the queen respects theirs, because of the greater cost to workers of damaging the queen (Section 7.3). Physical queen control of worker reproduction should also be easiest if colony size is small (e.g. E.O. Wilson 1971 p. 302; Ratnieks 1988; Ratnieks and Reeve 1992).

Aggressive queens are common in the monogynous bumble bees and wasps, which typically have small colonies and workers that reproduce only after the queen's death (e.g. Fletcher and Ross 1985; Ross and Matthews 1991). However, in monogynous ants, queen aggression towards workers is rare. This suggests that queens could instead control worker fertility with inhibitory pheromones. Various experiments show that queens produce chemicals in whose presence workers are sterile (e.g. Passera 1980a; Hölldobler and Wilson 1983). In addition, using pheromones arguably represents the most powerful means of queen control, especially in large colonies where aggressive control seems impracticable (E.O. Wilson 1971 pp. 299, 302).

The idea of pheromonal queen control of worker fertility used to enjoy wide acceptance (e.g. Brian 1979a, 1980; Passera 1984; Fletcher and Ross 1985; Hölldobler and Bartz 1985; Bourke 1988a; Hölldobler and Wilson 1990 p. 222). However, several authors have now suggested that the queen's pheromones are not instruments of force, but signals communicating the queen's presence (e.g. Seeley 1985 p. 30; Woyciechowski and Lomnicki 1987; Seger 1989, 1991; Crespi 1992b; Nonacs 1993b). The fullest case has been made by Keller and Nonacs (1993). They argued that pheromonal queen control, unless backed up by force, is not evolutionarily stable. Nothing seems to be stopping the workers from evading the effects of the pheromone if remaining sterile is contrary to their interests. Instead, Keller and Nonacs (1993) see queen pheromones as honest, evolutionarily stable signals indicating both the queen's presence and level of fecundity. In line with biological signaling theory (e.g. Grafen 1990b), stability is guaranteed both by the signal's honesty and by its benefiting both the queen and the workers (Keller and Nonacs 1993).

The queen benefits from the workers' response to her signal (sterility) because of her closer relatedness with offspring than with grandsons. Workers also benefit if the queen mates multiply, because they are then selected to be nonreproductive (Box 7.1). So queen signaling is not controversial in this case. But how do workers benefit when there is monogyny with single mating? Here the signaling hypothesis must assume that some of the productivity effects discussed above apply – for example, that worker reproduction is costly (Keller and Nonacs 1993). Workers then profit more from raising only queen-derived individuals, and so benefit from receiving the signal of the queen's continuing presence and health. The signaling hypothesis therefore denies not only that queen pheromones are coercive, but also that workers need coercing. However, queens must signal honestly, as Keller and Nonacs (1993) proposed, because workers only benefit if they judge queen productivity correctly.

One idea from the presignaling standpoint is that, as colony size increases, there is a trend from physical to pheromonal queen control (e.g. E.O. Wilson 1971 p. 299; Brian 1980). This was because physical policing of very many workers was believed impossible. It was also argued that one way in which a lineage changed from primitive to advanced eusociality was to acquire pheromonal queen control (E.O. Wilson 1971 p. 88), since advanced eusociality is associated with larger colonies. The signaling hypothesis explains the same phenomena. Larger, perennial societies, and greater queen–worker dimorphism, are all characteristic of advanced eusocial species (E.O. Wilson 1971 p. 88). These features also make it more likely that the queen's productivity outweighs the workers' and hence that, as the signaling hypothesis assumes, workers favor rearing her offspring rather than their own.

Is it possible to rescue pheromonal queen control from Keller and Nonacs' (1993) charge that it would be unstable? Perhaps it simply is unstable. West-Eberhard (1981) proposed that queens and workers run a continuous evolutionary arms race (Dawkins and Krebs 1979) over male production (see also Bourke 1988a). Queens acquire new pheromonal means of suppressing worker reproduction, and workers evolve fresh immunity to each new wave of chemical attack. However, the outcome of such a process should arguably be more arbitrary across lineages, whereas worker sterility in the queen's presence seems the norm in monogynous ants. West-Eberhard (1981) also suggested that the complexity of queen pheromones in honey bees supports the arms race hypothesis. However, signaling theory (Grafen 1990b) predicts that honest signals should be costly (when potential conflict exists between signaler and receiver). So perhaps queens evolve complex pheromonal blends because of their cost.

Another argument in defence of pheromonal queen control is that it could be stable if a queen–worker power asymmetry exists (Section 7.3). The queen can attempt to influence all workers from the egg stage onwards, but workers only know the queen as an adult. So queens may exert inhibition with pheromones more easily than workers resist it. In addition, workers could be constrained in their attempts to evade inhibition. In insects, single hormones commonly have multiple functions, with juvenile hormone providing a good example (e.g. Richards and Davies 1977 p. 277). Therefore, workers may be unable to resist the effects of queen pheromones on their ovarian development without costly disruption to their entire endocrine system.

Nevertheless, Keller and Nonacs (1993) have convincingly shown that the hypothesis of pheromonal queen control was uncritical. The alternative, signaling hypothesis is powerful, potentially correct, and deserves further investigation. It also agrees better with signaling theory (e.g. Grafen 1990b). Keller and Nonacs (1993) suggested some tests to discriminate between the hypotheses. One is that the honesty of queen signaling would be supported by a correlation between the pheromone output of queens and their fecundity. Another is that the signaling hypothesis would be undermined if queens producing only males were as strongly inhibitory as those producing both sexes. This is because, under a male-only producing queen, workers no longer benefit from remaining sterile, since (assuming monogyny and single mating) they are more closely related with sons and nephews than brothers.

The ponerine ant *Diacamma australe* has an extraordinary social system that provides an example of the second type of test. The chief reproductive of a colony is a single mated worker or "gamergate" (Peeters and Higashi 1989; Peeters and Billen 1991). All females have ovaries and a sperm receptacle and so are born potential gamergates, but individuals that have lost a pair of thoracic appendages ("gemmae") are not attractive to mates. In a colony without a gamergate, the first female to mature bites off the gemmae of later emerging females, so ensuring their virginity ("mutilation"). She keeps her own gemmae. She then dominates the other females (workers) with physical aggression, almost certainly to prevent them producing sons. When a foreign male enters the colony, the dominant female mates with him and so becomes the colony's gamergate. From then on, she no longer physically dominates the workers, yet these remain nonreproductive in her presence.

The crucial point is that before mating the gamergate-to-be must physically inhibit reproduction by the other females, but after mating she apparently does not have to. Peeters and Higashi (1989) interpreted this as a transition from physical to pheromonal control of nestmate fertility. But why the future gamergate cannot pheromonally inhibit the workers before mating is left unexplained. Alternatively, this case supports the signaling hypothesis. Before mating, when the future gamergate can produce only males, the other, mutilated females (which are probably her sisters) could challenge her in the role of male producer. Hence the future gamergate should exert physical inhibition to stop them. However, once the dominant female has mated, the mutilated workers cannot challenge her in her new role of female producer. In addition, any further attempts to replace her male eggs with their own might be costly, since female eggs might now be mistakenly destroyed instead of male ones (Nonacs 1993b). Therefore, once a gamergate is present, the workers are likely to benefit from not laying eggs and so should no longer require physical domination. In further support of the signaling hypothesis, there is also evidence that gamergates lose their inhibitory power as their fecundity declines (Peeters 1993).

Whether this is a correct interpretation requires more investigation. The main point is that the queen signaling hypothesis cannot be dismissed until it is critically tested in this and similar cases. Furthermore, phenomena in social mammals analogous to pheromonal queen control in ants – namely chemically induced reproductive suppression (e.g. Faulkes et al. 1991) – also bear reexamination from the standpoint of the signaling hypothesis.

3. SEXUAL DECEPTION

The final idea for why workers in colonies with single, once-mated queens are less reproductive than expected is the sexual deception hypothesis (Nonacs and Carlin 1990; Nonacs 1993b; Section 6.5). The theory of queen–worker conflict over male production (and sex allocation) assumes that workers can discriminate between queen-laid females and males in their brood stages. Only males, not the more closely related females, should be replaced by workers with their own male eggs. But if queens disguised brood so that workers could not tell sisters from brothers (sexual deception), workers would not be selected to attempt replacement. The cost in mistakenly destroyed sisters would exceed the gain in nephews, as detailed in Box 7.2. Since it makes worker reproduction costly, sexual deception represents a special case of productivity effects (see above) hindering the evolution of laying workers. Furthermore, sexual deception for only the early part of development could still protect brood from workers. This is because even identifiably male brood is expensive to replace if old, since more energy must be spent on raising the worker's egg to the same stage (Nonacs 1993b; Box 7.2).

The sexual deception hypothesis therefore predicts that the sex of brood should only become apparent late in development. Young ant larvae of both sexes certainly look the same to people. What about ants? Brian (1981) found that *Myrmica rubra* workers could recognize males at the third larval instar, but in this species most males appear to be worker-produced (Smeeton 1981), so perhaps sexual deception has failed in this case. The results of Aron et al. (1994, 1995) suggest that *Linepithema humile* (= *Iridomyrmex humilis*) and *Solenopsis invicta* workers preferentially eat haploid eggs (Section 6.4). Although workers in these species are nonreproductive, queens should arguably have used sexual deception to gain control in the conflict over sex allocation. By contrast, when presented with brood items of various ages, *Camponotus floridanus* workers treated only pupae differently by sex, suggesting they could not tell the sex of younger brood (Nonacs and Carlin 1990). Although no mechanism of sexual deception is yet known, these results indicate that Nonacs' (1993b) hypothesis deserves further investigation.

THE OCCURRENCE OF ACTUAL KIN CONFLICT OVER MALE PRODUCTION

Kin selection theory predicts both queen–worker and worker–worker conflict over male production (Box 7.1). Although potential conflict need not always be realized (Section 7.3), biologists expect evidence of conflict in some cases. Outside ants, this comes from fights over male egg-laying in bumble bees and wasps (e.g. Fletcher and Ross 1985; Ross and Matthews 1991), worker matricide in these groups (Bourke 1994a), worker policing in honey bees (Ratnieks and Visscher 1989; Ratnieks 1993), and the egg-laying rituals of stingless bees (Hamilton 1972).

Good evidence for overt conflict over male production in ants comes from three monogynous leptothoracine species. In these, workers are fairly exceptional in laying viable eggs in the queen's presence, and form dominance orders based on reproductive ability (reviewed by Sudd and Franks 1987 p. 19; Hölldobler and Wilson 1990 p. 222; Heinze 1993a; Heinze et al. 1994). The species are the nonparasitic *Leptothorax allardycei* from Florida (Cole 1981, 1986), the North American slave-

BOX 7.2 SEXUAL DECEPTION AND THE EVOLUTION
OF WORKER REPRODUCTION

This box describes the logic behind Nonacs and Carlin's (1990) and Nonacs' (1993b) sexual deception hypothesis (Sections 6.5, 7.4). Consider a colony with monogyny and single queen mating. Workers are expected to attempt male production, because their (life-for-life) relatedness with nephews (0.375) exceeds their relatedness with brothers (0.25) (Box 7.1). (Sons of individual workers are ignored, because they are assumed to form just a small fraction of all worker-produced males.)

Assume that the workers try to destroy queen-laid male eggs and replace them with worker-laid ones. (Replacement is necessary because colony productivity is taken to be fixed.) But workers sometimes mistakenly destroy sisters (0.75 relatedness) instead of brothers. Let the proportion of replacements that are errors be α. For example, $\alpha = 0.2$ means that one sister is destroyed for every 4 brothers. The gain from every replacement is the same (0.375 fitness units). But the loss from a correct replacement is 0.25, whereas from an incorrect one it is three times greater, 0.75. Therefore, if $\alpha > 0.25$ (more than one sister destroyed for every 3 brothers), replacement of queen brood is not favored (Nonacs 1993b). Workers would do better to avoid the destruction of queen brood altogether. A similar argument was used to work out the tolerable cost of manipulating sex allocation (Section 6.6).

Now imagine that queens disguise the sex of their offspring, such that, to workers, the queen's sons and daughters are totally indistinguishable (sexual deception). Then worker destruction of queen brood would be random ($\alpha = 0.5$), and so could not evolve. Sexual deception therefore helps explain why workers are not reproductive even under a single, once-mated queen (Nonacs 1993b).

As a brood item grows older, recognition errors during replacement become more costly. For example, mistakenly destroying a larval sister incurs a greater loss than destroying a sister as an egg, because it takes more energy to raise the substitute egg to the corresponding developmental stage. Let the sex of a brood item become apparent at a fraction x of the way through its development, so worker recognition is fully accurate after that time ($\alpha = 0$). When queen-laid males are replaced with worker-laid ones, the gain and loss to the workers are, as above, 0.375 and 0.25 fitness units respectively. But the gain must be devalued by a factor of $(1 - x)$, because, for each replacement event, x-worth of energy must be used to raise the worker's egg to the same developmental stage as the brother it replaced. For replacement to be favored, the net gain must exceed zero:

$$0.375(1 - x) - 0.25 > 0$$
$$x < 0.33.$$

BOX 7.2 CONT.

So, if the queen's brood is more than a third of the way through its development by the time its sex becomes discernible, workers are not selected to replace it. To do so would be too costly. Therefore, Nonacs (1993b) concluded, sexual deception need only operate during early development for queens to maintain their control over male production.

maker *Protomognathus* (= *Harpagoxenus*) *americanus* (Franks and Scovell 1983), and the European slave-maker *Harpagoxenus sublaevis* (Bourke 1988b; Bourke et al. 1988; Franks et al. 1990a).

The fertile workers in these species engage each other in ritual dominance bouts, with an alpha worker dominating a beta, beta dominating gamma, and so on. Worker rank correlates with the degree of ovary development or egg laying. In *Protomognathus americanus* the alpha worker is herself physically dominated by the queen. However, in *Harpagoxenus sublaevis* the queen is not aggressive, although her removal stimulates worker laying. Bourke (1988b) interpreted this as support for pheromonal queen control. But it could equally indicate queen signaling, assuming that workers do not respond to the queen's signal by total abstention from reproduction.

These phenomena demonstrate that queenright workers compete among themselves for egg-laying rights. The queen either opposes them with physical or pheromonal coercion, or chemically signals to them not to lay. So there is actual kin conflict between queen and workers, and among workers, over male production. However, mating frequency was only genetically investigated in *Harpagoxenus sublaevis*. Here, queens are singly mated, so workers are each others' full sisters (Bourke et al. 1988). Another uncertainty is over the exact level of worker male-production. Accurately determining what fraction of males in queenright colonies derive from workers is technically difficult (Cole 1981; Bourke et al. 1988). Confirming that a worker's production of adult sons depends on its individual dominance rank is even harder. At present, this is inferred from its ovary development or egg-laying rate. However, in future, more refined molecular methods may allow worker male-production to be measured directly.

Dominance behavior in workers producing male eggs also occurs outside the leptothoracines (reviewed by Heinze 1993a; Heinze et al. 1994). One case is in queenless workers of *Aphaenogaster cockerelli* (Hölldobler and Carlin 1989). But this phenomenon seems particularly prevalent in the ponerine ants, especially in colonies lacking either a queen or a mated worker ("gamergate"). Examples include two *Diacamma* species (Peeters et al. 1992; Peeters and Tsuji 1993), and three *Pachycondyla* species (Oliveira and Hölldobler 1990, 1991; Ito and Higashi 1991; Higashi et al. 1994). Although the kin structure of these ants is not definitely known, in *Diacamma* it appears that the one gamergate is singly mated (Peeters and Tsuji 1993), meaning worker nestmates would often be full sisters. Therefore, these cases further illustrate actual conflict between close worker kin over male production. Both leptothoracines and ponerines tend to have relatively small colonies (Heinze et al. 1994). This could contribute to the frequency of worker dominance in the two groups, since large colones seem unlikely to support dominance orders that are presumably based on some type of individual recognition (E.O. Wilson 1971 p. 302).

7.5 Kin Conflict in Multiple-queen Societies

Polygyny creates conditions for a further set of kin conflicts within ant societies. The parties involved are resident queens, resident workers, and queens seeking entry into the colony. Conflict arises because, as in previous cases, these parties need not be identically related with sexual brood (e.g. Rosengren and Pamilo 1983; Nonacs 1988; Pamilo 1991c; Herbers 1993). The outcome of conflict is again likely to be influenced by the relative power held by each group. Workers will probably be the most powerful party, through their greater numbers. Thus, workers should usually be able to prevent a queen entering the colony against their interests. Similarly, they should be able to adopt queens contrary to the interests of the resident ones. On the other hand, any resident party is unlikely to be capable of forcing a young queen to remain in the natal colony against her interests. Observations suggest that *Formica* workers sometimes prevent young winged queens from leaving the nest (e.g. Rosengren and Pamilo 1983). But this may happen because conditions for departure are unsuitable at that moment. If remaining in the nest is against their interests, young queens should evolve the appropriate dispersal behavior.

Many observations suggest that workers control queen number in ants (Fletcher and Ross 1985). For example, workers eliminate supernumerary queens in the monogynous form of the fire ant *Solenopsis invicta* (Fletcher and Blum 1983; Vargo and Porter 1993). In the polygynous form, workers kill all queens of a particular enzyme genotype, although the reasons for this are unclear (Keller and Ross 1993a;

Section 8.4). In other polygynous species workers also kill or expel nest-mate queens. Examples are the Argentine ant *Linepithema humile* (= *Iridomyrmex humilis*) (Keller et al. 1989; Passera and Aron 1993b), *Formica* ants (Fortelius et al. 1993; Yamauchi et al. 1994), and several *Leptothorax* species (e.g. Heinze and Smith 1990; Bourke 1991; Heinze et al. 1992b; Lipski et al. 1992; Heinze 1993b). Therefore, workers clear-ly have the potential to exercise power in queen–worker conflicts over queen number. This must also allow them the opportunity to control the relatedness class and quality of queens. Some kin conflicts arising in polygynous societies are now considered in detail.

QUEEN–WORKER CONFLICT OVER THE KIN CLASS OF ADOPTED QUEENS

The first is conflict over what kind of queen kin to adopt, assuming that recruiting fresh queens is favored. Several authors (e.g. Trivers and Hare 1976; Noonan 1978; Brockmann 1984; Nonacs 1988) have argued that queens and workers should each prefer their sisters as extra queens. Each caste is more closely related with the offspring of their full sisters than with the offspring of full sisters of the other caste. Hence there is queen–worker conflict over who to accept as a new queen. Polygyny is classified as primary if a polygynous colony arises directly from a group of foundress queens, and secondary if an established colony admits extra queens (Hölldobler and Wilson 1977; Section 8.2). Primary polygyny is rare, since foundress associations usually develop into monogynous colonies (Section 8.3). Brockmann (1984) and Nonacs (1988) attributed this to worker control of the adoption of new queens. Primary polygyny would represent workers accepting their mother's sis-ters as extra queens, whereas – as just discussed – workers prefer to recruit their own sisters, leading to secondary polygyny.

However, this reasoning assumes that queens in foundress associa-tions are relatives, which is not usually the case (Strassmann 1989). It also assumes that workers are responsible for foundress associations becoming monogynous, whereas sometimes the queens themselves fight (Rissing and Pollock 1988). The instability of foundress associations is instead better explained by the change from a closed to an open energy system when the first workers mature (Rissing and Pollock 1986; Section 8.3). Nevertheless, the origin of primary polygyny from groups of related queens is possible in principle. Therefore, the idea that work-ers favor sisters as extra queens could still partly explain the prevalence of secondary compared to primary polygyny.

Whether workers discriminate among classes of kin in granting queens admission is not known in any species. In laboratory introductions, *Leptothorax curvispinosus* workers usually rejected young, newly-mated queens if they were nonnestmates (Stuart et al. 1993). On the other hand, in a polygynous wood ant (*Formica lugubris*) with a uni-colonial social structure (workers from different nests mix freely), acceptance of new queens seemed to depend on whether they were mated or not (Fortelius et al. 1993).

QUEEN–WORKER CONFLICT OVER WHETHER TO ADOPT EXTRA QUEENS

Another kind of conflict arises when a monogynous colony adopts daughter queens to become (secondarily) polygynous, or when an already polygynous colony adopts an extra daughter queen. These cases were modeled by Nonacs (1988) and Pamilo (1991c). Not surprisingly, it was found that queen recruitment is favored if the colony afterwards becomes more productive. This is because, for all residents, adopting a queen who then reproduces in the colony dilutes relatedness with the average sexual, and this must be offset by gains in colony productivity or survivorship. However, the critical productivity threshold for the favoring of recruitment is always less for workers than for resident queens, because these queens experience a greater dilution of related-ness. For example, when a monogynous colony accepts daughter queens as reproductives, the resident queen trades offspring for grandoffspring, but the workers only trade sibs for nieces and nephews related via full sisters.

This situation creates a potential queen–worker conflict over whether to adopt new queens. Thus, if a post-adoption colony had a productivity greater than the workers' critical threshold, but below the queens', workers would favor adoption against the queens' interests (Nonacs 1988; Pamilo 1991c). However, no one has examined the extent and outcome of such conflicts in nature.

QUEEN–QUEEN CONFLICT OVER REPRODUCTION

A final type of potential kin conflict in multiple-queen societies is con-flict over reproduction among queens themselves. Although each queen is more closely related with her own offspring than with those of other queens, the level of overt competition usually seems very low. For example, fighting between queens is rare (reviewed by Heinze 1993a; Section 8.6). In some species, queens appear to inhibit each other's fecundity pheromonally (*Solenopsis invicta*: Vargo 1992). However, in others no such effect is detectable (*Leptothorax acervorum*: Bourke 1993), and the effects seen in *S. invicta* could stem from differential treatment of queens by workers (Keller and Nonacs 1993; Section 8.2). Queens may also compete by means of egg cannibalism (e.g. E.O. Wilson 1974a; Bourke 1991, 1994b).

One possible reason for the low levels of aggression in polygynous colonies is that the ecology of these species dictates a low stable reproductive skew (variance in reproduction), which might in turn discourage queens from fighting (Reeve and Ratnieks 1993; Bourke and Heinze 1994; Section 8.6). Overt queen fighting does occur in associations of colony-founding queens (e.g. *Iridomyrmex purpureus*: Hölldobler and Carlin 1985; *Messor* (= *Veromessor*) *pergandei*: Rissing and Pollock 1987), for reasons discussed in Section 8.3. Queen aggression also occurs in functionally monogynous species (in which only one of several mated queens per colony lays eggs) such as *Leptothorax* species A (Heinze and Smith 1990; reviewed by Heinze 1993a). In these societies, it could be linked to an ecology that promotes high reproductive skew (Reeve and Ratnieks 1993; Bourke and Heinze 1994; Section 8.6).

7.6 Other Kinds of Kin Conflict in Ants

This chapter has so far discussed kin conflict over nepotistic brood rearing, over male production, and in polygynous colonies (Sections 7.3–7.5). This section now considers other possible types of conflict. Some are speculative, because their existence is technically hard to establish.

QUEEN–WORKER CONFLICT OVER QUEEN QUALITY

Workers in both monogynous and polygynous colonies should be sensitive not only to the kin class of queens, but to their quality. Keller and Nonacs' (1993) queen signaling hypothesis argued that queens communicate their fecundity to workers, who are assumed to benefit from rearing queen offspring due to the queens' greater productivity. But if this declined, the workers' best course would be to replace the queen or start egg-laying themselves. Unless the queen's fecundity were very low, she would still favor the raising of her more closely related progeny. Therefore, an area of conflict exists in which workers seek the removal of the queen against her interests. In bumble bees and wasps there is evidence that such conflict leads to workers killing the queen (Bourke 1994a). However, there is no firm evidence for this type of worker matricide in ants. This could be because at most points in the life history of perennial societies, queens have a far greater future productivity than workers (Trivers and Hare 1976; Ratnieks 1988; Bourke 1994a; Section 7.4).

Workers in young colonies founded by groups of queens (foundress associations) are also probably sensitive to queen quality (e.g. Forsyth 1980; Section 8.3). Alexander et al. (1991) pointed out that in all types of eusocial society, helper individuals should continually monitor the health and fecundity of the reproductives, on which their future fitness depends.

CONFLICT AMONG DAUGHTER QUEENS OVER QUEEN REPLACEMENT

A few monogynous ant species have a social system in which the existing queen may be replaced by queens reared in the same colony (queen replacement). Examples include army ants (Franks and Hölldobler 1987) and ponerine ants of the genus *Diacamma*, with the queen caste being represented by gamergates or mated workers (Fukumoto et al. 1989; Peeters and Higashi 1989; Section 7.4). In this case, one expects conflict among daughter queens over replacing their mother, as each daughter should value herself more highly than her sisters. In honey bees, this situation is well known to lead to actual conflict. Newly-emerged queens engage in lethal fighting in the nest (Visscher 1993). Young queen bees also make "piping" sounds. These could be signals of queen quality designed to persuade the workers to let the most vigorous queen head a reproductive swarm (Seeley 1985 p. 65; Visscher 1993).

In ants, analogous phenomena occur. For example, the first female *Diacamma* to emerge mutilates the other individuals by removing their thoracic appendages (gemmae), so preventing their succeeding as the colony's single gamergate (Fukumoto et al. 1989; Peeters and Higashi 1989; Section 7.4). Among the army ants, overt conflict between queens is apparently absent, although it would be hard to observe. However, the equivalent to "piping" in honey bees may occur. Franks and Hölldobler (1987) proposed that both the existing army ant queen, and the handful of new ones, compete to head each daughter colony by advertising their fecundity to workers with pheromones (cf. Keller and Nonacs 1993). This represents another case where workers should be sensitive to queen quality (see above).

QUEEN-WORKER CONFLICT OVER COLONY MAINTENANCE VERSUS REPRODUCTION

In a perennial society of social Hymenoptera, a fraction of investment must each year be allocated to producing new workers (colony maintenance), and another fraction to producing new sexuals (reproduction).

Say there is a single colony queen. From her viewpoint, investment in workers is relatively attractive because it is investment in herself (it prolongs colony life), whereas investment in sexuals represents investment in her offspring. However, other things being equal, workers are indifferent between investment in their parents and investment in their sibs. Therefore, the queen favors relatively more investment in new workers than workers themselves, creating a potential queen–worker conflict over colony maintenance versus colony reproduction.

This idea was proposed by Pamilo (1991b). It demonstrates that even life history strategy is subject to kin conflict (Pamilo 1991b; Section 9.9). Pamilo's (1991b) model of this type of conflict predicted that, assuming worker control of resource allocation, the fraction of resources invested into sexual production should sometimes fall with rising colony survivorship (and hence size) under worker control (under queen control it should always rise). In polygynous colonies, no relationship between colony survivorship and the fraction of resources committed to sexuals is expected, because the conflict over growth versus maintenance is diminished. Sundström (1993b, 1995a) found evidence for both these predictions in monogynous and polygynous *Formica truncorum* colonies respectively. This supported Pamilo's (1991b) suggested conflict, and indicated that in this case the chief victors are workers.

Conflict over Investment in Individuals

Suppose that the expected fitness of adults depends on their body size and hence on their larval nutrition. A larva would then be selected to seek more and better resources for itself even if this involved costs to other parties. So a conflict would arise analogous to Trivers' (1974) original parent–offspring conflict (Section 7.2). In ants, queens plausibly benefit from increased body size because larger queens found new colonies more successfully (e.g. Keller and Passera 1989a). Similarly, males might gain because of sexual selection on males for large size (e.g. Davidson 1982).

The possibility of conflict over individual investment in social insects was raised by Trivers and Hare (1976). However, whether – as expected – larvae beg for more food than they need for basic survival is unknown. On the other hand, by examining with path analysis the relation between body size and colony demography, Backus (1993) detected conflict over individual investment in *Leptothorax longispinosus*.

Conceivably, a larva might also withhold resources from the colony to promote its personal survivorship. Selfish behavior of this sort was proposed by Wilson and Hölldobler (1980) in the weaver ant, *Oecophylla longinoda*. The nests of this species are woven by the workers using silk produced by the larvae (e.g. Hölldobler and Wilson 1990 p. 620). Wilson and Hölldobler (1980) found that male larvae had smaller silk glands, and provided less silk, than larval workers. They suggested that this occurred because male larvae are less closely related with sexual sibs on average than are female ones. In other words, they proposed that males are under kin selection to withhold communal resources (silk) for personal gain (such as improved growth). However, this argument is mistaken. Assuming monogyny and single mating by the queen, a male's (life-for-life) relatednesses with sisters and brothers both equal 0.5, and the corresponding relatednesses of a female are 0.75 and 0.25 (Grafen 1991; Table 3.1). Therefore, under a stable population sex ratio (cf. Box 3.2), males and females both value sibs equally, on average. (As others had done before [Section 3.5], Wilson and Hölldobler [1980] took male–sister relatedness to be 0.25, which explains how they reached their conclusion.) Thus, male weaver ant larvae must be producing less silk than worker larvae for other reasons. For example, worker larvae may have developed more efficient silk production because they are the only larval type present for much of the colony's lifetime.

QUEEN–WORKER CONFLICT OVER CASTE DETERMINATION

Another type of potential conflict is conflict among female larvae, queens, and workers over what caste the larvae should develop into (Pamilo 1991b; Nonacs and Tobin 1992; Ratnieks and Reeve 1992; Keller and Reeve 1994a). One reason why this might occur is queen–worker conflict over colony maintenance versus reproduction (Pamilo 1991b; see above). If this exists, workers might try to sexualize more female larvae than the queen favors (Pamilo 1991b).

Conflict would also arise if a female larva were selected to become one or other caste against the interests of other parties. For example, a female larva developing into a queen would oppose another larva doing so in her place, because this would mean trading her own offspring for the less closely related progeny of the other larva (Ratnieks and Reeve 1992). This raises the question of whether a female larva should prefer to be a queen or worker to begin with. The answer must depend on circumstances. For example, larvae should be indifferent to their caste in a population with a simple kin structure, where reproductives and workers are equally productive, and there is sex ratio equilibrium. This is because reproductives and workers in such a population achieve equal payoffs (Box 3.2).

Consequently, conditions in which larvae should prefer to be queens include the following: (1) when workers are far less productive than queens, as is likely in colonies of advanced eusocial species, especially as colonies grow larger and each new worker makes a smaller contribution to productivity (Keller and Nonacs 1993; Keller and Reeve 1994a); (2) in monogynous populations with multiple mating, since queens would then rear offspring but workers would rear kin less related than full sibs; and (3) in species with queen replacement, because a young queen might inherit the resources of the colony with a low risk of mortality.

Another option for a female larva is to become a reproductive worker. Nonacs and Tobin (1992) argued that, relative to becoming queens, female larvae of socially parasitic ant species favor this more than developing into sterile workers. Supporting evidence is that socially parasitic ants which have evolved from species with reproductive workers retain the worker caste more often (becoming slave-making species) than those that have evolved from species with worker sterility (which become workerless parasites) (Nonacs and Tobin 1992).

The possibility of any form of conflict over caste determination raises the issue of which party controls this trait – the workers, queens, or female eggs and larvae themselves. This is uncertain. In general, caste appears to be fixed in the egg ("blastogenic" caste determination) in some ants (e.g. Passera 1980b), whereas in others it is determined by the rearing conditions of larvae (reviews of Brian 1979a, 1980, 1983 p. 191; Wheeler 1986; Buschinger 1990b; Hölldobler and Wilson 1990 p. 348). These findings suggest control by the queen (or eggs) and workers respectively. In several polygynous species, the presence of colony queens pheromonally inhibits male production and the development of female larvae into new queens (e.g. *Plagiolepis pygmaea*: Passera 1980a; *Solenopsis invicta*: Vargo and Fletcher 1986a, b; *Monomorium pharaonis*: Edwards 1987, 1991; *Linepithema humile* [= *Iridomyrmex humilis*]: Passera et al. 1988a; Keller et al. 1989; Vargo and Passera 1991, 1992). This again suggests queen control of caste. However, in *S. invicta*, *M. pharaonis*, and *L. humile*, this inhibition occurs because workers in the presence of queens maltreat or destroy sexual larvae (Vargo and Fletcher 1986a; Vargo and Passera 1991; Edwards 1987, 1991). In another well known example, *Myrmica rubra*, workers in the presence of queens bite female larvae, so biasing their development towards workerhood (Brian 1973a). Therefore, in these cases, it is not always clear whether workers are behaving in their or the queens' interests.

The effects of queens and workers on caste determination were formerly interpreted in proximate terms as parts of a system for the regulation of sexual production (e.g. Brian 1980). However, as other investigators have shown (Ratnieks and Reeve 1992; Keller and Nonacs 1993; Keller and Reeve 1994a), it is illuminating to inquire into the ultimate reasons for these phenomena. One idea is that queens, acting in their interests, favor greater relative investment in workers, as predicted by Pamilo (1991b). But the intensity of queen–worker conflict over resource allocation between colony maintenance and reproduction should be low under polygyny (Pamilo 1991b). In addition, *Solenopsis invicta*, *Monomorium pharaonis*, *Linepithema humile*, and *Myrmica rubra* all frequently reproduce by colony budding (e.g. Elmes 1973a; Edwards 1987; Keller 1988; Vargo and Porter 1989). Queens and workers therefore have a common interest in producing workers to stock buds (Keller and Nonacs 1993). This would further reduce the intensity of Pamilo's (1991b) proposed conflict, and suggests the workers' behavior could be in their interests. Reproduction by colony budding also leads both queens and workers to favor male-biased sex allocation (Section 4.8). This could be why in some cases the presence of queens appears to inhibit the sexualization of female larvae more strongly than the production of males (e.g. *M. rubra*: Brian 1981; *L. humile*: Passera et al. 1988a; Vargo and Passera 1991). Finally, queens may inhibit the production of new queens because they oppose being replaced in their lifetimes.

However, when the existing queens die, all parties (the deceased queens, workers, and queen-potential larvae) favor their replacement (Keller and Nonacs 1993), which often occurs (e.g. *L. humile*: Passera et al. 1988b). This could explain why it appears that all species with pheromonal queen inhibition of new queen production are polygynous: monogynous species generally lack a mating system permitting replacement by related queens (Section 7.4), and so have no need to evolve the mechanism adopted by polygynous species to prevent queen replacement. In short, all the cases in which queens pheromonally inhibit sexual production stand to be reinterpreted in ultimate terms as cases in which one, some, or all parties in the colony have evolved a self-interested response to the presence of queens (Keller and Nonacs 1993; Keller and Reeve 1994a; Section 8.2).

Kin conflict over caste determination is clearly a promising area for future investigation. It would be particularly interesting to investigate systematically which party controls caste determination in relation to the expected severity of the conflict over investment in sexuals and workers suggested by Pamilo (1991b).

QUEEN–WORKER CONFLICT OVER WORKER ACTIVITY LEVELS

Reproductive workers should be inclined to work less hard than queens favor, because working hard decreases a worker's survivorship and hence its expected production of sons. Evidence for this idea is that reproductive workers commonly forage less than nonreproductive ones (reviewed by Ratnieks and Reeve 1992). Queens oppose worker reproduction because it involves the colony rearing grandsons in preference to more closely related sons (Section 7.4). They should oppose it even more if it also reduces colony productivity. Aside from attempting to inhibit worker reproduction (Section 7.4), queens might respond to lazy workers by attacking them and stimulating them to be active. There is evidence that queens behave this way towards potential replacements in both polistine wasps and naked mole-rats (Reeve and Sherman 1991; Reeve 1992). However, in ants, there are few observations of queens physically inciting worker activity. Queens of the slave-maker *Protomognathus americanus* preferentially attack the more ovary-developed workers (Franks and Scovell 1983), but this is probably direct physical inhibition of reproduction.

When workers are nonreproductive, worker inactivity might still be selected in some circumstances. For example, Schmid-Hempel (1990) showed that mutant genes for worker laziness could spread if worker bearers accumulated in colonies through low mortality, and if these workers preferentially reared sexuals also bearing the mutant gene. As Schmid-Hempel (1990) himself pointed out, this would again create a potential conflict over work levels between workers and queens (see also Schmid-Hempel 1991).

MALE–FEMALE CONFLICT

The interests of males are generally ignored in analyses of kin conflict in social Hymenoptera. One reason is that males usually die after mating and so have little opportunity to enforce their preferences. However, several authors have pointed out that males have their own fitness interests and conceivably attempt to realize them (e.g. Alexander 1974; Trivers and Hare 1976; Starr 1984; Brockmann and Grafen 1989; Ratnieks and Miller 1993). For example, under outbreeding, a queen's mate is unrelated with any males that she produces. So, if workers are sterile, males favor an all-female sex ratio. (Under monogyny with worker reproduction, the male's stable female : male sex ratio is $[1+p]/[1-p]$, where p equals the fraction of offspring males produced by queens. This is calculated from Equation 4.4 with $r_F = 1/[2k]$

and $r_M = [1 - p]/[2k]$, assuming queens use sperm from k males equally. The male's stable sex ratio is therefore independent of queen mating frequency.) In addition, because of male haploidy, a male's genetic interests are identical with those of his sperm (Starr 1984). These findings suggest that selection acts on sperm, in both the sperm's and the male's interests, to fertilize more eggs than queens favor (Alexander 1974; Trivers and Hare 1976; Starr 1984). However, power over fertilization (which is by sperm release from the sperm receptacle) presumably lies with the queen. So males seem unlikely to achieve their sex ratio goals in opposition to queens (Trivers and Hare 1976). Starr (1984) suggested that female bias (Section 5.2) could be evidence for partial male control of sex allocation. But unbiased allocation in slave-making ants (Bourke 1989; Section 5.3), and split sex ratios in *Formica truncorum* (Sundström 1994a; Section 5.4) indicate that male interests are rarely achieved. Males in these cases favor female bias in all colonies, but this does not occur.

Another way in which males might promote their interests would be to produce worker daughters capable of egg-laying. This would allow male genes to enter the progeny generation of males, from which they are otherwise absent (Lin and Michener 1972; Alexander 1974). However, it is hard to see how males could influence daughters that they never meet. In addition, males favor worker male-production at all queen mating frequencies, because they are always unrelated with the queen's sons. But, at least in honey bees, workers are nonreproductive at high mating frequencies because of worker policing (Ratnieks 1988; Section 7.4). This suggests that reproductive workers act in their own interests and not those of their fathers.

7.7 Conclusion

The main conclusion of this chapter is that potential kin conflict over reproduction is a universal feature of Hymenopteran societies. Therefore, to understand the social behavior and life history strategies of ants, it is essential to know the kin structure of their colonies. In many cases, actual conflict arises from potential conflict. Kin conflict theory helps explain the detailed nature of actual conflict. For example, it offers reasons why workers evolve dominance and male production in some societies but not in others (Section 7.4). This also constitutes evidence for kin selection theory. However, actual conflict is often absent or reduced despite the existence of potential conflict, with conflict between workers over nepotistic queen rearing being one example (Ratnieks and Reeve 1992; Section 7.3). Furthermore, although many types of potential kin conflict have been recognized (Sections 7.5, 7.6), technical difficulties have prevented the existence of several of them from being firmly demonstrated.

As social evolution proceeds, the kin structure of societies may alter, and with it the expected type of kin conflict. This conclusion has two implications. First, social evolution can be self-reinforcing, with changes in the kin and social structure precipitating further such changes. For example, the evolution of partial multiple mating may bring about facultative sex ratio biasing by workers, which in turn may lead to the evolution of higher queen mating frequencies and to worker policing (Ratnieks 1988; Boomsma and Grafen 1991; Pamilo 1991c; Queller 1993b; Ratnieks and Boomsma 1995; Sections 7.4, 11.4). Second, because evolutionary changes lag behind new selective pressures, species may exhibit in their current social organization traces of previous conflict (West-Eberhard 1981). For example, queen pheromones may now be signals of queen vigor (Keller and Nonacs 1993), but may previously have acted as forcible imposers of worker sterility (Section 7.4). Therefore, insect societies have a history, of which a changing pattern of kin conflict is almost certainly a major feature (West-Eberhard 1981; Bourke 1988a).

Lastly, kin conflict also influences the organization of work within colonies. A near-universal feature of insect societies is that young workers perform tasks in the nest, whereas old ones work outside it. Although the proximate reasons for this remain controversial (Section 12.3), its evolutionary origin may lie in the fitness interests of reproductive workers. Workers with a chance of replacing the queen or producing males would be selected to avoid risky tasks, at least until late in their lives when the costs of dying are smaller (West-Eberhard 1981; E.O. Wilson 1985a). Reeve and Ratnieks' (1993) models of the likelihood of performing risky tasks as a function of reproductive skew, and Schmid-Hempel's (1990) theory of the evolution of lazy workers (Section 7.6), also demonstrate how kin conflict might affect the division of labor.

7.8 Summary

1. Kin-selected conflicts occur because parties differ in their relatedness with the reproductive individuals reared within societies. Genes expressed in these parties for social acts towards reproductives will therefore be selected according to different criteria. For example, in parent–offspring conflict, selfish offspring demand extra parental investment at the expense of their less closely related sibs (Trivers 1974).

2. The main determinant of the nature of kin conflict in the social Hymenoptera is colony kin structure. In ants this varies with queen number, queen mating frequency, and the level of worker male-production. However, the potential kin conflict expected from kin structure need not always lead to actual conflict. Reasons for this are: (a) asymmetries in relative power, or in the fitness consequences of escalated fighting; (b) counter-manipulation by other parties, and costs to colony productivity of actual conflict; (c) relative unprofitability of actual conflict because of low reproductive skew (equitable reproduction); (d) worker inability to discriminate in favor of more closely related kin within colonies.

3. A principal kin conflict in social Hymenoptera is queen–worker and worker–worker conflict over who lays haploid, male-yielding eggs. Queens should oppose worker male-production because they prefer the colony to produce their more closely related sons. Under monogyny with single mating, or polygyny with single mating and unrelated queens, workers are more closely related with an average worker-produced male (e.g. nephew) than with an average queen-produced one, and so should be reproductive. However, under monogyny with multiple mating, and polygyny with many, highly related, singly mated queens, workers should be nonreproductive. Workers may be selected to prevent one another reproducing (worker policing [Ratnieks 1988]), because this ensures that the average worker-produced male is not highly related with a focal worker.

4. Unexpectedly, workers under monogyny with single mating are generally nonreproductive in the presence of queens. Explanations for this include: (a) costs to colony productivity; (b) queen inhibition of worker reproduction; (c) queens disguising their brood so that workers cannot tell valuable sisters from unwanted brothers (sexual deception [Nonacs 1993b]).

5. Queen inhibition of worker egg-laying has generally been assumed to be chemical (pheromonal queen control). However, queen pheromones may not be inhibitory, but rather honest, costly, stable signals of queen productivity, to which workers respond by being nonreproductive because this is in their fitness interests (Keller and Nonacs 1993).

6. In several leptothoracine and ponerine ants, reproductive workers form dominance orders in which rank correlates with ovary development. These cases demonstrate actual kin conflict over male production.

7. Polygynous societies induce kin conflict over: (a) the kin class of adopted queens (resident queens and workers each favor their full sisters as adoptees); (b) whether to adopt extra queens (the workers favor

this at lower levels of extra productivity than resident queens); and (c) reproduction among egg-laying queens. However, fighting among laying queens is rare in mature, polygynous ant colonies, possibly because of their low reproductive skew.

8. Other possible kin conflicts in social Hymenoptera occur over queen quality, queen succession, the relative amount of sexual and worker production, the level of investment in individual larvae, queen–worker caste determination, worker activity levels, and the fertilization of eggs. However, although some evidence exists for these conflicts, firmer support is lacking because of technical difficulties.

9. Some features of insect societies may stem from previous kin conflict, and present conflict could affect the division of labor. In sum, kin conflict is a powerful, all-pervading influence upon social organization, and its occurrence is further support for kin selection theory.

8 Evolution and Ecology of Multiple-queen Societies

8.1 Introduction

Societies of ants vary greatly in their fundamental social structure. A principal cause of this variation, as earlier chapters have mentioned, is variation in queen number. This exists at several levels – between colonies, populations, and species – but the basic difference is between colonies with one queen per colony (monogyny), and those with several (polygyny). This chapter reviews the evolution and ecology of queen number variation in ants (cf. Buschinger 1974a; Hölldobler and Wilson 1977, 1990; Rissing and Pollock 1988; Passera et al. 1991; Keller 1993a).

Polygyny is a major feature of ant biology, having evolved many times independently (Hölldobler and Wilson 1977; Ross and Carpenter 1991b; Elmes and Keller 1993). However, understanding polygynous ants is relevant not just to social insect biology, but also to behavioral ecology in general. Polygyny involves queens sharing reproduction with other queens, and workers raising brood less closely related than full sibs. It therefore represents a potential challenge to kin selection theory (Sections 8.3–8.5). Polygynous ants are also excellent subjects for the study of communal breeding in animals. In particular, they are useful for studying the influence of ecological factors on social structure (e.g. Hölldobler and Wilson 1977; Herbers 1993; Section 8.3–8.6), the occurrence of kin conflict inside complex societies (e.g. Nonacs 1988; Pamilo 1991c; Sections 7.5, 8.6), and the evolution of a stable reproductive skew (allocation of reproduction) among multiple breeders (Vehrencamp 1983a; Reeve and Ratnieks 1993; Keller and Reeve 1994a; Sherman et al. 1995; Section 8.6). These are areas of interest to both vertebrate and invertebrate biologists (e.g. E.O. Wilson 1975; Emlen 1991). Lastly, as pointed out by Rosengren and Pamilo (1983), Ross (1989), and Keller and Vargo (1993), the problem of why polygyny evolves is closely allied to that posed by the evolution of eusociality itself (Chapter 3). Therefore, studying either of these issues helps shed light on the other.

8.2 Types of Polygynous Society and their Features

There are several types of multiple-queen society. A key distinction is between primary and secondary polygyny (Hölldobler and Wilson 1977). Primary polygyny occurs when polygynous societies develop from groups of colony-founding queens (foundress associations). Secondary polygyny is when mature colonies admit extra queens (queen adoption). Although primary polygyny is rare, foundress associations are common, and their evolution forms a problem distinct from that posed by secondarily polygynous societies. Secondary polygyny is also common (Buschinger 1974a; Frumhoff and Ward 1992). As later described, it may itself be subdivided into two categories – the multicolonial type and the unicolonial one. Finally, colonies are said to be functionally monogynous if several mated queens coexist but only one lays eggs. Box 8.1 is a glossary defining these and related terms.

TRAITS OF SECONDARY POLYGYNY

Compared to monogyny, secondary polygyny is associated with changes in life history, dispersal behavior, and the mating system. It also involves alterations in the genetic structure of colonies and populations, and the individual traits of queens and workers (Table 8.1). The reasons for some of the traits of polygynous colonies are unclear. For example, it is uncertain why polygynous ants have less polymorphic workers than monogynous ones (Frumhoff and Ward 1992), although Keller (1995b) suggested that monogynous species, in the absence of colonies as genetically diverse as those of polygynous ants, evolve worker polymorphism as a defence against the spread of parasites. However, most features of polygynous ants are fairly obvious consequences of multiple queening.

To begin with, secondary polygyny involves queen adoption (by definition), which in turn entails a mixed mating and dispersal system in which some queens mate near the nest (and are adopted), whereas others leave the nest to mate. These features also mean that queens in polygynous colonies have several modes of colony foundation. They either found colonies independently after dispersal, or seek adoption (in their own or an unrelated colony), or found colonies by colony budding (Keller 1991; Table 8.1).

Queens in polygynous species are relatively smaller, and have fewer fat and glycogen reserves, than queens under monogyny (Table 8.1). Each of these features stems from the more limited dispersal behavior of polygynous species, and from their dependent colony foundation in

Table 8.1

Features Associated with Monogyny and (Secondary) Polygyny

Monogyny	Polygyny
a. Reproduction by emission of winged sexuals (or rarely by colony fission)	Reproduction by emission of winged sexuals and by colony budding
b. Mating away from nest in mating swarm; so queens more prone to fly from nest and less prone to shed wings in nest	Mating in, on, or near the nest, or in distant swarm; so queens less prone to fly from nest and more prone to shed wings in nest
c. Inbreeding rare	Inbreeding also rare (due to male dispersal or queens avoiding related mates?)
d. Wide dispersal	Wide dispersal and nondispersal (due to queen adoption), i.e. dispersal polymorphism
e. Independent colony foundation (usually claustrally by single queens)	Independent and dependent colony foundation (adoption or budding)
f. Colony dies soon after death of queen (queen replacement is rare)	Colony potentially immortal due to queen adoption
g. High within-colony relatedness	Lower within-colony relatedness
h. Little genetic structuring of population at colony level or above (neighboring colonies unrelated due to wide dispersal of sexuals)	Greater genetic structuring at or above colony level, e.g. neighboring colonies or nests related because of colony budding or polydomy
i. Monodomy more common than polydomy?	Polydomy common
j. Multicolonial	Multicolonial or unicolonial
k. Well-developed nestmate recognition	Less well-developed nestmate recognition
l. Workers larger and more frequently forming physical castes (polymorphism)	Workers smaller and forming physical castes less often (monomorphism)
m. Queens longer lived, larger, and with more fat and glycogen reserves, greater fecundity, and later first age of sexual production	Queens shorter lived, smaller, with fewer fat and glycogen reserves, lower fecundity, and earlier first age of sexual production
n. Wingless sexuals absent or rare in nonparasitic species	Wingless sexuals more common
o. Winged sexuals all of one size (unimodal)	Winged sexuals sometimes of two sizes (bimodal) – large or small (macro- and microgynes in queens, macra- and micraners in males)

NOTES: General references are E.O. Wilson (1971), Brian (1983), and Hölldobler and Wilson (1990). Detailed sources are as follows:

(a) Reproductive and mating system: Hölldobler and Bartz 1985. Examples of colony budding: Briese 1983; Vargo and Porter 1989; Stille and Stille 1993; reviewed by Keller 1991.

(b) Unusual monogynous species with mating near nest: Lenoir et al. 1988. Polygynous species with mating both near and far from the nest: Douwes et al. 1987; Cherix et al. 1991; Franks et al. 1991a. Polygynous species with mating mainly in nest: Passera et al. 1988b; Keller and Passera 1992; Passera and Keller 1994. Propensity of queens to fly or dealate: Fortelius 1987; Sundström 1995b.

(c) Rarity of inbreeding: Gadagkar 1991a; Keller and Passera 1993. Exceptions showing inbreeding: Pamilo and Rosengren 1984; Yamauchi et al. 1991; Pearson and Raybould 1993; Herbers and Grieco 1994. Males disperse between nests in polygynous species: Briese 1983; Kaufmann et al. 1992; Passera and Keller 1994. Queens avoid related mates: Keller and Passera 1993.

(d) Adoption of daughter queens: Elmes 1973a; Porter 1991; Satoh 1991; Stuart et al. 1993. Adoption of unrelated queens: Porter 1991; Stille and Stille 1992; Stuart et al. 1993. Queen adoption reviewed by Rissing and Pollock 1988; Herbers 1993.

(e) Mode of colony foundation: Keller 1991.

(f) Longevity of monogynous colonies: Tschinkel 1987a; Pamilo 1991d. Example of queen replacement under monogyny: Tschinkel and Howard 1978; Vargo 1990. Queen replacement in polygynous colonies: Passera et al. 1988b.

(g) Within-colony relatedness: reviews of Hölldobler and Wilson 1990 p. 187; Gadagkar 1991a; Herbers 1993; Rosengren et al. 1993.

(h) Genetic structuring above colony level in polygynous ants: Pamilo 1982c, 1983; Crozier et al. 1984; Stille and Stille 1993; Sundström 1993a; reviewed by Ross 1988a.

(i) Polydomy in monogynous species: Traniello and Levings 1986; Snyder and Herbers 1991; Buschinger et al. 1994. Polydomy in polygynous species: Alloway et al. 1982; Rosengren and Pamilo 1983. Polydomy commoner in polygynous species: Herbers 1993.

(j) Unicoloniality: Hölldobler and Wilson 1977.

(k) Poorer nestmate recognition in polygynous ants: Fletcher and Blum 1983; Keller and Passera 1989b; Provost and Cerdan 1990; Vander Meer et al. 1990; Fortelius et al. 1993; but see Stuart 1991.

(l) Worker size in polygynous ants: Elmes 1974b; Pisarski 1981; Ross and Fletcher 1985a. Physical castes: Frumhoff and Ward 1992; Keller 1995b.

(m) Queen longevity: Tschinkel 1987a; Hölldobler and Wilson 1990 p. 169; Keller and Passera 1990; Pamilo 1991d; Passera et al. 1991; Seppä 1994. Queen size and physiology: Keller and Passera 1989a; Passera and Keller 1990; Sundström 1995b. Queen fecundity: Fletcher et al. 1980; Vander Meer et al. 1992. Age of first reproduction: Keller and Passera 1990.

(n) Wingless sexuals in polygynous ants: Males only wingless: Kinomura and Yamauchi 1987; Stuart et al. 1987; Heinze et al. 1993a. Females only wingless: Briese 1983; Bolton 1986; Heinze and Buschinger 1987; Heinze et al. 1992a; Tinaut and Heinze 1992; Ohkawara et al. 1993. Both sexes wingless: Le Masne 1956; Yamauchi et al. 1991.

(o) Polymorphic winged sexuals in polygynous ants: Janzen 1973; Elmes 1976; Fortelius et al. 1987; Heinze and Hölldobler 1993a; reviewed by Bourke and Franks 1991.

which young queens do not require the energy stores needed for founding colonies singly (Porter et al. 1988; Keller 1991). In some cases, very strong selection for a mixed dispersal strategy has resulted in either wing loss or size polymorphism among the sexuals (Table 8.1), since both wingless queens and miniaturized queens are especially poorly equipped for independent colony foundation (e.g. Heinze 1989; Bourke and Franks 1991; Yamauchi et al. 1991; Rosengren et al. 1993; Section 8.6). Eventually, this process may lead to the evolution of interspecific social parasitism (Buschinger 1990a; Bourke and Franks 1991).

Polygyny also has notable genetic consequences. An obvious one is a reduction in within-colony relatedness (Table 8.1). Another is genetic structuring of the population above the colony level (Table 8.1).

BOX 8.1 GLOSSARY OF TERMS ASSOCIATED WITH POLYGYNY IN ANTS

Claustral colony foundation   Colony foundation by a nonforaging queen or queens, in which energy for rearing brood comes from queens' stored body reserves.

Colony budding   Colony foundation by a queen (or queens) plus workers leaving an established colony. Distinct from colony fission in monogynous colonies (Section 4.8).

Dealate queen   A standard morphological queen after she has shed her wings.

Dependent colony foundation   Colony foundation requiring the presence of workers (i.e. involves colony budding or queen adoption).

Ergatoid queen   A female specialized for reproduction but which is permanently wingless.

Facultative polygyny   Within populations, some colonies are monogynous and others are polygynous.

Foundress association   Group of colony-founding queens (in pleometrosis).

Functional monogyny   Several mated queens coexist, but only one lays eggs at any one time.

Gamergate   A mated worker, replacing the queen caste in some ponerines.

Haplometrosis   Colony foundation by a single queen.

Incipient colony   Newly-founded colony.

Independent colony foundation   Colony foundation by a queen or queens not requiring workers (e.g. haplometrosis).

Macraner   Large male.

Micraner   Small male.

Macrogyne   Large winged queen characteristic of monogynous colonies.

Microgyne   Miniature winged queen found as an intraspecific variant in some polygynous species.

Monodomy   Single colonies occupy single nests.

Monogyny   One mated, egg-laying queen per colony.

Multicoloniality   Population consists of many, distinct colonies, whose members are mutually hostile and do not mix.

Oligogyny   A few mated, egg-laying queens per colony; not clearly separable from polygyny with low queen numbers, though often used only if queens are mutually hostile.

Pleometrosis   Colony founding by several queens (in foundress associations).

Polydomy   Single colonies occupy several, neighboring nests.

Polygyny   More than one mated, egg-laying queen per colony.

Primary polygyny   Polygyny that develops from a foundress association.

BOX 8.1 CONT.

Queen adoption   Admission of new queens into colonies.
Reproductive skew   The allocation of reproduction among queens within multiple-queen colonies. High skew means an uneven sharing of reproduction (one individual produces most offspring), low skew means more equitable reproduction.
Reproductivity effect   Per capita decline in some measure of production (e.g. eggs, adults) as either queen number or worker number rises across colonies.
Secondary polygyny   Polygyny that develops from monogyny, e.g. by queen adoption.
Supercolony   A unicolonial population.
Unicoloniality   Population consists of nests without clear colony boundaries, whose members intermix.

However, whether this is just a consequence of polygyny, or whether it further affects polygynous evolution, has not been well studied. Polygyny is also associated with weaker nestmate recognition than under monogyny (Table 8.1), which presumably stems from an increased diversity in genetically-based odors among nestmates (Hölldobler and Michener 1980; Keller and Passera 1989b). The result is that members of different colonies often intermix, creating a trend towards unicoloniality and a reduction in genetic structuring at the nest level. Alternatively, conditions favoring polygyny or unicoloniality select for reduced nestmate recognition (Stuart 1991). Lastly, despite mating near the nest, polygyny rarely entails significant inbreeding. This seems to be either because males disperse before mating even if queens do not, or because queens deliberately avoid mating with related males (Table 8.1).

Two additional key features of polygyny are that queens almost certainly produce their first sexual offspring at a younger age than queens under monogyny, and have shorter lifespans (Table 8.1). The first of these traits is attributable to adopted queens exploiting the presence of a workforce to produce sexual offspring without initially producing any workers (Pamilo and Rosengren 1984; Keller and Passera 1990). The second could be a result of the first, since life history theory predicts a general association between reproducing early and dying young (Stearns 1992 p. 199; Section 9.3).

In sum, secondary polygyny is a syndrome of interconnected traits that together constitute a life history strategy differing from monogyny in many ways (Keller 1993b). This makes the question of why polygyny evolves a theoretically complex one. Polygyny clearly needs to be understood in its life history context (Keller 1993b). However, nearly all the relevant life history variables (e.g. average fecundity, longevity, age of first sexual production, and so on), are poorly known in both monogynous and polygynous forms. The gathering of these basic data remains a challenge for future investigations (Tschinkel 1991, 1993a). Nevertheless, as later sections describe, progress has been made in understanding polygynous evolution despite these uncertainties.

THE REPRODUCTIVITY EFFECT

Michener (1964) described how, in several taxa of social insects, there is a decline in productivity per worker as worker number rises across colonies. E.O. Wilson (1971 p. 338) termed this the *reproductivity effect*. In polygynous ants, analogous effects occur with respect to queen number. Thus, several studies have found that, as queen numbers rise, there is a fall in both the queens' per capita egg-laying rate (e.g. Mercier et al. 1985a,b; Keller 1988; Vargo and Fletcher 1989) and their per capita output of sexuals (Herbers 1984; Vargo and Fletcher 1987; Buschinger 1990b; Vargo 1993). (Brian [1989] describes an exception in which fecundity and queen number were linearly related.). Michener (1964) proposed that workers accidentally interfered with each other's labor in large colonies, leading to the effect he noted. Inefficiencies of this sort could explain some of the effects seen in polygynous ants, since queen number and colony size are generally correlated (Rissing and Pollock 1988; Elmes and Keller 1993). Another possible explanation in *Leptothorax* species is a rising level of egg cannibalism by queens as queen numbers increase (E.O. Wilson 1974a,b; Bourke 1991, 1993, 1994b).

On the other hand, the specific reasons why egg-laying rate and sexual production per queen fall with increasing queen number are not fully understood. A drop in egg-laying rate does not even necessarily entail a fall in sexual production, because queens apparently lay more eggs than are reared as adults (e.g. Elmes 1973a; Bourke 1993; Elmes and Keller 1993). Colony productivity is therefore set by the size and efficiency of the workforce, rather than by the number of eggs (e.g. Elmes 1989). At the proximate level, experiments suggest that in the fire ant *Solenopsis invicta*, queen fecundity falls as colonies grow more polygynous because queens produce pheromones that inhibit one another's egg-laying (Vargo 1992). Queens of *S. invicta* may also reduce the viability of the eggs of nestmate queens (Vargo and Ross 1989). By contrast, Bourke (1993) found no evidence for mutual pheromonal inhibition among queens in *Leptothorax acervorum*. Queens of several species pheromonally inhibit the sexualization of female larvae in their colonies (e.g. *S. invicta*: Vargo and Fletcher 1986a,b; Section 7.6). This provides a proximate reason for why the more polygynous colonies produce fewer reproductive daughters per queen.

At the ultimate level, reproductivity effects among polygynous ants could occur for several reasons. First, queens in colonies with high queen numbers may each produce fewer eggs and sexuals because workers adjust their treatment of queens to maximize brood production, which – since it does not depend on the egg supply – is largely independent of queen number (Keller and Nonacs 1993). In *Solenopsis invicta*, inhibitory queen pheromones could therefore be signals to workers facilitating the maximization of colony efficiency (Keller and Nonacs 1993). Second, colonies with many queens could produce fewer sexuals because they preferentially reproduce by colony budding. Third, queens may have evolved a lower sexual output per queen under polygyny as an energy saving, if this is more than offset by earlier sexual production (see above), or a higher survivorship of polygynous colonies and of recruited daughter queens (e.g. Herbers 1993; Keller 1993b; Section 8.4). Therefore reproductivity effects in polygynous colonies may not be inevitable costs of polygyny, as they are sometimes regarded, but evolved responses to it (Elmes and Keller 1993; Keller and Nonacs 1993).

MULTICOLONIAL AND UNICOLONIAL SECONDARY POLYGYNY

Although secondarily polygynous species share features distinguishing them from monogynous ones (Table 8.1), they themselves are a heterogeneous group. Some have fairly low queen numbers per colony, moderate to high within-colony relatedness, and clearly-defined, separate colonies (multicoloniality) (Table 8.2). Examples are *Leptothorax* and some *Myrmica* species (e.g. Douwes et al. 1987; Seppä 1994). But others have very many queens, low relatedness, and populations in which colony boundaries are blurred (unicoloniality) (Table 8.2). Examples are species of *Formica*, *Linepithema* (= *Iridomyrmex*), *Monomorium*, *Pheidole*, and *Wasmannia* (Hölldobler and Wilson 1977, 1990 p. 207). Multicolonial and unicolonial societies do not fall into totally distinct categories, but represent opposite ends of a spectrum of continuous variation. Within each type, separate traits are interconnected. For example, multicoloniality with low queen numbers and recruitment of mainly daughter queens clearly entails appreciable within-colony relatedness, whereas unicoloniality with many queens and adoption of non-relatives causes relatedness to fall.

Table 8.2

Traits of Two Types of Secondarily Polygynous Society

Multicolonial Type	Unicolonial Type
Moderate queen numbers on average (1–10 per colony)	High average queen numbers (10–100 or more per nest)
Moderate worker numbers per colony	Very large worker numbers per nest
Facultatively polygynous	Obligately polygynous
Some colony budding	Frequent nest budding
Mating near or far from nest	Mating often in nest
Some independent colony foundation	Nest foundation mostly dependent (adoption or budding)
Moderate to high between-queen relatedness	Frequently very low between-queen relatedness
Monodomous or polydomous	Polydomous (by definition)
Unrelated queens rarely adopted	Unrelated queens often adopted
Appreciable nestmate recognition	Almost absent nestmate recognition (by definition)

NOTE: For source references, see Table 8.1 and Sections 8.2, 8.5.

The phenomenon of unicoloniality has long been recognized (e.g. E.O. Wilson 1971 p. 457). However, the occurrence of two types of secondary polygyny is worth stressing (Rissing and Pollock 1988; Herbers 1993). For one thing, failure to distinguish the two classes has lead to confusion about the importance of kin selection in the evolution of polygyny. Several early allozyme analyses of polygynous ants found low within-colony relatedness levels (e.g. Pamilo 1982c; Pearson 1982), and these reinforced the view that polygyny represented a difficulty for kin selection theory (Section 8.1). But these studies were mostly of species tending to unicoloniality. Multicolonial species have moderate to high relatedness levels (e.g. Douwes et al. 1987; Stille et al. 1991). Therefore, polygyny *per se* need not be a problem for kin selection (e.g. Nonacs 1988), although unicoloniality might be (Section 8.5).

Moreover, the two sorts of secondarily polygynous society have different ecologies (Sections 8.4, 8.5). This means that separate environmental factors almost certainly underlie their evolution. Many unicolonial species thrive in new or disturbed areas, displacing other species (e.g. E.O. Wilson 1971 p. 451; Section 8.5). By contrast, multicolonial species seem to occupy more stable habitats, and are not noticeably good colonizers or competitors.

8.3 Evolution of Foundress Associations

In several ant species, young queens form groups in which they start a colony together. These are termed foundress associations, and communal colony foundation is referred to as pleometrosis. Examples include *Acromyrmex versicolor* (Rissing et al. 1989), *Lasius flavus* (Waloff 1957), *L. niger* (Waloff 1957; Sommer and Hölldobler 1992a), *L. pallitarsis* (Nonacs 1990, 1992), *Messor* (= *Veromessor*) *pergandei* (Rissing and Pollock 1986, 1987), *Myrmecocystus mimicus* (Bartz and Hölldobler 1982), and *Solenopsis invicta* (Tschinkel and Howard 1983). Foundress associations frequently mature into monogynous colonies through the death of all but one queen. This section examines why this occurs, and why foundress associations arise in the first place (cf. Pollock and Rissing 1988; Rissing and Pollock 1988; Strassmann 1989; Hölldobler and Wilson 1990 p. 217; Passera et al. 1991; Heinze 1993a; Herbers 1993; Itô 1993 p. 114).

FEATURES OF FOUNDRESS ASSOCIATIONS

An important feature of foundress associations is that queens in the same association are not related (e.g. Mintzer and Vinson 1985; Rissing and Pollock 1986; Hagen et al. 1988; Rissing et al. 1989), with *Lasius pallitarsis* being a possible exception (Nonacs 1990). In most cases, therefore, kin selection cannot explain why foundress associations form (Hagen et al. 1988; Pollock and Rissing 1988). A second trait is that all queens in a group typically lay eggs (e.g. Waloff 1957; Bartz and Hölldobler 1982; Mintzer and Vinson 1985; Rissing and Pollock 1986). This indicates that all queens can potentially become the colony queen. A third key feature is that, as just mentioned, foundress associations usually develop into monogynous colonies. This happens because, when the first workers emerge, the queens fight with each other until one is left (reviewed by Heinze 1993a). Alternatively, the workers themselves eliminate all queens but one, either by attacking them or by selective starvation (e.g. Bartz and Hölldobler 1982; Rissing and Pollock 1987; Nonacs 1990, 1992). Exceptions that show no reduction in queen number include *Iridomyrmex purpureus* (Hölldobler and Carlin 1985), *Atta texana* (Mintzer and Vinson 1985; Mintzer 1987), and *Acromyrmex versicolor* (Rissing et al. 1989). These species therefore represent rare cases of primary polygyny, although why they fail to revert to monogyny is unclear.

A CONDITION FOR THE EVOLUTION OF FOUNDRESS ASSOCIATIONS

Since most foundress associations become monogynous, a queen in an association of N queens has an average chance, other things being equal, of $1/N$ of becoming the colony queen. However, the success of a queen depends on the product of this probability and group survivorship. So, if group survivorship is much higher than that of singletons (more than N times higher), joining a foundress association is favored (Hamilton 1964b; Hölldobler and Wilson 1977; Bartz and Hölldobler 1982).

This simple observation is the key to understanding why foundress associations evolve, because – as will be seen – this occurs when queen groups have very high relative survivorship. It also shows that the lack of relatedness of associating queens does not represent a challenge to kin selection. The condition for a queen to join is that, on average, the benefit falls on herself (she has a higher expectation of offspring production by joining than by attempting to nest alone). So there is no true altruism. A possible exception occurs in *Acromyrmex versicolor*, where colony founding is, unusually, nonclaustral and one queen per group performs potentially risky foraging tasks (Rissing et al. 1989). As mentioned above, foundress associations in *A. versicolor* also fail to revert to monogyny, perhaps reducing the costs experienced by the foraging queen. However, although the foraging queen may on average increase her expectation of offspring (because colonies without foragers fail) (Dugatkin and Reeve 1994), it is not fully clear why any single queen acts with such apparent selflessness.

Several authors, noting the absence of relatedness among queens, and the partial subordinance of their individual interests to the good of the group, have interpreted foundress associations as clear examples of group selection (e.g. D.S. Wilson 1990; Dugatkin et al. 1992; Mesterton-Gibbons and Dugatkin 1992; but see Dugatkin and Reeve 1994). In particular, D.S. Wilson (1990) viewed the foraging queen in *Acromyrmex versicolor* as a "weak altruist" (Section 2.2). However, as these authors themselves made clear, a group selectionist view of foundress associations does not imply that queen selfishness is wholly absent, only that it is constrained by its effects on the group as a whole (cf. Sections 2.2, 2.3).

There are several ways in which queens might act selfishly. First, joining queens could be sensitive to the number of queens already in the group (e.g. Krebs and Rissing 1991). This is expected if some group sizes combine the highest survivorship and productivity benefits with the greatest chance of a group member becoming the eventual colony queen. In *Myrmecocystus mimicus*, the most common group size equals that expected from this argument (Bartz and Hölldobler 1982). Second, queens might selfishly withhold their stored energy reserves when feeding brood if this increased their individual longevity (Pollock and Rissing 1988). Evidence that this could be important is that, in *Lasius flavus*, queens in groups avoided the catastrophic loss of body weight experienced by single queens (Waloff 1957). Third, Nonacs (1989, 1990, 1992, 1993c) argued that queens should choose partners less likely than themselves to survive to head the mature colony. In support of this, a dynamic programing model showed that queens should be more influenced by competitive quality than kinship in partner choice (Nonacs 1989). This helps explain why associating queens are usually unrelated. In addition, experiments showed that *Lasius pallitarsis* queens were more likely to join a smaller (and therefore less competitive) partner than a larger one (Nonacs 1992).

FACTORS PROMOTING GROUP SURVIVORSHIP

A number of factors contribute to the high relative survivorship of foundress associations. For example, coexisting *Lasius flavus* queens groom each other and so prevent fungal infections that kill many solitary queens (Waloff 1957). In leaf-cutter ants, queens may be selected to associate because not all individuals carry a viable supply of the symbiotic fungus required for colony growth (Mintzer and Vinson 1985; Mintzer 1987). In weaver ants, only groups of queens can readily construct the nests of bound leaves that these ants inhabit (Peeters and Andersen 1989).

However, the factor that most frequently underpins the superior survival of foundress associations is clumping of incipient colonies (e.g. Tschinkel and Howard 1983; Rissing and Pollock 1987; Rissing et al. 1989). This occurs because colony-founding queens prefer areas unoccupied by mature colonies, which kill strangers entering their territories (e.g. Bartz and Hölldobler 1982; Tschinkel and Howard 1983). It also occurs when only some areas are physically suitable for colony foundation. For example, colony-founding *Solenopsis invicta* queens seek raised areas protected from flooding (Tschinkel and Howard 1983). Desert ants, which form a large fraction of pleometrotic species (e.g. Rissing and Pollock 1988; Herbers 1993), prefer to found colonies in shady patches (Rissing and Pollock 1988). In *Lasius pallitarsis*, Nonacs (1992) showed experimentally that queens were more likely to form groups when they were crowded.

Why does clumping of incipient colonies promote pleometrosis? The answer is that, when incipient nests are clumped, more colonies are started within patches than can eventually live there. This causes strong intraspecific competition for space. The colonies most likely to succeed are those that can monopolize a large territory by producing the greatest number of workers in the shortest time (Bartz and Hölldobler 1982). By pooling their stored energy reserves, queens in groups can achieve more rapid or more plentiful worker production than single queens (e.g. Waloff 1957; Bartz and Hölldobler 1982; Mintzer 1987; Nonacs 1990; Rissing and Pollock 1991; Sommer and Hölldobler 1992a; Tschinkel 1993b). Therefore clumping favors pleometrosis. In experimental colonies of *Solenopsis invicta*, the greater initial worker production of grouped queens also led directly to earlier sexual production (Vargo 1988).

Several pleometrotic species show intraspecific brood raiding, in which colonies steal workers as brood from their neighbors (e.g. Bartz and Hölldobler 1982; Rissing and Pollock 1987; Rissing et al. 1989; Sommer and Hölldobler 1992a; Tschinkel 1992a,b). This behavior could have originated as another way for young colonies to grow rapidly. It is also self-reinforcing. Since small colonies are most vulnerable to being raided (e.g. Tschinkel 1992a), raiding itself promotes early worker production and hence even more strongly developed raiding, and further pressure for pleometrosis (Bartz and Hölldobler 1982; Rissing and Pollock 1991; Tschinkel 1992b). Nonacs (1993c) also suggested that colonies may voluntarily submit to being "raided" because their queens have a chance of surviving to head the host colony. Certainly, colonies sometimes accept captured queens, and "raids" are often rather peaceful (e.g. Tschinkel 1992a,b).

Transition to Monogyny

One possible reason why most foundress associations become monogynous is that the lack of relatedness among queens makes the group unstable (Strassmann 1989). But the main reason seems to be that foundress groups change from being closed to open energy systems as they develop (Rissing and Pollock 1986). In claustral groups of queens, all available energy is in the queens' stored body reserves (closed system). A dominant queen would therefore endanger her own survival, since effort spent on aggression would reduce the energy available for rearing workers early. However, when the first workers mature and start to forage, energy enters the group from outside (open system). Lost energy can now be replaced, so nothing prevents the cooperation among queens from collapsing. This argument also explains why the reversion to monogyny usually starts when the first workers emerge.

In some species, queens fight among themselves (e.g. Waloff 1957;

Rissing and Pollock 1987; Sommer and Hölldobler 1992a). However, if queens are going to fight, each matriline (sibship) of workers should be selected to ensure that its mother survives. This is presumably why workers themselves sometimes actively eliminate queens (e.g. Bartz and Hölldobler 1982). In fact, one might expect from their greater power that workers should usually control which queen survives (cf. Section 7.5). Forsyth (1980) presented a model in which workers kill all but the most productive queen. Surprisingly, it is not known if this actually happens. However, when offered a set of queens, *Solenopsis invicta* workers killed those least swollen with eggs (Fletcher and Blum 1983). In addition, in foundress associations of *Myrmecocystus mimicus* and *Lasius pallitarsis*, heavier queens had the highest survival (Bartz and Hölldobler 1982; Nonacs 1990). A system in which only the most productive queen is spared would be stable from the workers' viewpoint, because it ensures that the average worker retains its own mother (Bartz and Hölldobler 1982). Forsyth's (1980) scheme might also select for queens that advertised their fecundity to workers, yielding another possible example of queen signaling (Sections 7.4, 7.6).

In sum, foundress associations are usually the product of specific ecological conditions (that cause clumping of incipient colonies). However, their stability is also affected by internal factors, namely relatedness (or the lack of it), and access to outside resources. As will be seen, both external and internal factors also affect the evolution of secondary polygyny (Sections 8.4, 8.6).

8.4 Evolution of Multicolonial, Secondary Polygyny

Multicolonial, secondary polygyny occurs when colonies acquire extra queens by adoption, but remain as separate entities (Section 8.2). It is a common type of social organization and its evolution has been discussed many times before (e.g. Buschinger 1974a; Hölldobler and Wilson 1977, 1990; Rissing and Pollock 1988; Passera et al. 1991; Herbers 1993; Keller 1993a; Bourke and Heinze 1994).

It is usually thought that monogyny is the primitive social system of ants, and secondary polygyny a derived trait (e.g. Hölldobler and Wilson 1977; Fletcher and Ross 1985). However, this is not proven, and other authors have suggested either that polygyny (Hamilton 1972) or within-species variation in queen number (Nonacs 1988; Ward 1989; Ross and Carpenter 1991b) is the primitive condition. Monogyny nevertheless represents the most stable type of ant society, since polygyny risks breaking down due to reproductive competition among queens

(Hölldobler and Wilson 1977). So this section takes the conventional approach of discussing reasons why polygyny evolves from monogyny, rather than the other way round.

A Condition for the Evolution of Secondary Polygyny

The basic axiom for understanding secondary polygyny is that a young queen should be selected to seek adoption if its expected fitness gain from doing so exceeds that from attempting to found a colony alone. Put another way, joining an existing colony is favored when nesting alone is relatively costly. This condition applies to the evolution of all types of communal breeding, including pleometrosis (e.g. Vehrencamp 1983a; Sections 3.2, 8.3).

If this condition holds, the question is whether the resident parties (existing queens and workers) should admit the young queen. Nonacs (1988) and Pamilo (1991c) presented kin selection models showing that, assuming young queens seeking adoption are daughters, the residents would be selected to accept them under some circumstances. These include when queens are short lived, colonies long lived, and polygyny does not greatly reduce productivity (Nonacs 1988). The result of adopting daughter queens is, of course, multicolonial, secondary polygyny. Therefore, provided the average adoptee is a relative, polygyny is in principle fully compatible with kin selection theory, contrary to early speculation in the literature (Section 8.2)

Nonacs (1988) and Pamilo (1991c) also predicted kin conflict between the resident queens and workers over accepting daughters (Section 7.5). Equally, there is a potential kin conflict between the residents and prospective daughter adoptees. Specifically, at some level of cost to nesting alone, daughter queens would favor joining their home colony, but residents would be selected not to admit them (Seger 1993). This is because each daughter is more closely related to herself than to the residents. At this level of cost, adoption would presumably not occur, because residents have the practical power to refuse queens admission. This is supported by findings that habitually monogynous species reject extra queens if they are experimentally offered to them (reviewed by Baroni Urbani 1968; Section 7.5). If residents do have the power to repel queens whose presence would be costly to them, the idea that adopted queens are intraspecific social parasites of polygynous colonies (Elmes 1973a, 1989, 1991) cannot be universally correct. A low proportion of queens may act parasitically (Section 8.6), but universal parasitism is inherently unstable.

When the cost of nesting alone rises beyond a certain threshold value, all parties should favor daughter adoption (Seger 1993). This is because a colony should adopt daughter queens that otherwise have no expectation of reproductive success. Pamilo (1991c) also found that workers should accept new queens more readily as queen number rises, since the dilution of workers' relatedness with sexuals caused by adoption grows proportionately less. Therefore, under some circumstances, the evolution of polygyny could be a self-reinforcing, runaway process.

Young queens conceivably obtain direct benefits from joining their home colony, as well as avoiding the costs of nesting alone. For example, joining permits a queen to achieve an earlier age of first sexual production than a queen on her own (Keller and Passera 1990; Section 8.2). In addition, as discussed later, polygyny may be favored because extra queens boost colony productivity or survivorship. There is presumably a trade-off between these benefits and the strength of the "reproductivity effects" (Section 8.2) observed in polygynous colonies (cf. Wenzel and Pickering 1991; Herbers 1993). Therefore, as argued earlier, these effects cannot be taken as automatic costs of polygyny. Instead, they could partly be evolved consequences of it (Nonacs 1988; Keller and Passera 1990; Elmes and Keller 1993; Section 8.2).

In sum, the factors that promote secondary polygyny are those that make solitary colony foundation costly and those that boost the productivity or survivorship of polygynous colonies. These factors can be further classified as ecological and social or genetic factors. These are now reviewed in turn.

ECOLOGICAL FACTORS PROMOTING SECONDARY POLYGYNY

Many ecological promoters of polygyny have been suggested to exist (e.g. Hölldobler and Wilson 1977; Herbers 1993), but some have limited applicability. These include species rarity, which conceivably selects for multiple queening because rising queen numbers increase effective population size (E.O. Wilson 1963; Hölldobler and Wilson 1990 p. 212); fragile nest sites, which could promote polygyny because polygyny minimizes the risk of queen loss in nest fragments (Hölldobler and Wilson 1977, 1990 p. 213); and interspecific social parasitism, which could select for polygynous colonies if these are better at resisting social parasites than monogynous ones (Rosengren and Pamilo 1983; Herbers 1986b). However, these factors lack generality: many polygynous species are common, inhabit stable nest sites, and occur in unparasitized populations (Herbers 1986b, 1991, 1993; Yamaguchi 1992; Bourke and Heinze 1994).

Instead, the main ecological factors promoting polygyny are those that make attempting to found colonies alone relatively costly (Crozier 1979; Rosengren and Pamilo 1983; Herbers 1993; Rosengren et al. 1993). These are referred to as ecological constraints on solitary breeding. The idea of ecological constraints has been used to explain communal breeding in many types of animal, including those with eusocial or helper-based societies (e.g. Emlen 1982; Vehrencamp 1983a; Koenig et al. 1992; Section 1.4). This illustrates both the generality of the idea, and the similarity of the forces affecting eusocial and polygynous evolution (Keller and Vargo 1993; Section 8.1).

In ants, founding a colony alone generally entails dispersal from the nest for the mating flight, followed by colony foundation itself. Therefore, factors that decrease survival during either of these phases will promote polygyny. An example of a factor operating during dispersal is heavy predation of sexuals during and after nuptial flights (e.g. Crozier 1979; Rosengren and Pamilo 1983; Bolton 1986; Herbers 1993). Another is the degree of habitat patchiness. Say a species only inhabits small patches scattered over a larger area of unsuitable habitat. Then young queens leaving a patch risk failing to locate another one, which will increase their dispersal costs (Rosengren and Pamilo 1983; Heinze 1992; Rosengren et al. 1993; Bourke and Heinze 1994; Section 8.6).

A major factor decreasing the probability that a queen, having dispersed, successfully founds a colony alone is habitat saturation or nest-site limitation (Herbers 1986c). This refers to the situation when most nest sites are already occupied. Evidence for its effects comes from Herbers' (1986c) finding that adding artificial nests to a population of *Leptothorax longispinosus* decreased the average queen number per nest. High absolute densities of existing colonies may similarly discourage solitary colony foundation (e.g. Nonacs 1993a). Some effects of climate also make solitary colony foundation harder. Several authors have suggested that a cold winter climate is more likely to kill hibernating queens if they are alone than if they are in a colony (Heinze and Buschinger 1988; Satoh 1989; Heinze 1992). Workers certainly seem to survive hibernation better in groups, although the exact reasons for this are unclear (Heinze 1992, 1993c; Heinze and Hölldobler 1994). Evidence also exists that whole colonies suffer less mortality during overwintering as queen number rises in leptothoracine ants, although the mechanism is again unknown (Herbers 1986b, 1993). In support of these climatic effects, the degree of polygyny across species and populations tends to rise with increasing latitude and altitude, at least in leptothoracines (Heinze and Buschinger 1988).

ECOLOGICAL FEATURES OF SECONDARILY POLYGYNOUS ANTS

If ecological factors promote polygyny, there should be consistent differences in the ecology of polygynous ants and their monogynous relatives. Surprisingly, little systematic attention has been paid to whether such differences exist. One reason is that factors such as cost of dispersal and habitat patchiness are hard to quantify (Keller and Vargo 1993). In addition, queen number can be tremendously variable at colony, population, and species level. It may also vary over time within populations (e.g. Elmes 1987a; Elmes and Petal 1990; Herbers 1990; Elmes and Keller 1993). These features make it difficult to characterize the gyny level of a population.

Existing data on the ecological correlates of multicolonial, secondary polygyny are reviewed below. Previous reviews of the ecology of polygyny are Brian (1983 p. 251), Hölldobler and Wilson (1977), Herbers (1993), and Rosengren et al. (1993). Some monogynous species or forms have a polygynous sibling species or form, but it is usually unicolonial. So the ecology of these pairs is discussed in Section 8.5.

1. *MYRMICA*

All species of *Myrmica* are potentially secondarily polygynous, in that all can recruit queens to varying degrees (Elmes 1980, 1991; Elmes and Keller 1993). Nevertheless, some species or populations have a very low average number of queens per colony and are effectively monogynous, whereas others are characteristically moderately or highly polygynous (Elmes 1980). In a few cases, differences in queen number are correlated with habitat. For example, colonies of *M. scabrinodis* on grassland contain more queens per worker than those on moorland (Elmes and Wardlaw 1982a). In a review, Elmes and Keller (1993) concluded that queen number in *Myrmica* is generally responsive to ecological factors, especially climatic ones (see below). However, the exact adaptive reasons for this responsiveness remain unclear.

Another puzzle is the cycle in queen numbers found in *Myrmica sulcinodis* (Elmes 1987a; Figure 8.1; Section 10.3). The causes of this phenomenon have not been resolved (Elmes 1987a). By contrast, in *M. rubra*, annual fluctuations in queen number over eighteen consecutive years were found to depend on annual variation in mean late summer temperature (Elmes and Petal 1990). In particular, if the late summer temperature in one year differed substantially from the average across years, queen number rose in the autumn of that year. Again, however, the reason why this effect occurred remains unclear (Elmes and Petal 1990).

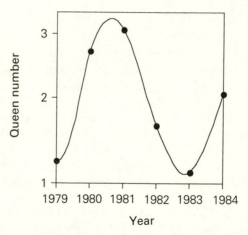

Figure 8.1 Cyclical variation in queen number (logarithmic scale, corrected for colony size) in two populations (combined data) of *Myrmica sulcinodis* in southern England. The curve is a fitted sine wave. The two populations, though 15 km apart, were in phase, and have continued to cycle since the phenomenon was first described (Elmes and Petal 1990; Elmes and Keller 1993). (From Elmes 1987a, by permission of Blackwell Scientific Publications Ltd)

2. *LEPTOTHORAX*

In *Leptothorax*, as already mentioned, polygynous species appear more prevalent as altitude and latitude rise. For example, the characteristically monogynous subgenus *Myrafant* is best represented in mild, Mediterranean habitats (Heinze and Buschinger 1988). This is consistent with high overwintering mortality promoting polygyny (Bourke and Heinze 1994). In addition, *Leptothorax* species typically nest in cavities, and their populations can be very dense (e.g. Heinze 1993b). Therefore, in northern areas the effects of nest site limitation (Herbers 1986c) probably add to those of the cold climate in selecting for polygyny. Lastly, facultatively polygynous *Leptothorax* species tend to occur in extensive, uniform habitats, whereas functionally monogynous ones, including some with wingless queens, prefer habitats of small, scattered patches (Heinze 1992, 1993b). This is as expected from the argument that very high costs of dispersal select for greater nondispersal and, as explained later, higher reproductive skew (Reeve and Ratnieks 1993; Bourke and Heinze 1994; Section 8.6).

3. *FORMICA*

In Finland, the wood ant *Formica rufa* is monogynous and monodomous and occurs in isolated nests on forest margins, whereas the highly polygynous and polydomous *F. aquilonia* and *F. polyctena* are typical of extensive uniform spruce forests (Rosengren and Pamilo 1983). Rosengren et al. (1993) proposed that, as the duration of occupancy of a habitat increases, so do nest density and the degree of habitat saturation, making polygyny more likely. Since spruce forests represent long lived and densely occupied habitat (Rosengren and Pamilo 1983), this could explain why they are inhabited by the polygynous *Formica*. Another possible example of this effect is *F. truncorum*. This species occurs in monogynous and polygynous populations on seemingly uniform islands in the archipelago off southwest Finland, but the polygynous populations appear to have been established for longer (Sundström 1993a).

4. PONERINE ANTS

Ants of the ponerine subfamily exhibit a remarkably complex set of social systems (reviewed by Peeters 1991a,b, 1993; Ito and Ohkawara 1994). In some, sexual reproduction is by specialized queens, which are either standard morphological queens (dealates), or wingless queens with a worker-like appearance (ergatoids). In others, a specialized queen caste is absent and only unspecialized, mated workers (gamergates) reproduce sexually. In others again, sexual reproduction is by both dealate queens and gamergates, although not necessarily in the same colonies. Species with dealate queens may be monogynous or polygynous. By contrast, species with ergatoid queens alone are always monogynous, whereas gamergate-only species are again either monogynous (there is one gamergate per colony) or polygynous (with several gamergates per colony) (Peeters 1991a, 1993; Sommer and Hölldobler 1992b; Peeters et al. 1994).

As in previous cases, little is known about the ecological causes of this variation. However, an important correlate of the possession of either ergatoid queens or gamergates is reproduction by colony fission or colony budding (Peeters 1993). A number of nonponerine species living in arid habitats have a proportion of wingless queens and reproduce by colony budding. The likely explanation for this is that, in resource-poor habitats, producing large numbers of dispersing queens suitably equipped for independent colony foundation is difficult (e.g. Briese 1983; Bolton 1986; Heinze et al. 1992a; Tinaut and Heinze 1992). Similar reasoning could explain an association in ponerines between having gamergates and living in arid habitats (Ward 1983a; Crozier and Pamilo 1986).

In addition, even when colony foundation is independent in ponerines, it is rarely claustral, since colony-founding queens generally perform some foraging (Peeters 1993). Villet (1989) therefore argued that in species specializing on prey which cannot be handled by single ants, independent colony foundation would be difficult. So these species are particularly prone to exhibit colony fission and hence queen ergatoidy (Villet 1989). On the other hand, ergatoid queens of the ponerine genus *Plectroctena* proved capable of handling prey on their own (Villet 1991). Moreover, all the above ideas still leave unanswered the question of why ponerines with ergatoid queens are obligately monogynous. What ecological factors underlie gyny variation in species with gamergates is also unknown, and it has even been suggested that none exist (Peeters 1990, 1993).

In sum, despite the many ecological factors that are believed capable of promoting polygyny, little information exists on the actual ecological correlates of queen number variation. Some existing field data support current ecological theory for polygynous evolution, but more are undoubtedly needed.

SOCIAL AND GENETIC FACTORS PROMOTING POLYGYNY

Secondary polygyny would be favored if adding queens to a colony somehow enhanced its survivorship or productivity through genetic or social effects. Several ways in which this could happen have been proposed. For example, multiple-queen colonies of wasps have been suggested to recover from predation better than monogynous ones (e.g. Itô 1993 pp. 32, 50), and the same could be true in ants. Since polygyny occurs in some species but not in others, such benefits presumably assist the evolution of polygyny when this is favored by the prevailing ecological conditions. The main suggested benefits are as follows.

1. REDUCED VARIANCE IN EXPECTED REPRODUCTIVE SUCCESS

Wenzel and Pickering (1991) proposed a very general advantage to group living, which can also be applied to polygyny. Say there is an optimum level of average per capita productivity for social groups in an ant species. Then, as queen number rises, the variance about this average falls, because large samples drawn from a statistical distribution have lower variances than small ones. If natural selection favored greater predictability of individual success, and hence reduced variance, group living (polygyny) could be favored. This factor could interact with ecological ones, since it suggests that group living is promoted by variable environments (Wenzel and Pickering 1991). Rosengren et al. (1993)

similarly proposed that polygyny represents a form of evolutionary "bet-hedging."

2. GENETICALLY MORE DIVERSE WORKFORCE

Herbers (1982) suggested that polygynous colonies exhibit a more efficient division of labor than monogynous ones, because their workforce is genetically more variable. Greater efficiency might increase productivity. (Similar advantages were proposed for multiple mating by Crozier and Page [1985; Section 11.4].) Workers' genotypes are associated with variation in their tendency to perform different tasks (Stuart and Page 1991; Snyder 1992, 1993; Carlin et al. 1993), so it is plausible that having a greater genetic mix of workers increases efficiency. However, Herbers (1982) and Herbers and Cunningham (1983) did not find any obvious behavioral differences between the workforces of monogynous and polygynous *Leptothorax longispinosus* colonies. In addition, whether associations between task performance and genotype are adaptive is controversial (e.g. Tofts and Franks 1992; Carlin et al. 1993; Section 12.4). There is evidence that genetically mixed colonies are more productive in honey bees (Fuchs and Schade 1994), but it is unknown if this is true for ants.

3. GREATER RESISTANCE TO PARASITES

Ants are subject to infection by various fungi, protozoa, and other parasitic organisms (e.g. Péru et al. 1990; Elton 1991; Sanchez-Peña et al. 1993; review of Hölldobler and Wilson 1990 p. 554), although little is known of their effects on colony growth and survival. If these effects are severe, polygyny might have evolved partly to increase genetic variation for disease resistance within colonies. This was proposed by Shykoff and Schmid-Hempel (1991c) and elaborated by Gadagkar (1992) and Schmid-Hempel (1994), following the suggestion that multiple mating confers similar benefits (Hamilton 1987b; Sherman et al. 1988; Section 11.4). Together, these ideas predict that monogynous ants should be more prone to multiple mating than polygynous ones (in which multiple queens already make colonies genetically variable). In a survey, Keller and Reeve (1994b) confirmed this prediction: 13 of 17 monogynous ant species or genera were polyandrous, compared to 9 of 25 polygynous species or genera (see also Bartels 1985; Pamilo et al. 1994). This association is also consistent with the idea (above) that colonies increase their genetic variability to promote efficiency. In addition, levels of polygyny and the mating frequency have yet to be correlated with differing levels of exposure to parasites.

4. PROTECTION FROM EFFECTS OF DIPLOID MALE PRODUCTION

A final idea is that polygyny reduces fitness losses due to the production of sterile, diploid males. (Again, multiple mating has been proposed to confer similar advantages [Page 1980, 1986].) In the case of polygyny, this hypothesis has been developed mainly for the fire ant *Solenopsis invicta* in a series of papers by K.G. Ross and colleagues (Ross and Fletcher 1985a,b, 1986; Ross et al. 1987b, 1993; Ross 1988a, 1989, 1993).

Diploid males in social Hymenoptera represent a cost to the colony and arise either from inbreeding or from low allelic diversity at the sex-determining locus (Box 8.2). In North America, to which it was introduced from South America earlier this century, *Solenopsis invicta* occurs in a monogynous and a polygynous form (Ross and Fletcher 1985a; Section 10.2). In the polygynous form, diploid males occur at high frequencies, but in polygynous colonies from South America they are far less common. An absence of inbreeding among the North American ants, and low levels of variation at their enzyme loci, suggest that the cause of the high frequency of diploid males is low allelic diversity at the sex-determining locus due to a founder effect occurring upon introduction (Ross and Fletcher 1985b; Ross et al. 1993; Box 8.2).

Diploid male production is a major source of colony mortality among incipient monogynous colonies from the North American population (Ross and Fletcher 1986). This is because, if singly mating queens produce diploid males at all, half their diploid offspring are male (Box 8.2). For a singly founding queen, this causes colony failure due to a lack of workers. By contrast, in the polygynous form, queens producing diploid males are buffered by the presence of workers produced by nestmate queens, and therefore survive.

These findings suggest that polygyny evolved in North American *Solenopsis invicta* to minimize colony mortality due to the high level of diploid male production (Ross and Fletcher 1986). However, polygynous colonies also occur in the South American population (Ross et al. 1993). Therefore, the special conditions in North America may just have reinforced an existing tendency to polygynous evolution. Ecological factors could also favor polygyny in *S. invicta*, as the next section describes. Furthermore, even if it has operated, the influence of diploid male production in promoting polygyny in *S. invicta* stems from the unusual history of this species in North America, and is unlikely to be general. Diploid males appear rare in most ant species, and are not particularly associated with polygyny (Ross and Fletcher 1985b; Pamilo et al. 1994; Box 8.2).

BOX 8.2 SEX DETERMINATION AND DIPLOID MALES IN THE HYMENOPTERA

In several Hymenopteran species, sex is determined in the following way (reviewed by Crozier 1975, 1977; Bull 1983; Ratnieks 1991b; Cook 1993). There is a single sex-determining locus with multiple alleles (e.g. *a, b, c* and so on). Individuals that are heterozygous at the locus (e.g. *ab*) become females, whereas homozygotes (e.g. *aa*) or hemizygotes (haploids, e.g. *c*) become males. Under outbreeding, each female is likely to mate with a male not bearing either of her alleles (e.g. *ab* × c). All diploid offspring of such matings will be heterozygotes (*ac, bc*) and therefore females. So, in the outbred case, diploids are females and haploids are males. This is the familiar result of haplodiploid sex determination in the Hymenoptera.

However, if there is inbreeding, a female is more likely to mate with a male bearing one of her sex-determining alleles (e.g. *ab* × a). Half the diploid offspring of such a "matched mating" would be homozygotes (*aa*) and therefore diploid males. Diploid males are nonviable or sterile (Bull 1983; Ross and Fletcher 1985b; Cook 1993). So their production is an entirely wasted investment, representing a "genetic load" of the Hymenopteran sex-determination and breeding system (e.g. Ross et al. 1993). This explains why, in honey bees, any diploid males are cannibalized by workers as larvae (Woyke 1963). These larvae apparently produce a pheromone alerting workers to their presence (Crozier 1975). This could be an act of altruism, since it would benefit the diploid drones' viable relatives by sparing the hive the cost of raising them (Crozier 1975; Crozier and Page 1985). In ants, destruction of diploid males as brood cannot be universal, since they occur as adults and their rearing has been observed in the laboratory (e.g. *Solenopsis invicta*: Ross and Fletcher 1985b, 1986; Vargo and Fletcher 1986b). However, selective elimination of diploid male brood has been suspected in *Linepithema humile* (= *Iridomyrmex humilis*) (Keller and Passera 1993) and some *Formica* species (Pamilo et al. 1994).

A fall in the number of alleles at the sex-determining locus would also make matched matings more likely and so increase the frequency of diploid males. It is believed that, when *Solenopsis invicta* was introduced from South to North America, diversity at the sex-determining locus fell because, by chance, the small founder population lacked some of the parent population's alleles (a founder effect). This would account for the high level of diploid male production in introduced *S. invicta* (Ross et al. 1993) (Section 8.4).

In ants as a whole, diploid males occur widely (Page and Metcalf 1982; Bull 1983; Cook 1993; Pamilo et al. 1994), but usually at low frequencies (Ross and Fletcher 1985b). This is consistent with the general occurrence

---

BOX 8.2 CONT.

of outbreeding in ants (e.g. Gadagkar 1991a) and selection for high allelic diversity at the locus for sex-determination (Crozier 1975). In species with habitual sib-mating (e.g. the *Epimyrma* social parasites; Section 5.8), it is assumed that a sex determination system has evolved that does not involve heterozygosity (Buschinger 1989a).

There is a further complication to the evolution of the polygynous form of *Solenopsis invicta* (Ross 1992; Keller and Ross 1993a, b; Ross and Shoemaker 1993; Ross and Keller 1995). In polygynous colonies from North America, queens of a particular genotype (homozygotes for an allele, *a*, at the enzyme locus, *Pgm*–3) are executed by the workers before they have a chance to become reproductive. However, the allele persists in polygynous populations because queens from these populations mate with *a*-bearing males from the monogynous ones, in which *Pgm*–3^a/3^a queens are not eliminated. (This gene flow is reinforced by the relative scarcity of fertile, haploid males in the polygynous populations.) Paradoxically, *Pgm*–3^a/3^a queens from polygynous colonies proved to have above-average fecundity in laboratory tests (Ross 1992; Keller and Ross 1993a).

This strange system implies that worker execution of queens need not be strictly kin-selected (cf. Section 7.5). Keller and Ross (1993a) also suggested that it could help maintain polygyny in *Solenopsis invicta*, since the execution of the most fecund queens would prevent polygynous colonies being dominated by one or a few queens and so reverting to monogyny. In sum, owing to the enormous detail in which *S. invicta* have been studied, we know that its polygyny is a complex phenomenon. Although some of its features could be unique, caution is therefore needed in drawing conclusions from far less well-known species.

8.5 Evolution of Unicolonial Polygyny

Some polygynous ant species live in so-called unicolonial populations in which there is extensive intermixing of queens and workers between nests, with colony boundaries being virtually abolished (Section 8.2). Examples include members of the genera *Formica*, *Linepithema* (= *Iridomyrmex*), *Monomorium*, *Pheidole*, and *Wasmannia* (Hölldobler and Wilson 1977, 1990 p. 207). A characteristic trait of unicolonial populations is a very low level of within-nest relatedness (Table 8.2). For example, allozyme studies have found relatedness levels close to or indistinguishable from zero in *Formica aquilonia* (Pamilo and Rosengren 1983; Pamilo et al. 1992), *Linepithema humile* (= *Iridomyrmex humilis*) (Kaufmann et al. 1992), *Solenopsis invicta* (Ross and Fletcher 1985a; Ross 1993), and *Lasius neglectus* (Boomsma et al. 1990).

Unicolonial species therefore represent a problem for kin selection theory. Their workers exhibit reproductive altruism towards brood that is no more related with them than the average individual, which is a situation that kin selection theory specifically forbids (Section 1.4). One might therefore expect unicoloniality to be unstable. For example, mutations making female larvae more likely to become queens rather than workers should rapidly spread, leading to loss of the worker caste and presumably extinction (Crozier 1979). Alternatively, mutations for nepotistic brood care by workers should be favored, leading to a reversion to multicoloniality.

This problem has received surprisingly little attention. Or, rather, secondarily polygynous species in general have sometimes been mistakenly viewed as embarrassing to kin selection theory, when it is actually the unicolonial species that represent the difficulty (Sections 8.2, 8.4). Of authors that have discussed the problem, Crozier (1977, 1979) proposed a solution based on interdemic group selection, whereas Rissing and Pollock (1988) suggested that unicolonial populations might in fact have some degree of internal genetic structuring favorable to kin selection. This last suggestion does not seem borne out by the estimates of within-nest relatedness. However, such estimates typically have large standard errors, so it is almost impossible to prove a total absence of relatedness. Nonacs (1993b) remarked that unicoloniality remains a paradox for kin selection theory.

Another solution to the problem of unicoloniality is that it arises when lineages prone to evolve it occur in the right ecological conditions. Workers of unicolonial species seem disproportionately likely to be totally sterile (e.g. *Iridomyrmex*: Halliday 1983; *Linepithema* [= *Iridomyrmex*]: Keller 1988; *Monomorium*: Oster and Wilson 1978 p. 93; *Solenopsis*: Ross 1993; review of Passera 1994). This suggests that a key feature predisposing a species or population to evolving unicoloniality is an inability of workers to respond to the kin structure of the nest by laying male eggs (Section 7.4). The ecological conditions that promote unicoloniality in such species or populations seem to occur in new, empty, or disturbed habitats. Unicolonial species thrive in such habitats (e.g. E.O. Wilson 1971 p. 451; Hölldobler and Wilson 1977; Passera 1994). As pointed out by several authors, this is due to some of the other traits associated with unicoloniality. These arguably confer advantages on unicoloniality that, by offsetting its inherent instability, permit it to evolve at least over the short term. The traits include rapid colony budding, a generalist diet, high worker numbers per nest, high nest densities, and mass recruitment of workers to food sources (Passera 1994; Table 8.2). All these make unicolonial species very strong interspecific competitors (e.g. *Solenopsis invicta*: Tschinkel 1987b; Porter et al. 1988; Porter and Savignano 1990). They also contribute to the success of unicolonial species in colonizing urban habitats and new countries to which they have been introduced by commerce. Thus, many are notorious "tramp" and pest species (e.g. *Linepithema humile* [= *Iridomyrmex humilis*], *Monomorium pharaonis*: E.O. Wilson 1971 p. 451; Hölldobler and Wilson 1977, 1990 p. 215; Passera 1994; Section 10.6). For these reasons, unicoloniality has been considered the ecological analogue of vegetative propagation in plants (e.g. E.O. Wilson 1971 p. 457; Brian 1983 p. 252), in which lateral spreading also appears to be favored in disturbed habitats (e.g. Fahrig et al. 1994). Furthermore, if it involves the occurrence of short-term benefits (ecological success) at the expense of long-term stability, the evolution of unicoloniality might resemble the evolution of asexual reproduction in certain animals, at least as viewed by some models (e.g. Maynard Smith 1984). As in these cases of asexual reproduction, the involvement of short-term benefits that are eventually outweighed by long-term costs is supported by the patchy taxonomic distribution of unicoloniality. Some unicolonial species occur within predominantly monogynous taxa (e.g. *Lasius*: Yamauchi et al. 1981; Boomsma et al. 1990), and, in general, it appears that unicoloniality is a derived trait that has evolved sporadically and independently in several lineages.

Several pairs of species or forms exist in ants in which one member is monogynous and the other is polygynous (reviewed by E.O. Wilson 1971 p. 457; Hölldobler and Wilson 1977, 1990 p. 215; Brian 1983 p. 252; Ward 1989; Bourke and Franks 1991; Passera et al. 1991; Herbers 1993). In several cases, polygyny is clearly of the unicolonial type (e.g. *Lasius neglectus*, *L. sakagamii*, *Solenopsis invicta*), and in the others much of the available information suggests the same. Therefore, the ecological features of unicoloniality may be further examined by looking for ecological differences between the members of such pairs. Although the relevant data are incomplete, and some come from studies that did not specifically seek ecological correlates of queen number (Table 8.3), they confirm that in several cases unicoloniality is favored in urban or in naturally disturbed habitats (e.g. by rivers). On the other hand, *Camponotus yamaokai*, *Formica lugubris*, *F. pallidefulva*, and *Pseudomyrmex venefica* do not obviously fit this pattern (Table 8.3). In the case of mound-building *Formica* ants, this could be because they have not in fact evolved to exploit novel or disturbed habitats. With their large, costly, and long lived nests, these species seem much more likely to be pursuing the "space-perennial" life history proposed by Nonacs (1993a), in which each colony occupies the same nest (or set of nests) over many generations in long lived, stable habitats (Sections 9.9, 10.5). This strategy is associated with polygyny (Section 9.9), and so high queen numbers per nest and low within-colony relatedness may have arisen by a different route than that found in other unicolonial species. In short, unicoloniality in mound-building *Formica* species and in those like *Linepithema humile*, with low-cost, frequently-changed nests, could be ecologically and evolutionarily widely different phenomena. Returning to monogynous and polygynous pairs of ants, it is clear that a consistent ecological basis for their occurrence is not always evident, as Herbers (1993) also concluded. To this extent, unicoloniality remains a puzzle. On the other hand, as in previous cases, better ecological information is required for a full understanding of unicolonial evolution.

8.6 Evolution of Functional Monogyny, Queen Aggression, and a Stable Reproductive Skew

Functional monogyny occurs when several mated queens coexist in a nest, but only one lays eggs at any given time (Buschinger 1968a). Reproductive skew describes the allocation of reproduction among members of a society. High skew is when one or a few individuals monopolize reproduction, and low skew is when it is more evenly shared (Vehrencamp 1983a,b). Thus, functionally monogynous colonies show high skew. This section examines why functional monogyny evolves, why it is associated with queen aggression within nests, and why the stable skew might vary. Skew evolution is a problem of general interest, because it occurs in all types of animal society (Vehrencamp 1983a,b; Reeve and Ratnieks 1993; Keller and Reeve 1994a; Reeve and Keller 1995).

Functional Monogyny and Skew Evolution in Leptothoracine Ants

The known cases of functional monogyny are mainly in the leptothoracine ants (Francoeur et al. 1985; Heinze and Buschinger 1988; Heinze 1993a), with rare, sporadic occurrences in other groups (Table 8.4). As well as functional monogyny, leptothoracines also exhibit facultative polygyny (multicolonial populations contain both monogynous and sec-

Table 8.3

Ecological Features of Related Monogynous and (unicolonial) Polygynous Species or Forms

	Monogynous Species or Form (M)	Polygynous Species or Form (P)	Mixed or Separate Populations?	Habitat Differences?
a.	Camponotus nawai	C. yamaokai	Separate but adjoining	P form is more northerly
b.	Conomyrma flavopecta	C. insana	No data	No data
c.	Formica argentea	F. podzolica	Mixed	M species is in logs in woodland, P species in sand in riverine areas
d.	Formica exsecta	F. exsecta	Separate	No data
e.	Formica lugubris	F. lugubris	Separate	M colonies in Finland and Ireland, P colonies in Central Europe
f.	Formica pallidefulva nitidiventris	F. p. schaufussi var. incerta	Mixed	Absent
g.	Lasius alienus	Lasius neglectus	Separate	L. neglectus is in urban habitat
h.	Lasius niger	Lasius sakagamii	Separate	L. sakagamii is in sandy areas subject to floods
i.	Myrmica ruginodis macrogyne form	M. ruginodis microgyne form	Separate and mixed	Scotland: M form in felled woodland, pine plantation, heather; P form in fields and grassland (Brian and Brian 1955). British Isles: P form more in wet cool areas with oceanic climate; M form commonest one in continental Europe (Collingwood 1958). Japan: M form in woodland, P form by rivers (Mizutani 1981)
j.	Pseudomyrmex ferruginea	P. venefica	Separate	P species in drier areas
k.	Solenopsis invicta	S. invicta	Separate, but P populations are inside M ones	Absent, but P form spread after introduction

NOTES: The third column refers to whether monogynous and polygynous colonies occur together or in separate populations. The fourth column gives information on any habitat differences between the two types of colony or population.

References: a. Satoh 1989, 1991; Terayama and Satoh 1990. b. Nickerson et al. 1975. c. Bennett 1987b. d. Pamilo and Rosengren 1984. e. Pamilo et al. 1992. f. Talbot 1948. g. Boomsma et al. 1990; Van Loon et al. 1990. h. Yamauchi et al. 1981, 1982. i. Brian and Brian 1955; Collingwood 1958; Mizutani 1981; in Finland, Seppä (1992) found no habitat differences between populations differing in gyny. j. Janzen 1973. k. Ross and Fletcher 1985a; Ross et al. 1987b.

ondarily polygynous colonies), and obligate monogyny. Functional monogyny is associated with a habitat of small, scattered patches, and facultative polygyny with extended, uniform habitats (Heinze and Buschinger 1987, 1989; Heinze 1989, 1992, 1993a,b). Bourke and Heinze (1994) recently explained these associations with a combination of ecological constraints hypotheses, game theory models of dispersal, and kin selection models of skew evolution. Their arguments are now summarized.

The starting point is the idea that multiple queening is promoted by high ecological constraints (high costs of dispersal and colony foundation for single queens) (Section 8.4). Therefore, as these costs increase, so should the propensity for daughter queens to seek adoption. In leptothoracines, major contributors to such costs are nest site limitation (Herbers 1986c), cold climate (Heinze 1992, 1993c; Heinze and Hölldobler 1994), and habitat patchiness (Heinze 1992). This last factor is effective because in a habitat of small, scattered patches, dispersing queens that leave a patch risk not finding another one (Section 8.4).

Game theory models predict that sedentary organisms will always

Table 8.4

Functionally Monogynous Species of Ants

Species	References and Notes
Leptothoracines	
Formicoxenus hirticornis	Buschinger 1979
Formicoxenus nitidulus	Buschinger and Winter 1976
Formicoxenus provancheri	Buschinger et al. 1980; Heinze et al. 1993b
Leptothorax acervorum	In Japan and Alaska: Ito 1990, Heinze and Ortius 1991. In Europe, *L. acervorum* is facultatively polygynous (Buschinger 1968a).
Leptothorax gredleri	Buschinger 1968a; Heinze et al. 1992b
Leptothorax species A	Heinze and Buschinger 1987, 1989; Heinze 1989, 1993b; Heinze and Smith 1990
Leptothorax sphagnicolus	Francoeur 1986; Buschinger and Francoeur 1991
Others	
Apterostigma dentigerum	Some colonies: Forsyth 1981
Myrmecina graminicola nipponica	Some colonies: Ohkawara et al. 1993
Pachycondyla sp.	Ito 1993a

NOTES: See reviews of Francoeur et al. (1985) and Heinze (1993a). *Formicoxenus* species are so-called "guest ants" that live inside the nests of other ant species (Francoeur et al. 1985). Tschinkel and Howard (1978) argued for functional monogyny in the monogynous form of *Solenopsis invicta*, but Vargo (1990) suggested they were mistaken.

produce some dispersing individuals among their offspring (Hamilton and May 1977). In other words, total nondispersal is not an ESS (evolutionarily stable strategy). This is because a population adopting a strategy of total nondispersal could be invaded by mutant individuals with a mixed strategy of dispersal and nondispersal: mutants could occupy sites occupied by the nondispersers, but nondispersers could not (by definition) occupy sites occupied by mutants, so eventually the whole population would become mixed strategists. These models also predict that – as intuition suggests – a rise in the mortality of dispersers among mixed strategists brings about an increase in the stable fraction of nondispersers (Hamilton and May 1977).

Applying these findings to leptothoracines suggests that species in extended uniform habitats with moderate costs of attempting solitary colony foundation should exhibit multiple queening, but with a relatively high fraction of dispersing (nonadopted) queens. By contrast, species in patchy habitats should show more pronounced multiple queening with a higher proportion of nondispersing, adopted young. These traits will also affect levels of within-colony relatedness. Assume that some queens disperse and then seek adoption by unrelated colonies (there is evidence for this in *Leptothorax acervorum* [Stille and Stille 1992]). Then, the greater the level of dispersal, the lower the level of within-colony relatedness compared to when the average adoptee is more likely to be a daughter queen (Bourke and Heinze 1994).

Both the severity of ecological constraints and the level of relatedness should affect the stable reproductive skew. This has been demonstrated by general models of skew evolution (Vehrencamp 1983a,b; Reeve and Ratnieks 1993; Keller and Reeve 1994a). Societies of multiple breeders are unstable if subordinate individuals could receive greater payoffs from attempting to breed alone. Therefore, a high skew should be associated with either severe ecological constraints, or high relatedness, or both. This is because a subordinate individual tolerates lower personal reproduction in the group if its expected reproductive success on its own is small, and if its kin-selected payoff via a related dominant's reproduction is high. Reeve and Ratnieks (1993) also predicted that high skew should be associated with high within-colony aggression. Under conditions of high skew, the payoff from becoming a dominant is also high (by definition). So subordinates should always be aggressively probing the dominants, which in turn should be constantly repressing the subordinates. By contrast, if skew is low, there is no great gain in becoming a (notional) dominant, so coexistence should be peaceful.

Combining all these arguments predicts that extended uniform habitats with moderate ecological constraints should be characterized by multiple-queening with (1) a mixed dispersal strategy with relatively many dispersers, (2) moderate within-colony relatedness, (3) low reproductive skew, (4) low within-colony aggression. By contrast, patchy habitats should promote multiple queening with (1) a mixed dispersal strategy with relatively many nondispersers, (2) high relatedness, (3) high skew, (4) high aggression (Bourke and Heinze 1994).

These associations of traits are shown by facultatively polygynous and functionally monogynous leptothoracines respectively (Bourke and Heinze 1994). In particular, the functionally monogynous (high skew) species inhabit small-patch habitats such as nests of other ant species (in the case of the "guest ant" genus *Formicoxenus*), forest edges, rocky outcrops, and moss hummocks (Heinze 1992, 1993a,b). Of the functionally monogynous ants in Table 8.4, the *Formicoxenus* species, *Leptothorax* species A, and *L. sphagnicolus* all have a fraction of wingless and therefore nondispersing queens (intermorphs) (Francoeur et al. 1985; Heinze and Buschinger 1987, 1989; Buschinger and Francoeur 1991). In addition, the *Formicoxenus* species, *Leptothorax* species A, and *L. gredleri* exhibit a high degree of within-colony aggression, with

Table 8.5

Measures of Reproductive Skew in Polygynous Ants

Species	Skew Index (range)	References and Notes
Leptothorax acervorum	0.12 (0.01–0.19)	Mean (range) for four colonies based on egg-laying rates (Bourke1991; Bourke and Heinze 1994)
Myrmica rubra	0.01 (–)	Value for one colony based on egg-laying rates (Brian 1986 Table VIII)
Solenopsis invicta	0.35 (0.12–0.51)	Mean (range) for six colonies based on analysis of parentage of daughter queens using allozyme data (Ross 1988b Table 3)

NOTE: The skew index was devised by H.K. Reeve and L. Keller (in Keller and Vargo [1993] and Reeve and Ratnieks [1993]) and equals $(N_b v + N_n)/(N_b + N_n)$, where N_b is the number of breeders in the colony, N_n is the number of nonbreeders, and v is the ratio, observed variance among breeders in proportionate share of reproduction within colony : maximum value of this variance. Hence v equals $\Sigma(p_i - 1/N_b)^2/(N_b - 1) : (1/N_b)$, with p_i being the fraction of offspring produced by the ith breeder. This index varies between 0 (no skew) and 1 (maximum skew, i.e. one individual produces all offspring).

an alpha queen that aggressively dominates her subordinate nestmates (Heinze and Smith 1990; Heinze et al. 1992b, 1993b). By contrast, neither wingless queens nor queen aggression occur in the facultatively polygynous species (Bourke and Heinze 1994). Specifically, peaceful queen coexistence has been noted in these species by E.O. Wilson (1974a), Herbers and Cunningham (1983), Bourke (1991, 1993), Lipski et al. (1992), and Heinze (1993b). In *L. curvispinosus* (E.O. Wilson 1974a) and *L. acervorum* (Bourke 1991), queens cannibalize eggs. However, in *L. acervorum*, experiments suggested that queens cannot recognize their own eggs, and hence that egg cannibalism does not affect reproductive skew (Bourke 1994b).

The arguments above predict that facultatively polygynous leptothoracines should have both lower within-colony relatedness than the functionally monogynous species, and also a low skew in their sexual progeny. However, as yet these predictions remain unconfirmed. This is because of a lack of relatedness estimates for functionally monogynous species, and because measuring skew in sexual output requires molecular methods of high resolution (Ross 1988b; Bourke and Heinze 1994). In *Leptothorax acervorum*, an index of skew devised by H.K. Reeve and L. Keller shows that skew in egg production is low, as is expected (Table 8.5). But this skew could be misleading because the fecundity of individual queens may not correlate with their final adult productivity (Ross 1988b). For example, despite indiscriminate egg cannibalism (Bourke 1994b), brood from different queens may still die at different rates. Stille et al. (1991) indirectly detected a degree of skew in *L. acervorum* worker production by using relatedness estimates derived from allozyme data in the method described by Queller (1993a) (see below). However, assuming workers are largely nonreproductive (Bourke 1991), skew in worker production is evolutionarily unimportant, since workers in polygynous ants probably show negligible within-colony kin discrimination (Carlin et al. 1993; Snyder 1993; Section 7.3).

Reeve and Keller (1995) pointed out that the pedigree structure of a group should also affect reproductive skew. In both diploids and haplodiploids, daughters are equally related (on average) with offspring and sibs, whereas mothers are more closely related with offspring than with grandoffspring. Therefore, other things being equal, daughters should be willing to cede all their reproduction to dominant mothers. (Charnov [1978a] made this point in the context of the evolution of eusociality [Section 3.6].) But in associations of sisters alone (semisocial groups), all individuals are comparatively more closely related with their own offspring than with those of group-mates. Thus, Reeve and Keller (1995) predicted that mother–daughter groups should exhibit greater reproductive skew than semisocial ones. Comparative data supported this view. Reeve and Keller (1995) also suggested that their hypothesis helped explain high skew in the functionally monogynous leptothoracines. However, the exact generational structure of these societies is not known. In addition, some colonies of the facultatively polygynous species almost certainly consist of mother–daughter groups, since they are founded by single queens (Bourke and Heinze 1994). Therefore, although Reeve and Keller's (1995) effect may assist, the very large difference in skew between the functionally monogynous and facultatively polygynous leptothoracines still seems best explained by ecological factors.

Finally, some level of skew may arise from proximate factors such as queen phenotypic quality and age (Brian 1988; Keller 1988), although very little is known about the effect of these. Equally, factors such as brood composition and food supply that are known to influence queen fecundity (e.g. Tschinkel 1988a; Brian 1989) may affect queens unequally at the proximate level. To conclude, gaps remain in our knowledge of skew evolution in leptothoracines. However, overall, the occurrence of facultative polygyny and functional monogyny in this group, as well as their ecological, morphological, and behavioral correlates, is well explained by the approach described above (Bourke and Heinze 1994).

FUNCTIONAL MONOGYNY IN ANTS OTHER THAN LEPTOTHORACINES

Above it was argued that functional monogyny with nondispersing, wingless queens evolves in the leptothoracines under conditions of particularly high dispersal costs (due to habitat patchiness). Similarly, in *Myrmecina graminicola*, one of the handful of functionally monogynous species found outside the leptothoracines (Table 8.4), Ohkawara et al. (1993) attributed wingless queens to habitat patchiness, and some colonies with such queens were functionally monogynous.

Another reason why queens might seek adoption, aside from rising ecological constraints, is if the nest structure is costly to build and therefore worth inheriting. This idea was suggested to apply to the mound-building *Formica* wood ants (Rosengren and Pamilo 1983; Rosengren et al. 1993; Section 10.5). It is also implicit in Nonacs' (1993a) proposal that polygynous *Formica* ants follow a "space-perennial" life cycle (Section 9.9; previous section). When within-colony relatedness is high, queen adoption for the purposes of inheriting the nest structure could also be associated with functional monogyny (see also Keller and Nonacs 1993). A dominant queen with a high chance of passing on her costly nest to readopted daughter queens when she dies, would lose little by denying offspring to these daughters during her lifetime.

Conceivably, this is why some colonies of the ant *Apterostigma dentigerum* are functionally monogynous (Table 8.4), since this species has potentially costly nests made from fungal mycelium (Forsyth 1981). In the functionally monogynous leptothoracine *Leptothorax gredleri*, Heinze et al. (1992b) observed that subordinates did inherit the alpha queen's position when she died, so similar phenomena seem likely in other taxa.

QUEEN INTERACTIONS, REPRODUCTIVE SKEW, AND INTRASPECIFIC SOCIAL PARASITISM IN POLYGYNOUS ANTS OTHER THAN LEPTOTHORACINES

As in the leptothoracines, polygyny in other ants is usually associated with a marked absence of between-queen aggression (e.g. *Neivamyrmex carolinensis*: Rettenmeyer and Watkins 1978; *Linepithema humile* [= *Iridomyrmex humilis*]: Keller 1988; *Camponotus yamaokai*: Terayama and Satoh 1990; Satoh 1991; reviewed by Heinze 1993a). This is again as predicted by the models of Reeve and Ratnieks (1993). A possible exception is *Myrmica rubra*, in which Evesham (1984) noted aggression among queens. However, other studies failed to report appreciable aggression in this species (Brian 1986, 1988; Sommeijer and Van Veen 1990; Elmes and Brian 1991). Mild aggression was also observed among *Camponotus planatus* queens, but its significance was unclear (Carlin et al. 1993). Another, as yet unexplained, exception is the ponerine ant *Odontomachus chelifer*, in which several mated, egg-laying morphological (dealate) queens in a colony were found to participate in a linear dominance order and engage in egg cannibalism (Medeiros et al. 1992). Similarly, Ito (1993b) found ritualized aggression among mated workers (gamergates) in a ponerine (*Amblyopone* sp.) with multiple gamergates in each colony. However, little appears to be known about the ecology of these species.

Conceivably, queens in polygynous colonies compete chemically rather than physically, by producing pheromones inhibiting one another's reproduction (e.g. Mercier et al. 1985b). Evidence for such pheromones was found by Vargo (1992) in *Solenopsis invicta*, but not by Bourke (1993) in *Leptothorax acervorum*. However, even in *S. invicta*, it remains controversial whether queen-produced pheromones are coercive agents or signals to which queens and workers respond voluntarily (Keller and Nonacs 1993; Sections 7.4, 8.2).

In polygynous ants, the society is arguably always at risk from destabilization by some queens concentrating on sexual production instead of producing a mix of sexuals and workers. Individual queens stand to benefit from such intraspecific social parasitism. But if all queens acted this way, hardly any workers would be produced and the society would collapse (Sturtevant 1938; Rosengren and Pamilo 1983). In addition, Reeve and Ratnieks (1993) predicted a low stable skew in the sexual production of queens in truly polygynous colonies. These arguments suggest that intraspecific social parasitism, if it occurs, is kept in check by between-group selection (Section 2.3), and reinforce the point (Section 8.4) that polygyny would not be stable if most queens imposed a net cost. Conceivably, however, sexual skew could be low but some queens might still produce very few workers. Such queens would in a sense remain intraspecific social parasites, if reducing their worker production allowed them to produce relatively more sexuals than they would otherwise have done. However, a parasitic strategy assumes that queens can bias the caste of their female offspring, and this ability may not be common (Hölldobler and Wilson 1990 p. 348; Section 7.6).

To investigate skew evolution and intraspecific social parasitism requires a knowledge of the reproductive skew among both sexual and worker offspring in polygynous ants. But, as in the leptothoracine case, technical difficulties have largely prevented this information from being collected. In *Myrmica rubra*, skew in egg-laying rates was low, as expected (Table 8.5). However, using allozyme analysis in this species, Pearson and Raybould (1993) found that relatedness among workers within colonies was sometimes significantly greater than zero, whereas relatedness among maternal queens was indistinguishable from zero. This would occur if some queens failed to contribute to the worker pool, suggesting a high skew in worker production (Pearson and Raybould 1993). But this conclusion is not certain, because (due to large standard errors) queen and worker relatedness never differed significantly from one another. Queller (1993a) proposed a quantitative method for investigating skew with relatedness data (see also Strassmann et al. 1991). Given a knowledge of the number of mother queens in a colony, their relatedness, and their variance in maternity (skew), it is possible to predict the relatedness level among their worker or queen offspring (cf. equations for r_F in Box 4.5). Therefore, one can equally predict the expected number of queens per colony given a knowledge of queen–queen relatedness and offspring relatedness and assuming zero skew. If the expected number falls below the observed number, skew must be greater than zero (Queller 1993a). Seppä (1994) used this method to assess the skew in worker production among *M. ruginodis* queens. Effective queen number was only slightly lower than the observed one, suggesting – again as expected – low skew.

In a pioneering study, Ross (1988b) investigated skew by genetic analysis of parentage in the polygynous (unicolonial) fire ant *Solenopsis invicta*. He set up artificial colonies in which the offspring of coexisting queens could be distinguished by their allozyme genotypes. The results showed that skew among daughter queens was fairly low (Table 8.5). Yet given negligible relatedness among queens in this species (Ross and Fletcher 1985a; Ross 1993), one might have expected even lower skew. However, it is unknown how the short-term skew measured by Ross compares with lifetime skew (Ross 1988b; Keller and Vargo 1993). Ross (1988b) also found that skew in worker production was lower than for queen production, suggesting that some mother queens specialized on producing sexual daughters at the expense of others. As discussed above, such queens would effectively be intraspecific social parasites.

Ross (1993) extended this study by using the methods of Queller (1993a) to investigate skew in polygynous *Solenopsis invicta* colonies from the field. Again, skew in worker production was low, although it rose with the number of colony queens. (The data do not permit the calculation of the actual skew index [Table 8.5].) However, skew in the production of new queens was higher (observed and expected colony queen numbers diverged more), although it was still not very pronounced. So these results again suggested that queens differ in the worker-to-sexual ratio among their brood (Ross 1993). In the weakly polygynous *Formica sanguinea*, Pamilo and Seppä (1994) found that relatedness among worker pupae was significantly lower than that among daughter queen pupae, also suggesting that some queens concentrated on queen production.

Another way in which queens could act as intraspecific social parasites would be to lay only unfertilized, male eggs. This is the only reproductive option open to uninseminated queens. Colonies containing unrelated male-producing queens derive neither workers nor kin-selected benefits from them. So, if workers can detect such queens, they should expel them. In keeping with this, Bruckner (1982) found that queenless *Leptothorax acervorum* colonies never accepted experimentally introduced virgin queens from other colonies, although they sometimes accepted foreign mated queens. In addition, even if uninseminated layers are relatives of their home colony, they still risk ejection because of conflict over male parentage (Section 7.4). In *Solenopsis invicta* and *Linepithema humile* (= *Iridomyrmex humilis*), young adult virgin queens in natal, queenright colonies are pheromonally inhibited from shedding their wings and developing their ovaries (Fletcher and Blum 1981; Willer and Fletcher 1986; Passera and Aron 1993a). This could be because both the colony queens and workers oppose their producing males (Keller and Nonacs 1993). Similarly, Elmes and Petal (1990) reported that *Myrmica* workers killed young virgin queens that had accidentally shed their wings in the nest due to disturbance.

Table 8.6

Frequency of Uninseminated Queens in Polygynous Ant Colonies

Species	% Uninseminated Queens	References and Notes
Camponotus nawai	4	Satoh 1989
Formica aquilonia	17	Rosengren et al. 1993; some were egg-layers
Formica polyctena	14	Rosengren et al. 1993
	45	Yamauchi et al. 1994
Formica uralensis	3	Rosengren et al. 1993
Lasius sakagamii	11	Yamauchi et al. 1981
Leptothorax acervorum	10	Douwes et al. 1987; a few (2% of all queens) were uninseminated egg-layers
Leptothorax ambiguus	22	Alloway et al. 1982; mean per colony figure; some were egg-layers
Leptothorax curvispinosus	0.05	Alloway et al. 1982; mean per colony figure; some were egg-layers
Leptothorax longispinosus	0.03	Alloway et al. 1982; mean per colony figure; all were egg-layers
	3	Herbers 1984
Linepithema humile (= Iridomyrmex humilis)	2	Keller and Passera 1992
Myrmica rubra	0	Elmes and Petal 1990
Myrmica ruginodis	0	Mizutani 1981
Myrmica sulcinodis	0	Elmes 1987a
Plagiolepis pygmaea	0	Mercier et al. 1985a
Solenopsis invicta	30	Ross 1989
Solenopsis geminata	40	Vargo 1993

NOTE: Uninseminated queens are dealate queens that dissections showed to be without sperm in their sperm receptacles. They could be either virgin queens that have shed their wings in the nest, or – less probably – queens that have mated and then exhausted their sperm supply (Alloway et al. 1982; Tschinkel 1987a,b). Egg-layers were detected by the presence of a combination of elongated ovarioles, mature eggs in the ovaries, or corpora lutea (ovarian structures formed when eggs are laid). See also review of Passera et al. (1991).

These arguments help explain why uninseminated, egg-laying dealate queens are in fact quite rare in most polygynous ants (Passera et al. 1991; Table 8.6). Exceptions include some *Formica* and *Solenopsis* species (Table 8.6). However, *Formica* queens seem to have the almost unique ability to mate as old, dealate virgins (Rosengren et al. 1993). Therefore, virgins might be tolerated because they are destined to remate with males from the nest that accepts them. This might also help explain the otherwise puzzling tendency of *F. lugubris* nests to admit uninseminated queens more readily than inseminated ones (Fortelius et al. 1993). In *Solenopsis*, uninseminated queens almost certainly achieve little reproductive success, since their eggs have very low viability (Ross 1989; Vargo 1993). Their abundance appears due a shortage of fertile, haploid males in polygynous populations, since their frequency falls with increasing proximity to monogynous populations, which produce plentiful males (Ross and Keller 1995; Section 8.4).

8.7 Conclusion

Ecological factors evidently play a large part in determining which of several possible social systems ant populations adopt. However, in some cases their exact influence is unclear, because the ecology of many species needs to be better understood (Sections 8.4, 8.5). Furthermore, as both the arguments concerning the evolution of pleometrosis (Section 8.3), and the stable skew models (Section 8.6) show, ecological factors need to be considered alongside internal ones such as relatedness. Both jointly determine the level of within-colony conflict, the stable social structure, and the stable skew. Explanations based on this insight have proved helpful in understanding leptothoracine social diversity (Section 8.6). Future work should be able to use increasingly sophisticated molecular techniques, such as the analysis of microsatellite DNA variation (e.g. Evans 1993; Hamaguchi et al. 1993), to find out what determines the stable skew at both the proximate and ultimate levels.

8.8 Summary

1. Polygyny, the occurrence of multiple queens within colonies, is common in ants. The major types are pleometrosis (the formation of associations of colony-founding queens), multicolonial secondary polygyny (populations consist of discrete colonies that acquire extra queens by adoption), unicolonial secondary polygyny (populations consist of freely mixing, secondarily polygynous nests), and functional monogyny (several mated queens coexist but only one lays eggs).

2. Polygynous colonies reproduce by a mixture of emission of winged sexuals and colony budding (emigrant queens and workers form new colonies). Queens mating near their natal colony and then reentering it (adoption) account for above-zero queen–queen relatedness under

polygyny. Due to colony foundation by adoption and lower dispersal, queens from polygynous colonies are smaller than their monogynous relatives and have fewer fat and glycogen reserves. Due to the presence of workers in the colonies adopting them, these queens also produce sexuals earlier and live less long. Monogyny and polygyny therefore constitute distinct life history strategies.

3. As their numbers rise within colonies, the per capita fecundity and sexual productivity of queens fall (reproductivity effect). Proximate explanations include increasing worker inefficiency in larger colonies, rising levels of egg cannibalism, and increasing amounts of queen pheromones that inhibit fecundity and the sexualization of female larvae. Possible ultimate causes include worker action to maximize brood production, more frequent colony budding as queen number rises, or energy savings made possible by other benefits of polygyny.

4. Foundress associations evolve if they have especially high survivorship relative to single foundresses. The main ecological factor promoting foundress associations is clumping of incipient colonies. This encourages rapid worker production to ensure competitiveness, which favors several queens raising eggs together. In most species, foundress associations revert to monogyny when the first workers emerge. This is probably because a dominant queen would no longer harm herself by being aggressive once energy enters the colony from outside. Workers favor the survival of the most productive queen, since she will be the average worker's mother.

5. Multicolonial secondary polygyny evolves when the costs of dispersal and colony foundation are particularly high for single queens (severe ecological constraints). Queens should then seek adoption in their home colony, which should be selected to admit them. Severe ecological constraints arise from predation of sexuals, habitat patchiness, nest-site limitation, and cold climate. Existing data confirm the influence of these factors in some taxa (e.g. *Leptothorax*, *Formica*), but in others the ecological basis of polygyny is unclear (e.g. *Myrmica*).

6. Polygyny is also favored if extra queens allow colonies to survive predation better, or improve their overwintering success, or if a genetically more variable workforce boosts colony efficiency or protection against parasites. The presence of several queens per colony also reduces the variance in their expected reproductive success. In *Solenopsis invicta* in North America, polygyny may buffer queens against colony mortality due to diploid male production.

7. Nests in unicolonial populations are characterized by zero or very low within-nest relatedness, which poses a problem for kin selection theory. However, unicoloniality could evolve for short-term ecological gain in new, urban, or disturbed habitats. Some, but not all, pairs of

forms or species in which one member of the pair is monogynous and the other unicolonial and polygynous match this interpretation.

8. In leptothoracines, functionally monogynous societies evolve when extreme habitat patchiness makes the costs of dispersal very high. Consequently, a fraction of queens are wingless nondispersers and within-colony relatedness rises. Both severe ecological constraints and high relatedness promote high skew, which in turn favors high between-queen aggression. By contrast, facultatively polygynous lepto-thoracines have nonaggressive queens and live in nonpatchy habitats, but whether their skew in sexual production is low is unconfirmed.

9. In polygynous species from other taxa, queen coexistence is gener-ally peaceful. In rare cases where it has been measured, skew is low as expected. *Solenopsis invicta* queens appear to have unequal skew in worker and sexual production, suggesting a degree of intraspecific social parasitism. The production of males by uninseminated queens is rare in polygynous ants.

10. Overall, there is reasonable evidence that ecological factors pro-mote the different forms of multiple queening in ants as theory predicts. However, in many cases the ecological correlates of polygyny, and asso-ciated levels of skew, remain poorly known and require further investi-gation.

9 Life History Theory in Ants

9.1 Introduction

Both this chapter and the next are about life history evolution in ants. This chapter is concerned mostly with theoretical issues and the next with case studies. Since monogyny and polygyny constitute different life history syndromes (Section 8.2), the previous chapter also dealt with a number of life history issues in ants.

To set the stage, consider a hypothetically "typical" ant life history. This will provide a baseline for later evaluation of theoretical issues, and a comparison for the empirical case studies. This typical ant has colonies with one queen (monogyny), who is singly mated (monandry). The whole society lives in a single nest site (monodomy). The queen is the sole source of reproductives during her lifetime, but the workers may produce their own sons after the queen's death. Mating occurs between winged males and females from different colonies in a nuptial flight, after which the males die. The colony is perennial and passes through four stages. The first, the *foundation stage*, is characterized by solitary queens that, after mating, attempt to start their colonies without any help from workers or other queens. Successful colonies then enter an *ergonomic stage* in which they rapidly grow by producing all-worker broods for a number of years. Next, in the *reproductive stage*, the colony each year produces a mixture of workers and sexuals (new queens and males). If, during the year, workers are produced first, then sexuals, a colony may undergo an ergonomic followed by a reproductive "substage" within each season. The overall reproductive stage continues for the rest of the queen's life. Lastly, after the death of the queen, workers start to produce their own sons (the *orphanage stage*). This, and the colony as a whole, ends with the eventual death of the workers, but could last for more than one year, depending on the workers' longevity. The founding, ergonomic, and reproductive stages were originally defined by Oster and Wilson (1978 p. 28). The significance of the orphanage period is discussed in Franks et al. (1990a).

In reality, almost all ant societies deviate from this "canonical" society in at least one of their attributes (Chapters 8, 10, 11). For example, in some species colonies can be founded by several queens (Section 8.3). In others, colonies are obligately monogynous but reproduce by fission (Section 10.8), whereas in others again, colonies are perpetually polygynous and reproduce in part by budding, or grow into super-colonies (Sections 8.5, 10.5, 10.6). Some species have obligately sterile workers, whereas others have workers that produce sons in the presence of the queen (Bourke 1988a; Sections 7.4, 10.2–10.4). There is even a species of ant with neither queens nor males, whose workers produce other workers parthenogenetically (Section 10.7).

A comprehensive life history theory for social insects should aim to explain, first, why the basic ant life cycle is perennial and iteroparous (with repeated breeding episodes). Next, it should explain the nature, timing, and scale of each part of the basic life cycle, and especially how resources are partitioned between investment in growth (worker production) and investment in reproduction (sexual production). Lastly, it should uncover the reasons why variants on the basic pattern evolve (cf. Chapters 7, 8, 11). All this should be done with regard to the possibility of kin conflict within the colony (Pamilo 1991b). This chapter aims to suggest some of the ideas that may be important in this as yet almost unattempted enterprise. The following section considers the basis of life history theory in general, and Oster and Wilson's (1978) pioneering analysis of life history strategy in social insects. Subsequent sections review additional principles provided by life history theory (developed mostly for other organisms) that are likely to be of value in the study of life history in ants. These concern issues such as the evolution of perenniality and iteroparity (Section 9.3), dispersal (Section 9.4), trade-offs and propagule size (Section 9.5), modular growth (Section 9.6), reaction norms (Section 9.7), and lineage-specific effects (Section 9.8). The final section discusses special issues in social insect life history evolution.

Traditionally, there are two approaches in life history biology. One is to examine general principles and patterns (the subject, mostly, of this chapter), and the other is to look at particular case histories in quantitative detail. Chapter 10 takes the second approach, highlighting – where possible – what evidence supports or contradicts the theoretical framework presented here. The present chapter also asks what extra dimensions sociality adds to the study of life history in ants. Modular growth is one answer (Section 9.6), and kin conflict within the colony is another (Chapters 6, 7; Section 9.9).

9.2 Life History Theory in General and in Social Insects

SOME BASIC PRINCIPLES IN LIFE HISTORY THEORY

The theory of life history evolution is based on the notion that a phenotype consists of demographic traits, connected by constraining relationships that lead to trade-offs. For ants, the phenotype in question is that of the individual and also the extended phenotype (Dawkins 1982a) represented by the colony. The idea of a trade-off is that, due to constraints such as limited resources, investment in one trait can only be made at the expense of a reduced investment in another. According to Stearns (1992 p. 10), the principal life history traits are size at birth, growth pattern, age at maturity, size at maturity, number, size, and sex ratio of offspring, age-specific and size-specific reproductive investment and mortality schedules, and length of life. These traits are bound together by numerous trade-offs, including those between current reproduction and survival, current reproduction and future reproduction, and number, size, and sex of offspring. In fact, at least 45 trade-offs are readily defined between life history traits (Stearns 1992 p. 72). Of these, the most studied are those between current reproduction and growth, reproduction and condition, and the number and quality of offspring.

Life history analysis is about understanding the diversity of reproductive allocation strategies. Interpretations must take into account both selection pressures and constraints. Stearns (1992 p. 115) summed up the problem admirably: "To understand a pattern of variation in life history traits, whether interspecific or intraspecific, we need information on phylogenetic effects, phenotypic variation, genetic variation, trade-off structure, and the selection pressures generated by demography." Stearns (1992 p. 115) concluded that such a multidimensional analysis of life history evolution has not yet been achieved for any group of organisms. The study of life history strategy in social insects is at a particularly early stage. Furthermore, few connections have been made between social insect life history theory and the general study of life histories. For example, in his splendid book, Stearns (1992) did not consider life history evolution in social insects at all!

OSTER AND WILSON'S (1978) LIFE HISTORY ANALYSIS FOR SOCIAL INSECTS

A pioneering and detailed treatment of life history evolution in social insects was presented by Oster and Wilson (1978). They focused on per-

302 • CHAPTER 9

haps the most central issue, the scale and timing of the allocation of resources between worker and sexual production. Oster and Wilson (1978 pp. 50–69) considered both annual and perennial species. As discussed later, there are no annual ants. In the annual bumble bees and polistine and vespine wasps, Oster and Wilson (1978) predicted that the life history policy maximizing colony fitness (sexual productivity) was a "bang-bang" strategy. It is called "bang-bang" to indicate that there is a complete switch from the exclusive production of one form of progeny (workers), to the exclusive production of another (sexuals). In other words, annual colonies should produce workers for as long as possible and then switch completely to the production of sexuals. In this way they can rear the greatest possible number of workers to raise, just before the end of the annual cycle, the greatest possible number of sexuals.

For the perennial species, such as ants, (Oster and Wilson 1978 p. 65) also predicted a "bang-bang" strategy within each season. Thus, when they are sufficiently large, perennial colonies should alternate (as is often observed) ergonomic and reproductive phases within each year for a number of years. Just as in annual colonies, the ergonomic phase within each year in a perennial colony should be as long as possible so that more workers are available to help raise the largest number of sexuals during the second part of the year. However, perennial colonies also have to invest in workers to ensure that the colony survives the winter and can continue to reproduce in subsequent years. Oster and Wilson (1978 pp. 65–67) found that if the probability of the colony surviving the winter is high, it should start to produce sexuals relatively late in its life cycle. This makes sense in that colonies which risk dying early should not delay reproduction. Similarly, within a season, the switch to producing sexuals occurred later as colony overwintering survivorship rose.

In *Solenopsis invicta* colonies, a bang-bang strategy may be occurring after only a year or two dedicated to growth through the exclusive production of workers (Tschinkel 1993a; Figure 10.1). The observed oscillations in worker numbers are probably a consequence of repeatedly switching to sexual production. The issue of resource allocation between growth and reproduction is further considered later in this chapter, and in the next one (e.g. Sections 10.2–10.4, 10.8). In perennial societies, an important new dimension was introduced by Pamilo's (1991b) model of kin conflict over resource allocation (Sections 7.6, 9.9).

9.3 The Evolution of a Perennial Life Cycle in Ants

THE ADVANTAGE OF EARLY BREEDING, AND AGE AND SIZE AT MATURITY

A general principle in life history theory is that early reproduction is frequently advantageous (L.C. Cole 1954; Lewontin 1965; Sibly and Calow 1986; Stearns 1992). For example, in an expanding population, genes for early breeding should spread much more rapidly through a population than genes for delayed breeding (Stearns 1992 p. 26). This line of reasoning poses the question of why it is that there are no annual ant colonies. We return to this enigma below.

The benefit of early reproduction can be considered as the advantage of faster returns on compound interest. It immediately suggests a downwards selection pressure on worker size in social insect colonies. Small workers consume fewer resources during development and, all else being equal, should become adults more quickly than large workers. Therefore, they should be able to help raise other workers, and ultimately sexuals, faster than if the queen had produced fewer large workers each with a longer development time. This argument helps explain why the first workers produced by colony-founding queens, the so-called nanitics or dwarf workers (Hölldobler and Wilson 1990 pp. 157–159), are often far smaller than workers in mature colonies of the same species. This is most clearly the case when there is claustral colony foundation by single queens, since claustral queens do not forage and their first workers are raised purely from the queen's on-board resources. By first producing the maximum number of small workers, claustral queens attempt to maximize the rate of growth of their colony as it moves from the foundation phase into the ergonomic phase.

COLE'S PARADOX AND THE ADVANTAGE OF ITEROPARITY

Life histories may be classified as semelparous (with one breeding event per lifetime) or, as already described, iteroparous (with repeated breeding events per lifetime). They may also be annual (lasting one year) or perennial (lasting many years), with perenniality and iteroparity often going together. In his classic life history paper, L.C. Cole (1954 p. 118) built a simple model to compare an annual with a perennial and found that "For an annual species, the absolute gain in intrinsic population growth which could be achieved by changing to the perennial reproductive habit would be exactly equivalent to adding one individual to the average litter size." Since increasing litter size by one seems easy to achieve, why then are there any perennials at all? This is known as Cole's paradox.

In fact, Cole's result is not paradoxical because it ignored mortality rates (Stearns 1992 p. 187). As Charnov and Schaffer (1973) showed, a perennial, iteroparous life history will be favored if adult mortality is low compared with that of juveniles. In the case of ants, the adult form can be regarded as the established colony, and the juvenile form as the colony-founding queen. Seen in this way, low "adult" mortality and high "juvenile" mortality are almost ubiquitous in ants, and most other social insects. The distinction between annual and perennial life cycles is of direct interest to students of social insects because many wasps and bumble bees are annual and semelparous, whereas all ants are perennial and iteroparous (Hölldobler and Wilson 1990 p. 143).

Whatever its causes, the perenniality of ant societies is a key point in understanding their life history evolution. Because of its longevity, robustness, and productivity, the adult ant colony has been aptly described as a factory within a fortress (E.O. Wilson 1971 p. 342). By contrast, an isolated new queen starting a colony is a largely defenseless morsel for predators. These frequently include neighboring factory–fortresses. (As Auguste Forel remarked, the greatest enemies of ants are other ants [E.O. Wilson 1971 p. 447].) This is because many ant colonies maintain territories in one form or another, including foraging territories (Hölldobler and Lumsden 1980; Hölldobler and Wilson 1990 p. 400). In addition, mature colonies should be selected to destroy smaller ones of the same or closely related species in their neighborhood, before the smaller ones significantly deplete local resources (Levings and Franks 1982). Thus, even established colonies of herbivorous species such as *Atta* will destroy newly-founded colonies that are close to them (Lewis 1975; Levings and Traniello 1981).

The result is that, in certain habitats, such is the density of long lived, territorial ant colonies that there may be few, if any, ant-free gaps. In turn, this means that in many cases the life history strategies of ants are dominated by gap dynamics. That is, as discussed below, they are sensitive to the availability in time and space of gaps between existing colonies where young colonies have a chance of becoming established. They are also characterized by adaptations that enable young queens or colony propagules to find and survive in such gaps.

GAP DYNAMICS AND ITEROPARITY AS BET-HEDGING

One of the disadvantages of an annual life cycle, and other forms of semelparity, is the risk of the single reproductive event occurring at a particularly bad time, or in a particularly bad year. Iteroparity spreads the risk. In other words, it is a form of bet-hedging (Seger and Brockmann 1987).

Isolated, young ant queens are greatly at risk compared to their parental colony. For example, imagine an ant colony releasing all its sexual progeny on a single afternoon that ends with a torrential thunderstorm. All its daughter queens attempting solitary colony foundation could be wiped out. As just mentioned, dispersing young queens also frequently face the risk of detection and destruction by mature colonies. Thus, following a nuptial flight, queens of species that found their colonies independently must try to find gaps between established colonies in which to start their own colony. Iteroparity may help to ensure that at least some sexual progeny are produced in years in which gap availability is better than average.

This scenario has not been explicitly modeled for ant colony establishment. However, an analogous life history pattern occurs in other organisms in which the mature stage is sessile (ant colonies can generally move nest site, but in many ways they resemble totally sessile organisms) and the young can disperse. These are trees. Stearns and Crandall (1981) presented a model to explore bet-hedging by iteroparity in tropical canopy trees (see also Stearns 1992 p. 195). Like ant colonies, these produce dispersing propagules (seeds) that have to find uncolonized gaps (in this case tree-fall gaps or light-gaps) to survive and grow to reproductive maturity. Stearns and Crandall (1981) modeled the life history traits of trees on a two dimensional plane where patches suitable for offspring occurred unpredictably. In the models, the density of offspring dispersing from the parental organism fell off exponentially with distance from the parent. Dispersal interacted with the random pattern of patches that were suitable for colonization in each generation. This produced a relationship between the number of patches colonized and the reproductive effort of the parent. In addition, the parental organisms were assumed to have an accelerating cost of reproduction, in that their mortality increased exponentially as their reproductive effort increased.

The general conclusion of Stearns and Crandall's (1981) models was that juvenile mortality was of almost paramount importance in influencing fitness, and that gap-colonizers are under strong selection to produce high-quality offspring that disperse over a wide area. Ant colonies do indeed produce high quality offspring, in the form of well-provisioned queens (Keller and Passera 1989a; Section 9.5), or even, if juvenile mortality is exceptionally high, in the form of foundress associations and colony buds (Sections 8.3, 9.5). They also often have widely dispersing young (Sections 9.4, 11.5).

These predictions are at variance with the traditional view of colonizers as classic "*r*-strategists" (Section 9.8) that are short lived and produce many small propagules. This is because the traditional view was based on two major assumptions, namely (1) that the environment colonized deteriorated equally rapidly for juveniles and adults, and (2) that fitness is always best defined as *r*, the intrinsic rate of natural increase of a population or genotype. The first assumption might well apply to short-lived habitats, but it is not applicable to the colonization of gaps that are simply patches where the probability of surviving to maturity is high compared to occupied habitat.

Concerning assumption (2) above, Stearns and Crandall (1981) explored the consequences of using two different definitions of fitness in their models. In the first definition (the bet-hedging definition), maximizing fitness involved maximizing the rate at which a clone increases in an unpredictable environment. This was equivalent to maximizing *r*, the intrinsic rate of natural increase for that genotype. In the second (the persistence definition), it involved minimizing the probability that a clone will disappear from the population. These two fitness definitions led to different predictions from the models. Nevertheless, at the optimum under either fitness definition, the trees were always iteroparous. In addition, a tree could have a seed output that optimized one fitness measure without suffering a significant decrease in the other.

Another important result of Stearns and Crandall's (1981) model was that, if the abundance of patches was reduced, there was a concomitant reduction in the reproductive effort of the adult organisms. Consequently, the organisms lived longer and hence even fewer gaps became available. This created a cycle of positive feedback selecting for organisms of yet greater longevity. Indeed, many sessile organisms, from trees to marine invertebrates such as clams, corals, and sponges, are very long lived (Stearns 1992 p. 196), and so are most ant colonies. However, Stearns and Crandall (1981) only obtained this result for the first of their fitness definitions. When they sought to minimize the probability of extinction they found that a decrease in patch frequency led to increased seed output and shorter life. Clearly, which definition of fitness is employed is important. Stearns and Crandall (1981) and Stearns (1992) favor the fitness definition based on maximizing *r*.

Gap Dynamics, Territoriality, and Perenniality in Ants

We propose that the gap dynamics described above help to explain why ant colonies are perennial, iteroparous, and long lived with widely dispersing, relatively high-quality offspring. This is in contrast with the annual, semelparous, short lived societies of wasps and bumble bees. In short, ants may be perennial due to established colonies holding and defending territories, thereby creating the kind of life history dominated by gap dynamics that Stearns and Crandall (1981) modeled.

This argument raises the question of why ant colonies should defend exclusive territories in the first place. The answer seems intimately connected with the fact that ants have wingless workers and are (typically) soil-nesters. This means that they are required to be central place foragers limited to exploiting resources relatively close to their nest. As a consequence, the area around the nest needs defending, not least – as stressed earlier – against conspecific neighbors. By contrast, bees and wasps have flying workers that can track rapidly-shifting, distant food sources such as flower patches, making territory defense unnecessary and uneconomical. Furthermore, established bee and wasp colonies only rarely if ever kill neighboring colonies. This could be why there is no territoriality in these groups, and hence no life history characterized by gap dynamics and perenniality. For these reasons, ant colonies are like trees and annual wasps and bumble bees are like weeds. Moreover, just as certain trees (particularly the North American Black Walnut, *Juglans nigra*) engage in contests for a gap in the canopy and produce chemicals that harm neighboring plants (Harborne 1988 p. 280), ant colonies fight for defensible territories and kill incipient colonies that are too close.

There are also self-reinforcing effects in the evolution of ant perenniality. First, defending territories requires numerous workers and large colony size is most easily achieved if the colony can grow over more than one year. Second, big colonies can make larger and better protected nests and so can survive winters better (Section 10.5). Third, with some notable exceptions (Sections 10.5, 10.6, 10.8), big colonies are less likely to abandon their large and valuable nests. Being less mobile, they tend to fill up the habitat more, so making gaps even more valuable and large colony size and perenniality even more advantageous. Seeley (1985 pp. 57–60) and Nonacs (1993a) also discuss the evolution of perennial and annual life cycles in social insects.

Although, there are no annual ant colonies, certain ants have evolved ways of life that resemble, in certain regards, an annual life cycle (P. Nonacs, personal communication). For example, in *Linepithema humile* (= *Iridomyrmex humilis*) a unicolonial species, 90% of queens are killed by the workers each year (Keller et al. 1989; Section 10.6), so many queens are unlikely to live for more than one year. Thus, although colonies are perennial, individual queens may be annual and semelparous. This is also suggested as a possibility for *Solenopsis invicta* queens in Section 10.2. In the genus *Epimyrma* there are socially parasitic queens that utilize *Leptothorax* host colonies, kill the host queens, and produce no workers of their own (Buschinger 1989a). The longevity of such parasitized colonies, and the *Epimyrma* queens they support, must be greatly reduced and some of these queens may therefore also exhibit an approximately annual life cycle (Section 10.4).

Another issue that is closely related to the question of whether life histories should be annual or perennial is the evolution of reproductive lifespan. (Seeley [1985 pp. 57–60] discusses lifespan evolution in honey bee queens). Stearns (1992 p. 205) suggested that the selection pressures determining reproductive lifespan are largely associated with the relative magnitude of both the mean and the variance of adult and juvenile mortality. In general, an increase in the mean and the variance of adult mortality should decrease the reproductive lifespan, eventually producing semelparity. By contrast, an increase in the mean and variance of juvenile mortality can lengthen the reproductive lifespan, producing very long lived organisms with high investment in somatic structures and their repair. As discussed above, mortality rates of incipient ant colonies are likely to be high and variable. This should select for persistence among established colonies. In addition, viewing an ant colony as a superorganism (Seeley 1989; Wilson and Sober 1989; Section 2.6), it is clear that almost all ant colonies have some form of massive somatic investment, such as workers, nest structures, and foraging pathways. Also, the modular construction (Section 9.6) of colonies from large numbers of workers and brood is highly suited to rapid repair of this somatic investment. It also permits a division of labor in ant colonies which is robust in the face of perturbations (Section 12.4). So, in general, ant colonies do seem to be selected for a long reproductive lifespan.

SEED DORMANCY

An additional reason for iteroparity is that ants do not have anything equivalent to seed dormancy in plants. Plant seeds commonly show dormancy in conditions seemingly favorable to germination (Watkinson and White 1986). This is presumably an adaptation for temporal bet-hedging, since it provides a safeguard against unsuitable conditions in a variable environment (Léon 1985). Thus, an annual plant that produces seeds which emerge from dormancy and germinate in different years is, from the point of view of bet-hedging, iteroparous.

We are unaware of any phenomenon in ants closely analogous to seed dormancy. Certain insects do enter prolonged diapause (Denlinger 1985), so in principle something equivalent to dormancy is possible. For example, ant queens could leave the parental nest, perform a nuptial flight, and then hide for variable periods before attempting to start their colony. One reason why dormancy might not have evolved in ants is that dormant queens are likely to be more vulnerable to predation than dormant seeds. This is because established colonies are liable to seek out and kill queens that are too close to them. By contrast, the predators of plant seeds are not other plants, and, instead of removing a potential competitor, simply obtain the direct rewards of eating each seed they destroy. Therefore, ant queens, rather than lying dormant, might be better off trying to found a colony as quickly as possible so it can grow to a size at which it cannot be overwhelmed without a disproportionate cost to the aggressor (Franks and Partridge 1993).

Perhaps the nearest thing to seed dormancy in ants is found in some *Leptothorax* species. Young queens, instead of attempting colony foundation in the summer of their emergence as adults, hibernate in the parental nest and leave to start a colony the following spring (Heinze 1992; Heinze and Hölldobler 1994). These queens are potentially flexible in the timing of their departure (they could conceivably wait more than one year), and so in unfavorable conditions may delay beginning an independent life until a better time. However, it is more likely that they would try to remain in the parental colony as members of a polygynous society.

9.4 Dispersal in Stable Habitats

Why do so many ant species produce winged sexuals that participate in large-scale nuptial flights followed by wide-ranging dispersal (Section 11.5)? This seems paradoxical given that established colonies are likely to have very high year-to-year survivorship (e.g. Keeler 1988, 1993; Porter and Jorgensen 1988; Thurber et al. 1993). It also seems unexpected given that isolated queens and small new colonies suffer tremendous mortalities, and that secondary polygyny allows colonies to be potentially immortal (Sections 8.4, 8.5). An alternative to wide dispersal is propagation by colony budding, though this appears to depend on permanent polygyny (e.g. Section 10.5). However, Hamilton and May (1977) showed that there is strong selection for dispersal even in predictable and uniform habitats. This is because, unlike universal dispersal, total nondispersal is not evolutionarily stable (Section 8.6). Hamilton and May's (1977) reasoning explains the nuptial flights and dispersal of the majority of ants (Section 11.5). More intriguingly, it helps explain the mixed dispersal strategy found in polygynous ants such as *Leptothorax* and various *Formica* species in which there are nuptial flights (e.g. Franks et al. 1991a), but in which some queens are readopted by the parental nest (Sections 8.6, 10.4, 10.5, 11.5). This mixed strategy of partial dispersal also represents a classic case of bet-hedging (Rosengren et al. 1993).

9.5 Trade-offs, Propagule Size, and Modes of Colony Foundation

EVOLUTION OF PROPAGULE SIZE

Previous sections have suggested that the reason ant colonies are perennial and iteroparous is that the growth conditions for small colonies are extremely poor compared with those for well-established colonies. This section examines in greater detail the effects this might have on propagule size among ants, where "propagule" means a product of reproduction such as a young queen, a group of queens, a colony bud, or a daughter colony resulting from fission.

The general assumption in models of propagule size, as in other areas of life history analysis, is that a system of trade-offs operates. The larger the investment in any one propagule, the fewer propagules can be produced. The greater the overall investment in reproductive effort, the lower the growth rate and survival of the parental organism. However, demonstrating such trade-offs in models is immeasurably easier than detecting them in field populations. Van Noordwijk and de Jong (1986) explained why the negative correlations predicted by life history theory are often not found in practice. Negative correlations are expected precisely because life history theory in couched in terms of trade-offs. That is, an organism either invests in reproduction *or* growth and survival, with limited resources being partitioned one way or the other. But many empirical results show that organisms that invest most one year also invest most in the next.

Van Noordwijk and de Jong (1986) suggested an analogy to explain why. Consider the application of simple life history theory to human families in the industrial world. Here, one might also expect trade-offs, for example a negative correlation between such major investments as the cost of the home and the cost of the family car. However, in reality the human families that invest most in houses are the wealthiest ones and so also tend to spend most on cars. Thus, the expected trade-off is not seen because family incomes are very variable. For ant colonies, which are largely rooted in one place and reliant on a local environment in which resources are likely to fluctuate in space and time, there should be a huge variation in "family income." Therefore, trade-offs might be even harder to show in ants than in other organisms.

Sibly et al. (1988) used a simple but robust graphical method to determine optimal trade-off strategies for animals in which reproduction occurs at a given time (e.g. a breeding season) rather than at particular body sizes. The trade-off is between producing many small reproductive propagules, or fewer, larger ones. Sibly et al. (1988) showed that, under specified conditions, larger propagules should be produced when conditions for juvenile growth are poor. The reason is that, assuming the juvenile growth period has a fixed duration, adult size will depend to a large extent on propagule size, because bigger propagules get a head start. In addition, bigger adults can gather more resources for reproduction, and so have a fecundity advantage. Thus, there is a selection pressure for producing fewer, larger, and more fecund offspring, which is further exacerbated when the conditions for growth are particularly bad for juveniles compared with adults, because juveniles from bigger propagules survive the poorer conditions better. Sibly et al. (1988) further suggested that increased adult density favors larger propagules and adults if resources are obtained by "contest competition", but not if they are obtained by "scramble competition" (Begon et al. 1986 p. 220). In scramble competition, it is supposed that all individuals obtain some resources, so each obtains less if the population density is higher. The optimal strategy under these circumstances is to produce the smallest viable adults, and hence small propagules, irrespective of population density. By contrast, in extreme contest competition all the resources are obtained by the larger animals, which get their full requirements, while the smaller animals get effectively nothing. Therefore, in contest competition, increased density favors larger adults and larger propagules.

Many ant species have fixed breeding seasons, reproducing at a particular time of year (Hölldobler and Wilson 1990 pp. 145–157). This might be more important in their life history evolution than previously considered. Because of the difficulty of knowing the age of colonies in the field, it is not known, in general, if colonies first reproduce at a fixed size or, after a set juvenile period, at a fixed age. However, it may be the case that almost all colonies in a population start to reproduce in, say, their third year. Such colonies may vary considerably in size, but none would gain by delaying reproduction until the fourth year. Colonies with these characteristics would have a rather constant, though not strictly fixed, juvenile period. Therefore, the Sibly et al. (1988) model, even though it is based on a fixed juvenile period, may have some relevance to ant biology.

In addition, we suggest that ant colonies generally engage in contest competition rather than scramble competition. This is because, as already discussed, there are many species in which colonies hold exclusive territories (or at least deny space to smaller colonies), and hence in which there is only a limited and roughly constant number of possible winners. Therefore, in most ant populations, the scale of ecological competition is tipped towards a contest, rather than a scramble, for resources. If so, the Sibly et al. (1988) model would predict increasing colony and propagule size among ants as colony density rises.

MODES OF COLONY FOUNDATION

The evolution of propagule size is closely connected with the mode of colony foundation in ants. For example, ant species have either independent or dependent modes of colony foundation (e.g. Keller 1991). In the first case, young queens found colonies without the help of workers (propagules are small), whereas in the second case they are accompanied by workers (propagules are large). Hence, models of the evolution of propagule size could help explain the diverse modes of colony foundation in ants.

The colony founding stage for solitary queens can be very long. In her classic study of *Lasius flavus*, Waloff (1957) showed that for solitary foundresses the average time to production of the first adult workers was 206 days and the limits were 96 and 347 days. These figures indicate the importance of the on-board resources of colony-founding queens, since these resources must support the queen and her first brood over all this time. Not surprisingly, therefore, Keller and Passera (1989a) have shown that a relationship exists between the mode of colony founding and the physiology of queens. They analyzed young queens of 24 ant species. Young queens utilizing independent colony founding had a far higher relative fat content than queens of species employing dependent colony founding (Figures 9.1, 9.2). These resources are built up between the time a queen emerges from the pupa as an adult and the time of mating, and serve as vital energy supplies for the queen and the first brood she rears during colony foundation. Queens of species founding independently but nonclaustrally (meaning that they forage for new food after they have started a nest) had intermediate relative fat contents. This makes sense in that these queens require some on-board resources, but not as many as independent claustral foundresses that do not forage at all.

We suggest that the findings of Keller and Passera (1989a) point to the existence of five classes of selective regime influencing the mode of colony foundation.

1. Independent, nonclaustral colony foundation might be favored in sites with a low density of parental colonies and hence relatively little competition between nests. In these circumstances, foundress queens

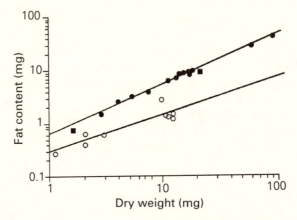

Figure 9.1   Fat content of young queens in relation to their dry weight and mode of colony foundation. The queens are from species in which colony foundation is either independent and claustral (filled circles), dependent (open circles), or independent and nonclaustral (filled squares). (From Keller and Passera 1989a, Figure 1, by permission of Springer-Verlag and the authors. Copyright © 1989 by Springer-Verlag)

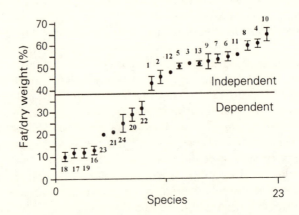

Figure 9.2   Mean (SD) relative fat content of young queens in relation to mode of colony foundation. Numbers 1–13: Queens of independently-founding species of the genera *Aphaenogaster, Camponotus, Formica, Lasius, Solenopsis,* and *Tetramorium.* Numbers 16–24: Queens of dependently-founding species of the genera *Cataglyphis, Formica, Lasius, Linepithema,* and *Myrmica.* (From Keller and Passera 1989a, Figure 2, by permission of Springer-Verlag and the authors. Copyright © 1989 by Springer-Verlag)

would have a small chance of discovery by other established nests if they left their nest site to forage. They would also not require the presence of workers to defend the nest site, or a large body size to house on-board energy resources. Hence, colony foundation could be inde-

314 · CHAPTER 9

pendent and nonclaustral by relatively small queens, and the best strategy for a colony would be to make propagules of minimum size and maximum number. However, suitable circumstances seem unusual, since independent and nonclaustral colony foundation is probably rare among ants, and is mostly restricted to members of the relatively primitive ponerine subfamily (Hölldobler and Wilson 1990 p. 157).

2. Extending this reasoning, independent but claustral colony foundation is a plausible strategy for habitats with moderate competition between colonies. Parental colonies would then be selected to produce queens who avoid detection through claustral colony foundation, and who would therefore have to be individually better-endowed and hence fewer in number.

3. If there are few gaps between established colonies, many more foundresses may aggregate in an available gap than the gap could sustain mature colonies. In these circumstances claustral colony foundation is again required to minimize detection by foragers from established colonies. In addition, there is selection for pleometrotic foundress associations in which unrelated queens may cooperate temporarily, because the larger and more quickly produced worker populations that result may lead to greater success in the competition for living space (Strassmann 1989; Section 8.3).

4. In situations with very high colony densities and hence extreme competition, parental colonies might be selected to invest relatively little in on-board resources for individual queens, and instead to invest massively in a few, large propagules by providing a small number of queens with worker support. In other words, there would be selection for colony foundation through colony budding. An example is in the study of *Formica montana* by Henderson and Jeanne (1992). Colony budding is associated with polygyny (Table 8.1), which is also promoted by habitat saturation (Section 8.6). The models of Sibly et al. (1988) discussed earlier would predict larger propagules (colony buds) as growth conditions for young colonies worsened, and as colony density rose.

5. A further alternative in ant-saturated habitats is for colonies to produce vast numbers of poorly-endowed queens who adopt the socially parasitic strategy of entering already established, foreign nests. This circumvents the need to invest in colony buds and represents the ultimate in dependent colony foundation (e.g. Bourke and Franks 1991; Nonacs 1993a).

9.6 Modular Growth

One of the basic differences between social insect colonies (considered as superorganisms) and many other solitary animals is that insect colonies have modular growth (Harper and Bell 1979; Harper 1980; Mackie 1986). Modular organisms are ones made up of discrete units. In the case of insect societies, these are represented by the eggs, larvae, pupae, workers, queens, and males. Although it might be argued that all advanced metazoans are made up of modules, namely cells, these are so small and numerous relative to the whole organism that growth in advanced metazoans resembles a continuous rather than a discrete process. In addition, many large plants and animals depart from simple modular growth because some of their cells form tissues and organs that are shared by numerous other parts of the organism. For example, plants that have modular growth among their leaves have nonmodular root and vascular systems. Similarly, even though an ant colony is populated by modules such as workers, eggs, and larvae, there is usually a shared infrastructure in the form of a nest and foraging trails. Nevertheless, even though strict modular growth is usually only a theoretical possibility, the idea of modular growth still has conceptual advantages. Thus, it helps explain variable size in modular organisms, by readily suggesting how such organisms can decrease markedly in size whilst remaining both as healthy, and as potentially vigorous in growth, as similar organisms that have only just grown to the same size. Such plasticity and robustness is not seen in organisms with determinate, nonmodular development.

Modular growth also has some advantages for an experimental approach to life history analysis. For example, an ant colony can easily be taken apart and put together again. However, although such experiments have contributed greatly to the study of the division of labor in social insect colonies, they have rarely been used for the empirical investigation of ant life history strategies. Hasegawa's (1993) excellent study of *Colobopsis nipponicus* provides an exception. In this species, the major workers have specialized heads that they use to block the nest entrance. In field experiments, Hasegawa (1993) showed that artificial colonies from which the majors had been removed could not survive, whereas colonies with at least one major per nest entrance could defend their nest site from being usurped by conspecific colonies. One factor preventing more work of a similar type is the difficulty of following experimentally manipulated colonies in the field over meaningfully long timescales. In many cases, it might take years for manipulated colonies to mature and produce sexuals.

In the hope of stimulating such work, this section examines life history analyses of organisms such as certain plants that have at least some aspects of a modular growth pattern. Traniello and Levings (1986) and López et al. (1994) also draw parallels between the foraging strategies of modular plants and ants.

Watkinson and White (1986) suggested a broad definition of modular growth that encompasses all vascular plants. In almost all these plants, the zygote develops into an organism in which one or more structural tissue units are iterated by one or (usually) many growing points, each capable of self-perpetuation. Examples of such structural units include a leaf with its axillary meristem, a bud, a shoot, or a branching system.

The advantage of an analysis of life histories in modular organisms is that it permits a population dynamics or demographic approach to be applied to the growth of the organism at the level of modules (Harper 1980). Watkinson and White (1986) therefore explored the nature and life history consequences of modular construction in plants in terms of growth, reproduction, and survival. For example, changes in size, whether positive or negative, depend on the birth and death rates of modules. If births continue to exceed deaths, plants have the capability of attaining enormous size, especially if they are clonal.

This very familiar Malthusian concept of the dramatic consequences of birth rates exceeding death rates immediately suggests how unicolonial ant societies so rapidly achieve enormous size (Section 10.6). A modular growth viewpoint also helps explain why oscillations may occur in subpopulations of modular units competing for resources (Brian et al. 1981; Section 10.3). Also of general importance, as pointed out by Watkinson and White (1986), is that plants of the same age may show large variation in individual size if individuals differ in their relative growth rates. Therefore correlations in plants between age and size are often weak. The same is almost certainly true of many ant colonies.

The above points, although obvious to anyone with training in population dynamics, are not trivial, because they suggest some useful and different approaches to thinking about the life histories of modular organisms. For example, Watkinson and White (1986) refer to competition between modules for a limited supply of resources yet, in general, there has been little work on this kind of competition in ants. For example, few people have studied resource competition between different brood stages or worker castes in ant societies, although Box 9.1 describes a possible application of this approach.

Care needs to be taken, however, to distinguish between the almost inevitable competition that can occur among modules when resources are limiting, from competition that is the result of kin conflict (Chapter 7). For example, at the proximate level, larvae as a whole and workers as a whole may compete for food when it is in short supply, because members of these two groups have different rates of assimilation of food, and different responses to starvation. The regulation of such competition could even be important at the ultimate level, because in a starving colony it could, say, be better for a small number of weak larvae to die, than for all colony members to be harmed. By contrast, competition among reproductive workers for food, with the winners being able to lay more viable eggs, occurs at the ultimate level as the outcome of kin-selected conflict even when the colony as a whole has sufficient resources (e.g. Cole 1981; Franks and Scovell 1983; Section 7.4).

The potential importance of looking at the effects of modular growth on life history is also highlighted by Samson and Werk's (1986) analysis of reproductive effort in plants. These authors (1986 p. 670) found "strikingly linear relationships between absolute allocation to reproductive biomass and vegetative plant size . . . in many annual species and at least some perennial species." Stearns (1992 p. 88) suggested that this study's findings imply that, over a broad range of plant sizes, reproductive effort is fairly constant per leaf or per branch within a species. One explanation of this finding is that plants respond to environmental heterogeneity by varying the number of modules out of which they are built, but keep the level of reproductive effort within each module roughly constant. In stressful environments, the number of modules is low and the influence of capital expenditures on structures shared by all modules (e.g. tap root and main stem) should be more evident.

These findings suggest, by analogy, that in social insects the number of reproductives raised per worker (the social insect equivalent of a leaf) might also be rather constant, allowing for ecological and phylogenetic constraints. For example, Cole (1984) showed that, in colonies of *Leptothorax allardycei*, the total number of brood items increased linearly with the number of workers. However, in a variety of social insect species reviewed by Michener (1964, 1974), most species attained their highest efficiency at small colony sizes, as measured by the production of immature stages per worker (the "reproductivity effect": E.O. Wilson 1971 p. 338; Section 8.2). On the other hand, the contribution to colony costs of capital expenditure on shared structures (such as the nest fabric) is hardly ever taken into account in assessing worker efficiency.

Finally, in a new approach that falls within the general area of modular life history dynamics, Houston et al. (1988) considered how the foraging strategy of social insect workers might influence the growth and life history strategies of colonies. Foraging workers are the foraging modules of the colony, and the different strategies they adopt may have a large impact on their own mortality rates and hence on the colony's growth rate. Houston et al. (1988) showed in a model that strategies that reduce mortalities of foragers may be selected in preference to strategies maximizing the net rate of gain of new resources. This approach is important because it combines foraging theory and life history theory for social insects, allowing foraging strategies to be evaluated in terms of fitness.

9.7 Reaction Norms

One of the major areas of current interest in life history evolution is phenotypic plasticity (West-Eberhard 1989). This is also of central importance for ant biology. There are two main levels of phenotypic plasticity within colonies. First, plasticity occurs at the level of the individual worker. For example, since caste determination is environmental, larvae with no average genetic differences can develop (in certain species) into either minor or major workers. Second, there is phenotypic plasticity at the colony level. This has already been introduced in the previous section in the discussion of modular growth.

Phenotypic plasticity in life history strategies is common to a wide variety of organisms, and has led to the concept of the reaction norm. This was defined by Stearns (1992 p. 223) as describing "the full set of phenotypes that [the] genotype will express in interaction with the full set of environments in which it can survive." We are not aware of a single thorough study of reaction norms in ant colony life histories. In the future this should be a highly important focus for investigation.

Stearns and Koella (1986) highlighted the usefulness of the reaction norm perspective by using it to address the long-running discussion over whether organisms mature at a predetermined size or at a predetermined age (Sibly et al. 1988; Section 9.5). A reaction norm analysis suggests that they are more likely to do neither. The advantage will lie with organisms that have flexibility. Thus, many organisms are likely to have evolved a reaction norm in which maturation is plastic yet predictable over a range of values that are determined both by age and size. Over part of the reaction norm, usually associated with low growth rates, organisms may appear to mature at a fixed size. Alternatively, in other environments, organisms may appear to mature at a fairly constant age even though they may have highly variable growth rates. At intermediate growth rates, both age and size at maturity are likely to change rapidly with changes in growth rates.

Reaction norms were first put forward to explain plasticity within the life histories of unitary organisms. However, the level of plasticity that one might expect to observe among ant colonies is even greater, because of their modular growth. In future it will be important to combine the two approaches and examine both modular growth and reaction norms. For example, in ant biology it will be necessary to explore colony growth experimentally under different regimes of resource availability. This should help to answer the following key questions. How flexible are the life history patterns of ant colonies? Can colonies swiftly reallocate resources? One possible mechanism for resource reallocation is brood cannibalism (Box 9.1). Is such cannibalism widespread? If so, it suggests that when they are presented with different food supplies, some colonies may risk high levels of foraging and have high work loads (both possibly leading to high rates of mortality among adult workers). This would allow colonies to build up brood reserves that they can rear or cannibalize depending on whether food supplies become better or worse. Thus brood cannibalism and the flexibility it yields could enhance the long-term survival of established colonies. This also illustrates the propensity in social insects for long-term time-lags in their life history dynamics (Box 9.1; Section 10.3).

9.8 Lineage-specific Effects and the Concept of r and K Selection

Twenty years ago, life history theory was considered to have provided one of the most successful broad frameworks in ecology. One general principle within this framework was the idea of r and K selection (MacArthur and Wilson 1967). The letters r and K refer to parameters of the logistic equation. Organisms are classified as r-selected if they reproduce rapidly (and so have a high intrinsic rate of natural increase, r), whereas K-selected organisms are those most likely to make a contribution to the next generation of a population which is at its carrying capacity (K). The r and K concept is therefore based on there being two contrasting types of habitat, r-selecting and K-selecting.

In this scheme, K-selected individuals are predicted to have a large size at maturity, to delay reproduction, to be iteroparous, and to invest relatively less of their resources in reproduction and more in their own maintenance and survival. They are also predicted to invest strongly, possibly with parental care, in producing few, well-endowed progeny. Nevertheless, due to intense competition, many K-selected organisms will have short and, in terms of reproduction, completely unsuccessful lives. At the other extreme lie r-selected organisms. These are predicted to have a small size at maturity, to reproduce early (sometimes being semelparous), and to invest a high proportion of their resources in reproduction and little in their own survival and maintenance. They should also produce high numbers of relatively poorly-endowed

BOX 9.1 BROOD CANNIBALISM IN ANTS

Brood cannibalism in ants is a phenomenon that illustrates the importance of considering ant colonies from the point of view of modular growth (Section 9.6), with an emphasis on phenotypic plasticity and variation in reaction norms (Section 9.7). This is because brood cannibalism permits ant colonies a large degree of flexibility in their life history strategies.

With exceptions such as harvester ants, whose granaries have been known since biblical times, *Cerapachys* ants, which store larvae stolen from other ant colonies (Hölldobler 1982), and honeypot ants, which store liquid food in the bodies of specially modified "replete" workers (Hölldobler and Wilson 1990 p. 333), ants do not hoard food in their nests. Many ants are carnivorous (though many are also opportunistic omnivores). This means that their food supply may fluctuate from feast to famine over short time-scales, so suggesting that food storage would be highly advantageous. However, insect prey is difficult to store because it putrefies. One important advantage of modular growth is that energy supplies can be reabsorbed from certain modules and redirected for use in others. Therefore, subpopulations of modular units may serve as a type of food store. Such a phenomenon may occur quite commonly in ant colonies, as a substitute for storing prey, because ants can cannibalize their brood. In *Camponotus floridanus*, Nonacs (1991b) showed that there is no simple relation between food supply and colony growth. Nonacs suggested that this stemmed from the occurrence of high levels of brood cannibalism in response to the nutritional status of the colony (see also a recent field study on *Pogonomyrmex desertorum* by Munger 1992).

Mature colonies of *Camponotus floridanus* have a single queen and a polymorphic workforce consisting of majors and minors. The workers have the potential to lay eggs, but do so only very rarely when their colony has a viable queen (Nonacs 1991b). Given this social structure, variability of growth among queenright *C. floridanus* colonies cannot arise from competition among queens within colonies, and is unlikely to be due to competition between workers over egg-laying opportunities.

Camponotus floridanus colonies in the field seem subject to a widely fluctuating availability of food both in quantity and composition, with a variable abundance of plant products and insect prey. Protein-rich insect prey would appear to be most valuable of all. However, as mentioned above, dead insects cannot be stored for long before they are consumed by microbes. The only method available to most ants for storing such material is to convert it into their own brood.

Against this background, Nonacs (1991b) measured the growth of six colonies of *Camponotus floridanus* under three regimes of insect–prey and protein availability. He found a linear relationship between a rising abundance of insects in the diet and changes in a number of colony traits

BOX 9.1 CONT.

that were presumably under the queen's control. These changes included increases in the number of eggs, total egg clutch biomass, and larval number, and a decrease in the weight of individual eggs. This last trend probably resulted from a higher rate of egg-laying and shorter maturation of the eggs within the queen's ovaries. Thus the queen responded to an increase in insect prey with a linear rise in productivity. In contrast, traits such as larval and total brood biomass and pupal numbers, which were presumably under the workers' control, changed nonlinearly with greater insect prey availability. Specifically, the colonies grew fastest under the intermediate regime of insect protein availability.

Nonacs (1991b) proposed that these results were due to colonies with intermediate levels of protein availability working harder in foraging and maintaining brood, to have larvae to cannibalize if their relatively intermittent protein supply fell even more. This would not necessarily mean that these colonies would have the highest long-term growth rates in the field, because in the laboratory environment the extra risk of predation from having hard-working foragers was artificially absent. But it does point to the possible importance of using larvae as living food storage vessels to safeguard the colony against environmental variation (Nonacs 1991b). In addition, Nonacs argued that colonies with the high-protein diet had no need to shunt resources from dead prey into storage in larval tissue, and this accounted for their lower growth rates in the laboratory. These possibilities remain to be tested in the field. Apart from being important because they show how brood cannibalism can serve as a mechanism in a colony bet-hedging strategy, these conclusions also point to the existence of reaction norms due to modular growth in social insect colonies (with the flexible response of colonies to different nutritional environments constituting the reaction norm). They also suggest the occurrence of considerable and variable time-lags both in the effect of external change, and in the colony's response, in ant life histories.

As Nonacs (1991b) pointed out, the idea that cannibalism of offspring may be adaptive in social organisms is not new. The consumption of eggs and brood in social insects is very common (e.g. Dlussky and Kupianskaya 1972; Crespi 1992b). For example, *Solenopsis invicta* colonies cannibalize larvae in periods of hunger stress and may also show low levels of brood cannibalism even under moderate stress (Sorensen et al. 1983). Brood cannibalism may be particularly important during colony foundation by monogynous *S. invicta* queens (Tschinkel 1993b).

Lastly, by buffering colonies against fluctuations in the food supply, brood cannibalism within ant colonies may help to reduce the variance in reproductive success. This may be advantageous in certain circumstances (e.g. Kozlowski and Stearns 1989; Wenzel and Pickering 1991).

progeny. Furthermore, the actual survivorship of *r*-selected individuals will vary considerably depending on the (unpredictable) environment in which they find themselves (Begon et al. 1986 pp. 521–522).

For these reasons, it is tempting to consider all ants as *K*-selected. However, this is a misleading use of the concept, because *r* and *K* selection can only be used as comparative terms. This is worth exploring in detail because it suggests some important cautionary points for the study of life histories in ants.

Although the *r* and *K* hypothesis was originally intended to make comparisons between closely related species, it has often been used at higher taxonomic levels. But support for the hypothesis from the level of families, orders, or even phyla is partly illusory. This is because of biases caused by ancient lineage-specific effects. For example, when Stearns (1983) analyzed vertebrate life histories, he found that *r* and *K* patterns in the unmanipulated data accounted for 68% of the covariation in ten life history traits. However, the fraction of the variation that could be explained in terms of *r* and *K* selection was progressively reduced when the effects of body weight, order membership, and family membership were statistically removed. Therefore much of the evidence for *r* and *K* selection was an artefact of making comparisons at a high taxonomic level (Stearns 1992 p. 105).

This kind of analysis provides a warning for studies of social insects in general and ants in particular. Not only are there few data on colony size and reproductive allocation in social insects (Tschinkel 1991), but the phylogenetic relationships of social insects are much more poorly understood than in many vertebrate groups. In general, a comparison of differences in strategies is only meaningful if the organisms involved could conceivably use the same strategies. Therefore, returning to the earlier theme, ants should not be labeled as *K*-selected in the absence of a group with which to compare them. In addition, we suspect that lineage-specific effects will continue to haunt social insect biologists in their attempt to understand life history evolution in ants at levels above the species. Certainly, recently recognized difficulties with the comparative method must be taken into account (Harvey and Pagel 1991; Section 5.2).

9.9 Special Issues in Social Insect Life History Evolution

This section considers a number of theoretical issues, some unique to social insects, that have a direct bearing on their life history evolution. Some of these issues arise because of the genetic structure of social insect colonies. For example, this structure means that the fitness interests of colony members are often expected to diverge (Chapters 6, 7). In addition, if colony kin structure varies within populations, colonies are selected to have variable patterns of sex allocation (Boomsma and Grafen 1990; Section 6.2). This may promote variation in the life history strategies of colonies within the same population.

KIN CONFLICT AND LIFE HISTORY IN ANTS

There are very few studies, even theoretical ones, that explore the relationship between kin conflict and life history patterns in ants. One of the most recent and thorough is that of Pamilo (1991b; Section 7.6). Others were presented by Bulmer and Taylor (1981) and Bulmer (1983b), who analyzed a special case of conflict and its effects on colonies that reproduce by fission (Sections 4.8, 10.8).

The central insight is that, in perennial colonies, queens and workers may have different interests in how resources are allocated between sexual production and the colony's maintenance (worker production). Specifically, the queen should favor investment in colony maintenance to keep herself as a reproductive individual, whereas the workers should prefer to invest more in sexual sibs (Section 7.6). In modeling this conflict, Pamilo (1991b) assumed that a perennial colony can split its resources three ways, investing one fraction in daughters, another in sons, and the third in workers. Pamilo also assumed, for simplicity, that the combined reproductive success of the sexual offspring is a linear function of investment, and that colonies have one, singly mated queen producing all offspring. The model further assumed that every colony has the same amount of resources to invest in total (although this is unlikely ever to be the case: Section 9.5), and that the sex ratio is not affected by resource allocation between sexuals as a whole and colony maintenance.

The results of Pamilo's (1991b) model showed that the greater the success of colony-founding queens, and the greater the survivorship of established colonies, the lower the expected fraction of investment in worker production. These findings make sense in that more queens should be produced if each survives well, and similarly fewer workers need to be produced to maintain the colony if colonies are unlikely to die. Therefore, this modeling demonstrates the importance of the quantitative relationship between investment in colony maintenance and the probability that the colony and its daughter queens will survive. The model also demonstrated that the exact form of this relationship differed under queen and worker control. This confirmed that conflict should occur between queens and workers over resource allocation.

Evidence for the existence of this conflict in *Formica truncorum* colonies in the field was presented by Sundström (1993b, 1995a; Section 7.6).

RELATEDNESS ASYMMETRIES AND SPLIT SEX RATIOS

When the workers' relatedness asymmetry varies among colonies, workers are selected to produce colony sex ratios of mainly one sex, leading to split population sex ratios (Boomsma and Grafen 1990; Sections 4.5, 6.2). This phenomenon is potentially of direct relevance to understanding sources of variability in ant colony life history patterns. First, relatedness asymmetry may add another level of complexity in any attempt at understanding reaction norms for social insect colonies (Section 9.7). Second, colonies selected to raise exclusively males may be constrained in their worker production schedules, since there could be fewer diploid eggs available for rearing as workers. The potential for significant consequences of variable relatedness asymmetries for life history traits again illustrates the need to know the genetic structure of ant colonies to understand their social evolution.

THE EFFECTS OF POLYGYNY ON LIFE HISTORY STRATEGIES

In a recent review, Nonacs (1993a) suggested that there are three major life history strategies in polygynous ants. He termed these the alate dispersing, fissioning, and space-perennial strategies. Alate dispersing refers to the production of winged queens with the ability to disperse to distant habitat patches (Sections 8.6, 11.5). By fissioning, Nonacs (1993a) meant the division of the parental colony in both true colony fission, as shown by the army ants, and in the separate process of colony budding, as found among wood ants (Sections 4.8, 10.5, 10.8). Space-perennial is a useful new term for colonies that are selected to maximize their longevity in a given nest site. Even though all three life histories may be exhibited within a single species, these categories are helpful because a number of predictions can be made from them, particularly concerning the effects of polygyny. Nonacs' (1993a) main predictions for the alate dispersing strategy and the fissioning strategy concerned sex allocation and were discussed in Chapter 4 (Sections 4.7, 4.8). An important aspect of the fissioning strategy that has life history consequences is as follows. During colony division, more workers are expected to stay with their mother than join a daughter colony because of their higher relatedness with the offspring of their mother (cf.

Bulmer 1983b; Pamilo 1991b; Section 4.8). Polygyny tends to eliminate this relatedness difference, so the division of the worker population between daughter colonies should be more equitable.

The space-perennial strategy, as its name suggests, also has important life history consequences. Nonacs (1993a) proposed that space-perenniality should be highly favored among colonies that compete for limited and saturated habitat patches, since mortalities of young colonies are likely to be extremely high due to the low availability of gaps (cf. Section 9.3). But a colony can only realistically monopolize the same site indefinitely if it polygynous, since queen adoption allows colonies to be potentially immortal. In addition, habitat saturation independently promotes polygyny (Herbers 1986c; Section 8.4). Therefore space-perenniality and polygyny go together (Nonacs 1993a). Space-perennial species should also invest much more in workers than in reproductives. However, the last generalization refers to a population average, and certain colonies might invest heavily in sexual production in any given year or across several seasons. Good examples of space-perennial, polygynous species that occur, as expected, in very dense populations are the *Formica* wood ants (Sections 8.6, 10.5). A further consequence of space-perenniality is that dispersing females may often attempt to invade or parasitize existing colonies in preference to founding their own nests independently. Therefore, it should lead to intraspecific and interspecific social parasitism (Nonacs 1993a); the latter is certainly prevalent in *Formica* ants (Section 10.5).

9.10 Conclusion

Life history theory is potentially one of the most important unifying approaches in evolutionary biology and ecology. Ideally it should explain such basic features as associations between different genotypes and their environments (e.g. reactions norms), patterns of dispersal and growth, resource allocation between growth and reproduction, and patterns of juvenile and adult mortality. Ultimately, it should explain how population dynamics and population genetics are linked. All of these issues are fundamental to understanding why certain organisms are fitter than others in the same population. However, life history theory has arguably not yet provided broad, well-tested unifying concepts. In fact, Stearns (1992 p. 208) concluded that "There are virtually no general predictions in life history theory. . . ." Nevertheless, life history theory provides many pointers to areas of importance.

The application of life history analysis to social insects in general and ants in particular is still in its infancy. On the one hand, ant colonies look like ideal experimental material for understanding, for example, the role of modular growth in life history evolution, because the structure of colonies can be experimentally manipulated. On the other hand, the existence of within-colony kin conflict in ants adds another layer of complexity to life history theory. In addition, the study of life history evolution in ants has been hampered by practical difficulties with collecting basic demographic data on the age structure, productivity schedules, and survivorship of individuals and colonies (Tschinkel 1991). These arise from the great longevity of colonies, and the near impossibility of following the lives of individual queens and workers in the field over the long term. All these features make life history analysis within the social insects particularly challenging. These challenges should be taken up, however, since a full understanding of kin conflict will only be achieved in the context of life history analysis. They should also be addressed because ants have a dominant role in most terrestrial ecosystems (e.g. Sudd and Franks 1987; Hölldobler and Wilson 1990), and this role depends to a large extent on their life history patterns (Chapter 10). Finally, studying life history evolution in ants promises to yield many findings of wide applicability to the field as a whole.

9.11 Summary

1. Ant colonies follow a perennial life history with repeated episodes of reproduction (iteroparity). Typically, a colony starts with a single foundress queen (founding stage), grows by producing workers only over several years (ergonomic stage), undergoes a period of mixed worker and sexual production (reproductive phase), and declines after the death of the queen, when workers may produce their own sons (orphanage phase). However, there are many variations on this basic pattern.

2. The life history model of Oster and Wilson (1978) predicted that the optimal policy for perennial colonies is, after a period of worker-only production, to produce workers and sexuals in a "bang-bang" pattern each season, involving mostly worker followed by mostly sexual production.

3. As a soil-nester with wingless workers, the typical ant colony occupies a territory in which workers both forage for food and destroy any incipient colonies that arise. Therefore, available space tends to become filled by mature colonies and juvenile mortality (the mortality of young colony-founding queens) is very high. This creates a life history dominated by gap dynamics, which probably underlies the perenniality and iteroparity of ants. Colonies must produce sexuals repeatedly so that

some will have a chance of finding a rare, unoccupied gap (a form of bet-hedging), which also means they must monopolize their nest site over many years. This interpretation is supported by models developed for life history evolution in trees. At very high mortalities of foundress queens, colonies may evolve both perenniality and polygyny, allowing them to monopolize their nest site indefinitely in a "space-perennial" strategy.

4. As the mortality of young queens increases, for example with an increasing density of nests, the mode of colony founding should also change. It should switch from independent and nonclaustral (queens found nests alone using energy obtained on foraging trips), to independent and claustral (queens found nests alone using metabolized energy reserves), and finally to a dependent mode. This last step may involve the evolution of colony fission, colony budding, or even colony foundation by social parasitism. Associated with these changes are changes in the relative body size and fat content of young queens.

5. Ant colonies are characterized by modular growth, meaning that they grow or shrink according to changes in the sizes of subpopulations of the discrete units of which they are composed (individual adults and brood items). Ideas about modular growth, developed largely for plants, may contribute to an understanding of ant life histories. For example, the relative ease with which colonies can reallocate resources from one set of modules to another without overall harm to the colony (as when workers cannibalize brood in starving colonies) buffers them against environmental heterogeneity. This flexibility on the part of colonies also suggests that the idea of a reaction norm (the mapping of the genotype onto the phenotype across different environments) is important for ant life history analysis.

6. Comparative studies of life history evolution in ants are likely to be incomplete until their phylogeny is better understood, because of the need to eliminate lineage-specific effects.

7. A model by Pamilo (1991b) predicts kin conflict between the queen and workers over the allocation of resources between investment in worker production, and investment in sexual production. The queen favors relatively greater investment in worker production, since this maintains her as a reproductive for longer. The occurrence of kin conflict adds a uniquely new dimension to the study of life history evolution in ants.

10 The Diversity of Life Histories in Ants

10.1 Introduction

This chapter considers the social structure and life history patterns of seven groups of ants. These are the species *Solenopsis invicta*, *Linepithema humile* (= *Iridomyrmex humilis*), *Pristomyrmex pungens*, and *Eciton burchelli*, the genera *Myrmica* and *Formica*, and the tribe *Leptothoracini*. We focus on some of the species that have been most intensively studied over the last few years. The seven case studies also show some of the remarkable diversity and complexity in life history (sometimes within a single species) found in ants. Thus, they nearly all deviate in interesting ways from the "typical" ant life history outlined in Section 9.1. Equally importantly, they show that it is only possible to understand the evolutionary ramifications of such processes as kin selection and kin conflict with a detailed knowledge of a species' ecology, social structure, and life history. On the other hand, their very complexity means that a complete understanding of all the phenomena uncovered in these case studies remains elusive.

10.2 The Imported Fire Ant, *Solenopsis invicta*

The ant *Solenopsis invicta* is a small myrmicine known as the imported fire ant because of its accidental introduction to the United States of America from Brazil in around 1940 (Lofgren et al. 1975), and because of its vicious sting. *S. invicta* is now a major agricultural pest in the U.S.A. It is also one of the best understood ants in the world, partly because to control pest organisms it is often necessary for biologists to have a deep understanding of their basic biology.

One of the most fascinating things about *Solenopsis invicta* in North America is that it occurs in a monogynous and a polygynous form. Several polygynous populations have apparently evolved independently in only a few decades from an originally monogynous founder population (Glancey et al. 1973, 1975; Jouvenaz et al. 1989; Ross et al.

1987a, b, 1993). The recent and rapid evolution of a polygynous form of *S. invicta* may provide special insights into social evolution in action (Sections 8.4, 8.5).

RELATEDNESS AND SOCIAL STRUCTURE

Two important features that the monogynous and polygynous forms of *Solenopsis invicta* have in common are that queens appear to be always singly inseminated, and workers possess no ovaries and so are obligately sterile (Ross and Fletcher 1985a; Ross et al. 1988). Ross and Fletcher (1985a) also showed that functional queens in polygynous colonies are no more closely related with nestmate queens than with other queens. Consequently within-nest relatedness of workers in polygynous nests is also low, although it is greater than zero (Ross 1993).

Colonies of *Solenopsis invicta* in monogynous populations may be founded either haplometrotically or pleometrotically, that is by one queen or by several queens. However, if there is pleometrosis, the first workers that emerge kill all but one of the foundresses (Tschinkel and Howard 1983). Thus mature colonies are monogynous (Section 8.3). By contrast, the polygynous colonies appear to reproduce by budding, so that a new colony may contain several thousand workers and sometimes even a hundred or more queens (Fletcher 1983; Greenberg et al. 1985). The workers in polygynous colonies are generally smaller than those in monogynous nests (Greenberg et al. 1985; Ross and Fletcher 1985a). Monogynous colonies also have workforces that can be distinctly polymorphic, with a conspicuous major and minor caste (Calabi and Porter 1989).

ECOLOGY AND COLONY GROWTH

Tschinkel (1988b) examined *Solenopsis invicta* colonies ranging in age from incipient ones to some that were 12 years old. The study population consisted of monogynous nests. Tschinkel's method was to look at colonies in sites that had been disturbed at known times in the past. Even though colony size showed strong seasonal variation, he demonstrated that colonies grew approximately logistically, reaching half size between 2.5 and 3.5 years and reaching a maximum size of 220,000 workers after 4 to 6 years. There was some circumstantial evidence that colony growth rate varied with food density. Tschinkel (1988b) also showed that incipient colonies have small monomorphic workers and that colony growth is associated with a strong skewing of the worker size frequency distribution. Thus, in full grown colonies the distribution of worker sizes resembled two slightly overlapping normal distributions, namely a narrow one of minor workers and a much broader one of majors.

Recently, Tschinkel (1993a) reported an exhaustive study of colony demography in monogynous *Solenopsis invicta* populations. A key finding was that the proportion of investment in sexuals (as a fraction of total annual production) increased rapidly in colonies of between 20,000 and 50,000 workers. This proportion remained at about 0.33 over each subsequent year of colony life. Hence the transition from the ergonomic to the reproductive stage in colony development was abrupt (Tschinkel 1993a), as predicted by Oster and Wilson's (1978) life history model (Section 9.2). In addition, colony size appeared to fluctuate as worker production annually gave way to sexual production (Figure 10.1).

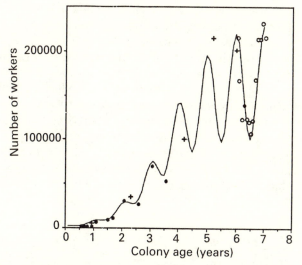

Figure 10.1 Growth of *Solenopsis invicta* colonies. The curve is fitted to data from S.D. Porter (unpublished data) (open circles), Markin et al. (1973) (filled circles), and Tschinkel (1988b) (crosses), and is a logistic function with a logistic increase of seasonal variation. (From Tschinkel 1993a. Copyright © 1993 by the Ecological Society of America. Reprinted by permission)

A major worker in a monogynous colony of *Solenopsis invicta* is a magnificent creature, and can be up to 20 times the weight of a minor (Calabi and Porter 1989). Tschinkel (1988b) showed that the increased relative abundance of majors in larger colonies came about because the growth of the subpopulation of majors is faster than that of the worker population as a whole. (This is an excellent example of the modular growth of two subpopulations within the same nest: Section 9.6.) In full-sized colonies, about 35% of the workers are majors. Total biomass investment in majors increases as long as colonies grow, beginning at about 10% at 2 months and reaching about 70% in mature colonies. Why mature colonies should put such massive investment into the production of majors is not yet clear. However, from a life history perspective, majors could represent the equivalent of somatic investment in a long lived perennial organism.

Calabi and Porter (1989) made a very thorough examination of the relative costs of minors and majors in *Solenopsis invicta*. They calculated energetic costs for three sizes of workers from data on biomass, longevity, respiration rates, and the energy content of worker tissue. A strong feature of this study was that it included calculations for two different temperature regimes, reflecting the variation in temperatures that colonies might encounter in nature. Calabi and Porter (1989) showed that large workers lived considerably longer than small workers (50–140% longer depending on the temperature) and respired more slowly per unit weight of tissue (40% more slowly). It was therefore possible to estimate that, overall, large workers cost about 30% less, in energy terms, than an equal weight of small workers. However, on an individual basis, large workers were four times more expensive than small workers because of the six-fold weight difference. Thus, each large worker would have to provide services equivalent to a least four small workers to justify the colony's energy investment (Calabi and Porter 1989).

This consideration of a colony's investment patterns poses the problem of what mechanisms are responsible for changing priorities in allocating investment. There are two obvious candidates: patterns of food flow within colonies and social control of egg laying.

Food Flow and Egg Laying within Colonies

Tschinkel (1988a) discovered an intriguing potential interaction between food distribution and processing by different members of the colony on the one hand, and the regulation of the queen's egg-laying on the other. (For a pioneering study of food flow in ant colonies, see Wilson and Eisner [1957]. For other studies of food flow in *Solenopsis invicta*, see Howard and Tschinkel [1981] and Sorensen et al. [1985].) Tschinkel (1988a) showed that there was a significant, positive relationship between the logarithm of the number of fourth instar larvae and the logarithm of the queen's egg-laying rate. So it seems that fourth instar larvae are needed to stimulate and maintain oviposition. For example, the larvae may form a digestive and metabolic caste that processes protein for egg production by the queen. Tschinkel (1988a) thus suggested that the queen's egg-laying rate and the abundance of fourth instar larvae are linked in a positive feedback loop. It would follow that either the logarithmic relation of fecundity to larval numbers, or physical limitations of the queen, set the maximum egg-laying rate and thus, at the proximate level, determine the upper colony size (Tschinkel 1988a).

THE RELATIONSHIP BETWEEN MONOGYNOUS AND POLYGYNOUS POPULATIONS

As mentioned above, the polygynous form of *Solenopsis invicta* differs from the monogynous form not only in its social organization, but also in the diminutive size of its workers and the likely use of budding as a mode of colony foundation. Because of these differences, investigators have tried to determine if populations of the two forms are reproductively isolated from one another. However, genetic work shows that in fact they seem to interbreed rather freely (Ross et al. 1987a,b; Ross 1992; Ross and Shoemaker 1993; Ross and Keller 1995; Section 8.4). Queens of each form are readily adopted into colonies of the alternative form in the laboratory. In addition, sexuals of the two forms exhibit virtually identical morphologies, and similar frequencies of queens producing diploid males occur in each (Ross and Fletcher 1985b, 1986; Ross et al. 1987b). The adaptive advantage of polygyny in *S. invicta* is unclear, but could lie in the protection it affords against the damage to colony survival caused by diploid male production (Sections 8.4, 8.5).

CONFLICTS IN POLYGYNOUS COLONIES AND THE CONTROL OF PRODUCTIVITY

Competition among the queens in polygynous *Solenopsis invicta* colonies, in particular over the production of sexuals, is expected because such queens are generally not related to one another. Indeed, a decrease in individual reproductive output with increasing number of reproductive individuals is believed to be a general feature of social insect colonies (Michener 1964; Section 8.2). Such "reproductivity effects" are well documented in *S. invicta* (Vargo and Fletcher 1986a,b, 1987, 1989). However, whether they are a direct aspect of reproductive competition is unclear; they could instead represent an adaptive response of workers to the presence of extra queens (Keller and Nonacs 1993; Keller and Reeve 1994a; Section 8.2). The same applies to Vargo's (1992) finding that *S. invicta* queens pheromonally inhibit one another's reproduction (Keller and Nonacs 1993; Section 8.2).

Some individual queens in polygynous *Solenopsis invicta* colonies

may dominate the production of sexual progeny and produce few workers, but they may only do so at the cost of low survivorship (Ross 1988b;
Section 8.6). This intriguing association of sexual production with a
short life raises the possibility that there are two kinds of queen strategy. The first consists of long lived queens producing few daughter
queens per season but doing so repeatedly, and the second of short
lived queens that opt for more massive reproduction in a single season
(Keller 1993b). In other words, it is possible that individual queens in
polygynous *S. invicta* nests exhibit either annual or perennial reproductive strategies.

There are also genetic influences on the outcome of reproductive
competition in *Solenopsis invicta*, and on which queens are reared in
polygynous colonies. Workers preferentially kill queens of a particular
genotype (at the *Pgm–3* locus: Section 8.4). Such queens are potentially
so productive that if they were raised, and successfully invaded polygynous colonies, they might cause them to revert to monogyny (Ross
1992; Keller and Ross 1993a). Thus the so far unexplained behavior of
the workers may help to preserve the polygynous form of *S. invicta* in
North America (Section 8.4).

Given the somewhat exotic genetics of *Solenopsis invicta*, possibly
associated with its status as an introduced species (Ross et al. 1993), it
may be dangerous to generalize from the North American populations
of this species to the evolution of polygyny in other ants (Section 8.4).
Nevertheless, *S. invicta* does reveal the extent to which there can be very
different social phenotypes in an ant with little genetic differentiation.

CULTURAL TRANSMISSION OF SOCIAL STRUCTURE

Keller and Ross (1993b) made the intriguing discovery that the social
organization of *Solenopsis invicta* colonies can be culturally transmitted.
New queens reared in monogynous or polygynous colonies of *S. invicta*
differ hardly at all in fat content and overall weight at the pupal stage
and when they first become adults. However, by the time the queens
are ready to participate in a nuptial flight, large-scale differences
between queens reared in the two types of nest have appeared. Mature
queens from monogynous colonies are, on average, 63% heavier in
terms of dry weight than those from polygynous colonies. Cross-foster-
ing experiments suggested that it is the social environment in which pre-
flight queens are reared, rather than their genetic constitution, that is
the major determinant of their phenotypic traits. For example, queen
pupae transferred from polygynous to monogynous colonies were fat-
tened up just like the resident young queens. Therefore, this pioneering
study shows that the social environment of ant colonies can cause

changes that probably feed forwards across generations. In other words, it suggests that, at the proximate level, monogyny and polygyny can be self-perpetuating. For instance, light queens would be poor solitary foundresses and so may have little option but to try to infiltrate established polygynous colonies, thus perpetuating polygyny. This raises fascinating issues of the interplay between cultural transmission and genetic change in the evolution of ant societies. The work of Keller and Ross (1993b) also highlights the considerable phenotypic plasticity that may exist within colonies of social insects (West-Eberhard 1989; Section 9.7).

10.3 The Red Ants, *Myrmica*

Myrmica is one of the most extensively investigated of all ant genera. In Britain, it has been studied in detail since the 1940s by M.V. Brian and colleagues, most notably in recent times by G.W. Elmes. The following account is largely drawn from the excellent review by Elmes (1991), with some changes in emphasis. For example, *M. rubra* had been studied intensively for 30 years before Smeeton (1981, 1982a,b,c) showed that most males are produced from worker-laid eggs. This worker male-production is one fundamental difference between *Myrmica* and *Solenopsis invicta* (where workers are completely sterile). Factors such as the parentage of males are central to understanding conflicts between workers and queens in these societies, and these conflicts may in turn influence life history patterns.

Myrmica are extremely common in north temperate habitats. They occur in open pine woodland, chalk downlands, acid heathland, and peat bogs. According to Elmes (1991), more than 600 species and varieties of *Myrmica* have been described. The workers of *Myrmica* are always monomorphic and may be up to 10 mm in length. As in other polygynous ants, the queens in *Myrmica* are not much bigger than their workers.

A particular strength of the work of Brian and Elmes and their colleagues on *Myrmica* is that they have investigated the social physiology (how colonies function) of *Myrmica* in the context of the ecology of this genus (Elmes and Keller 1993; Section 8.4). Elmes (1991) suggested that all species of *Myrmica* are nonterritorial (Brian 1956) and have "submissive" foragers (Savolainen and Vepsäläinen 1988, 1989; Vepsäläinen and Savolainen 1990). In other words, they run away from encounters with larger or more numerous foragers from other ant genera. *Myrmica* colonies are found, for example, at decreasing density with increasing proximity to nests of the aggressive wood ant *Formica polyctena* in Finnish forests. Species with submissive foragers can avoid competition with more aggressive species by habitat partitioning (Brian 1955a), and by concentrating foraging activity at different times of day or in different micro-sites (Savolainen and Vepsäläinen 1989), or even at different seasons within the year (Elmes 1982). *Myrmica* ants may also compete among themselves for nest sites (Elmes and Wardlaw 1982b,c; Elmes 1991). The picture that emerges from these studies is of a genus of ants that must attempt, rapidly, to make a living in the gaps between more aggressive colonies. *Myrmica* is therefore an ecological "guerrilla strategist" (Begon et al. 1986 p. 195), meaning a hit-and-run exploiter of ephemeral opportunity. This is in marked contrast to the "phalanx strategy" of, for example, *Linepithema humile* (= *Iridomyrmex humilis*), whose large and potentially overwhelming numbers allow it to dominate the communities it invades (Section 10.6). Elmes (1991 p. 20) concludes: "one expects that *Myrmica* colonies should be mobile, highly defensive of their nest-site, should be able to rapidly colonize new nest sites and to be flexible in their reproductive strategy."

Myrmica colonies can be extremely variable in terms of the population sizes of both workers and queens, even within a species. Mature colony sizes range from fewer than 50 workers to more than 5000 (Elmes 1991). Within-population studies generally reveal a clear correlation between the numbers of queens and workers inside colonies (e.g. Elmes 1973a). But, according to Elmes (1991), there is no reason to assume that either queen or worker number is dependent on the other. Across species, queen number ranges from an average of less than one per colony, to more than fifteen. (An average of less than one implies a population of monogynous colonies of which a fraction are orphaned, queenless colonies.) For example, *Myrmica schencki* is basically monogynous, whereas *M. rubra* has a large but highly variable number of queens per colony. Most intriguingly, however, the shapes of the frequency distributions of numbers of queens per colony in nine *Myrmica* species can all be described by the same mathematical function (Elmes 1991). Elmes (1991) therefore suggested that queen number in all these species is generated by a common process of queen recruitment. Elmes and Petal (1990) showed that queen numbers within a population can vary markedly both in time and space (Section 8.4). For example, the average number of queens in colonies of *M. rubra* varied significantly between years and between different grassland sites. Cyclical variation in queen numbers was also found in two separate populations of *M. sulcinodis* (Elmes 1987a; Figure 8.1). *Myrmica* queens may only live for three years or less whereas polygynous colonies may continue almost indefinitely (Elmes 1991). According to Elmes (1991), the short lifespans of queens, combined with frequent queen recruitment, may help explain both the observed frequency distributions of queens per colony in different species, and the cycles in queen number.

Elmes (1973a) made the controversial suggestion that *Myrmica* queens act as intraspecific social parasites upon the worker population. Queens in *Myrmica* colonies are periodically recruited into colonies (Elmes 1980), and even though related queens may be favored, non-related queens may infiltrate the colony during this process (Elmes 1991). However, in deciding whether queens should be regarded as parasites, what matters is not that some unrelated queens invade nests, but (1) the average relatedness of recruited queens, which is almost certainly above zero in *Myrmica*, and (2) whether all queens contribute to the production of workers (Section 8.6). In short, if the average adoptee imposes a net cost, adoption is expected to be unstable (Section 8.4). Therefore, the label of social parasite might be best reserved for queens who only produce sexual offspring in the colonies they invade (Section 8.6). As described below, this condition is strongly suggested, so far, only for *M. rubra* microgynes.

Before considering sexual production by *Myrmica* ants, it is worth recalling that these are typically north temperate species living in highly seasonal environments. Life history strategies are largely about the timing of resource allocation (Section 9.2). *Myrmica* seasonality is therefore a central feature of life history strategy within the genus. However, ant biologists simply do not know if the complex response to seasonality shown by *Myrmica* ants occurs in other ant species, because there are few detailed studies of the social physiology of other ant genera. It would therefore be unwise to dismiss *Myrmica* as an unusually complex case.

One set of seasonal effects influences caste determination in *Myrmica*. For example, eggs laid by queens in early summer tend to become workers, whereas ones laid later in summer overwinter and emerge the next year as either queens or "spring" workers (e.g. Brian 1955b; Elmes 1991). There is also a "queen effect" on caste determination in the genus, which results from workers (in the presence of a colony queen or queens) treating female larvae in such a way that they are more likely to develop into workers than young queens (Brian and Carr 1960; Brian 1973a; Elmes and Wardlaw 1983; Section 7.6). The proximate cause appears to be a pheromone produced in the abdomen of fertilized queens (Brian 1973b). Queen effect seems universal in its action among *Myrmica* species (Elmes and Wardlaw 1983; Elmes 1991). When a queen is not present workers care differentially for the biggest larvae. These larvae then have the best chance of successfully developing into a young queen (if diploid) or male (if haploid). So it appears that workers can recognize which larvae have the greatest potential to develop into young queens (Brian 1975). However, the ultimate significance of the queen effect is not yet clear (Section 7.6).

The queen effect notwithstanding, the workers in *Myrmica* maintain some independent control because they can alter larval growth rates through their control of the feeding of the larvae and the temperature at which they are reared (larvae may be moved to or away from warmer chambers). The queen effect also has a role in the production of male eggs by workers. In the presence of queens, workers eventually switch from producing reproductive eggs to laying highly nutritious trophic (nonviable) eggs (Brian 1969; Brian and Rigby 1978). These may be directly consumed by the queens.

This phenomenon is not sufficient, however, to prevent substantial production of males by workers. Under natural conditions, *Myrmica rubra* colonies produce a large number of male eggs in late summer, and almost all of these may be laid by young workers that emerged in early summer (Smeeton 1981, 1982b). Male-egg production by workers seems to be inversely correlated with the abundance of larvae in the nest (Smeeton 1982a). This might be directly caused by nests with few larvae having surplus resources, allowing workers themselves to have sufficient resources to be able to lay eggs. Another suggestion is that egg laying by workers is determined in part by the availability of food in the nest when they themselves were larvae (Smeeton 1982c). These densely interdependent relationships of production and inhibition may give rise to complex cycles in the production schedules of *Myrmica* colonies (Brian et al. 1981; and see below).

The social system of *Myrmica* can rapidly adapt to changing conditions, but only at the possible expense of further plasticity. For example, a colony might overwinter with brood that could develop into large numbers of males and sexual females, but if conditions turned bad in the spring many of the diploid larvae would develop into workers rather than queens. Such workers are, so to speak, failed young queens and would be larger than average and more likely to produce many of their own male offspring (Smeeton 1982b). The colony would therefore tend to produce a male-biased sex ratio during the summer following the bad spring, and also to have a higher potential male production the following year (Elmes 1991). In this way the results of one poor spring could feed forwards for a considerable period of time. This in turn means that the effect of social factors can only be detected with long-term population censuses, and when climatic and other physical variation can be eliminated statistically (Elmes 1991).

The flexibility in *Myrmica* that stems from plastic queen–worker caste determination, polygyny, and worker production of males can be

interpreted as a possible colony-level adaptation in an ant that is fundamentally an opportunist (Elmes 1991). Its use of unpredictable and
ephemeral resources creates a selection pressure that favors rapid
changes in resource allocation rather than a steady, perennial pattern of
investment and expenditure. In *Myrmica* much of the flexibility results
in a biased investment towards males rather than female alates.
Nevertheless, sex ratios in *Myrmica* are not always male-biased (e.g.
Elmes 1987b; Sections 5.5–5.7), suggesting that the constraints of social
design in *Myrmica* colonies can frequently be overcome by selective
processes.

A QUANTITATIVE LIFE HISTORY MODEL FOR *MYRMICA*

One of the only life history models that has been specifically designed
to explore the characteristics of a particular social insect genus or
species is that developed by Brian et al. (1981) for *Myrmica*. In this
regard, the model has one peculiarity: it is specifically for a monogynous
society. Monogyny is uncommon in *Myrmica* (but see Elmes 1974a;
Snyder and Herbers 1991), and its encoding in the model was clearly to
avoid the extra complexity associated with possible conflict among
queens in polygynous societies. Nevertheless, the model of Brian et al.
(1981) makes some important general predictions.

The *Myrmica* model is numerical and deterministic. It considers two
discrete time periods per year. In this way the model can take into
account the observation that more female eggs are laid in the first half
of the year and all male eggs are laid by workers in the second half
(Smeeton 1981). The model assumes that the system is fundamentally
food-limited and that available food is partitioned between the sexual
female and worker offspring of the queen and the male offspring of the
workers. Control "switches" in the model regulate the growth of different populations within the society. As worker numbers build up, production is shifted into the output of reproductives to prevent further
growth and resulting inefficiency. For example, the production of both
males and females is regulated by a control system that assesses the
worker : queen ratio. The setting of the controls can, in theory, be opti-
mally tuned so that maximum numbers of sexuals are produced with an
equal investment in male and female reproductives.

Equal investment in sexual progeny was taken as a constraint
deduced from theoretical considerations (Brian et al. 1981). (The stable
sex ratio for queens and nonlaying workers is indeed 1 : 1 in monogynous colonies with all males coming from workers [Table 4.1],
although Brian et al. [1981] assumed polygyny in their discussion of sex

allocation.) Since queen weight is roughly double that of males, this investment necessitates two males for each queen. As Brian et al. (1981 p. 396) state, "Thus the problem is to partition the eggs the queen lays into gynes [young queens] and workers in such a way that the males produced by the young workers . . . are in the correct ratio to gynes, whilst the total weight of sexuals is maximal."

The *Myrmica* model colony typically grows sigmoidally and then passes into a period of stable or convergent oscillations. There seemed to be two major causes of the oscillations. These were (1) competition for resources between sexual brood and worker brood, and (2) the time-lags that are inherent in a system where different subpopulations interact both in the short and long term. For example, in the model, there was a more sensitive relationship between the production of new queens and the colony's food supply than there was for worker or male production. This is a direct result of the queens being double the size of workers and their need for food in two successive growth intervals.

In one of the first populations of *Myrmica ruginodis* studied by M.V. Brian, some colonies produced only males. These colonies were smaller than those producing both sexes, and colonies producing only females were very rare (Brian and Brian 1951; Brian 1957). Colonies producing more brood produced more males. This was taken to imply that colonies might pass through a stage of producing males only before they matured and produced both sexes. However, in the study population, such male-only production might also have been associated with small colonies tending to be more common in restricted woodland glades in which food was rare and consequently often further from the nest. The life history model of Brian et al. (1981) supported this interpretation. Undernourished "model" colonies were also small and yielded many males but few young queens, whereas both types of sexuals were produced in large numbers in colonies that were not food-limited (see also Sections 5.2, 5.6, and 6.2 for effects of resource shortage on sex allocation).

The model of Brian et al. (1981) assumed that colonies emit flying sexuals. But *Myrmica* colonies also reproduce by colony budding (Section 4.8), although its relative frequency is unknown. Brian et al. (1981) suggested that budding may be favored in habitats where densities of both *Myrmica* ants and other ants are high. Budding may be the major mechanism for successful interference with and displacement of established competitors. (See Franks and Partridge [1993] for a model that shows the potential importance of colonies having large numbers of workers in such situations.) This aspect of the ecology and reproductive behavior of *Myrmica* again suggests that gap dynamics (Section 9.3) are central to the diversity of ant life history strategies.

Colony budding might also be associated with the evolution of strong "queen effects" (see above). If these were encoded in the model of Brian et al. (1981), only a few young queens were formed compared to many workers and many males. In addition, when strong queen effects were incorporated, the model predicted that colonies might divide into two equal parts, each of 300 workers, and then resume growth to the 600 worker steady state. The average size of colonies would then be roughly 450 workers. This equals the average size of colonies emitting both sorts of sexuals. Thus the model predicted that in habitats promoting both colony budding and reproduction by the emission of sexuals (most probably heterogeneous habitats), colonies of similar size might either emit just males (and bud) or emit both types of sexual. Hence the reproductive strategy of *Myrmica* could provide a good example of phenotypic plasticity within populations (West-Eberhard 1989).

Microgynes, Macrogynes, and Social Parasites in *Myrmica*

There can be substantial size variation among *Myrmica* queens within a population. Small queens tend to produce fewer offspring and relatively fewer summer workers and more spring workers. There is also a clear positive correlation between the size of new queens and their mothers. This strongly suggests a genetic component to queen size and fitness (Elmes 1991).

In certain species of *Myrmica* queen size has a clearly bimodal distribution. The small queens are termed microgynes and the larger ones macrogynes (Pearson 1981; Brian 1983 p. 255; Sudd and Franks 1987 p. 144; Bourke and Franks 1991). An example is provided by the two forms of *M. ruginodis*, *macrogyna* and *microgyna* (Brian and Brian 1949, 1955). In Britain, most colonies have only one type of queen, though there are many mixed colonies (Elmes 1991). The macrogyne colonies produce sexuals that participate in mating flights followed by independent colony foundation (although some polygynous colonies contain multiple macrogynes: Bourke and Franks 1991). By contrast, the microgyne colonies propagate by budding (Brian and Brian 1949, 1955; Elmes and Clarke 1981). In the colonies containing both types of queen, it is conceivable that the microgynes behave as social parasites producing their own sexual offspring and very few workers. However, the small queen forms in *M. ruginodis* live mostly in independent colonies where they produce workers. Seppä (1992) made a genetic study of *M. ruginodis* in Finland which challenged early suggestions that there is habitat specialization in the two forms (e.g. Brian and Brian 1955; Table 8.3). However, Seppä's study confirmed Elmes' (1991) recent view that queen size variation is not a clear cut phenomenon and that the two forms are not ecologically well segregated.

There is little overlap in the size range of microgyne and macrogyne forms of *Myrmica rubra* (Elmes 1973b, 1976, 1991). Also unlike the case of *M. ruginodis*, the small forms are almost always found only in colonies with the larger queens. Allozyme analysis (Pearson and Child 1980; Pearson 1981) and rearing experiments (Elmes 1976) showed that *M. rubra* macrogynes produce macrogynes, workers, and few or no males, whereas microgynes produce microgynes, few or no workers, and males. In addition, microgynes are much more queen-productive than macrogynes, producing nearly forty times as many sexual female offspring per queen as macrogynes. Lastly, microgynes exert as strong a queen effect as macrogynes (thereby promoting worker production), but microgyne-produced female larvae themselves escape the queen effect, explaining why they develop preferentially into sexuals. It is, therefore, almost certain that *M. rubra* microgynes act as social parasites of macrogynes (Pearson 1981; Bourke and Franks 1991).

An unsolved issue is whether *Myrmica rubra* microgynes are a separate parasitic species. Pearson and Child (1980) and Pearson (1981) found highly dissimilar gene frequencies in micro- and macrogynes in their electrophoretic study, concluding that parasite and host were therefore separate species. However, this conclusion is uncertain, since these authors only studied allozyme variation at a single locus (esterase). *M. rubra* microgynes are therefore either intraspecific inquilines (workerless social parasites), or a separate inquiline species. There are also many fully parasitic *Myrmica* species that produce no workers of their own and utilize the economies of free-living species in the genus (Elmes 1978, 1983; Bolton 1988; Bourke and Franks 1991). That the social systems of *Myrmica* are literally the breeding ground for interspecific parasites is perhaps hardly surprising given the extent to which many colonies are highly polygynous and potentially rife with internecine conflict (Bourke and Franks 1991).

10.4 The Leptothoracines

The tribe *Leptothoracini* has assumed importance in the study of ant sociobiology in the last 25 years both because colonies are relatively easy to collect and observe, and because leptothoracine ants display a huge variety of social structures and life history traits (Buschinger 1987; Section 8.6). These features are found in members of each of the two principal subgenera, *Leptothorax* (*Myrafant*) and *Leptothorax* (*Leptothorax*).

Leptothoracines are easy to collect and study because their colonies generally consist of at most about 500 very small workers, with one or more queens and associated brood. In addition, they nest in preformed natural cavities such as hollows in decaying plant material or fissures in rocks. This nesting habit has important consequences for leptothoracine ecology and life history. First, nest sites may be short lived relative to the lifespan of colonies, which may live for 12 or more years (Franks et al. 1990a). Thus frequent nest emigrations are likely even in mono-domous species. Second, since leptothoracines generally do not build or repair their nest sites (but see Franks et al. 1992), and since the availability of cavities may be limited, competition for nest sites may be widespread (Herbers 1986c). This may have selected for the common leptothoracine trait of colony foundation by queen adoption, along with polygyny and polydomy (Section 8.4).

Another characteristic trait among the leptothoracines is that queens and workers are similar in size. In the *Leptothorax* subgenus this is particularly true (*L. [L.] acervorum* queens weigh only 1.4 times as much as their workers: Franks et al. 1990a; Chan and Bourke 1994). But even *Myrafant* queens may only weigh about three times as much as their workers (G. Orledge, personal communication). Possibly associated with their close physical similarity (as in *Myrmica*), queens and workers have lifespans that are less divergent than in many other species. For example, *Harpagoxenus sublaevis* queens probably live as adults for ten years, and their slave-maker workers for three years (Franks et al. 1990a). This means that the last broods of workers in a monogynous *Leptothorax* colony may outlive their queen, and that the orphanage period after the queen's death can represent a substantial fraction of the overall lifetime of the colony. Worker reproduction in leptothoracines is very common both in the queenright phase and in the orphanage period (Bourke 1988a). Worker reproduction may have been maintained because of the high risk of workers losing their queen, or of losing contact with the rest of their colony when the society emigrates to a new nest site. Finally, another major trait associated with worker reproduction in leptothoracines is the formation of stable worker dominance hierarchies in which high ranking workers produce their own sons while inhibiting their subordinate sisters (Cole 1981; Franks and Scovell 1983; Bourke 1988b; Section 7.4).

STUDIES OF *LEPTOTHORAX LONGISPINOSUS* IN NORTH AMERICA

Leptothorax (Myrafant) longispinosus is locally common in broadleafed woodland sites across large areas of eastern North America. *L.*

longispinosus workers and queens are minute and often nest in fallen acorns that have been hollowed out by other insects. In similar habitats, two other free-living *Myrafant* species, *L. ambiguus* and *L. curvispinosus*, also occur. All three of these species can be hosts to the slave-makers *Protomognathus* (= *Harpagoxenus*) *americanus* and *L. duloticus*. There is even a rare workerless social parasite, *L. minutissimus*, that parasitizes *L. curvispinosus* (Hölldobler and Wilson 1990 p. 440). In common with *Myrmica* species and many other species of ants, the free-living North American *Myrafant* ants are secondarily polygynous (e.g. Alloway et al. 1982; Herbers 1984; Section 8.4). In *L. curvispinosus*, Stuart et al. (1993) found that the acceptance of new queens into nests was extremely rare if the queen was not reared by that nest, whereas acceptance of related queens occurred intermittently. Some nests reaccepted queens readily, and some did not. The proximate and ultimate reasons for this are, as yet, unknown (Stuart et al. 1993).

Aggression among queens in *Leptothorax longispinosus* seems very rare and has never been observed to lead to monogyny secondarily (Herbers 1982, 1986c). A lack of aggression between queens may be normal in polygynous ants (Bourke 1991; Section 8.6). Herbers (1986c) suggested that there is no morphological distinction between queens from monogynous and polygynous nests in *L. longispinosus*. This is probably typical of facultatively polygynous ant species, since microgyny (as found in certain *Myrmica* species: Section 10.3) is generally rare.

Colonies of *Leptothorax longispinosus* frequently occupy many acorns (polydomy), although nests of several queens and up to about 100 workers can also be found in a single acorn. Despite the ease with which nests of *L. longispinosus* can be sampled, polydomy makes it extremely difficult to examine the colony structure of this species. Nests occur densely in some areas, so it is not always obvious which nests belong together in the same colony. For this reason, most studies have taken nests as the unit of sampling with little or no information on colony size or structure (but see Alloway and Del Rio Pesado 1983). This is problematical, especially as colonies seem to be highly seasonal in their polydomy. That is, colonies disperse into several acorn nests during the spring, summer, and early fall, and coalesce back into one or a few nest sites to overwinter (Herbers 1984, 1990; Herbers and Stuart 1990). Since eggs are laid during the summer and most brood overwinters, seasonal polydomy leads to some loss of information about which parties in the colony invest most in growth or sexual production within the colony. It seems unlikely that queens, workers, and the brood they begin to raise in one year all return as a cohesive group to a single nest in the next year.

To investigate the effects of nest availability on polygyny and poly-domy in *Leptothorax longispinosus*, Herbers (1986c) artificially increased the abundance of nest sites in an experimental plot in New York State. The major result was that both queen number and worker number per nest were reduced relative to controls. In addition, the queen : worker ratios were not different between control and experimental nests. This suggested that the observed reductions were caused by colonies becom-ing more polydomous. This in turn suggested that nest sites were limiting in the population, and hence that polygyny could itself be a response to habitat saturation (Herbers 1986c, 1989; Section 8.4).

Herbers (1990) analyzed an impressive ten years of data on reproduc-tive investment and sex allocation ratios for two study populations of *Leptothorax longispinosus*, in Vermont and New York State respective-ly (Sections 5.6, 6.2). The populations were similar with respect to their nest sites and overall level of polygyny, but differed strongly in that the Vermont ants reared smaller sexuals and produced broods that were more male-biased than those in New York. In addition, the pattern of polygyny was not consistent from year to year within each site (Table 5.7). Worker number also showed considerable annual variation in each site. Herbers' (1990) data therefore provide yet another warning about deducing the life history parameters of social insect populations from information from only a few seasons.

Herbers (1990) investigated the allocation strategies of *Leptothorax longispinosus* using path analysis, a multivariate statistical technique that attempts to estimate which of a number of relevant factors most influence traits of interest. This suggested that differential rearing costs of the two sexes, local resource competition, and local mate competi-tion had little or no influence on sex allocation (Sections 5.7–5.9). Instead, Herbers (1990) proposed that queen–worker conflict helped explain the sex investment patterns observed in *L. longispinosus*. However, these patterns conceivably also resulted from resource abun-dance effects, or from sex ratio splitting due to variations in relatedness asymmetry (Section 6.2). The path analysis further suggested that much variation among nests and between populations in resource allocation remained to be explained. A factor possibly accounting for this residual variation is allocation to colony growth in terms of worker production, which was not measured. Along with sex allocation, this might depend upon resource–acquisition ability (Herbers 1990; Backus and Herbers 1992; Backus 1993). Resource availability may have differed between the Vermont and New York sites, with more competition from both other *L. longispinosus* nests and other ant species leading to fewer available resources in Vermont (Section 5.6). This could explain why Vermont nests had smaller workers and routinely put less energy into reproduction than their counterparts in New York. Site-specific sizes of winged reproductives may also have represented a response to local resource levels (Herbers 1990).

In many ant populations, greater food availability boosts sexual production. Backus and Herbers (1992) cite several cases of nests in food-rich areas, or which receive protein-enriched food, rearing more sexuals (especially females). They include *Leptothorax curvispinosus* (Wesson 1940), *Lasius niger* (Boomsma et al. 1982), *Formica aquilonia* (Luther 1987), *Linepithema humile* (= *Iridomyrmex humilis*) (Passera et al. 1987), *Leptothorax acervorum* (Buschinger and Pfeifer 1988), *Myrmica rubra* (Brian 1989), and *Solenopsis invicta* (Porter 1989). Both Nonacs (1986b) and Herbers (1990) suggested a link between resource availability and sex allocation. In ants a direct link could occur simply because underfed diploid larvae mature to become workers instead of queens (Sections 6.2).

Backus and Herbers (1992) therefore attempted to find out if the difference in sex allocation between their Vermont and New York *Leptothorax longispinosus* study populations was determined by the availability of food. However, experimentally adding extra food in both the field and laboratory did not alter the sex ratios of nests (Backus and Herbers 1992; see also Backus 1993). On the other hand, although the nest site and resource manipulation experiments of Herbers (1986c) and Backus and Herbers (1992) were undoubtedly pioneering, it is possible that much of the unexplained residual variation in allocation patterns in *L. longispinosus* stemmed from using nests instead of colonies as the unit of sampling. The seasonal polydomy and possible brood mixing by *L. longispinosus* nests during overwintering cast doubt over the independence of their nests in terms of determining sex allocation and other investment patterns.

LIFE HISTORY CONSEQUENCES OF WORKER REPRODUCTION IN LEPTOTHORACINES

The production of haploid, male-yielding eggs by workers is very common among the leptothoracines (Bourke 1988a). It may occur both in the presence of the queen or, more commonly, in orphaned colonies. Indeed, in ants, worker dominance hierarchies associated with high ranking workers that produce sons and physically prevent reproduction by low ranking workers were first shown in leptothoracine colonies (Cole 1981). This level of worker conflict is only likely in small colonies (Heinze et al. 1994), where individual workers can dominate the social system and in which worker longevity is not trivial compared with queen longevity. Relative longevities of queen and workers are important because they are likely to be correlated with relative reproductive potential. A dominant worker may live long enough to lay male-producing eggs over more than one reproductive season, even when the colony no longer has a queen. By contrast, in many other species, workers live only for part of one reproductive season, and accordingly have a limited opportunity to reproduce. Indeed, worker dominance orders have now been demonstrated in ponerine ants that also have relatively small colonies and long lived workers (e.g. Oliveira and Hölldobler 1990; Ito and Higashi 1991; Section 7.4).

This overt conflict and the use of scarce resources in worker reproduction undoubtedly affects colony life history evolution. For example, Cole investigated the impact of worker dominance on the time budgets of workers and hence on the efficiency of the colony as a whole in *Leptothorax allardycei*. He found that dominance behavior constrained food exchange in the colony, limited the movement patterns of workers, and reduced the time spent on brood care (Cole 1981, 1986, 1988). In a model, Cole (1986) also found that provided worker reproduction does not impair colony productivity by more than about 20%, it can spread through a population (Section 7.4). This was consistent with the occurrence of reproductive workers in *L. allardycei*, because, in terms of loss of colony efficiency, Cole (1986) estimated the cost of worker reproduction to be between 13% and 15%.

Cole's (1986) study focused on naturally queenless colonies of *Leptothorax allardycei*. This was appropriate because approximately 20% of *L. allardycei* colonies lack a queen (Cole 1984, 1986). Similarly, in other leptothoracine species (e.g. *Harpagoxenus sublaevis*), the orphanage period is likely to be important in determining colony life history and the inclusive fitness of individuals, as discussed below.

LIFE HISTORY STRATEGIES IN THE SLAVE-MAKING ANT *HARPAGOXENUS SUBLAEVIS*

Franks et al. (1990a) conducted an intensive study of life history strategies in the slave-making ant *Harpagoxenus sublaevis*. One of the issues they explored was the impact of worker reproduction and dominance orders on the inclusive fitness of all members of the society.

In social insects, life history analysis should take into account queen–worker and worker–worker kin conflict (Sections 9.1, 9.9). One problem with doing this is determining how much more productive, for the colony as a whole, a worker could be if it were not trying to produce its own sons. In slave-making ants, this problem is simplified because slave-maker workers have only one altruistic task, namely participating in slave-raids to bring in more "slave" workers. *Harpagoxenus sublaevis* workers form dominance orders in which high ranking individuals gain better opportunities to lay their own eggs. These workers do little or no slave-raiding, and it is therefore relatively easy to determine how many slaves are lost to the society by their selfishness. Therefore, curiously, the habit of slave-making facilitates the analysis of life history strategy.

Another advantage of *Harpagoxenus sublaevis* for life history analysis is that quite a lot is known about its social behavior and life history parameters (Buschinger 1966a,b, 1968b, 1974b; Buschinger and Winter 1978; Bourke 1988b; Bourke et al. 1988). Colonies are monogynous and monodomous, and the queens mate singly (Bourke et al. 1988). Queen and worker longevities have been estimated at ten and three years respectively (Franks et al. 1990a). There is little reproduction by *H. sublaevis* workers during the lifetime of the queen. However, a few dominant workers produce their own sons during the three year orphanage period (Bourke 1988b; Bourke et al. 1988).

The major host species of *Harpagoxenus sublaevis* is *Leptothorax acervorum*. In the mixed colonies, the slaves of this species carry out all the foraging, and rear all the new slave-maker queens, males, and workers. Franks et al. (1990a) estimated the per capita productivity of slaves from the productivity of free-living *L. acervorum* workers in neighboring colonies. *H. sublaevis* queens found colonies by invading host colonies and driving off the queens and adult workers. Franks et al. (1990a) also estimated that only 0.4% of new *H. sublaevis* queens successfully complete this process.

The model was constructed in the following way. The queen chooses the proportion of total excess energy, provided by the slave workers, that is taken to raise slave-maker workers or sexuals in any particular year. In so doing, the queen attempts to maximize her fitness by appropriate choice of "policy vector." The queen's fitness is a function of the number of her sexual daughters, sons, and grandsons (produced by orphaned workers), weighted by her relatedness with these progeny. Using a numerical optimization technique, the model calculated the life history strategy ("policy vector") of the queen that maximized her fitness measured in this way.

The model found that if the original queen followed her optimal strategy, the number of colonies descended from her would take about 10.5 years to double – a realistic figure for relatively long lived perennial organisms. Franks et al. (1990a) also used variants of the original model to explore plausible alternative evolutionary scenarios. These involved (1) all the slave-maker workers being totally sterile, and (2) there being male-production but no slave-raiding by orphaned colonies, contrary to the natural situation. The results showed that in both cases the queen's fitness was reduced. For example, when the workers were sterile, the queen could not alter her investment policy sufficiently during her lifetime to compensate for the loss of grandson production by her orphaned colony. One of the reasons was that the queen's sons and sexual daughters take a long time to develop (one and two years respectively). Therefore, the eggs laid just before the queen's death need large numbers of slaves to rear them, and the queen must achieve this by producing slave-maker workers rather than more sexuals. Thus, the queen evidently benefits from the high level of worker reproduction by the orphaned colonies (Alexander 1974; Franks et al. 1990a).

Lastly, the model also showed that even the subordinate slave-raiding *Harpagoxenus* workers, which take all the risks of slave-raiding in the orphanage period but produce no sons, might benefit from raiding. This was because raiding allows the queenless colony to produce many more extra males, which are the subordinates' nephews. In other words, orphaned slave-maker workers should fight hard to be the dominant egg layer, but if they lose they should promote their inclusive fitness by slave-raiding. This result also suggested that the division of labor between slave-raiding and nonraiding workers observed in *H. sublaevis* colonies stems from worker–worker reproductive conflict (cf. West-Eberhard 1975, 1981; Section 12.3).

ADVANTAGES OF EARLY VERSUS LATE SEXUAL PRODUCTION IN THE SOCIALLY PARASITIC GENUS *EPIMYRMA*

The socially parasitic ant species within the leptothoracine genus *Epimyrma* provide good examples of divergent life history strategies. In particular, they exemplify the evolution of early "big bang" reproduction on the one hand, and delayed reproduction with iteroparity on the other. This could be in response to differing environmental conditions, with a special feature being that the environment in question is that provided by the host species.

Epimyrma ants, together with the closely related *Myrmoxenus gordiagini*, have been investigated more thoroughly than almost any other group of socially parasitic ants (Winter and Buschinger 1983; Buschinger and Winter 1983, 1985; Buschinger et al. 1986; Buschinger 1989a,b; Douwes et al. 1988). Detailed studies have been conducted on eight species from southern Europe and North Africa. All the species utilize hosts in the *Leptothorax* (*Myrafant*) group, with little overlap among the parasites in the species they exploit. The parasites are also phylogenetically very closely related. For life history analysis, this avoids the problems associated with comparisons among species with diverse phylogenetic origins (Section 9.8).

All the *Epimyrma* species except one (*E. algeriana*) are monogynous, and all utilize monogynous host species (with the exception of the host species *Leptothorax tuberum*, which can be facultatively polygynous). They also all have nonindependent colony foundation. Newly-inseminated females enter foreign host nests, sting some of the host workers (who usually recover), and somehow gain acceptance by the rest of the host workers. The parasite queen then sets about eliminating the host queen by climbing on her and slowly throttling her.

The important result of this process is that the *Epimyrma* queen is faced with a host workforce that will dwindle in numbers. Consequently, as time passes, the parasitized colony will be able to rear progressively fewer new parasites, unless the parasitic queen produces slave-raiding workers who can capture more host workers. Slave-making is in fact believed to be the primitive condition in *Epimyrma* (Buschinger 1989a). However, as in *Harpagoxenus sublaevis*, there must be a trade-off each season between producing more slave-maker workers (to have more slaves to produce more sexual offspring later) and producing more sexual offspring (which can enhance the original's queen fitness by founding new colonies sooner). In some *Epimyrma* species, this trade-off has apparently been taken to an extreme, and queens produce no workers at all, but only sexuals (Table 10.1). In these species the host colony is utilized for as long as the original host workers survive. Reproduction therefore consists of a "big bang" of sexual production over only a couple of seasons. By contrast, in the slave-making species, the trade-off has evidently been resolved the other way, and sexual production almost certainly occurs over many years after a period of colony growth brought about through slave-raiding.

There is also an association between these different life history patterns and the rate of brood development. The workerless species have fast-developing ("rapid") brood in which females develop from eggs laid in the same year, without larval hibernation. The slave-making species have brood in which sexual females develop from hibernated larvae and which therefore only emerge as adults in the next season. *Epimyrma kraussei* (Sections 4.9, 5.8) represents a transition between the slave-making and nonslave-making species. It can produce low numbers of workers (that apparently do not raid in the field) but has rapid brood. Mating behavior in *Epimyrma* also shows a transition between the two sets of species. Most of the slave-makers have sexuals that disperse on the wing and participate in mating swarms, while the nonslave-makers have within-nest mating and disperse on foot.

Buschinger (1989a) suggested that the loss of the slave-making habit is brought about by selection for reduced dispersal from the local popu-

Table 10.1

Life History Traits of *Epimyrma* and *Myrmoxenus* Social Parasites

Parasite Species	Monogyny/ Polygyny	Colony Founding with Throttling	Mating Behavior	Sexual Brood	Slave-raiding	Parasite Worker Number	Host Species
M. gordiagini	M	+	swarming	slow	+	high	Leptothorax lichtensteini, L. serviculus
E. ravouxi	M	+	swarming	slow	+	high	L. unifasciatus, L. nigriceps, L. affinis, L. interruptus, and others
E. stumperi	M	+	swarming	slow	+	high	L. tuberum
E. bernardi	M	+	intranidal	slow	+	high(?)	L. gredosi
E. algeriana	P	+	intranidal	slow	+	high	L. spinosus (and 3 other species)
E. kraussei	M	+	intranidal	rapid	±	low	L. recedens
E. corsica	M	+	intranidal	rapid	−	none	L. exilis
E. adlerzi	M	+	intranidal	rapid	−	none	L. cf. exilis

NOTES: Colony foundation with throttling involves the parasite queen invading a host colony and killing the host queen or queens. "Slow" and "rapid" refer to the rate of development of the sexual brood of the parasite species. (From Buschinger 1989a, by permission of Birkhäuser Verlag, Basel.)

lation of host colonies (philopatry). A reduction in dispersal is favored because queens that disperse far risk leaving the host population. It is achieved by evolving the system of queens mating in the nest and dispersing on foot. This in turn leads to a reduced need to carry out slave-raids, because by allowing neighboring colonies to survive unmolested (at least temporarily), an established colony preferentially benefits its own philopatric queens when the time comes for them to found colonies (Buschinger 1989a). Therefore, there is selection for "big bang" early reproduction by the parasites. As these ideas suggest, the workerless *Epimyrma* parasites certainly do occur in concentrated patches of parasitized colonies scattered inside dense host populations (Buschinger 1989a). The density of the hosts ensures that host colonies destroyed by parasitism are rapidly replenished by colony foundation by fresh host sexuals (Buschinger 1989a). This rapid turnover of hosts would itself facilitate a "big bang" life history. The corollary of this reasoning is that, if host colonies are more thinly spread, or have slower rates of reestablishment, the parasites favor the perennial, slave-making life history. The less dense distribution of the hosts would promote dispersal among the parasite sexuals (as occurs in the slave-makers). The slower replacement of parasitized colonies would select for parasites that exploit existing colonies for longer, over a perennial life cycle. Of course, slave-makers need some nearby host colonies (and hence fairly dense host populations) from which to obtain slaves. But raided colonies survive being pillaged (Buschinger 1989a), and therefore do not require replacement, whereas colonies invaded by a parasite queen are eventually destroyed.

More data on the demography and distribution of *Epimyrma* and their hosts are necessary to test these ideas. However, the life histories of these parasites are certainly fascinatingly dissimilar, and the likeliest explanation for why some are practically annuals and others perennials is that their life history strategies have evolved in response to their hosts' biology.

10.5 The Wood Ants, *Formica*

The genus *Formica* is of major ecological importance in boreal ecosystems, dominating many northern temperate habitats, especially coniferous forests. *Formica* species are very diverse in their life history patterns. This applies both within and between the two major groupings in the genus, namely (1) *Formica sensu stricto*, the wood ants of the *Formica rufa* group, and (2) the *Formica fusca* group (Sudd and Franks 1987 p. 38). The *F. rufa* group is characterized by very big, mound-building colonies with large numbers of aggressive workers. Ants of the *F. fusca* group, by contrast, form smaller colonies under stones or in rotting wood, and have timid foragers. Because of their ecological dominance and economic importance in forestry, the species of the *F. rufa* group have been intensively studied. For this reason, this review focuses on the wood ants.

Species in the *Formica rufa* group often inhabit seasonally very cold environments where the annual period when high-tempo activities can be undertaken is short. Hence colonies need to be very large to produce nests big enough to allow thermoregulation. Individual nest mounds are often huge. We have seen some in southern Sweden that resemble haystacks, 1.6 m high at their summit, containing several cubic meters of nest material. Nests also have a sophisticated internal structure, and include excavations penetrating over two meters below the soil surface (Bristow et al. 1992).

Rosengren et al. (1987) showed that, in nests with more that a million workers, the wood ants can generate enough warmth to melt the snow off the mound early in the spring. In their study colony, the temperature inside the nest remained at a stable level (about 30 °C) in late spring and summer, even when the outside temperature dropped below freezing point. This extraordinary ability to thermoregulate probably has significant effects on fitness. Rosengren et al. (1987) found that those nests which later produced sexuals maintained a significantly higher inner temperature than spring nests that later produced only worker broods. These authors suggested that the nest-warming effect of insolation (sunshine) is more important for small and weak colonies, whereas large nests have, through their size, much more control over their thermal environment.

The importance of large nests for thermoregulation was also illustrated by a study of an island population of *Formica truncorum* by Rosengren et al. (1985). The population formed a very large polydomous colony occupying a three hectare pine-forest. Approximately 1.2 million workers occupied 50 to 60 nests in summer, but during the fall 60% of these nests were abandoned as the ants withdrew to shared hibernation nests. This high level of seasonal polydomy seems to have two major functions. First, expansion into multiple nests during the summer allows the wood ants to monopolize patches. Second, retreating to a smaller number of large nests for the rest of the year may permit the rapid development of brood and reduce the risk to adult colony members of predation. In sum, the construction and maintenance of a large nest is an important feature of the life history strategy of wood ants. It is also associated with their habit of dominating and improving resource availability in the local forest patch through aphid "farming" (see below). In addition, huge nests and large worker populations are themselves a long-lived and heritable resource. For example, Pamilo (1991d) showed that the nest (and queens) of the monogynous form of *F. exsecta* survive on average for 20 years.

Foraging Systems and Habitat Patches

The large, long lived nests of wood ants are supported by foraging on a massive scale. These ants typically consume aphids and insect larvae, but the most important food source both in terms of dry weight and energy is aphid honeydew. To obtain these resources, the foraging range of individual mounds may exceed 100 m, and a significant effect on insect prey occurs up to 50 m from the nest (Rosengren 1977; Rosengren et al. 1979). For these reasons, Rosengren et al. (1979) suggested that the greatest ecological impact of wood ants is only realized if there are at least two to three inhabited mounds per hectare. At these densities, wood ants even have an economic impact because they significantly depress populations of the defoliating insect pests of forestry. Even in England, where wood ant colonies are typically rather small and at low density, they can have a large local effect on other forest insects (Skinner 1980). In addition, wood ants effectively "farm" aphids by protecting them against predators, so building up their numbers. This both creates another heritable resource and causes local enrichment of the ants' environment, so amplifying the patchiness of the habitat (Rosengren and Sundström 1991). Such a socially induced habitat patchiness could contribute to the variability in the life history strategies of wood ants.

SUPERCOLONIES AND THE DISTRIBUTION OF MONOGYNY AND POLYGYNY

Colony foundation in the *Formica rufa* group is rarely, if ever, independent. It is either through temporary social parasitism of members of the *F. fusca* group (in which queens usurp the existing colony queen), or through colony budding. Established colonies may use budding as a strategy to spread from the favorable habitat they have formed around themselves into less favorable habitat (Rosengren and Pamilo 1983).

The genetic structure of mature *Formica* colonies is potentially very complex. Nests can be either monogynous or polygynous (and in the latter case colonies are frequently polydomous as well), and queens may be either singly or multiply mated (e.g. Pamilo 1982b; Sundström 1993a, 1994a). In polygynous nests, queens may or may not be related (e.g. Pamilo 1982c).

In addition, certain species in the *Formica rufa* group sometimes form enormous polygynous and polydomous "supercolonies" (unicolonial populations: Section 8.5). This is further evidence of social plasticity in these ants. The largest example of a supercolony is an enormous polydomous population of the Japanese *F. yessensis*, which covered an area of 2.7 km^2 and contained an estimated 306 million workers and 1 million queens (Higashi and Yamauchi 1979; Higashi 1983). In Russia, a *Formica* supercolony consisted of 140 million workers. In these cases, forager densities may exceed 100 per square meter and in certain areas there can be 100 mounds per hectare (Savolainen and Vepsäläinen 1988). Cherix et al. (1991) described a long-term study of a supercolony of *F. lugubris* in Switzerland. The supercolony occurred on the northwest slope of the Jura Mountains at an elevation of 1400 m (Cherix 1980), and consisted of about 1200 interconnected nests occupying an area of 70 ha. However, why polygynous *Formica* ants form supercolonies rather than remaining multicolonial (polygynous with mutually hostile colonies) is unclear (Sections 8.4, 8.5).

Rosengren et al. (1993) have provided a recent review of monogyny and polygyny in the *Formica rufa* group in relation to different dispersal tactics (Sections 8.4, 8.6). Of the eight Palaearctic species that have been studied in detail, only three (*F. polyctena*, *F. aquilonia*, and *F.*

yessensis) are obligately polygynous. By contrast, none are obligately monogynous in the Palaearctic region, though *F. pratensis* and *F. rufa* may be mostly monogynous in central and northern Europe. In southern Finland, *F. lugubris* and *F. truncorum* are monogynous and monodomous, but there are highly polygynous and polydomous populations in Switzerland and Germany. Monogyny may also occur in *F. uralensis* and *F. pressilabris*. Most of the species are known to exist in both monogynous and polygynous populations and can be classified as facultatively polygynous (at least at the population level). In short, polygyny is by far the most prevalent condition in the majority of the *Formica rufa* group species.

There is a strong, proximate association between dealation (wing-shedding) behavior in queens and the occurrence of monogyny and polygyny. Rosengren et al. (1993) state that virgin queens of monogynous and monodomous *Formica rufa* group species rarely drop their wings if they have not participated in a mating flight. Females from polygynous colonies, in marked contrast, often shed their wings without flying (e.g. Fortelius 1987; Sundström 1995b). Nevertheless, this may still be followed by successful copulation and may therefore be a tactic for staying in the parental nest. The mixed dispersal strategy that such tactics allow could be associated with the habitat patchiness created by the behavior of colonies through their aphid farming (cf. Section 8.6). Rosengren et al. (1993) also attributed polygyny and polydomy in *Formica* wood ants to long lived, stable habitats (Section 8.4). Certainly wood ants practice, par excellence, the "space-perennial" strategy that Nonacs (1993a) argued should be favored in such habitats (Section 9.9).

10.6 The Argentine Ant, *Linepithema humile*

Linepithema humile, formerly known as *Iridomyrmex humilis* (Shattuck 1992), is a dolichoderine ant that has become a significant pest in many parts of the world. Its ability to outcompete several different kinds of local ant fauna, which seems to be based on its extraordinary social structure, makes it an important case study in the investigation of ant life history strategies. *L. humile* provides another example of unicolonial polygyny (Section 8.5). Nests contain numerous queens, and the relatedness among these queens is close to zero (Kaufmann et al. 1992). The genetic studies of Kaufmann et al. (1992) also showed that inbreeding is absent despite within-nest mating (Markin 1970). This is almost certainly due to dispersal of queens and males between nests, and to queens avoiding matings with related males (Keller and Passera 1993;

Passera and Keller 1994). The existence of unicolonial ants is enigmatic from the perspective of kin selection theory (Section 8.5). This section focuses on the life history and social behavior of *L. humile*. It also considers related *Iridomyrmex* species, certain of which have some similarities with *L. humile*.

Linepithema humile is called the Argentine ant in recognition of its origin in South America. It has now been introduced to, and thrives in, many parts of the world with Mediterranean-like climates (Newell and Barber 1913; Marchal 1917; Skaife 1955). In such environments, it often ousts the native ants. It has also been recognized as a pest species because it invades human habitations, and damages plants by tending herbivorous homopterans such as aphids.

A major competitive advantage of unicoloniality comes from the associated habit of colony budding. This avoids the high-risk part of the life cycle experienced by isolated colony-founding queens. Another possible advantage is that unicolonial species make foraging more efficient. Thus, in a study of the Australian "meat ant" *Iridomyrmex sanguineus*, McIver (1991) showed that a single colony may have widely dispersed nest holes. Different sets of specialist foragers were faithful to different groups of entrances. This habit, coupled with the frequent transport of larvae between nest holes, implies that these ants take their brood to the food, whereas most ants do the reverse and transport food back to a single nest site. Therefore, since a unicolonial social system allow nests to be dispersed, it partly avoids some of the costs of central place foraging. This foraging advantage may contribute to the ecological success of unicolonial species (Section 8.5), and to the evolution of polydomy in general (Traniello and Levings 1986).

QUEEN EXECUTION

A macabre aspect of the social life of *Linepithema humile* is the periodic mass execution of queens by workers. Keller et al. (1989) showed that about 90% of queens in a French population are executed in May, at the beginning of the reproductive season. There was no correlation between the percentage of queens executed and their weight and fecundity. Furthermore, just prior to the mass execution, nearly all queens were of the same age (less than one year old). So it seems unlikely that the age of queens determines whether the workers select them for execution. In addition, Keller and Passera (1992) excluded the possibility that the executed queens had exhausted their sperm supply. (However, it is not known if these queens have an unusual genotype as in *Solenopsis invicta*: Keller and Ross 1993a [Sections 8.4, 10.2].) Queen execution imposes a large energetic cost on the colony. Keller et al.

(1989) estimated that replacing the queens would take 7% of the total energetic cost invested in the production of all individuals. This makes the occurrence of queen execution all the more puzzling.

Keller et al. (1989) argued that, since *Linepithema humile* queens often change nests (Newell and Barber 1913; Keller 1988), it is likely that the level of relatedness among nestmate queens, as well as between queens and workers, decreases after spring when sexuals are produced. So these authors hypothesized that, to prevent the level of relatedness among nestmates becoming low, and hence to safeguard their inclusive fitness, workers execute queens to which they are less related. However, this would only benefit the workers if queen execution occurred just before the production of eggs giving rise to the sexuals (L. Keller, personal communication). But whether this is the case is unknown. In addition, against a background of low relatedness, high queen mobility, and weak nestmate recognition (Keller and Passera 1989b; Kaufmann et al. 1992), it is hard to understand how workers could use queen killing to favor close relatives. This possibility might be examined by using sensitive molecular techniques.

Keller et al. (1989) and Vargo and Passera (1992) also suggested that, at the proximate level, queen killing facilitates the production of sexuals. This is because the presence of queens inhibits the production of new queens (Vargo and Passera 1991, 1992) and males (Passera et al. 1988a). For example, male production is inhibited by queens because the queens apparently consume food resources that would otherwise have gone to male larvae (Passera et al. 1988a). However, if queen execution is just a proximate device to inhibit sexual production, and so happens year after year, why have the ants not evolved a more efficient, less wasteful mechanism? Thus, queen execution in *Linepithema humile* remains a largely unsolved problem. At the ultimate level, it possibly results from a form of bet-hedging (Seger and Brockmann 1987). Perhaps large numbers of queens are produced in case conditions are particularly favorable, and are executed if conditions turn out to be poor. This interpretation would be supported if fewer queens were executed in years with plentiful resources.

An important way to resolve this issue would be to make an intensive ecological and genetic study of the ancestral population of *Linepithema humile* in South America. The crucial question is: to what extent do Argentine ants practice queen execution in their original habitat? It is possible that this phenomenon is a remnant of an ancestral behavior pattern and is not now adaptive in introduced populations of *L. humile*. Alternatively, queen execution could be an example of a novel trait brought about by the rapid spread of a species into a new environment (cf. polygyny in *Solenopsis invicta*; Section 10.2).

10.7 The Obligately Thelytokous Ant,
Pristomyrmex pungens

The small Japanese myrmicine, *Pristomyrmex pungens*, has a life history pattern that is unique among ants (Itow et al. 1984; Tsuji 1988a,b, 1990a,b, 1994). In colonies of *P. pungens*, which are common in Japan, there are no morphological queens, and males are produced only very rarely (Itow et al. 1984; Tsuji 1988b). Instead, colonies consist entirely of workers, none of whom have a sperm receptacle. However, all workers appear to have the ability, at least early in life, to produce diploid eggs yielding daughter workers by thelytokous parthenogenesis (Box 3.1). *P. pungens* is thus the only known completely thelytokous ant (Itow et al. 1984; Tsuji 1988b, 1990b).

Since it lacks a standard reproductive division of labor, Tsuji (1992) argued that *Pristomyrmex pungens* is also the only free-living, noneusocial ant (see also Furey 1992). However, arguments about whether *P. pungens* should be defined as eusocial or not distract from biologically more important matters. The species is clearly highly social and derived from orthodox eusocial ants, as evidenced by its vestigial sexual production. It is the life history and genetic system of *P. pungens* that are highly unusual and demand explanation.

At present, the precise genetic structure of *Pristomyrmex pungens* colonies is unknown. Worker thelytoky implies that colonies form clones. But preliminary molecular evidence suggests that this is not strictly true, and that colonies may instead be composed of several clonal worker subgroups (Tsuji 1994). Colony propagation is inevitably by budding, implying that neighboring colonies should also be closely related. However, in two field sites in Nagoya City, Japan, the average nest longevity in summer was only 16 or 17 days (Tsuji 1988a). Therefore, between-nest relatedness must be limited by the high mobility of nests. One reason that colonies might move so frequently is to track their ephemeral food resources. For example, the experimental removal of weeds upon which *P. pungens* workers were tending aphids seemed to promote nest relocation (Tsuji 1988a).

Since all *Pristomyrmex pungens* workers are apparently able to reproduce, much attention has been paid to the division of labor within colonies. It seems that all workers pass through a phase early in their lives when they remain in the nest and lay eggs (the intranidal phase), and a post-reproductive phase when they forage for their nestmates outside the nest (the extranidal phase) (Tsuji 1988b). Why should any workers go out and forage for food only to benefit other workers, rather than invest for longer in their own reproduction? Tsuji (1994) recently suggested the answer to be that colonies with noncooperative, nonforaging mutants would suffer from reduced productivity. In short, he proposed that the proportion of egg-laying workers in *P. pungens* colonies is maintained by a balance of within-colony selection for selfish reproduction, and between-colony selection for cooperative foraging (cf. Sections 2.2, 2.3).

Why has obligate thelytoky evolved in *Pristomyrmex pungens* in the first place? In general, thelytoky is sporadic in queen and worker ants (E.O. Wilson 1971 p. 325; Crozier 1975; Bourke 1988a; Choe 1988; Hölldobler and Wilson 1990 p. 189). Apart from *P. pungens*, the only other species with routine thelytoky appears to be *Cataglyphis cursor*, but here it is not obligate (Lenoir and Cagniant 1986; Lenoir et al. 1988). E.O. Wilson (1971 p. 327) suggested that worker thelytoky evolves to enable rapid expansion into new sites in variable environments. This is reminiscent of explanations both for unicoloniality and the inbreeding strategy of *Technomyrmex albipes* (Sections 5.8, 8.5). In other words, thelytoky, unicoloniality, and inbreeding all involve a sacrifice in dispersal power, and in the maintenance of genetic diversity through sex, in return for increased lateral spreading power. Lenoir et al. (1988) proposed that worker thelytoky in *C. cursor*, which is monogynous, is a mechanism to permit queen replacement. This species has a mating system, unusual for a monogynous ant, in which queens mate with nonnestmate males outside their home nest and then found a new colony by budding (Lenoir et al. 1988). So orphaned colonies can produce a new queen by thelytoky and then adopt her as their colony queen after she has mated. This neatly overcomes the difficulty that monogynous species otherwise have in replacing a dead queen with a relative (Section 7.4). However, it is not clear why, apparently alone among monogynous species, *C. cursor* has evolved this mating and genetic system and hence its ability to replace queens. In *P. pungens*, somewhat similarly, worker thelytoky is suggested to be for maintaining units that can reproduce given a nomadic lifestyle and the consequent risk of nest fragmentation and queen loss (Tsuji 1988a,b, 1990b). Insurance against queen loss was also proposed to underlie the regular occurrence of worker thelytoky in the Cape race of the honey bee, *Apis mellifera capensis* (Moritz 1986a, 1989).

10.8 The Army Ants, *Eciton*

The army ants are at another extreme among the great variety of life history strategies in ants. Although nearly always monogynous, colonies are huge and propagate exclusively through colony fission, making a massive investment in a minimum number of new colonies. The army ants are almost certainly a polyphyletic assemblage (Bolton 1990). They are also a diverse group, with perhaps 150 or more species occurring in the New World (Watkins 1976), and many others in parts of the Old World tropics. *Eciton* is an exclusively New World genus with about 12 species (Watkins 1976). Reviews of army ant diversity and behavior can be found in Gotwald (1982), Franks (1989a), and Hölldobler and Wilson (1990). There are few quantitative data on colony sizes, growth rates, and reproductive investment in species other than *E. burchelli*. The most intensively studied population of *E. burchelli* is on Barro Colorado Island, Panama (e.g. Rettenmeyer 1963; Willis 1967; Schneirla 1971; Franks 1982a,b, 1985, 1989a,b; Franks and Bossert 1983; Franks and Fletcher 1983), and the following account refers mostly to this population.

THE ARMY ANT SYNDROME

The whole army ant way of life depends on having very populous colonies. *Eciton burchelli* colonies, for example, contain 300,000 to 600,000 workers (Franks 1985). Army ants tend to predate large or numerous prey such as crickets, cockroaches, scorpions, and social insect colonies (Schneirla 1971; Gotwald 1982; Franks 1989a), and these prey can only be overwhelmed by many ants raiding in concert. In addition, because their prey occur unpredictably in space, army ants raids have to be very large to sample extensive areas. Hence army ants produce very substantial, densely populated raid systems. Since these locally deplete the prey populations, the ants also have to move regularly to find new feeding areas. Nomadism is characteristic of army ants (see below), and is itself facilitated by large colony size. For example, nomadic colonies have to pass through the territories of other ants, where protection through large numbers is advantageous.

All army ants probably build temporary nests ("bivouacs") partly from their own living bodies. In *Eciton burchelli*, bivouacs have a high and almost constant internal temperature (Franks 1989b). This seems to help *E. burchelli* colonies rear their worker broods very quickly and over a precisely set period, which is another army ant trait (see below). Complete worker development, from egg to adult, occurs within a fixed period of just 45 days.

Many army ants exhibit stereotyped cycles of activity and brood development that they repeat throughout their lives and which are unique among ant life histories. They seem to have evolved as an integral result of the nomadic lifestyle of army ants. In *Eciton burchelli*, these cycles each last 35 days and correlate with the production of discrete cohorts of new ants. A cycle consists of two phases (Schneirla

1971). The 15 day nomadic phase coincides with the development of the larvae. Since these require feeding, the ants raid each day and emigrate to a new bivouac site almost every night (so a fresh area can be then be raided the next day). The 20 day "statary" phase starts when the larvae pupate, and lasts the entire period of pupal development. In addition, in the middle of the statary period, the queen lays the eggs, numbering 100,000 in total. Since the colony has no larvae to feed during this phase, it raids on only 13 of the 20 days and remains in the same bivouac site. The eggs hatch in synchrony with the emergence of all the new adults from their pupal cases. Since the colony has a new batch of brood to feed, the statary period ends and once again the colony becomes nomadic.

Eciton burchelli colonies produce about 50,000 new workers per 35 day cycle. During each day of raiding, 100,000 or more workers go into the field in a huge predatory swarm (Rettenmeyer 1963; Willis 1967; Schneirla 1971), and more than 30,000 prey items may be retrieved (Franks 1989a). However, if only 1% or 2% of the foragers are killed each day, colonies cannot grow (Franks 1985). Recently, Franks and Partridge (1993) used mathematical models to show that the very large numbers of ants present at the leading edge of a raiding swarm may be essential to minimize such losses. Only by vastly outnumbering their large insect prey can army ants overwhelm them without suffering catastrophic casualties.

LIFE HISTORY CONSEQUENCES OF COLONY FISSION

As a result of reproduction by colony fission, *Eciton burchelli* colonies start life at half their final size, which is then reached in three years. Army ants probably reproduce by fission because their colonies could not economically gather prey if they were small (Franks and Hölldobler 1987). In fission, the largest colonies split into two, with either the old queen and one of her daughter queens, or two new daughter queens, taking over the resultant colonies. Queens appear to compete to head the new colonies (Section 7.6). Queens are never winged, presumably because of the lack of mating swarms and independent colony foundation. However, outbreeding occurs because the males have wings and fly to other colonies, which they enter in search of mates. Males may therefore have been selected to resemble queens and, unlike males in other ant species, to have wings that they eventually shed (Franks and Hölldobler 1987; Section 11.5).

During colony fission, the colony produces about six new queens and 4,000 males (Franks and Hölldobler 1987), as expected from sex allocation theory for fissioning species (Pamilo 1991b; Section 5.7). In addition, each parental colony divides at about the size at which it can produce two daughter colonies whose combined growth rate just exceeds its own current growth rate (Franks 1985). In this way, the parental colony produces daughter colonies that should themselves most rapidly succeed in producing daughter colonies. This is as predicted by a model for the optimum timing of fission based on minimizing the time between reproductive events (Franks 1985). However, this model did not take into account possible conflicts of interest arising from asymmetric degrees of relatedness between the workers and the different queens (mothers or sisters) they may accompany (e.g. Seeley 1985 pp. 60–67).

The evolution of resource allocation when there is colony fission was considered in detail by Pamilo (1991b; Section 4.8). One prediction was that, when a daughter colony retained the old queen, it should always be larger than the other daughter colony containing a new queen (Section 4.8). The findings for *Eciton burchelli* are at variance with this prediction because colonies appear to divide into two equal parts (Franks 1985, 1989a). However, this observation has to be treated with caution, because what proportion of fission events involve the retention of the old queen is unknown. In addition, datasets for army ant reproduction are typically based on small sample sizes. Nevertheless, some old queens almost certainly head daughter colonies, and highly uneven colony fissions have not been recorded (Franks 1985, 1989a). By contrast, in honey bees, another social insect group with obligate colony fission (Seeley 1985), very uneven colony division appears common. About 70% of workers accompany their mother queen in the prime swarm that leaves the nest (Getz et al. 1982). In honey bees, the mother queen always departs in the daughter colony (the swarm) that leaves the parental hive. Therefore workers, unlike army ant workers, have some knowledge of which daughter colony contains their mother. In line with Pamilo's (1991b) model, this could be precisely why more than half the workers join the prime swarm.

Clearly, a consequence of reproduction by colony fission in army ants is that it provides circumstances where a maternal queen can be superseded by one of her daughters. This means that colonies do not have an orphanage period, which might select for workers to retain the ability to produce their own sons (Franks et al. 1990a; Sections 7.4, 10.4). *Eciton* workers are indeed sterile (Oster and Wilson 1978 p. 93). It also means that the old queen may have to compete with her daughters to head one of the new colonies (Section 7.6). The result of this competition is likely to be decided by the workers because of their vastly greater numbers.

In *Eciton*, the old queen may periodically remate with a new, incoming male, possibly even once per year (Rettenmeyer 1963; Section 11.3). However, there are no firm genetic data on the number of matings and the pattern of sperm utilization. The possibility of remating suggests genetic circumstances that would favor workers joining a daughter colony headed by a sister rather than by their mother, since these arise if the mother queen switches to using sperm from a male other than the workers' father. The reason is that, in this case, the (life-for-life) relatedness of the workers with new daughters of the queen would tend to 0.25 (and with her sons would remain at 0.25). But if the workers were sure to work for a full sister queen their relatedness with both her sons and daughters would be 0.375. However, favoring working for a sister for this reason is implausible, because it assumes that the old queen changes her pattern of sperm use at exactly the moment of colony fission. Therefore, in general, workers should often seek to retain the old queen as head of one of the daughter colonies, unless her presence would seriously damage the survival and reproductive chances of that colony, for example if she were unhealthy.

In honey bees, replacement of a dead queen can occur at almost any stage of the colony life cycle, so long as diploid eggs are present from which workers can rear a new queen (Michener 1974 p. 109). This would explain why one of the daughter colonies resulting from fission is always headed by the old queen. Even if she were unhealthy, workers could replace her provided she laid some diploid eggs in the new hive. By contrast, in army ants, there is no evidence that colonies can replace a dead queen unless her death precisely coincides with a fission event. Therefore, retention of a mother queen who is potentially low in fecundity is likely to be fatal to a colony. This could be why, unlike the case in honey bees, army ant colonies sometimes split into two parts each headed by a daughter queen.

Connected with this, there is evidence that army ant colonies that have lost their queen will fuse with another colony after only a few days (Schneirla 1971). Since colonies are highly nomadic, it is unlikely that the two fusing colonies are close relatives. Such colony fusion is not, however, an evolutionary enigma (see also Lin and Michener 1972; West-Eberhard 1975). The workers in a queenless colony are at an evolutionary dead end, being both sterile and unable to replace their queen. But a queenright colony that is able to acquire extra workers effectively undergoes a growth spurt enabling it to reproduce all the sooner. Thus, there should be strong selection on queenright colonies to accept unrelated orphaned workers, and little counteracting selection on these workers to resist adoption.

10.9 Conclusion

All the case studies considered in this chapter show that there is immense variability and flexibility in life history patterns in ants, sometimes within single species. In addition, they show that genetic studies are essential for interpreting life history patterns. Long-term field data for each species, preferably from more than one site, are also of unrivalled value. An obvious lack is the absence of basic demographic data on the age structure and survivorship of individuals and colonies (Tschinkel 1991). Furthermore, as the last chapter discussed, it is clear that there is as yet no all-embracing, general life history theory for social insects. Therefore, new developments in theory, and long-term empirical studies, are now at an equal premium in the life history analysis of ants.

10.10 Summary

1. The seven case studies reviewed in this chapter show the extreme diversity and plasticity that exist in the social systems of ants. For example, species, populations, and colonies vary in queen number, queen mating frequency, the mating system, the mode of colony reproduction, and the level of worker reproduction. The case studies also show that variation in these traits has many consequences for life history evolution. However, several of these consequences remain to be explained or elucidated.

2. Monogynous colonies of *Solenopsis invicta* exemplify the sharp switch from the ergonomic (worker producing) to the reproductive (sexual producing) phase of ant life history as colony size and age increase. Complex interactions, which are not fully understood, occur between queen number, larval number, queen fecundity, and sexual productivity. Individual queens within polygynous colonies conceivably follow either annual or perennial strategies. Monogyny and polygyny may be perpetuated, at the proximate level, by a "cultural" influence of colony type on queen size.

3. *Myrmica* species may be predominantly monogynous or polygynous. There is a high degree of worker male-production. Again, a complex set of interactions within colonies determines the pattern and timing of worker and sexual production. Specifically, the presence of colony queens inhibits (at least at the proximate level) the sexualization of female larvae and male-production by old workers ("queen effect"). These traits make *Myrmica* ants very plastic in their life history patterns. A life history model for *Myrmica* by Brian et al. (1981) predicted

that mature colonies should oscillate in size, and that food-stressed colonies should produce mainly males and reproduce by budding.

4. Herbers' (1984, 1990) long-term study of *Leptothorax* (*Myrafant*) *longispinosus*, which is both polygynous and seasonally polydomous, reveals a high level of variation between colonies, populations, and years in sex allocation and other reproductive parameters. Experiments did not confirm resource differences as an explanation of the between-population variation. In the monogynous leptothoracine slave-maker *Harpagoxenus sublaevis*, a life history model by Franks et al. (1990a) showed that the queen loses fitness if orphan workers fail to reproduce, because she cannot compensate by additional sexual production late in her life. Members of the socially parasitic leptothoracine genus *Epimyrma* exhibit life histories involving either exclusively sexual production for a year or two ("big bang"), or a slow build-up of slave workers by slave-raiding followed by iteroparous sexual production. The particular strategy that evolves seems to depend on the distribution and abundance of the host species.

5. The wood ants of the *Formica* group exhibit the "space-perennial" life history strategy. Colonies occupy for long periods large, populous, thermoregulated mounds, and amplify the value of their local patch by "farming" aphids. These factors explain, in part, the high levels of polygyny, polydomy, and colony budding in this group. However, why some species go on to form unicolonial populations ("supercolonies") is less clear.

6. In the Argentine ant, *Linepithema humile* (= *Iridomyrmex humilis*), unicoloniality is associated with a tendency to spread as an introduction in new areas. It may also help to minimize the costs of central place foraging. Workers exhibit the bizarre habit, which has yet to be explained, of periodically executing most of their queens.

7. *Pristomyrmex pungens* colonies are unique in consisting entirely of workers that reproduce by thelytokous parthenogenesis, yielding more workers. Thelytoky may have evolved as a response to the risk of queen loss in a nomadic species. The fraction of egg-laying workers in *P. pungens* could be maintained by a balance of between- and within-colony selection.

8. *Eciton* army ants live in enormous monogynous colonies that reproduce by colony fission, following a series of regular, synchronized cycles of brood production. Reproduction by colony fission stems from colonies needing to be large to exploit large prey. This in turn enforces nomadism and brood production cycles. Workers may not always favor remaining in their mother's half of the colony at fission, because they cannot replace an old queen that dies between fission events.

11 Mating Biology

11.1 Introduction

The mating biology of ants is a little explored topic, but one that never-theless holds several general lessons for the evolutionary study of mat-ing frequency, sex roles, sperm competition, and related issues in the field of sexual selection. This chapter reviews the natural history and evolutionary significance of mating biology in ants. A previous, influen-tial review of this topic was that of Hölldobler and Bartz (1985), and it was also discussed by Hölldobler and Wilson (1990). Thornhill and Alcock (1983) described mating systems and the operation of sexual selection in insects in general, and Stubblefield and Seger (1994) review sexual selection issues in the Hymenoptera as a group. Andersson (1994) is a comprehensive monograph dealing with sexual selection across all animals and plants.

The chapter begins by considering what aspects of sexual selection theory are applicable to the mating biology of ants, and what traits in the behavior of male and female ants they help to explain (Section 11.2). It then examines sperm use and sperm competition in ants (Section 11.3). The following section (11.4) discusses the evolution of multiple mating in ant queens. Finally, Section 11.5 explores the diversi-ty of mating systems found in ants, including the more unusual ones such as those in which males are wingless and fight for possession of females in the nest.

11.2 Sexual Selection and Ant Sexual Behavior

SEXUAL SELECTION THEORY

Sexual selection promotes the evolution of traits that are associated purely with increasing mating success. According to classical sexual selection theory, such traits take two forms. First, there are those that result from intrasexual selection and enhance the ability of members of one sex to compete directly with one another for fertilization opportunities. Weapons for fighting among males to gain access to females, such as antlers in deer, are well-known examples. Second, there are traits that result from intersexual selection and influence mate choice. The time-honored example here is the peacock's tail, an apparently useless ornament used by males to attract potential partners. Intrasexual and intersexual selection may operate simultaneously (e.g. Krebs and Davies 1993 p. 183).

In most animals, males are the competitive sex, and females the choosy one. The exact reason for this has been much debated. Recently, Clutton-Brock and Vincent (1991) and Clutton-Brock and Parker (1992) proposed that what affects the strength of sexual selection, and the role each sex adopts, is the relative potential rate of reproduction of either sex. This refers to the rate at which individuals can produce gametes, or gametes plus surviving offspring if there is parental care. The sex with the highest potential rate of reproduction should be the competitive sex, and the other one the choosy sex. This is because the potential rate of reproduction affects the *operational sex ratio*, which is defined as the ratio of receptive females to sexually active males. In the sex with the highest potential rate of reproduction, some individuals will obtain disproportionately more partners, so biasing the operational sex ratio among the individuals without partners towards the same sex. This bias in the operational sex ratio is the immediate cause of the greater competitiveness of that sex. For example, if sexually active males outnumber receptive females, then males will be the competitive sex. However, if the operational sex ratio is biased towards females, the females should be competitive (Emlen and Oring 1977; Clutton-Brock and Vincent 1991; Clutton-Brock and Parker 1992). This turns out to be an important argument for understanding the sexual behavior of ants.

ANT SEXUAL BEHAVIOR

There are four widespread aspects of ant sexual behavior that demand explanation in terms of sexual selection theory. They are now considered in turn.

1. To begin with, male ants are notable for their general lack of adaptations for intrasexual fighting (Starr 1984). (Exceptions such as *Pogonomyrmex* and *Cardiocondyla* males are described below and in Section 11.5.) For example, they typically have small mandibles, and lack any other structures that could act as "antlers". Instead, male ants are characterized by adaptations for finding queens on the wing (Heinze and Hölldobler 1993b). These include large compound eyes and prominent ocelli (simple eyes) for visual orientation, a voluminous

thorax to house the large wing muscles, and an otherwise lightweight and flimsy build (Heinze and Hölldobler 1993b). Males ants also typically have an elaborate genital architecture (e.g. Donisthorpe 1927 pp. 13–15; Collingwood 1979 p. 16), which presumably functions in clasping the female's cloaca during copulation and allows the male to remain attached to his partner even if she is in flight. In fact, a male's grip on his mate may be so tight that, while mating, he remains attached to her by his genitalia alone (e.g. Buschinger and Alloway 1979; Robertson and Villet 1989). Lastly, male ants live notably briefly as adults. For example, adult *Harpagoxenus sublaevis* males live for an average of only six days (Winter and Buschinger 1983).

2. A widespread, but again not ubiquitous, behavioral trait of male ants is single mating. Thus, males in many species typically discharge all their sperm in a mating with just one female (Nonacs 1986a). They often also leave their genitalia stuck to the female's cloaca (e.g. Hölldobler 1976; Nonacs 1986a; Fukumoto et al. 1989), apparently as a "mating plug" that hinders her remating.

Strangely, males throughout the Hymenoptera exhibit degeneration of the testes prior to sexual maturity (Heinze and Hölldobler 1993b). In a number of ants from diverse phylogenetic backgrounds, it has been shown that this occurs even before the males emerge as adults from the pupal stage, with the only known exception being *Cardiocondyla* (Heinze and Hölldobler 1993b; Section 11.5). In the absence of the testes, the male's entire batch of sperm is stored in an expanded vas deferens (Hölldobler and Bartz 1985; Hölldobler and Wilson 1990 pp. 154–155; Passera and Keller 1992). An important consequence is that, unlike males in many other organisms, sexually mature ant males cannot replenish their sperm supply. Arguably, this explains why they typically mate just once; perhaps they do not have enough sperm to donate to more than one female (Hölldobler and Bartz 1985; Passera and Keller 1992; Heinze and Hölldobler 1993b). However, in principle, males have the choice of distributing their limited sperm over several queens (Hölldobler and Bartz 1985), or of evolving continuous sperm production as in *Cardiocondyla* (Heinze and Hölldobler 1993b). At the proximate level, female ants often terminate copulation by turning round and biting the male. This can be fatal (Nonacs 1986a). But this behavior also fails to explain single male mating because, at the ultimate level, males could evolve to separate from the females themselves. Single mating by male ants, therefore, still warrants explanation.

3. The third feature of ant mating behavior that needs explaining is that females do not seem to discriminate among conspecific males. In other words, there seems to be hardly any female choice. This is arguably because no one has systematically looked for it. However, sup-

port for this assertion comes from the absence of any obvious adaptations in male ants for advertising themselves to females; they have no "peacock's tails" (Starr 1984).

Female choice may exist in ants in one respect, namely over whether or not to mate with a relative. This type of choice seems likeliest to evolve in species with mating in the nest, where the chances of encountering relatives are obviously high. Keller and Passera (1993) presented evidence for females discriminating against related males as mates in such a species, *Linepithema humile* (= *Iridomyrmex humilis*), where unrelated males compete with nestmate males by entering foreign nests (Section 11.5). However, in species with mating outside the nest, in mating swarms, selection for this type of female choice should be weak, because the chances of a female meeting a relative are very small. Consistent with this, Mintzer (1982) found that females mated with both nestmates and nonnestmates in experiments with the swarm-mating *Pogonomyrmex californicus*. It still needs explaining why, throughout ants as a whole, female choice for other male qualities seems absent.

4. The last aspect of ant sexual behavior that needs to be explained is variation in the mating frequency of queens (e.g. Page 1986; Keller and Reeve 1994b). This has been the subject of much attention because it influences the genetic structure of colonies (Sections 4.5, 5.4, 11.3).

We return to the evolution of queen mating frequency in Section 11.4, and to the other three aspects of ant sexual behavior described above in the remainder of this section. First, we outline some basic features of ant mating systems. Thus, we now consider the sexual and social environment in which ant sexual behavior occurs.

Key Aspects of Ant Mating Systems

A very conspicuous feature of sexual behavior in ants is that mating often takes place during a large-scale nuptial flight, or mating swarm, of very limited duration (Sudd 1967; Brian 1983; Hölldobler and Bartz 1985; Section 11.5). In the classification proposed by Thornhill and Alcock (1983 p. 263), ants therefore approximate a mating system involving "scramble competition polygyny". Afterwards, queens never remate except in perhaps one or two cases (see below). Note, however, that large nuptial flights are unlikely to represent a primitive trait in ants. Thus in the primitive ponerine ants they are almost unknown (Peeters 1991a; Section 11.5). Furthermore, even in other ant subfamilies, nuptial flights are not always on a large scale (e.g. Hasegawa and Yamaguchi 1994).

Nevertheless, extensive, synchronized, short-lived nuptial flights are common in ants. They presumably evolved because of the seasonality exhibited by many species in their sexual production, and through selection to swamp predators, to maintain reproductive isolation from related species, and to promote outbreeding within species (Sudd 1967 p. 137; Brian 1983 p. 230; Starr 1984; Hölldobler and Bartz 1985; Hölldobler and Wilson 1990 p. 152). The need to satiate predators is clear from the observation that swarming ant sexuals are subject to heavy predation (e.g. Robertson and Villet 1989; O'Neill 1994). The importance of outbreeding is suggested by the finding that, as far as is known, in no species of ants do all representatives of both sexes lack wings, although in many species one sex is wingless (Wheeler 1910 p. 182; Starr 1984). Even in one very unusual case, *Technomyrmex albipes*, in which there are many wingless males and females, there are also winged forms of both sexes (Yamauchi et al. 1991; Section 5.8). However, some of the socially parasitic *Epimyrma* ants appear to inbreed routinely (Buschinger 1989a).

Connected with the occurrence of nuptial flights, another important trait of ant mating biology is that all reproduction (egg laying) by queens occurs during the period of the queen's life after the single episode of mating, with the use of sperm stored in a sperm receptacle next to the ovaries. Thus, unlike many other organisms, ant queens do not alternate bouts of mating with bouts of raising offspring. This is a basic aspect of ant social organization (Section 11.3).

A final, critically important feature of ant reproductive biology is that all sperm from any one male are genetically identical because, due to male haploidy, they are produced by mitosis. This trait, of course, leads to the haplodiploid relatedness values that are so crucial to understanding eusocial evolution and kin conflict in the Hymenoptera (Chapters 3–7). It also influences facets of sexual selection in ants. For example, it means that competition between individual sperm to fertilize eggs (sperm competition) is not expected between sperm from the same male (since they form a clone). Instead, it should only occur, if at all, between the ejaculates of different males, and hence is only expected if queens are multiply mated. Furthermore, since many fertilized eggs become sterile workers, sperm competition in ants may involve competition to fertilize queen-yielding eggs (Starr 1984; Section 7.6).

Explanation of the Major Traits of Ant Sexual Behavior in Terms of Sexual Selection Theory

1. Single Mating and Other Typical Traits of Ant Males

The concept of the potential rate of reproduction (Clutton-Brock and Vincent 1991; Clutton-Brock and Parker 1992) is not readily applicable to ants, because of the use of stored sperm by queens. Applying this concept presupposes that individuals can alternate periods of mating, egg laying, and care (and can therefore have a reproductive rate that limits the activities of the opposite sex), whereas in ants this is not the case. However, the concept of the operational sex ratio (Clutton-Brock and Parker 1992) is highly relevant, and applies to ants in a very straightforward manner. In ants with mating swarms, the operational sex ratio simply equals the *numerical sex ratio* (Box 4.1) in the swarm, since swarms consist exclusively of sexually active individuals.

For a fixed degree of female–male sexual size dimorphism, the numerical sex ratio will be set by the stable investment sex ratio and hence by the kin structure and mating pattern of colonies (Chapter 4). For monogynous ants, which typically have large females (for independent colony foundation), random mating, and female-biased sex investment (Table 5.1), the numerical sex ratio will often be male-biased. (In other words, even with a stable sex investment ratio of $3:1$ females : males, male numerical bias should often occur because in monogynous ants females are frequently more than three times as costly as males [Table 5.1].) Many polygynous species have male-biased sex investment (Table 5.1) and relatively small queens (because of nonindependent colony foundation), and so should also have numerically male-biased sex ratios. Lastly, species with colony fission and budding are also expected to have male-biased numerical sex ratios (Section 4.8). In fact, as expected, male bias in the numerical sex ratios of ants seems a frequent occurrence (Nonacs 1986a; Keller and Passera 1992; Table 5.1). Examples of numerical male bias being observed in swarms in the field are provided by Brian and Brian (1955), Chapman (1969), Markin et al. (1971), Hölldobler (1976), Davidson (1982), and O'Neill (1994). Male bias may frequently be even greater than expected from the population-wide numerical sex ratio. Many ants exhibit the so-called "male aggregation syndrome" in which groups of males attract incoming females (Hölldobler and Bartz 1985; Section 11.5). Since females appear to stagger their entry to the swarm, and remain in it for only a short while (e.g. Brian and Brian 1955; Markin et al. 1971), male bias at any one time is likely to be very high.

As earlier discussed, the predicted consequence of a male-biased numerical (operational) sex ratio is that males should be the competitive sex (Clutton-Brock and Parker 1992). In some ants, this is borne out in an obvious way. For example, in *Pogonomyrmex* species, whose swarms are numerically male-biased, males form a tight ball around queens, competing directly for matings with other males. This results in sexual selection for a large body size, strong mandibles, and a robust build (Hölldobler 1976; Davidson 1982). However, no male–male competition or large-male advantage was observed in a *Formica* species with a male-biased numerical sex ratio (O'Neill 1994). The reason for this difference appears to lie in the site of mating. *Pogonomyrmex* sexuals mate on the ground, whereas the *Formica* ants mated on shrubs and grass after a period of overhead flight (O'Neill 1994). Males that mate on the ground can be large and combative because by being so they do not risk the loss of the queen (Davidson 1982; Hölldobler and Wilson 1990 p. 146). (Similarly, male aggression is seen in some species with mating in the nest [e.g. Keller and Passera 1992], reaching extremes in species with wingless males [Section 11.5].) By contrast, species with partner location and mating in the air, or on unstable vegetation, are less likely to evolve large male body size. The top priority of males in these species is, after finding a female, to remain attached to her in copula. For this, small size could be an advantage, because it would help the copulating pair to remain airborne, or on their perch.

Another conceivable outcome of a male-biased operational sex ratio would be a system in which males defend harems of females, or hold territories to which females are attracted. However, it seems unlikely that ant males ever have the opportunity to defend resources that females need, since females either already have these resources stored in their bodies (in species with claustral colony foundation) or will find them in the colonies that adopt them (in species with dependent colony foundation). This is probably why male ants, unlike males of many vertebrates and of other insects such as damselflies (Waage 1979), are non-territorial.

We suggest that, in species with aerial mating (possibly the majority of ant species), a male-biased numerical sex ratio still makes the males the competitive sex. However, for the reasons just given, the result is not males that compete with one another overtly, but instead males that show the typical traits of ant males listed earlier (traits [1] and [2]). These include a small body designed for locating queens and for flight (Heinze and Hölldobler 1993b), single mating, the retention of the Hymenopteran trait of limited sperm production, the use of the detached genitalia as mating plugs, and a short adult lifespan.

The reason for these features is that male-bias in the numerical sex ratio means that a male has only a small chance of finding even one free partner, and hardly any chance of finding two. This is especially so given that nuptial fights are, as earlier mentioned, typically both subject to heavy predation and of limited duration. These features mean that males risk either being eaten between mating attempts (Nonacs 1986a), or failing to find available females at all. The effect also applies in species with multiply mating queens, because queen mating frequencies are usually low (Pamilo 1993; Queller 1993b; Keller and Reeve 1994b). Therefore a male should be designed to devote all his reproductive effort first to locating a female, and then to discharging all his sperm in a single mating with her. Furthermore, the male should also be selected to leave a mating plug in the queen to help guarantee his paternity, since she is likely to be approached by many other suitors. The male should leave a mating plug even if it involves suicidally detaching his genitalia, since the improbability of his having two partners would make the benefits of this behavior outweigh its very low cost. The elaborate structure of the male genitals almost certainly facilitates not only the male's clasping of the female in copula, but also the use of the severed genitals as a mating plug. Finally, given such a strategy, a male would have no need for continuous sperm production or anything but a brief adult life.

The argument that a male-biased numerical sex ratio affects the chance of a male finding more than one mate, and hence selects for males to mate once (and suicidally), was made for honey bees by Bulmer (1983b) and Thornhill and Alcock (1983 p. 235). An influence of the numerical sex ratio on ant sexual behavior was also suggested by Davidson (1982), Yamauchi et al. (1991), and Keller and Passera (1992). Honey bee drones are well known for leaving their detached genitalia in the queen after mating, and killing themselves as a result. This "mating sign" was previously interpreted as a mating plug. However, in fact it fails to prevent further matings by the queen, leading Woyciechowski et al. (1994a) to suggest that its real function is to prevent sperm leaking from the queen's genital tract. This is conceivably also the case in some ants, although few ant queens mate as frequently as queen honey bees (Page 1986; Keller and Reeve 1994b). A possible example is the ant *Carebara vidua*, in which Robertson and Villet (1989) found queens containing up to four male structures resembling mating plugs. However, suicidal mating for any reason still underlines the point that, in both honey bees and many ants, mating opportunities for males are so rare that complete investment in one mating is advantageous.

Another reason why males might be selected to mate once, which would reinforce the effects of their outnumbering the females, arises specifically when there is multiple mating by queens. In this case, males should arguably maximize the sperm load delivered to females, to ensure that, in the sperm competition with her other mates, at least some of their sperm fertilize the sexual-yielding eggs (Starr 1984; Hölldobler and Bartz 1985; Nonacs 1986a).

Since single mating by males appears common, all its proposed advantages presumably usually offset one of its possible costs, which is the risk to a male of donating all his sperm to a single queen who may herself be predated (Hölldobler and Bartz 1985). However, at present the data on numerical sex ratios, predation rates, and reproductive success that would allow this to be verified do not exist.

As earlier mentioned, it has sometimes been suggested that male ants mate once simply because they lack sufficient sperm to inseminate more than one queen (e.g. Hölldobler and Bartz 1985). However, this seems a poor explanation because it presupposes that males cannot evolve larger sperm loads (Section 11.4). Therefore, if male ants typically have only enough sperm for one queen (Hölldobler and Bartz 1985), this could be because they evolved to be this way, through most having experienced conditions in which single male mating is selectively advantageous.

2. MULTIPLE MATING IN ANT MALES

As with all general rules in biology, some male ants violate the norm and mate multiply. But, as also often happens, the exceptions can be accounted for by the same underlying logic. Thus, we propose that multiple mating in ant males occurs for two main reasons. First, it occurs when there is a *female-biased* numerical sex ratio. In this case, the previous argument is reversed, and males are no longer selected to mate just once. Second, multiple mating by males can evolve when the risk of predation on males is unusually low, due to mating occurring in the nest (Yamauchi et al. 1991). Note that these two factors covary: several species with mating in the nest have a degree of local mate competition among males, and hence female-biased numerical sex ratios (Section 5.8), both of which favor multiple male mating.

To test these proposals would require knowing both the typical male mating frequency and the numerical sex ratio in many species. These data do not yet exist. However, the available evidence tentatively supports the predicted links. Thus, with two exceptions, there appears to be an association between single male mating and a male-biased numerical sex ratio, and between multiple male mating and a female-biased numerical sex ratio (Table 11.1). These associations even occur intraspecifically, as shown by the case of *Technomyrmex albipes*

(Yamauchi et al. 1991; Table 11.1). The effect of the numerical sex ratio on male mating frequency appears to override that of the site of mating. For example, *Linepithema humile* has numerical male bias and single male mating, yet mating is in the nest (Table 11.1). Conversely, *Formica aquilonia* and *F. polyctena* have numerical female bias and multiple male mating, but mating is outside (or possibly on) the nest (Table 11.1). It is also worth noting that multiply mating males and mating in the nest are also associated with an extended adult male lifespan (e.g. *Epimyrma kraussei*: Winter and Buschinger 1986; *T. albipes*: Yamauchi et al. 1991; *Cardiocondyla nuda*: Heinze and Hölldobler 1993b).

One of the two exceptions to the expected association between the numerical sex ratio and male mating frequency is *Harpagoxenus sublaevis*, which has numerical male bias in the sex ratio and males that mate several times (Table 11.1). A possible reason is that *H. sublaevis* has a mating system that does not involve swarming or large concentrations of both sexes. Instead, it shows "female calling" behavior (Section 11.5). Females walk a short way from the nest and then attract males from a stationary position on the ground by releasing sexual pheromones (Buschinger 1968b, 1983). Therefore, a single male may occasionally encounter virgin females in locally high concentrations (say, around an isolated nest), which are worth exploiting before the arrival of other males. In essence, the *local* numerical sex ratio could often be very female-biased, and this would favor a male ability to mate multiply. By contrast, males in species where males gather to await arriving females (species with male aggregation syndrome) must nearly always find themselves in the company of large numbers of other males. The argument used for *H. sublaevis* could also apply to other species with female calling, such as the ponerines, although whether males mate singly or multiply does not appear to have been recorded in this group (Peeters 1991a). Similarly, in fireflies, females call males with light signals, and those males that can outrace their rivals to reach them may get multiple mating opportunities (Lloyd 1986).

3. THE LACK OF FEMALE CHOICE IN ANTS

Finally, it needs explaining why female choice appears absent in ants (trait [3] in the earlier list). To start with, females are almost certainly very limited in the number of male properties they could choose. Since males do not participate in rearing offspring, these properties potentially include only a favorable genetic make-up ("good genes"), a lack of transferable parasites, and sufficient plentiful, healthy, and long lived sperm. Conceivably, they also include accessory gland secretions (transmitted with the sperm during insemination) that might be required to

Table 11.1

Male Mating Frequency, Site of Mating, and the Numerical Sex Ratio in Ants

Species with Single Male Mating:
Linepithema humile (= *Iridomyrmex humilis*) (Keller and Passera 1992).
Numerical sex ratio is male-biased and mating is in nest (Keller and Passera 1992).
Technomyrmex albipes winged male (Yamauchi et al. 1991).
Numerical sex ratio among winged sexuals is male-biased and mating is outside nest
(Yamauchi et al. 1991; Tsuji and Yamauchi 1994).

Species with Multiple Male Mating:
Cataglyphis cursor (Lenoir et al. 1988).
No data on numerical sex ratio.
Cardiocondyla nuda ergatoid male (Heinze and Hölldobler 1993b).
No data on numerical sex ratio; mating is in nest and sperm production is continuous
(Heinze and Hölldobler 1993b; Heinze et al. 1993a).
Epimyrma kraussei (Winter and Buschinger 1983).
Numerical sex ratio is female-biased and mating is in nest (Winter and Buschinger
1983).
Formica aquilonia (Pamilo et al. 1978).
Numerical sex ratio is female-biased (Pamilo and Rosengren 1983).
Formica polyctena (Yamauchi et al. 1994).
Numerical sex ratio is female-biased (Yamauchi et al. 1994).
Harpagoxenus sublaevis (Buschinger and Alloway 1979; Winter and Buschinger 1986).
Numerical sex ratio is male-biased (Bourke et al. 1988).
Meranoplus peringueyi (Robertson and Villet 1989).
Numerical sex ratio is male-biased (Robertson and Villet 1989).
Technomyrmex albipes ergatoid male (Yamauchi et al. 1991).
Numerical sex ratio in wingless sexuals is female-biased and mating is in nest
(Yamauchi et al. 1991; Tsuji and Yamauchi 1994).

feed either the sperm in the queen's sperm receptacle, or even the queen herself during colony foundation (Wheeler and Krutzsch 1992). However, it is not obvious how a queen could assess any of these attributes. In addition, as stressed earlier, nuptial flights are typically brief and frantic because of heavy predation on ant sexuals. Therefore, males would probably not be selected to spend time in advertising their qualities to queens, and queens (after mating their specified number of times) are probably most concerned to leave the swarm to found a colony as quickly as possible. In short, it appears that female choice has not evolved because it would be difficult, time-wasting, and costly. However, some kind of female choice seems more likely in species with mating on the ground or in the nest, where both male advertisement and female choice would carry less risk of curtailment by predators.

Earlier, it was argued that female choice with respect to inbreeding avoidance was expected in species with mating in the nest. Note that this would also be facilitated by the relative protection from predators.

In sum, it seems that the sexual behavior of ants is largely dictated by the fact that both sexes typically mate in just one episode at the start of their lives, in large, short lived swarms, subject to predation, in which males outnumber females. As we have argued, these features allow the basic idea that sex roles are determined by the operational sex ratio to be applied very simply to ants. The typical products are males that display little overt male–male competition, mate once, and die young, and females that have hardly any opportunity or incentive to display female choice of a partner. Hence, although sexual selection affects ants in a predictable manner, its products are not the conventional dominant, flamboyant males and coy, selective females of the textbooks.

11.3 Sperm Use and Sperm Competition in Ants

Multiple Mating, Sperm use, and the Relatedness Structure within Colonies

In many species of ants, queens mate multiply (e.g. Page 1986; Hölldobler and Wilson 1990 p. 156; Keller and Reeve 1994b; Section 11.4). But how does multiple mating actually affect the relatedness structure of colonies? There are two issues. First, if a queen mates multiply, do all her mates achieve equal paternity? If, for example, one male's sperm fertilized most eggs, then the effect on within-colony relatedness would be less than if males achieved equal fatherhood. Second, do multiply mated queens change their pattern of sperm use over time? That is, over their lifetimes, do they always draw on the different males' sperm within the sperm receptacle in proportion to the sperm's abundance, or do they use first one male's sperm, then a second's, and so on? Trivers and Hare (1976) raised this possibility by suggesting that an individual male's sperm clumps in the sperm receptacle. At one extreme, if sperm does clump, then workers might usually rear full sisters and the effects of multiple mating will be mitigated. At the other, if multiply mated queens utilize sperm randomly through time, for example through complete mixing of sperm in the sperm receptacle, the effect of multiple mating on relatedness will be fully felt.

Some genetic data exist on both of these issues (reviewed by Crozier and Brückner 1981; Starr 1984; Page 1986). On the first, they suggest that males often do differ in their share of the offspring, with some males being better represented than others. This was found in polyandrous honey bees (Page and Metcalf 1982; Moritz 1986b; Estoup et al. 1994), vespine wasps (Ross 1986), polistine wasps (Metcalf and Whitt 1977), and several *Formica* ants (Pamilo 1982b, 1993; Sundström 1993a). On the other hand, no significant inequalities in paternity were found in mature *Lasius niger* ant colonies headed by doubly mated queens (Van der Have et al. 1988). It is not clear whether unequal paternity results from males delivering different amounts of sperm, or from sperm competition, or simply from differential survival of sperm or brood after insemination. On the second issue, the data suggest that the real situation lies between the two possible extremes, but is closer to the random usage case than the total clumping one. That is, a high level of sperm clumping is absent, but sperm usage over time is not totally random. Thus, for example, Page and Metcalf (1982), Laidlaw and Page (1984), and Estoup et al. (1994) found fairly uniform temporal patterns of queen usage of sperm from different males in honey bees, and Ross (1986) obtained similar results in vespine wasps. Therefore, the general finding is that multiple mating does appreciably reduce relatedness within a queen's brood, though not always by as much as the number of queen's mates alone suggests. This is why it is often best to refer to the *effective number of matings* to describe the phenomena of multiple mating, where this means the mating frequency taking into account any biases in the pattern of sperm use (e.g. Starr 1984; Page 1986; Queller 1993a).

THE EFFICIENCY OF SPERM USE BY ANT QUEENS

As described in the previous section, nearly all ant queens mate in just a single nuptial episode prior to colony foundation. This means that queens must store very large numbers of sperm for use during the whole of their reproductive lives. Large numbers of sperm are required because queens are typically long lived and productive, and need to fertilize the eggs that will develop into their entire complement of diploid offspring. These eggs include those that will be lost through natural mortality, egg eating, or brood cannibalism (e.g. Crespi 1992b; Box 9.1). The economics of sperm use are therefore worth examining.

To see the scale of the problem, consider the case of the army ant, *Eciton burchelli*. Assuming a colony takes 3 years to grow between fission events, it will pass through 31 brood-rearing cycles, in each of which the queen may lay 100,000 eggs, of which perhaps only half develop into workers. In other words, in 3 years the *E. burchelli* queen will lay roughly 3 million eggs and produce about 1.5 million workers. Since some queens may live 6 years, they would need at least 6 million sperm (Franks 1989a; Section 10.8). However, the amount of sperm that males deliver is unknown.

The efficiency of sperm use has been most thoroughly investigated in the monogynous form of *Solenopsis invicta*. The queen can live nearly 7 years. She mates with just a single male, and is known to start her reproductive life with about 7 million sperm. Over her lifetime, the queen uses these to produce 2.5 million workers and 0.1 million new queens (Tschinkel 1987a; Tschinkel and Porter 1988). Therefore, supposing each diploid egg reached maturity, a queen would have to release only about 3 sperm to fertilize each egg (Tschinkel and Porter 1988). Since some brood mortality is inevitable, it is likely that *S. invicta* queens can inseminate an egg with almost every sperm released from the sperm receptacle. This represents an astonishingly efficient use of sperm (Tschinkel and Porter 1988). For example, it is estimated that honey bee queens release 20 to 30 sperm per fertilized egg (Tschinkel and Porter 1988). In ants, extreme efficiency of sperm use could be a consequence of sperm being expensive for queens to store (Tschinkel 1987c; Section 11.4).

WHY DO MOST QUEENS NEVER REMATE?

The typical ant queen, after mating, sheds her wings and founds a colony. Clearly, such a queen cannot participate in another nuptial flight, and it would make little sense for her to depart from the protection of her workforce to mate again. This is why, in earlier discussion, we took a single, early mating episode as a given fact of ant reproductive biology. However, it is still worth asking why some males do not seek out established females, allowing them the opportunity to remate. This is at least feasible, since, in certain ponerines, mating occurs when foreign males enter nests to seek out the virgin queens (Peeters 1991a). Males that mated with an established queen might benefit because they would be contributing sperm to a "going concern," and not just to a highly vulnerable young queen who had yet to found a colony. However, the reason remating by established queens does not occur is almost certainly because it would nearly always be against the interests of the workers. This is because it would dilute their relatedness with future brood. For example, if the queen were originally singly mated, her remating with an unrelated, incoming male would involve the workers' trading full sisters for half sisters. Hence, instead, we see the usual system in which queens mate on just one occasion and thereafter use stored sperm.

Very few exceptions have been proposed to the general rule that mature ant queens do not remate. These were suggested by Rettenmeyer (1963) in *Eciton* and other army ants. However, the evidence is anecdotal and partly based on laboratory observations in circumstances that might have been unrealistic. Thus, genetic data and field observations are required to verify or refute the possibility that queens remate.

SPERM COMPETITION

One set of ideas for why some ants have evolved multiple mating among their queens holds that a greater genetic diversity in the workforce improves the colony's survivorship or productivity (e.g. Crozier and Page 1985; following section). If this is so, then multiple mating by queens would be favored by her mates as well, even though it means they must share their paternity. This could be why, as far as is known, male ants do not possess structures for removing the sperm from previous mates of their partner, as are found in some solitary insects (e.g. damselflies: Waage 1979). Equally, in species with single mating by the queen, males could lack such structures because they rarely encounter already-mated queens, whose priority is to leave the swarm. However, as already discussed, males apparently do sometimes attempt to safeguard their paternity by leaving their genitalia in the queen as a mating plug.

Another possibility, if queens mate multiply, is sperm competition between the ejaculates of the different males (see previous section). Starr (1984) suggested that this might lead to the evolution of what might be termed "soldier sperm." Since the sperm from a single male are genetically identical, some might easily evolve a morphology and behavior designed for suicidal attacks on the sperm of the other males in the sperm receptacle. So, in ants with multiply mating queens, one might expect sperm of unusual morphology. However, there are no reports of such sperm in any species (Starr 1984). Similarly, the idea of altruistic "kamikaze" sperm in mammals has not received empirical support (Harcourt 1991). In ants, an additional expected outcome of sperm competition between ejaculates would be that some of a queen's mates were genetically better represented in the offspring than others. Although this is the case in several species (e.g. Pamilo 1982b), it could result from unequal delivery or survivorship of sperm (see above).

If it is genuine, a possible reason for the lack of sperm competition in ants is that the multiple mates of ant queens themselves benefit from the genetically variable workforce, as just discussed. Since a multiply mating queen presumably always herself benefits from polyandry (or otherwise she would take only one partner), the queen may also deliberately mix the sperm in her sperm receptacle to stop sperm competition occurring. Certainly such mixing seems to exist, at least in bees and wasps, because polyandrous queens use sperm from the different males approximately randomly over time (e.g. Page and Metcalf 1982; see above).

11.4 The Evolution of Multiple Mating in Ant Queens

As we have had occasion to say already, multiple mating by queens (polyandry) is relatively common in ants. Over half of the species investigated show it to some degree (Page and Metcalf 1982; Cole 1983a; Starr 1984; Page 1986; Hölldobler and Wilson 1990 p. 156; Keller and Reeve 1994b). However, even in these, some queens may remain singly mated. In addition, the overall mating frequency of the other queens may be low (with two to three mates being typical), and their effective mating frequency even lower (e.g. Van der Have et al. 1988; Pamilo 1991d, 1993; Keller and Reeve 1994b; Section 11.3). Nevertheless, multiple mating affects many facets of social evolution, including sex allocation, conflict over male parentage, and the level of genetic diversity within the colony (e.g. Hamilton 1987b; Ratnieks 1988; Boomsma and Grafen 1990; Sections 4.5, 7.4). In addition, it presumably has advantages that outweigh its costs to queens, which include a higher risk of queens' contracting disease from their partners, a greater exposure to predators, and the metabolic costs of storing extra sperm (Starr 1984; Sherman et al. 1988; Tschinkel 1987c). Therefore, this section reviews the major hypotheses that have been proposed for the evolution of polyandry in ants. Previous reviews of this topic, in ants and other social Hymenoptera, include those by Page and Metcalf (1982), Cole (1983a), Starr (1984), Crozier and Page (1985), Page (1986), Hölldobler and Wilson (1990 p. 155), Pamilo (1991c), and Moritz and Southwick (1992 p. 293). The evolution of multiple mating has also been investigated in solitary insects (e.g. Thornhill and Alcock 1983; Ridley 1988) and vertebrates (e.g. Madsen et al. 1992).

Hypothesis 1

Multiple mating evolves because single males have insufficient sperm for a queen's requirements (Parker 1970; West-Eberhard 1975; Trivers and Hare 1976; Cole 1983a).

This hypothesis has been advocated most strongly by Cole (1983a). He suggested that multiple mating evolves to provide queens with enough sperm to produce large colonies. To support this, Cole (1983a) documented an association between large colony size and multiple mating in ants. However, the sample sizes were small (Pamilo 1991c), and the data on mating frequency came mostly from behavioral observations. Male ants may copulate with queens without actually transferring sperm (e.g. Keller and Passera 1992), so genetic techniques provide the only sure way to establish the (effective) mating frequency. Moreover, as pointed out by Ross et al. (1988), *Solenopsis invicta* has very large colonies (even in the monogynous form) and singly mating queens. Therefore, the proposed connection between mating frequency and colony size is uncertain.

In addition, this hypothesis presupposes that individual male ants cannot evolve to produce sufficient sperm to serve the needs of a single queen, whatever they may be. It is not clear why this should be so. A possible example is *Atta sexdens*, in which each virgin male has 40 to 80 million sperm, whereas newly-mated queens contain 200 to 310 million (Hölldobler and Wilson 1990 p. 154). However, the actual lifetime requirements of an *Atta* queen for sperm are unknown. She may mate multiply to obtain genetically diverse sperm, and fail to use all the sperm she acquires. Furthermore, this species appears exceptional in that the queen's mating frequency is unusually high (Keller and Reeve 1994b).

Tschinkel (1987c) suggested that sperm might be expensive for queens to store, because they must be kept alive in the sperm receptacle over many years. In dissections of queens of 25 ant species, Tschinkel (1987c) found a positive relationship between the logarithm of the number of sperm stored in the sperm receptacle, and the logarithm of the number of ovarioles in the ovaries (Figure 11.1). (Ovarioles are the tubes in which eggs are produced in the ovaries, and their number is almost certainly linked with the lifetime number of eggs produced by queens.) That is, for each increase in the number of ovarioles, there is a disproportionate increase in the number of sperm stored. For example, queens of a *Hypoponera* species have six ovarioles and store 11,000 sperm, whereas *Solenopsis invicta* queens have 200 ovarioles and store 7 million sperm. As expected if keeping sperm alive is costly, this suggested that the queens store no more sperm than their potential fecundity demands. But Tschinkel's (1987c) findings also suggest that the queen's reproductive system evolves to suit her life history. The same could be true of ant males. Certainly, in bees and wasps, male sperm loads rise with female storage capacity (Crozier and Page 1985), suggesting that males evolve the sperm load appropriate to their mating biology.

Finally, it seems unlikely that male ants fail to evolve larger sperm stores to avoid impairing their flight performance, as has been suggested in honey bees (Starr 1984). For example, in *Solenopsis invicta*, males may remain airborne for hours at altitudes of over 90 m (Markin et al.

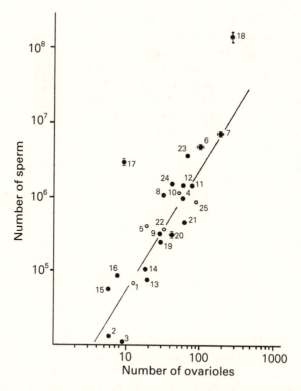

Figure 11.1 The number of spermatozoa in the sperm receptacle of ant queens regressed on their ovariole number (both axes are logarithmic scales). Each numbered point represents a different species of ant queen. Numbers 17 and 18 (*Acromyrmex versicolor* and *Atta texana* respectively) were outliers and were not used in the regression. In *Linepithema humile*, a species not included in the sample, Keller and Passera (1992) estimated that queens store about ten times more sperm than they use in their lifetimes. (Reprinted from Tschinkel 1987c; by permission of the Entomological Society of America)

1971), and yet every male contains more than enough sperm for all the requirements of a queen (Tschinkel and Porter 1988).

Hypothesis 2

A genetically more diverse workforce displays a more efficient division of labor (Starr 1984; Crozier and Page 1985).

Multiple mating means that the workforce in a colony consists of a genetically diverse set of sibships, each arising from a different father. Such sibships are termed patrilines (if they stem from different mothers, they are termed matrilines). This and each of the remaining hypotheses for multiple mating propose that its advantages lie in increasing the genetic diversity within the colony, although the specific benefits involved vary. Some of these hypotheses have also been put forward to help explain the evolution of polygyny (Section 8.4).

There is evidence in both honey bees (Frumhoff and Baker 1988; Robinson and Page 1988) and ants (Stuart and Page 1991; Snyder 1992, 1993; Carlin et al. 1993) that different worker patrilines or matrilines may become associated with the performance of different tasks. Arguably, then, multiple mating evolves because it provides queens with a spread of worker genotypes fitted to different tasks, so raising the efficiency of the colony, and hence its productivity (Frumhoff and Baker 1988). However, why some worker sibships have a greater tendency to perform certain tasks is unknown. This could be just a result of the way the division of labor develops (Tofts and Franks 1992; Franks and Tofts 1994; Section 12.4). In addition, crucially, there is no suggestion in the literature that key tasks are not performed if certain genotypes are absent from the colony. Moreover, although Fuchs and Schade (1994) found that honey bee colonies with multiply mated queens outproduced those artificially started from a singly mated queen, it is unknown if the same would be true in ants.

HYPOTHESIS 3

Colonies with greater genetic diversity are less prone to damage from diseases and parasites (Hamilton 1987b; Sherman et al. 1988; Schmid-Hempel 1994).

This hypothesis suggests that, by mating multiply, queens acquire a workforce that will have a greater mix of genes for resistance to parasites and disease. Therefore, any one pathogen might damage only one patriline, rather than infect the whole colony. Singly-fathered Hymenopteran societies may be particularly vulnerable to pathogens, since they have a long generation time and contain very closely related offspring at high densities (Hamilton 1987b; Sherman et al. 1988).

In bumble bees, Shykoff and Schmid-Hempel (1991b, 1991c) showed experimentally that workers contracted a trypanosomal infection more readily from related individuals. This implied that, as this hypothesis for multiple mating predicts, high within-colony relatedness increases susceptibility to disease. However, Woyciechowski et al. (1994b) detected no difference in protozoan infection between workers in honey bee colonies with high and low genetic diversities. Keller and Reeve (1994b) found an association in ants between monogyny and polyandry, and polygyny and single mating (Section 8.4). This is consistent with all the theories that multiple mating serves to enhance genetic diversity. In addition, although ants are indeed subject to attack by parasites (Hölldobler and Wilson 1990 p. 554; Section 8.4), it has not yet been investigated whether multiple mating confers any protection.

HYPOTHESIS 4

Multiple mating leads to a reduction in the variance of diploid male production, which means that each queen has a lower risk of heading an unproductive colony (Page 1980, 1986; Page and Metcalf 1982; Crozier and Page 1985; Pamilo et al. 1994).

In the Hymenoptera it is believed that sex is often determined by the genotype at a specific sex determining locus (Cook 1993; Box 8.2). Heterozygotes at the locus become females, whereas homozygotes and hemizygotes (haploids) become males. The diploid, homozygote males are sterile and therefore represent a burden to any colony producing them (Box 8.2). The fourth hypothesis for multiple mating assumes that this sex determination system is widespread. Then, if all the queens in a population of a typical ant species are singly mated (and there is monogyny), most colonies will produce no diploid males, but some will have half of their diploid offspring becoming diploid males. However, if queens mate a large number of times, nearly all colonies are likely to produce a small fraction of diploid males (because it is likely that one – but no more than one – of a queen's many mates will bear one of her alleles at the sex determination locus). Thus, multiple mating reduces the variation among colonies in the frequency of diploid males.

This effect could lead to selection for multiple mating if there were a nonlinear relationship between colony fitness and the fraction of diploid males produced (Page 1980; Page and Metcalf 1982; Crozier and Page 1985). For example, say colonies are unaffected by the presence of diploid males if these make up fewer than 25% of the diploid offspring, but die of starvation if they are any more plentiful. Then, in the above example, all the colonies with diploid males would die under single mating, but all could survive under multiple mating. Therefore, in such a population, queens could be selected always to mate multiply. In other words, a queen might accept the cost of producing a fairly predictable but low number of diploid males, rather than risk total failure if she happened to mate with a male bearing one of her sex determination alleles.

In the monogynous form of *Solenopsis invicta* in North America, colony mortality due to diploid male production is high, and single mating is the rule (Ross and Fletcher 1986; Section 8.4). This arguably goes against the present hypothesis for multiple mating (although, instead, *S. invicta* may have evolved polygynous colonies to avoid the problem caused by diploid males: Section 8.4). However, with other assumptions concerning the way colony fitness changes with the frequency of diploid males, or concerning the timing of sexual production and the detection (if any) of diploid males, it is possible to predict that single mating should be the result (e.g. Crozier and Page 1985; Ratnieks 1990b; Pamilo et al. 1994). This makes the diploid male hypothesis for multiple mating difficult to test. For example, Pamilo et al. (1994) were unable to confirm or refute the hypothesis with data on the frequency of diploid males in *Formica* ants.

HYPOTHESIS 5

Queens evolve multiple mating to reduce kin conflict with the workers over sex allocation and male parentage (Starr 1984; Moritz 1985; Woyciechowski and Lomnicki 1987).

This last hypothesis was proposed because, if there is multiple mating, the workers favor a less female-biased population sex ratio and are less likely to produce their own sons (Sections 4.5, 7.4). Therefore conflict with queens, who favor a 1 : 1 sex ratio and worker sterility, would be reduced (Starr 1984; Moritz 1985; Woyciechowski and Lomnicki 1987). However, Ratnieks (1988) and Ratnieks and Reeve (1992) pointed out that multiple mating, to evolve, must provide immediate benefits to queens. But if queens started to mate multiply, any worker response could only appear gradually as genes for the appropriate worker behavior were slowly selected and spread through the population. Therefore, a rare, mutant polyandrous queen in a population of singly mating queens would experience no benefits, because her workers would not alter either their sex allocation strategy or lessen any inclination to lay male eggs. This suggested that multiple mating would not evolve to reduce kin conflict (Ratnieks 1988; Ratnieks and Reeve 1992).

Suppose, however, that workers were already rearing split sex ratios due to partial multiple mating in a population (Pamilo 1991c; Ratnieks and Reeve 1992; Queller 1993b; Ratnieks and Boomsma 1995). Thus, say a population exists in which some queens are singly mated (and so have workers who favor female-biased colony sex ratios) and others are multiply mated (and so have workers who favor rearing males) (Boomsma and Grafen 1990; Section 4.5). Then queens that switched to multiple mating would obtain colony sex ratios that were more male-biased and hence more favorable to them (Pamilo 1991c; Queller

1993b; Ratnieks and Boomsma 1995). (Although queens are equally related with sons and daughters, queens heading colonies producing male-biased broods would have greater inclusive fitness than those in colonies with female-biased ones, because, under a worker-controlled female-biased population sex ratio, sons would have a higher mating success than daughters.) The workers' response to multiple mating would be immediate because they would already be under selection for facultatively biasing the colony sex ratio according to their queen's mating frequency. It has been suggested that, in this way, queens may evolve multiple mating to enhance their inclusive fitness directly (Pamilo 1991c; Ratnieks and Reeve 1992; Queller 1993b; Ratnieks and Boomsma 1995). One consequence would be to reduce conflict over population sex allocation and male parentage, but this would be incidental.

This idea would still leave unexplained why partial multiple mating had evolved in the first place. In addition, if double mating spread through a population through the above process, the workers' stable population sex ratio would grow relatively more male-biased, and so the incentive for queens to mate twice would decrease (Ratnieks and Boomsma 1995). Therefore, the hypothesis is best suited to explaining low queen mating frequencies; but this is not a problem, given that in polyandrous ants they usually are low (Keller and Reeve 1994b). Other brakes on the further spread of multiple mating could be its costs to queens, for example in terms of extra predation (Queller 1993b).

In conclusion, why multiple mating occurs in ants is not yet firmly established. All the hypotheses have some empirical support. Moreover, in particular cases, multiple mating may have evolved through a combination of the reasons proposed, since the hypotheses are not mutually exclusive. For example, some queens may first have evolved multiple mating to decrease the colony's vulnerability to disease (Hypothesis 3), but the habit could then have spread further through direct effects on the queens' inclusive fitness (Hypothesis 5). Once again, better data are required, particularly on sperm loads and queen productivities, on the effect of multiple mating on colony efficiency and productivity, on genetic resistance to parasites, on the shapes of the functions relating colony fitness to the frequency of diploid males, on the levels of facultative sex allocation and worker policing, and lastly on the effective mating frequencies of queen ants in natural populations.

11.5 The Location of Mating: Causes and Consequences

This section considers the diversity of the mating systems found in ants. Although the categories considered are not always distinct, our classification is designed to help clarify the evolutionary forces affecting where and how ants mate.

LARGE-SCALE NUPTIAL FLIGHTS

Many ant species stage large-scale nuptial flights in which numerous colonies over a large area simultaneously release all their young winged queens and males (e.g. Hölldobler and Maschwitz 1965; Markin et al. 1971; Boomsma and Leusink 1981; Section 11.2). Examples include species in the genera *Solenopsis*, *Atta*, *Myrmecocystus*, *Lasius*, *Formica*, *Messor*, *Myrmica*, and *Pogonomyrmex*. Indeed, since female calling is known mainly from the ponerines and leptothoracines (Hölldobler and Bartz 1985), a mating system involving a nuptial flight may be the commonest type in ants, although it needs repeating that not all nuptial flights are large and extensive (e.g. Hasegawa and Yamaguchi 1994). One purpose of these flights might be to ensure outbreeding (Hölldobler and Bartz 1985), which is a high priority in Hymenoptera given that inbreeding leads to the production of sterile diploid males (Box 8.2). Inbreeding is in fact fairly rare in ants (e.g. Gadagkar 1991a; Table 8.1), and, as mentioned earlier, there are no ant species in which both sexes are exclusively wingless (Starr 1984). The typically large scale and infrequency of nuptial flights (roughly two or three per year) are further evidence for their role in inbreeding avoidance. The synchronized release of sexuals may also serve as a selfish herd phenomenon (Hamilton 1971b). Thus, the vast numbers of sexuals present in a flight may satiate predators, giving a colony a good chance that at least some of its sexuals will survive (Starr 1984). Nuptial flights are also associated with dispersal of the ant sexuals, since they often occur far from nests (e.g. Markin et al. 1971; O'Neill 1994).

Nuptial flights frequently involve gatherings of males, either in the air or on the ground, which are then visited by the females. In some cases, the males are known to attract the females with pheromones (e.g. Hölldobler 1976), although once the sexuals are swarming together it appears that sex pheromones produced by females serve to stimulate mating (Hölldobler 1976; Walter et al. 1993). These systems were therefore referred to by Hölldobler and Bartz (1985) as displaying the "male aggregation syndrome." They also have some resemblances to vertebrate leks, especially in the ground-swarming species of *Pogonomyrmex* (Hölldobler 1976; Davidson 1982; Rust 1988).

FEMALE CALLING SYNDROME

The "female calling syndrome" is the term applied to the system in which colonies release small numbers of virgin females, who emit sexual pheromones that attract or "call" males (Hölldobler and Bartz 1985). The females typically disperse a short way from the nest, then call from the ground or from low vegetation by raising their abdomen and releasing pheromones onto the breeze. However, inbreeding is probably still avoided in this system by males dispersing from their colonies before mating (e.g. Bourke et al. 1988).

The female calling syndrome is typical of species with relatively small, scattered colonies. Good examples are the ponerines and leptothoracines (e.g. Hölldobler and Haskins 1977; Buschinger and Alloway 1979; Peeters 1991a; Bourke and Heinze 1994), although female calling (or very similar behavior) also occurs sporadically outside these groups (e.g. *Chelaner*: Briese 1983; *Carebara*: Robertson and Villet 1989). In the ponerines and leptothoracines, female calling probably has several advantages over staging nuptial flights. First, populations often do not develop large numbers of sexuals that could form an extensive nuptial flight or lek. Thus, by calling, females probably maximize their chances of finding a mate when both sexes are at low density. Second, small scattered colonies may be associated with a patchy habitat. If so, then the low dispersal that goes with female calling means that females do not risk leaving the patch of suitable habitat (Bourke and Heinze 1994; Section 8.6). This helps explain why female calling is so prevalent among the socially parasitic leptothoracine ants (e.g. Buschinger 1968b, 1971, 1972), where each host colony or group of colonies constitutes the favorable patch (Heinze and Buschinger 1989). If the costs of dispersal are high enough to favor polygyny or functional monogyny, female calling also facilitates the adoption of queens into their natal colonies after mating (Bourke and Heinze 1994; Section 8.6; and see below).

THE MIXED STRATEGY OF HIGH AND LOW DISPERSAL

Hamilton and May (1977) used game theory to show that having entirely nondispersing progeny is not an evolutionarily stable strategy (Sections 8.6, 9.4). However, if dispersal is costly, having a fraction of progeny that stays at home and shares the local resources of the habitat patch, or inherits the infrastructure of the maternal nest, is advantageous (Bourke and Heinze 1994; Section 8.6). Thus, one might expect many ants to show a mixed strategy of high and low dispersal. This would imply a mating system in which not all males and females participated in a large and distant nuptial flight, because it is highly unlikely that after such a flight an ant queen could find its way back to its natal patch or colony.

The facultatively polygynous ant *Leptothorax acervorum* provides a nice example of a mixed dispersal strategy (Douwes et al. 1987; Franks et al. 1991a; Bourke and Heinze 1994). (Others are given in Table 8.1.) In the *Leptothorax* subgenus to which it belongs, female calling is almost certainly the primitive mating system (Buschinger and Alloway 1979). However, in *L. acervorum*, observers have seen large nuptial flights in which, interestingly, females do not mate on the wing in the swarm, but alight on the ground and show classic calling behavior. The males then also descend from the swarm to mate (Buschinger 1971; Franks et al. 1991a). This suggests that nuptial flight behavior evolved from the ancestral mating system involving female calling alone. It also raises the possibility that other queens fail to participate in the swarm, and instead call for males from near their home nest before being readopted by it. Although this has not been observed, it almost certainly happens because, in *L. acervorum* colonies, coexisting queens are related on average (Douwes et al. 1987; Stille et al. 1991; Heinze et al. 1995). (Again, it is unlikely that queens can return home from a faraway swarm.) Thus, the suggestion is that, because selection favored facultative polygyny in *L. acervorum*, there has been a *de novo* evolution of a mixed mating strategy (Franks et al. 1991a; Bourke and Heinze 1994; Section 8.6).

Another trait that is intimately linked with a mixed mating and dispersal strategy in polygynous ants is the production of polymorphic sexuals (Table 8.1). For example, Fortelius et al. (1987) showed that, in the wood ant *Formica exsecta*, there are two distinct male size classes (see also Agosti and Hauschteck-Jungen 1987). These occurred within single nests, but each morph was also characteristic of a particular population type. *F. exsecta* tends to occur in monogynous, monodomous populations with even or female-biased sex ratios, and polygynous, polydomous populations with male-biased sex ratios (Pamilo and Rosengren 1983, 1984; Table 5.6.). Fortelius et al. (1987) found that small males (micraners) were produced more commonly in dense populations of polydomous colonies, whereas large males (macraners) were produced by the monodomous populations. Micraners also seemed to be the better dispersers. Fortelius et al. (1987) therefore suggested that colonies start to produce micraners as their population makes the transition from the monogynous to the polygynous type. This is because the mating success of sedentary males declines as the sex ratio grows more male-biased. A related interpretation is that, if polygyny is favored in a population, the females should display a mixed mating strategy (as in the case of *Leptothorax acervorum*). This creates two sites of mating for males – near and far from the mounds – so favoring the emergence of a small, dispersing morph because on the mound males of below-average size would be outcompeted for matings by the large ones (Bourke and Franks 1991). Although the exact dispersal pattern of *F. exsecta* queens appears unknown (Fortelius et al. 1987), dispersal polymorphisms have long been suspected in *Formica* queens in general, with some queens being reluctant to mate before flying, whereas others are prone to shed their wings before mating (Rosengren et al. 1993; Sundström 1995b).

MATING IN THE NEST

1. THE PONERINES

Among the ponerines there is probably a greater diversity of mating systems than in any other subfamily of ants (Peeters 1991a). This may be associated with the ponerines being phylogenetically the most primitive ant subfamily. However, ponerines also have a notably diverse set of social systems (Peeters 1993; Section 8.4), and this has evidently affected how their mating biology evolves (Peeters 1991a). One of the most primitive species is *Amblyopone pallipes*. This exhibits the classic female calling syndrome (Haskins 1978), which appears to be ancestral in the subfamily (Peeters 1991a). Only very rarely do ponerines produce large mating groups. An example is *Ponera pennsylvanica* (Haskins 1970). Certain species have an extremely unusual social and mating system. For example, some have wingless queens with a worker-like external morphology (ergatoids), which males locate by following foraging trails, whereas others have workers capable of being mated (gamergates), which males find by entering nests (Peeters 1991a; Section 8.4; Figure 11.2).

The possession of ergatoid queens and gamergates seems to be a derived feature of ponerine colonies, with the ancestral trait being reproduction by orthodox, dealate queens (Peeters 1991a, 1993). Although the factors favoring the evolution of colonies with ergatoid queens and gamergates are unclear (Section 8.4), it is clear that the switch to these systems has been facilitated by the fact that female calling was the ancestral mating system in ponerines. Female calling involves males dispersing to find virgin females near the females' nests of origin. Thus, first, it is a small behavioral step for males to change from flying to foreign nests to mate with females on the ground outside, to flying to nests to mate with females inside them. Second, if males take this step, sexually reproductive females can evolve winglessness without suffering any costs of inbreeding (Peeters 1991a). A number of other features that enable, or are associated with, the evolution of gamergates are listed in Table 11.2. Their co-occurrence in the poner-ines provides another reason why gamergates are only known in this group.

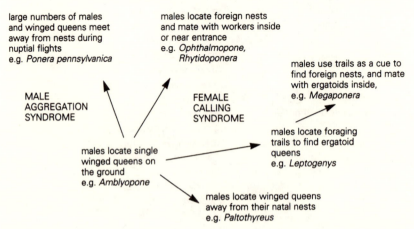

Figure 11.2   The variety of male mating behavior patterns among ponerine ants, start-ing with the ancestral pattern (simple female calling) exhibited in the most primitive genus *Amblyopone*. (From Peeters 1991a; by permission of Academic Press Ltd)

2. THE ARMY ANTS

Army ant queens are exceptionally large, permanently wingless erga-toids, and mating occurs in the nest at the time of colony fission (Section 10.8). (As mentioned in Section 11.3, remating by old queens has also been suggested.) The production of a small number of new queens just before colony fission (Sections 4.8, 5.7) means that daughter queens are in competition among themselves, or with their mother, to head one or other of the new colonies. It is very likely that the workers determine the outcome of this competition, and they may even abandon their own mother during fission. The proximate mechanism that the workers use in their selection of queens may be based on the relative strength of their pheromonal signals. Thus, *Eciton* army ant queens have batteries of chemical secreting glands along the dorsal surface of

Table 11.2

Occurrence Outside the Ponerine Ants of Biological Traits that Enable Workers to
Reproduce Sexually

Characteristics	Occurrence outside the Ponerinae
1. Do workers have a sperm receptacle?	Generally absent, except in *Myrmecia*, and a proportion of workers in *Formicoxenus*.
2. Limited difference in fecundity between queens and workers?	True in some *Myrmecia*, Pseudomyrmecinae, various leptothoracines (often social parasites). But queens are very fecund in other Myrmicinae, and many Formicinae and Dorylinae.
3. Do males seek other nests?	No mass nuptial flight in several *Myrmecia* species. In army ants, males find nomadic colonies with wingless queens. Mating occurs outside or inside nest in socially parasitic Myrmicinae.
4. Can workers attract males?	Unknown at present since it has been assumed that workers never mate. [This must be the case in species where workers lack a sperm receptacle.] In *Formicoxenus provancheri* some workers show female calling.
5. Do mechanisms of colony fragmentation exist?	Throughout the ants [excluding obligately monogynous, monodomous species with reproduction by sexual emission]. Exclusive mode of colony reproduction in army ants.

NOTE: From Peeters (1991a); by permission of Academic Press Ltd.

their abdomen, which the workers are constantly licking (Franks and Hölldobler 1987; Section 7.6).

Male army ants, unlike queens, start life with wings and it is likely that they always fly to other colonies for mating (so promoting outbreeding). However, in several respects, they are very unusual for ant males. To begin with, they are extremely large and robust, being of similar build to the oversize queens, whereas the typical ant male is small and flimsy (Section 11.2). In fact, the first army ant males to be described by Linnaeus were classified as wasps, and in Africa their large size and elongate abdomens have earned them the name "sausage flies" (Franks and Hölldobler 1987). In the African driver ant *Dorylus nigricans*, both the queen and the males are the largest known ants (Gotwald 1982).

The other remarkable things about male army ants are that, uniquely among ant males, they have deciduous wings, which they shed after flying to a foreign colony; and they are richly endowed with glands in the abdomen that open at sites corresponding to the openings of the glands of the queens. Franks and Hölldobler (1987) proposed that the reason for all these traits is that male army ants have been selected to resemble the queens. Males entering an alien colony must run the gauntlet of the workers to gain access to the queen, who is always surrounded by a worker entourage. This allows the workers the opportunity to be selective about the mate of the queen who may soon head their own colony. The workers can afford to be selective because of the very male-biased sex ratio in army ants (Franks and Hölldobler 1987). (This also means that a male has little chance of remating, so shedding his wings is unlikely to be costly. Moreover, it means competition between males is potentially high, so that the ten-hour duration of copulation recorded for *Eciton* [Schneirla 1949; Rettenmeyer 1963] could represent a form of mate guarding.) The workers may simply apply the same criteria for selecting males as they do for picking queens. This would explain why, when a male enters a colony, the workers lick his abdomen as they do that of a queen, presumably sampling the corresponding glandular products. In short, a form of sexual selection for mate choice acting via the workers apparently forces the males to use similar channels of communication as the queens use to gain the approval of the workers (Franks and Hölldobler 1987). Since the males bring no resources to the colony, and die shortly after mating, this implies that their pheromones act as honest signals to the workers of their "good genes."

A role for workers in selecting both their queen and her mate is practically unknown in other social insects. For example, honey bee workers may pick their new queen (e.g. Starr 1984; Section 7.6), but because mating occurs outside the nest they cannot select her mates. In the unicolonial species *Linepithema humile* (= *Iridomyrmex humilis*), mating occurs in the nest between resident females and incoming males (Keller and Passera 1992, 1993; Passera and Keller 1994). In experiments, workers killed nearly all strange males outside the period when female sexuals were available, but during this period few or no males were killed (Passera and Keller 1994), suggesting that workers do not select which males gain access to the queens.

3. *CARDIOCONDYLA*: SPECIES WITH WINGLESS, FIGHTING MALES

A number of species in this genus of tiny-bodied ants have a particularly unusual and interesting mating system. Their males, instead of being the typical, winged, inoffensive type (Section 11.2), are wingless, fighting, worker-like individuals (ergatoid males). The best-known examples occur in *Cardiocondyla wroughtoni*, in which only some males are ergatoids (Kinomura and Yamauchi 1987; Stuart et al. 1987), and *C. nuda*, in which all are (Heinze et al. 1993a; Heinze and Hölldobler 1993b). The ergatoid males are equipped with either very robust (*C. nuda*) or especially long pincer-like mandibles (*C. wroughtoni*), with which they engage one another in lethal combat in the nest. In both species, the ergatoids pick on other males in the vulnerable soft-bodied stage just after emergence from the pupa. In *C. wroughtoni*, they also smear rival males with a secretion that causes these rivals to be attacked by the workers (Yamauchi and Kawase 1992). The result is that, of a group of ergatoid males, only one typically survives, and he then mates multiply in the nest with any available females. In *C. wroughtoni*, in which winged males exist alongside the ergatoids in the same nest, the winged males are typically not attacked, and nor do they fight among themselves. Instead, they disperse from the nest to obtain matings (Kinomura and Yamauchi 1987; Stuart et al. 1987).

Recently, Heinze and Hölldobler (1993b) showed with both behavioral and histological studies that ergatoid *Cardiocondyla* males are able to continue manufacturing sperm after becoming sexually mature. This is in strong contrast to winged conspecific males, to the males of all other ants so far investigated (Section 11.2), and indeed, as far as is known, to all males throughout the Hymenoptera (Heinze and Hölldobler 1993b). Continuous sperm production is obviously highly adaptive for a male morph that will encounter multiple mating opportunities due to its elimination of rivals (and due to a female-biased numerical sex ratio: Kinomura and Yamauchi 1987). Connected with this, *Cardiocondyla* males are also notably long lived (Heinze and Hölldobler 1993b; Section 11.2).

Exactly why wingless, fighting males evolve in some species and not in others is unclear. *Cardiocondyla wroughtoni* and *C. nuda* are both polygynous, and their "tramp ant" habits suggest that they tend towards a unicolonial organization (Kinomura and Yamauchi 1987; Stuart et al. 1987; Heinze et al. 1993a). "Tramp" behavior is also found in another ant with fighting ergatoid males, *Hypoponera punctatissima* (Hamilton 1979). Unicoloniality often involves mating in the nest, and appears to be associated with the exploitation of ephemeral, disturbed habitats (Section 8.5). In these circumstances, dispersal polymorphisms among males may be favored, just as they are among females under polygyny (see above). Indeed, female dispersal polymorphisms almost automatically entail dispersal polymorphisms in males, since the nondispersing females provide an alternative mating opportunity for males. Male ergatoidy could therefore be an extreme outcome of selection for male dispersal polymorphisms in species that already have in-nest mating. The *Formicoxenus* "guest ants," which live in the nests of other species, also have ergatoid males, although little appears to be known about their behavior (Francoeur et al. 1985). Again, these could have evolved through selection for dispersal polymorphism linked with the extreme patchiness of their habitat (Section 8.6). Similarly, in solitary insects, wingless fighting males appear linked with restricted habitats with high concentrations of females (Hamilton 1979; Stubblefield and Seger 1994).

However, it is still not clear why *Cardiocondyla nuda* has exclusively ergatoid males, whereas *C. wroughtoni* retains both morphs (Heinze and Hölldobler 1993b). In addition, in some species males are wingless, but apparently do not fight (e.g. *Hypoponera eduardi*: Le Masne 1956; *Technomyrmex albipes*: Yamauchi et al. 1991). In *T. albipes*, this could be because nestmate males all descend (via inbred parents) from a single foundress, and so are highly related with one another (Yamauchi et al. 1991; cf. Hamilton 1979). By contrast, when ergatoid males occur in polygynous colonies, they may be less related and so more aggressive to each other (Yamauchi et al. 1991). However, in the nonsocial fig wasps, there is no clear link between the level of fighting among wingless males and their relatedness (M.G. Murray 1989). In fact, according to a general model, the most important determinant of whether fatal fighting evolves between conspecifics is not relatedness, but the value of the contested resource compared to the value of remaining alive (Enquist and Leimar 1990). This suggests that, in ants, wingless males evolve to be fatal fighters more readily when the period of female emergence is brief, because the value of surviving beyond the period of mate availability is obviously small. But, in general, not enough is currently known about the ecology, relatedness structure, and colony life history of ant species with male ergatoidy to unravel its evolutionary basis fully.

4. RAPID DEVELOPMENT, NEOTENY, AND MATING IN THE NEST

In ants with mating in the nest, there is an obvious advantage to sexual larvae of developing rapidly to adulthood. Slow developers may not only obtain fewer mating opportunities, but they may also be killed before having any chance to mate at all. Of species discussed above, this applies to *Eciton* army ant queens, in which the workers have to select which adult queen will head a new colony (Franks and Hölldobler 1987), and to *Cardiocondyla* ergatoid males, who are liable to be killed by rival males as they emerge from their pupae (Heinze et al. 1993a). In addition, in the ponerine *Diacamma australe*, late-emerging females are "mutilated" and forced to become nonmating, nonreproductive workers by the reigning mated worker or gamergate (Peeters and Higashi 1989; Section 7.4).

A consequence of this sexually-selected rapid development could be the retention of some juvenile features by reproductively competent adults (neoteny). This may have important consequences. For example, typical ant males manufacture sperm as pupae, but cease as adults (Hölldobler and Bartz 1985). But, if *Cardiocondyla* males were selected to undergo rapid development, they might have retained the "juvenile" trait of sperm manufacture into their adult lives. This would have benefited them by allowing them more mating opportunities, and, as discussed above, it does occur (Heinze and Hölldobler 1993b). Thus the principle of neoteny may have facilitated, at the proximate level, the evolution of the unusual mating biology of ergatoid *Cardiocondyla* males. Since ant workers typically develop more quickly than queens, rapid development of female sexuals could also lead to some traits being shared with workers. This could provide a proximate explanation for the characteristically worker-like morphology of *Eciton* queens (involving a relatively small, wingless thorax, and small eyes).

A particularly striking example of neoteny occurs in *Anergates atratulus*, a workerless social parasite of *Tetramorium caespitum* (e.g. Dumpert 1981 p. 174). The males are wholly wingless, and so all mating occurs in the nest. The females, by contrast, are winged, and disperse after mating. The males are probably the offspring of one or a few *Anergates* foundresses, and the likelihood of female-biased sex ratios within nests (Donisthorpe 1927 pp. 98–101) suggests that male numbers have been kept low by the effects of local mate competition on sex allocation (Section 5.8). A high relative abundance of females, resulting in a low value of fatal fighting over matings (see above), seems to have lead to a form of male–male competition that does not involve male ergatoidy. Instead, the males resemble walking pupae rather than adults (Dumpert 1981 p. 174). This trait has presumably evolved through selection on males to achieve sexual maturity early in order to gain access to virgin females before their nestmates. The result is that *Anergates* males now show the neotenous combination of a pupal body form and sexual competence. In the light of the occurrence of continuous sperm manufacture in ergatoid *Cardiocondyla* males, it seems reasonable to expect that *Anergates* pupoid males also have fully functional testes. The same is true of the ergatoid males in other genera discussed earlier. In the solitary parasitoid Hymenoptera with local mate competition, one might also expect males to mate multiply, and so perhaps to have continuous sperm production as well. Evidence for multiple mating by males in these species does exist, although sperm depletion has been reported too (Godfray 1994 pp. 269, 281).

11.6 Conclusion

The classical concepts of sexual selection cannot be transferred in a straightforward way to the case of the ants. For example, the idea that the relative cost of producing gametes determines the roles of the sexes has little relevance to insects in which females lay eggs fertilized with stored sperm. Instead, we have argued that a recent extension of sexual selection theory, the idea that sex roles are largely determined by the operational sex ratio (Clutton-Brock and Parker 1992), can be very easily applied to ants. This is because, in a typical species, the operational sex ratio simply equals the numerical sex ratio in the mating swarm. From this, and from the fact that they often mate in large-scale nuptial flights with a male-biased numerical sex ratio, flow many of the characteristic features of male and female ants (Section 11.2).

Another conclusion is that the lack of continuous sperm production by adult ant males (Hölldobler and Bartz 1985) should not be viewed as an innate constraint of Hymenopteran design. Instead, most species may have retained this ancestral Hymenopteran trait because it suits a selective regime encouraging single male mating (Section 11.2). The example of *Cardiocondyla nuda* shows that continuous sperm production can evolve under suitable circumstances (Heinze and Hölldobler 1993b; Section 11.5).

Finally, where ants show atypical mating systems, such as in the army ants and species with ergatoid males, this is often more or less easy to link with an atypical social system (in these examples, respectively, reproduction by colony fission, and probably unicoloniality) (Section

11.5). However, exactly why queens vary in their mating frequencies remains another conundrum for future research (Section 11.4).

11.7 Summary

1. Typically, ants mate during brief, synchronized large-scale nuptial flights in which sexuals from many colonies participate, so promoting outbreeding, predator satiation, and wide dispersal. After swarming, male ants die, whereas queens attempt colony foundation. Queens never mate again (with the possible exception of army ants), and instead produce fertilized eggs from stored sperm for the rest of their lives.

2. In many ant species, males cannot replace their sperm supply as adults, because the testes degenerate early leaving the sperm to be stored in the vas deferens.

3. Female choice of male quality appears absent in ants, except, it is suggested, with respect to inbreeding avoidance in species with mating in the nest. In species with nuptial flights, female choice is probably absent because the chance of mating with a relative is small, and exercise of any type of choice would be too costly in terms of predation risk.

4. In ants, the operational sex ratio (ratio of receptive females to sexually active males) equals the numerical sex ratio of the mating swarm. The numerical sex ratio seems to be nearly always male-biased. This is ultimately a consequence of the stable sex investment ratio, and the level of sexual size dimorphism. Therefore, applying in a simple way the recent idea that the operational sex ratio determines sex role, it is argued that males should usually be the competitive sex in ants.

5. In particular, the excess of males means that the average male has only a small chance of finding even a single mate. This may well account for many typical traits of male ants. These include a body designed for finding queens in flight (well-developed eyes and wing muscles, along with a flimsy build), lack of adaptations for fighting, single mating, the inability to replenish the sperm supply, elaborate clasping genitalia that may be suicidally detached and left in the female as mating plugs, and a brief adult lifespan. However, in species that swarm on the ground rather than in the air, a male-biased numerical sex ratio may promote aggression among males and hence a robust build. Consistent with these arguments, species known to have female-biased numerical sex ratios (or mating in the nest, where males are less likely to be predated between mating attempts) tend to have multiply mating males.

6. Sperm competition is expected in ants only between the ejaculates of different males in the sperm receptacle of multiply mated queens.

This is because any one male's sperm cells are genetically identical, since males are haploid. There is little evidence that sperm competition occurs, although a queen's mates may be unequally represented among her offspring.

7. Queens mate multiply in over half of ant species surveyed, although mating frequencies are usually low. The exact reasons are unknown. Proposed advantages of multiple mating include: a genetically more diverse, and hence more efficient workforce; a workforce that is genetically better protected against disease; a reduction in the variance of diploid male production; and an inclusive fitness gain for multiply mating queens in species where workers bias the colony sex ratio towards males when under queens with above-average mating frequencies.

8. Variants of the typical ant mating system include: female calling syndrome, in which females pheromonally attract males on the ground; mixed mating and dispersal systems with mating both near and away from the nest; systems in which males enter strange colonies to find mates, as in ponerines and army ants; and systems in which males are wingless and fight to the death for the possession of multiple females in the nest (e.g. *Cardiocondyla*). All these may be explicable, to a greater or lesser extent, as products of the prevailing social system (for example, polygyny, reproduction by colony fission, unicoloniality). In *Cardiocondyla* species, males have evolved the unusual ability to produce sperm continuously.

12 The Division of Labor

12.1 Introduction

This chapter evaluates recent theoretical and empirical treatments of the organization of work and division of labor in ants and other social insects, and presents an alternative way of interpreting these phenomena. A division of labor among the workers of insect societies is a conspicuous feature of their biology. It is also fundamental to the incredible ecological success of ants (E.O. Wilson 1987, 1990). Arguably, social insects do not generally exploit resources that are not also consumed by solitary insects. Instead, their ecological success almost certainly stems from the efficiency that derives from a division of labor (Oster and Wilson 1978 p. 9). In addition, division of labor is an important topic because, as the following section describes, many of the principles involved apply to complex biological systems in general.

Previous reviews of division of labor and caste evolution in social insects include those of Oster and Wilson (1978), E.O. Wilson (1985a,c), Calabi (1988), Jaisson et al. (1988), Gordon (1989a), Hölldobler and Wilson (1990), Schmid-Hempel (1991), and Robinson (1992). In ants, it has often been assumed that the division of labor is fundamentally determined by centralized hierarchical control of the workforce's demographic structure (e.g. E.O. Wilson 1985a,c), even though some of the day to day running of the colony may be decentralized through being within the control of the active workforce at the worksite (e.g. Wilson and Hölldobler 1988). However, in this chapter we argue that decentralized control may have far wider significance for the longer-term organization and evolution of the division of labor in ant societies, because it promotes flexibility and robustness. In particular, we conclude that the division of labor in ant colonies is based on extensive behavioral flexibility on the part of individual workers, and on surprisingly simple and robust self-organized systems. Self-organization has already been shown to be important in diverse aspects of social insect behavior, from the development of foraging systems to nest building (Deneubourg and Goss 1989; Deneubourg et al. 1989; Seeley 1989; Camazine 1991; Franks et al. 1991b, 1992).

12.2 The General Significance of a Division of Labor

The division of labor, with its associated enhancement of efficiency, is fundamental to the biological evolution of complexity and diversity. Division of labor is not a term invented by students of social insects. The modern usage of the term comes from studies of human societies and economics:

> This great increase of the quantity of work which, in consequence of the division of labor, the same number of people are capable of performing, is owing to three different circumstances; first, to the increase of dexterity in every particular workman; secondly, to the saving of the time which is commonly lost in passing from one species of work to another; and lastly, to the invention of a great number of machines which facilitate and abridge labour, and enable one man to do the work of many. (Adam Smith 1776 p. 112)

This chapter is about the evolution and development of the division of labor in social insects and in ants in particular. In it, we consider each of Adam Smith's explanations for the benefit of a division of labor, namely skill acquisition, spatial efficiency, and mechanical specialization. The existence of a division of labor in insect societies has long been recognized. Aristotle was clearly aware of it in honey bees:

> . . . they differentiate their work; some make wax, some make honey, some make bee-bread, some shape and mould combs, some bring water to the cells and mingle it with the honey, some engage in out-of-door work. (Aristotle in Smith and Ross 1910 p. 627)

Aristotle also knew that the tasks of honey bees were associated with their age (though he wrongly supposed that the bees got hairier with age):

> The elder bees do the indoor work, and are rough and hairy from staying indoors; the young bees do the outer carrying, and are comparatively smooth. (Aristotle in Smith and Ross 1910 p. 626)

These passages from Adam Smith and Aristotle illustrate the general importance of a division of labor in many kinds of organization.

Biologists now recognize that selection pressures for greater efficiency through divided labor have driven both the evolution of the eukaryotic cell, probably through symbiotic association (Margulis 1981), and the evolution of metazoan life in general (Bonner 1988). The benefits of a division of labor also underlie morphogenesis (the development of form). Thus, the study of the division of labor in social insects can illuminate many forms of biological integration, not by mere analogy but because the fundamental selection pressures are likely to be the same (cf. Leigh 1991; Sections 2.5, 2.6).

One of the reasons that students of social insects have almost come to regard themselves as the inventors of the term "division of labor" is that a reproductive division of labor is explicitly included in the standard definition of eusociality (E.O. Wilson 1971 p. 4; Section 1.3). This can lead to the mistaken impression that explaining, in terms of genetic factors alone, why some individuals forgo reproduction to help raise relatives provides all the needed insight into eusocial evolution. But this is not the case (e.g. Seger 1991). Division of labor must deliver significant quantitative benefits for altruism and eusociality to evolve by kin selection, by making the benefit to cost ratio high enough for Hamilton's rule to be satisfied (Section 1.4). Thus, the division of labor is not just about partitioning reproduction, but also about maximizing the production of young.

As recognized in E.O. Wilson's splendid metaphor of the ant colony as a factory within a fortress (E.O. Wilson 1971 p. 342), biological systems never display a division of labor more obviously than in insect societies. In ant colonies, the queens and workers can be both easily recognized and observed. Thus, the division of labor (in both the reproductive and the maintenance activities of the society) is achieved by identifiable units. Furthermore, because the units are largely autonomous, they form unrivalled experimental material. The colony can have parts of its workforce removed and replaced in so-called "sociotomy" (Lenoir 1979a, b; Lachaud and Fresneau 1987) or "pseudomutant" (E.O. Wilson 1980a, b) experiments. In sociotomy, the colony is partitioned to see how different parts function. In the pseudomutant technique, the colony is taken apart and then reconstructed in new ways. So alternative social designs (which might plausibly have evolved), and the developmental biology of whole societies, can be investigated experimentally. In other words, the "sociogenesis" of insect societies is experimentally far easier to study than the equivalent process of cell and tissue differentiation in whole organisms, morphogenesis (E.O. Wilson 1985a).

For these reasons, it is unfortunate that there has been so little dialogue between students of social insects and others studying developmental biology and physiology. A possible cause is the wide acceptance of an apparently straightforward explanation of the division of labor among social insect workers. This is the idea that the labor is divided according to the age of individuals (consider the earlier quotation from Aristotle). This immediately seems to set the division of labor in social insects apart from that within whole organisms. However, we will suggest that there is in fact no direct causal link between age and task in social insect workers, only a correlation (e.g. Tofts and Franks 1992; Franks and Tofts 1994; Robinson et al. 1994). Therefore, sociogenesis and morphogenesis may be even more tightly linked than previously thought.

The idea of a system of division of labor in which changes in worker age drive changes in worker tasks may have stifled the possibility of even deeper and more general lessons being learnt from social insect studies. For example, because of the inflexibility inherent in such a system, this idea may have led to an underappreciation of the need for plasticity and fault tolerance in biological organization. In insect societies, this need is very evident, since colonies may often be exposed to unpredictable catastrophes such as the destruction of most of the foragers in a storm. A colony with a deterministic relationship between the age and role of its workers could not quickly recover from this type of event.

This chapter argues that surprising simplicity may underlie much of the complexity and plasticity in the organization of work in social insects. Recent work in many branches of science is beginning to show how interactions between large numbers of relatively simple entities can lead to collective patterns of behavior of considerable sophistication, flexibility, and robustness (Nicolis and Prigogine 1977). In systems exhibiting such *self-organization*, control of the overall process is not hierarchical, but decentralized. The role of decentralized self-organization in the division of labor in social insects is the main theme of this chapter (Deneubourg and Goss 1989; Franks 1989a; Seeley 1989; Camazine et al. 1990; Camazine 1991).

The approach we will advocate partly contradicts widely held theories about the organization of social insect colonies. Thus, we suggest that the division of labor in ant colonies is not based on a causal link between age and task, but instead, at the proximate level, on individuals responding in the short term to local events and workloads in the light of their experience (Section 12.4). At the ultimate level, work may also be organized in the context of selection pressures for personal reproduction (e.g. West-Eberhard 1981; Sections 7.7, 12.3).

To begin with, the following section discusses the strengths and weaknesses of earlier work on the division of labor. Section 12.4 then introduces an alternative way of understanding the division of labor, the foraging for work algorithm. Section 12.5 reviews empirical studies in the light of our treatment, Section 12.6 extends it to the spatial structure within nests, and Section 12.7 proposes that there are lessons to be learnt from the division of labor in social insects for general organizational theory (Morgan 1986) and vice versa. In Section 12.7, then, we return to consider the wider significance of a division of labor, as foreshadowed in the quotation from Adam Smith with which this section began.

12.3 Temporal Polyethism and a Confusion of Cause and Effect

It is well known that workers in social insect colonies tend to change the tasks they perform with time. This is termed *temporal polyethism*. If workers change their tasks specifically as a function of their age, the phenomenon is termed *age polyethism* (Hölldobler and Wilson 1990 p. 635). Temporal polyethism involving a correlation between age and task was first demonstrated in honey bees (Free 1965; E.O. Wilson 1971 p. 174; Seeley 1982; Robinson 1992; Van der Blom 1993a,b), and has also been shown to occur in a large number of ants (Otto 1958; E.O. Wilson 1976a; Lenoir 1979a,b; McDonald and Topoff 1985; Tsuji 1990b; Dejean and Lachaud 1991; Hölldobler and Wilson 1990 pp. 312–317).

Typically, young newly-emerged workers ("callows") remain in the nest in the close proximity of the queen and the brood, and act as nurses. Slightly older workers also remain in the nest, but may live closer to the entrance and act in nest maintenance (clearing debris, repairing, and building). The oldest workers act as foragers outside the nest. These individuals run the gauntlet of the external environment, with its aggressive competitors and predators. For example, in the field the life expectancy of *Cataglyphis bicolor* workers foraging for the first time is about six days, whereas in the laboratory foragers can live several months (Wehner et al. 1983; Schmid-Hempel and Schmid-Hempel 1984). In *Oecophylla smaragdina*, the oldest workers patrol the colony's territorial boundaries and engage in deadly fights with hostile neighbors (Hölldobler 1983). It is the frequently observed correlation between age and task in social insects that has led to the view that aging provides the mechanism underlying the division of labor.

ADAPTIVE DEMOGRAPHY

E.O. Wilson's (1975 p. 14, 1976b, 1985a) adaptive demography hypothesis demonstrates how influential has been the idea of a causal link between age and task in social insects (Oster and Wilson 1978 p. 159; Hölldobler and Wilson 1990 p. 307). Adaptive demography is the idea that the ageing schedules of workers are so tightly coupled to the effectiveness of the division of labor, and hence to colony efficiency and productivity, that the entire demographic structure of the workforce within the colony is best regarded as a colony-level adaptation. In contrast, the demographic structure of noneusocial populations is likely to be an epiphenomenon of selection acting on individual life histories. This is because it will be determined by the senescence of individuals, which current evolutionary theory explains as the result of genes that are beneficial to individuals before the age of reproduction, but detrimental to them afterwards (Medawar 1957; G.C. Williams 1957; Hamilton 1966).

Social insect workers should also be subject to similar processes of senescence. In fact, in species with reproductive workers (Bourke 1988a; Choe 1988), senescence should evolve by exactly the same means as in solitary animals. And even if workers are sterile through kin selection, senescence theory should still apply, but with the qualification that the critical age that triggers senescence is the age at which helping occurs. That is, a pleiotropic gene that made a helper particularly good at raising relatives early in life, and die after the age of helping, would be positively selected. This is because it would be likely to be present in the beneficiaries of the aid (Alexander et al. 1991).

However, the idea behind adaptive demography is that an individual worker's schedule of senescence is determined mainly by its conformity to a hypothetical age distribution which, by promoting colony efficiency, benefits the colony as a whole. It is certainly feasible that between-colony selection influences the age and caste profile of the workforce (Section 2.3). But implicit in the conventional idea of adaptive demography is a rather inflexible causal link between age and task allocation. This is because, without such a link, the age of a worker will not map closely onto its job, and hence the age frequency distribution of workers will not determine the colony's efficiency.

CENTRIFUGAL POLYETHISM

Centrifugal polyethism describes the tendency of social insect workers to start their employment at the center of the nest, and later to perform each successive task closer and closer to the periphery of the colony's activities (e.g. E.O. Wilson 1985a). For example, as already described, workers often begin as nurses in the nest and end as foragers outside it. West-Eberhard (1979, 1981) suggested that centrifugal polyethism stems from selection for reproduction by workers. Thus, when a worker is young and has a high (age-specific) reproductive value, it is selected to lay eggs and so to remain in the safety of the nest. But, as senescence commences and reproductive value declines (so death becomes less costly), a worker's best strategy is to help raise related brood by taking on risky occupations outside the nest such as defense and foraging. This scheme also generates an age-related division of labor, because young workers are found in the nest, and old workers outside it.

West-Eberhard's idea provides a cogent, ultimate explanation of how an age-related division of labor might first have evolved in social insect workers (Section 7.7). It should still apply in social insect species with reproductive workers, where studies have confirmed that ovary-developed workers tend to remain in the nest (examples among ants are in Otto [1958], Bourke [1988b], Breed and Harrison [1988], and Tsuji [1988b]). However, it clearly does not apply to those species in which workers are nonreproductive in the presence of queens, or are totally sterile, unless in these it is a mere evolutionary relic (Bourke 1988a; Choe 1988). In addition, at the proximate level, it does not provide (nor did it seek to provide) any form of algorithm or organizational procedure for generating the intricate "production lines" (multi-staged division of labor) seen within advanced social insect colonies. However, it is these that most require explanation.

AGE POLYETHISM

The traditional explanation for the most elaborate forms of division of labor is age polyethism (E.O. Wilson 1976a, 1985c; Oster and Wilson 1978; Calabi 1988). As discussed above, this is the idea of an age-based division of labor. As we have also mentioned, this concept implies a causal relationship between age and task. Thus, it suggests that each worker has an internal clock that ticks off the days of its life and directs it into one role after another. However, this is problematic, because all the existing data merely show a correlation between age and task, and, moreover, indicate that this correlation is often weak (e.g. Calabi and Rosengaus 1988; Corbara et al. 1989; Dejean and Lachaud 1991; Sendova-Franks and Franks 1993, 1994). Of course, assuming an internal ageing clock, it is almost impossible to show anything more than a correlation. To show causation would require taking a young worker and turning its clock forwards, or taking an old worker and turning its clock back, and then determining if the workers changed roles. A number of other factors have been shown to covary with age and task in social insect workers. These include titers of juvenile hormone in honey bees (Rutz et al. 1976; Fluri et al. 1982; Robinson 1985; Seeley 1985 p. 31; Robinson and Page 1989; Robinson et al. 1989; Robinson 1992), and changes in stimulus sensitivity with age in ants (Topoff et al. 1972; Cammaerts-Tricot 1974; Jaisson 1975; Topoff and Mirenda 1978; MacKay 1983; Jaisson et al. 1988). However, none of these covariates substantiates a causal relationship between ageing and task allocation.

Furthermore, a growing body of evidence challenges the hypothesis of a causal link between age and task. For example, in honey bees, workers move very quickly to new tasks, or revert to old ones performed earlier in their lives (Huang and Robinson 1992; Robinson 1992). In addition, "behavioral reversion" from foraging to brood care was observed in honey bee colonies from which young bees were experimentally removed (Robinson 1992). In unmanipulated colonies, this type of reversion could routinely occur when the social organization is perturbed by natural events in the life cycle, such as overwintering or colony fission (Robinson 1992). Fukuda (1983) also provided evidence of natural flexibility in the behavior of honey bees. Similarly, in ants, studies have found that a significant proportion of old workers may remain within the nest, and that some callows rapidly become foragers (Calabi and Traniello 1989a,b; Corbara et al. 1989; Dejean and Lachaud 1991; Sendova-Franks and Franks 1993; Section 12.5). In fact, both young foragers and old nurses may occur simultaneously in the same colonies (Sendova-Franks and Franks 1993, 1994). Calabi (1988) and Gordon (1989b) also presented evidence for behavioral flexibility and task switching in ants. All these findings go against the idea of a relatively rigid, age-determined division of labor.

Recent work on honey bees (Frumhoff and Baker 1988; Robinson and Page 1988, 1989; Robinson et al. 1989; Robinson 1992) has also shown that correlations exist between juvenile hormone concentrations, patriline membership, and age and task clusters within the workforce of the hive. But even this meticulous work has yet to determine the direction of any arrows of causality linking these factors.

In view of these findings, we advocate the need for a new and deeper understanding of the division of labor. Workers may vary in many ways, but cataloging correlations between their traits and what they do is likely to lead to only a phenomenological understanding of their behavior. Instead, we suggest that the link between age and task is not causal, and that the best way to interpret the looseness of this link, along with behavioral flexibility, is to consider the fundamental algorithms that might drive a self-organizing, self-tuning, and self-correcting division of labor (Deneubourg et al. 1987; Tofts and Franks 1992; Tofts 1993). For this reason, as the following section describes, Tofts and Franks (1992) and Tofts (1993) presented an algorithm that not only overcomes, but also capitalizes upon, the natural variability of workers to create a robust yet flexible division of labor.

The traditional hypothesis of a causal relationship between age and task does suggest one algorithm for a division of labor. This is that the colony rears the right number of eggs, at the appropriate rates, to set up a suitable age structure and hence division of labor among the workforce. But this algorithm is implausible, both in terms of its intrinsic inflexibility and because of contradictory empirical findings (Sendova-Franks and Franks 1993; Tofts 1993; Franks and Tofts 1994: Robinson et al. 1994).

In sum, the new approach adopted here asks how an efficient division of labor might be organized that would permit highly flexible responses to changing task demand which would not be possible if there were a direct causal link between age and task. In addition, it asks whether a nonage-based scheme could produce, purely as a by-product, a correlation between age and task in workers. These themes are the subject of the following section.

12.4 An Algorithmic Approach to the Division of Labor

> It is clear that the egg contains not a description of the adult, but a programme for making it, and this programme may be simpler than the description. Relatively simple cellular forces can give rise to complex changes in form: it seems simpler to specify how to make complex shapes than describe them. (S. Wolpert 1974 p. 16)

THE FORAGING FOR WORK ALGORITHM

Tofts (1993) developed the *foraging for work algorithm* as a candidate algorithm for robust and flexible task allocation and division of labor among workers. It was designed to meet the essential conditions of any such algorithm: (1) if the workforce is of sufficient size, all the required work will be done; (2) no individual is overloaded; and (3) individuals do not waste time needlessly looking for work to do. In other words, as a good task-allocation algorithm, it aimed to arrange the agents carrying out the work in proportion to the amount of work. As will be seen, Tofts' (1993) model is highly abstract, because it was designed to have predictive generality. By representing work implicitly (involving abstract "tokens") rather than explicitly, the algorithm is also easier to describe and to evaluate in mathematical terms (Tofts 1993).

A fundamental assumption of the model is that all the ants are initially identical and each is unaware of its age. It envisages the organization of work as a production line housed in a series of linked rooms containing the ants (Figure 12.1). "Tokens" pass along the production line, with each ant receiving a token performing some kind of transformation on it (work) before passing it down the line. The successful completion of the work depends on the availability of sufficient "tokens" from the upstream room. Thus, another assumption is that tasks are (linearly) ordered. This plausibly arises in real nests because of their internal structure. For example, the ordering of different tasks might be based on their distance from the center of the brood pile, or on similar positional cues (e.g. Camazine 1991; Franks and Sendova-Franks 1992; Section 12.6). The importance of spatial structure within nests has often been overlooked in division of labor studies, although Seeley (1985 pp. 32–35) argued for the importance of the spatial separation of tasks for the efficiency of work in honey bees.

The algorithm attempts to maximize the throughput of tokens along the production line. The use of the term "tokens" reflects the abstract nature of the model, but possible examples are food items, different brood stages, building materials, or even some form of information. In the case of food, one part of a production line in a real ant colony could be the chain along which food passes from foragers to nurse workers to

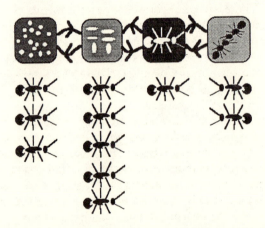

Figure 12.1 A diagram illustrating the foraging for work algorithm. A sequence of four tasks is shown. From left to right, they are care of eggs, care of larvae, care of callow workers, and trophallaxis resulting from foragers exchanging food with workers who stay in the nest. Below are shown the hypothetical numbers of ants in each of the worksites. Workers can move to a neighboring worksite in response to changing task demand. (From Tofts 1993. Copyright © 1993 by kind permission of Elsevier Science Ltd)

brood. Maximizing throughput along this line would help to maximize the production of new ants.

In more detail, the algorithm works in the following way. If the ants in a certain room are "starved" of a sufficient supply of tokens coming downstream, or if tokens are being taken at too low a rate from their room, they become "restless". Perhaps one or more of the ants becomes so restless that it moves to a neighboring room to find work there. Logically, in the case of "starvation", ants move upstream, whereas if tokens are accumulating they move downstream. The basic algorithm that an individual employs is therefore as follows:

1. Attempt to take a token from upstream (work on it) and pass the product downstream.

2. If this succeeds, then stay where you are and remember that you found work.

3. Keep a cumulative score of each time a direction fails to give or take work.

4. If the number of failures for either direction exceeds a critical amount, move in that direction with some probability.

In short, in this scheme, *workers do not allocate tasks, but tasks allocate workers*. Furthermore, because unemployed workers seek work (by moving towards zones where tasks are waiting), the label "foraging for work" is appropriate (Sendova-Franks and Franks 1993; Franks and Tofts 1994). These two catchphrases recognize the major role played by the working environment in organizing the division of labor. The rule of thumb is that workers seek gainful employment in the local environment. That is, they forage for work and changing task demand continuously tunes the distribution of labor. These characteristics lead to the conclusion that, if it is based on the foraging for work algorithm, the division of labor is self-organized.

In the algorithm, the threshold number of failures an ant might accept before switching tasks could be made extremely small. But this would make the system as a whole very sensitive to short-term fluctuations in work loads, making workers change tasks with unnecessarily high frequencies. Thus, some inertia on the part of workers (hesitation before changing tasks) is more economical even at the cost of longer delays in the time the system takes to equilibrate (Tofts 1993).

Tofts' (1993) model was formulated in a "process algebra," a mathematical language developed to facilitate formal reasoning about algorithms for concurrent systems such as parallel-processing computers (e.g. Milner 1989). A detailed explanation is provided by Tofts (1993). Having formally proved that an algorithm behaves in a certain way, it is then possible to use it as the basis for simulation. In this way Tofts

(1993) showed that, as anticipated, the foraging for work algorithm is sufficient to distribute workers efficiently and flexibly according to the changing abundance of tasks. Tofts (1993) went on to consider different levels of difficulty associated with the performance of different tasks (the forwarding of different tokens). Under this and other changes, the basic properties of the algorithm were robust: its efficiency at distributing workers according to needs remained, or in some cases was enhanced (Figure 12.2).

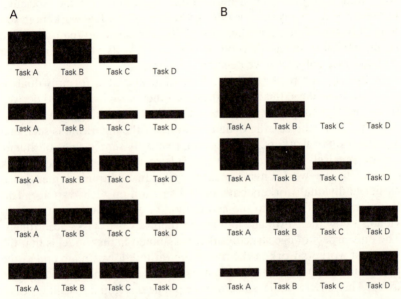

Figure 12.2 Results from executing the foraging for work algorithm for hypothetical ants and tasks. The heights of the bars indicate the number of ants performing a task. A. Eight ants evenly allocating themselves among four tasks. The algorithm was sampled after 1, 7, 14, and 20 execution steps. The final state (bottom row) shows the optimal arrangement, with workers allocated to tasks according to task demand. B. Eight ants allocating themselves to an uneven task demand. Individual ants can perform 3, 2, 2, or 1 units of work per step, in Tasks A, B, C, and D (the hardest) respectively. The algorithm is sampled after 1, 8, 20, and 45 turns to demonstrate how allocation proceeds. (Reprinted from Tofts 1993. Copyright © 1993 by kind permission of Elsevier Science Ltd)

The flexibility of the algorithm allows a colony to operate efficiently if lazy individuals opt out of the production lines, for example due to their being selfish, reproductive workers (Sections 7.4, 7.6). Thus, as long as sufficient workers behave altruistically and "forage for work," the society can still flourish. In addition, as described in more detail below, the colony can cope if certain individuals fixate on a particular task through learning, or through being physically specialized for a particular task (as is likely with physical worker castes). These findings suggest that in a stable environment a colony could function even if it contained a minority of ants that performed only one particular task. However, in a variable environment, the colony would be more efficient if these specialized ants could perform other work. Bigger colonies are likely to be better buffered against environmental variability by their larger infrastructure and greater supply of resources. This reasoning implies that worker flexibility should be more common in small societies and worker specialization more common in large societies. However, since many ant societies grow from just a few workers in size to hundreds or thousands, there should be a widespread requirement to retain flexibility in the division of labor even if its value is no longer obvious once colonies have become very large.

In the algorithm, the global information available to individuals is minimal. In a sense, the only information they have is local and derived from the particular task they are currently undertaking or waiting to perform. Nevertheless, the dynamic nature of the algorithm is such that a population of workers, each obeying relatively simple rules of thumb, can be precisely tuned to the workloads of the colony and can track changes in the work environment closely. In addition, as mentioned earlier, individual workers have no information on their own age. This is in complete contrast to the assumption of age polyethism, that individuals know their own age and assign themselves to tasks accordingly. The only aspect of age structure that is assumed in the model is that the youngest workers work in the first room (at the downstream end of the production line), and that older workers have a higher probability of dying. This is reasonable, since worker pupae develop into callow adults at the center of the nest and so are likely to seek their first work in the brood pile.

Significantly, the foraging for work algorithm, in conjunction with the spatial arrangement of tasks, generates several of the classical phenomena of the division of labor. Most important are the weak correlation between worker age and role on which the idea of age polyethism was based (Figure 12.3), and the tendency for young workers to work in the nest and old ones outside (centrifugal polyethism). These both arise because workers tend to be passively displaced from task to task away from the nest's center, where they first emerge, as new workers emerge at the center and as fresh tasks are encountered away from it. However, both phenomena are by-products of the more fundamental self-organization process, coupled with the spatial ordering of tasks. Thus, the algorithm shows that there can be an association between age and task even though it is not explicitly encoded in each worker's behavioral development. The emergence of structure (here, the association between age and task) through relatively simple, local rules of behavior is characteristic of self-organization in social insects. In addition, if a shortage of workers to perform tasks at the nest center arises, foraging for work predicts that old workers from the periphery should return to carry them out. Therefore, it also accounts for the behavioral reversions and individual flexibility now known to be characteristic of workers.

Much recent work has shown that honey bee workers belonging to different genetic subgroups (patrilines) tend to perform different tasks in the colony (Frumhoff and Baker 1988; Robinson and Page 1988,

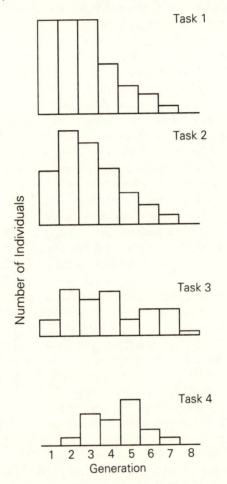

Figure 12.3   The age frequency distributions of workers in four unspecified tasks using the simulated foraging for work algorithm, illustrating how worker age distribution can change over tasks. (From Tofts and Franks 1992; by permission of the authors and Elsevier Trends Journals)

1989; Robinson 1992; see also Tofts and Franks 1992). However, the exact significance of this finding is controversial. One idea is that multiple mating in honey bee queens has evolved precisely to provide the colony with an optimal mix of worker genotypes (e.g. Frumhoff and Baker 1988; Robinson and Page 1988; Section 11.4). But it seems highly unlikely that the queen could choose to mate with just the right combination of males to achieve this. Therefore, a danger exists in assuming a direct causal link between genes and tasks in honey bee colonies. Instead, like the link between age and task, the association could follow from the foraging for work algorithm. In the algorithm, a tendency based on genotype for particular workers to be biased in their association with certain tasks might leave gaps in the production lines elsewhere and so draw other workers into those roles. For example, a worker of a particular genotype specializing in removing its dead nestmates (Robinson and Page 1988) may be doing so not through an innate preference but simply because such work tends to be neglected by others. Initial biases in task allocation would also be perpetuated or even amplified, because workers have a tendency to fixate on tasks (see below). In fact, the foraging for work algorithm suggests that *any* initial variation among workers, both phenotypic and genetic, could influence the allocation of tasks in the future. So rather than assume that age or genetic differences determine the division of labor, the foraging for work algorithm looks for a deeper understanding of how a colony can function effectively despite these hard-to-control influences.

The foraging for work algorithm generates a self-adjusting temporal polyethism, which is flexible, robust, and decentralized. The advantage of such organization is that it is event-driven and not necessarily based on a genetically preprogramed solution to the problem of how to divide labor. "Hardwired" programing in a dynamic system like an ant colony, which must continuously adjust to a capricious external environment, would be more likely to add to the problem than provide a solution. For example, as mentioned in the previous section, it does not easily allow recovery if a colony suffers a catastrophe. Thus, if a large proportion of old foragers or young nest workers are killed, the resulting inefficiencies would persist for years because they would make it all the harder for the colony to redevelop its optimized worker age structure. The alternative of decentralized control through event-driven self-organization is both simpler and more resistant to perturbations.

The foraging for work algorithm can also take account of the possibility of learning in workers. According to the algorithm, a naive young worker does the first job it encounters. Since similar jobs are likely to be in the same area because of the spatial organization of the nest interior, there is a high probability (reinforced by positive feedback) that the worker will repeat the task and so stay in the vicinity. This might entrain workers to certain tasks and help them acquire skills through practice. Thus task fixation and learning may be intimately associated both with one another and spatial efficiency. Although learning has been repeatedly demonstrated in ants (Schneirla 1943; Jaisson 1975; Wehner et al. 1983), we are not aware of any demonstration that workers learn to be more skilful or quicker at particular tasks within nests, as the above argument predicts. However, this could be simply due to the technical difficulties of demonstrating this phenomenon.

Lastly, Franks and Tofts (1994) suggested a way in which the foraging for work algorithm could be experimentally tested. Robinson et al. (1994) claimed that temporal polyethism in social insects is fundamentally a developmental process. To show this, it would be necessary to demonstrate that physiological changes occur in individuals prior to a change in those workers' tasks, when task demand is held constant. By contrast, in the foraging for work model, a worker should not change its task if task demand is held constant. Instead, it should only switch its location and task in response to changing demand (and might accordingly undergo physiological changes on recognizing the change in demand, and then move). Thus, showing that developmental transitions occur even in the face of constant external conditions would strengthen the case that temporal polyethism is a developmental process, and weaken the case for the foraging for work algorithm.

THE FORAGING FOR WORK ALGORITHM AND PHYSICAL POLYMORPHISM

Physical worker castes (physical worker polymorphism) occur in a number of ant genera (e.g. E.O. Wilson 1953; Oster and Wilson 1978). Such castes are classes of workers, occurring within single colonies, that are distinct in size and shape (Figure 12.4). Physical castes show behavioral differences that are qualitative, and not just attributable to quantitative variations in size. In addition, they are generated not by genetic differences among workers, but developmentally, by the differential (allometrical) growth rates of body parts (e.g. Huxley 1932; E.O. Wilson 1953; Wheeler 1986; Franks and Norris 1987). For detailed reviews of the developmental basis of worker polymorphism, see Nijhout and Wheeler (1982), Wheeler and Nijhout (1984), Wheeler (1986, 1991), and Hölldobler and Wilson (1990 pp. 348–354).

The use of specialized individuals to perform particular tasks has many advantages. The basic organizational strategy seems straightforward: a worker might simply perform the task to which it is physically suited. Further, it is possible to produce individuals whose degree of

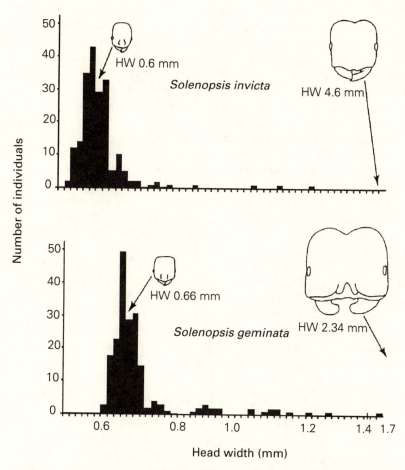

Figure 12.4 Size frequency distribution of worker castes in two fire ant species, *Solenopsis invicta* and *S. geminata*. Both distributions are based on random samples of 200 ants taken from laboratory colonies. Both distributions are skewed to the right, especially that of *S. geminata*, compared with the normal distributions found in monomorphic species. The outlines shown are minor and major workers at the two extremes of the size variation. (From E.O. Wilson 1978; by permission of the Kansas Entomological Society)

specialization is such that each can replace many generalists, thus freeing those individuals for other tasks. (This recalls Adam Smith's endorsement of physical specialization: ". . . the invention of a great number of machines which facilitate and abridge labour, and enable one man to do the work of many." [Section 12.2].)

In order to exploit these organizational principles, the ratio of various specialists must be controlled to provide an efficient workforce.

Specialization of individuals leaves them highly unsuited to perform any task but their speciality. Thus if demand for workers of a particular type is reduced, the workers cannot effectively work on any other task. Since, in ants, the morphology of a specialist is already determined by the time it emerges as an adult, a system based on physical castes is therefore unable to respond rapidly to environmental change requiring a large-scale adjustment in the caste profile of the colony. This could be why physical polymorphism is in fact relatively rare, being only found in about 20% of ant genera. In these, moreover, the number of physical castes per species is almost always just two or three (Oster and Wilson 1978 p. 181; Tofts and Franks 1992), although four castes have been reported in *Atta sexdens* (E.O. Wilson 1980a) and *Eciton burchelli* (Franks 1985).

What is the relationship between physical castes and the foraging for work algorithm? A physically specialized ant can be regarded as an extreme case of task fixation. However, when such individuals are introduced into the foraging for work algorithm, no detrimental effect is observed. They cause no major disruption, provided that there are sufficient generalist ants (ones that have no fixed task) to deal with fluctuations in task demand caused by environmental variability (Tofts 1993). Clearly, if the environment changed so radically that all the individuals engaged in a particular task became unnecessary, the presence of task-fixated individuals would represent a cost to the colony. However, whenever the environment is reasonably stable, it is possible to accommodate the presence of a minority of specialists, while maintaining an accurate assignment of individuals to tasks. In nature, colonies of morphologically different ants also contain an abundant caste of generalist individuals. As long as the number of generalists is sufficient to cope with any changes, nests can allocate individuals to roles efficiently.

Franks and Norris (1987) explored the development of physical worker castes by using computer graphics to carry out two dimensional Cartesian transforms of worker morphology of the type first suggested by D'Arcy Thompson (Figure 12.5). They showed with greater resolution than traditional allometrical analysis (e.g. E.O. Wilson 1985c) that castes cannot be formed in isolation from one another. For example, the development of the morphology of major workers is severely constrained by that of the minor workers and vice versa, because the same functions describe the transformations of the different castes in a colony, meaning that only a limited number of shapes can be produced. Such constraints suggest that extreme physical polymorphism is fairly haphazard and that physical castes are not precisely tuned machine parts. Thus, contrary to the view that advanced ant societies are based on hardwired organization, it seems that even physically polymorphic

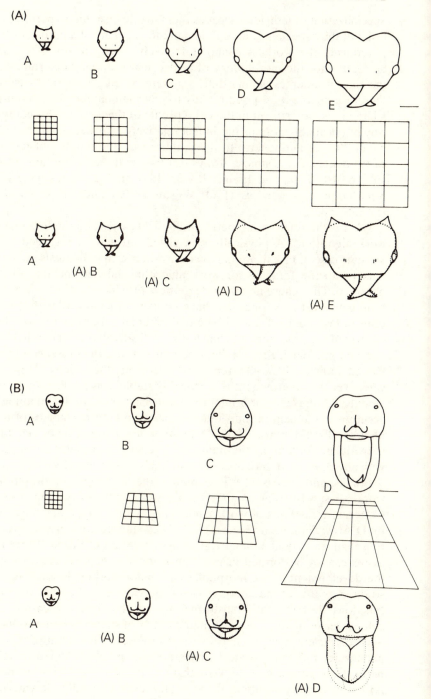

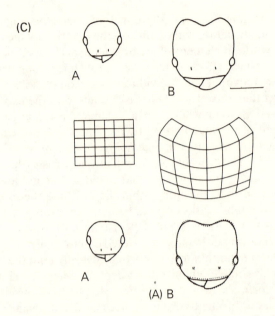

Figure 12.5 A (*facing page*). The top row of outlines are accurate drawings of a nest series of *Atta columbica* workers. The grids represent the changes in shape (Cartesian transforms) required to transform the minor (A) into each of the larger workers. The bottom row shows as solid outlines the resulting shapes of the modified minor, as it appears when transformed into each of the large workers. The unmodified outline of these larger workers is represented by the dotted lines. The scale bar to the right of E represents 1 mm. B (*facing page*). Cartesian transforms for the heads of the *Eciton burchelli* minor, medium, submajor, and major worker. The scale bar to the right of D represents 1 mm. C. Cartesian transforms for heads of the minor and major of a *Pheidole* species. The scale bar to the right of B represents 1 mm. (From Franks and Norris 1987; by permission of Birkhäuser Verlag, Basel)

workforces would require flexibility in their basic task allocation algorithm. In other words, because physical castes cannot evolve independently of one another (Franks and Norris 1987), there is all the more reason to expect that physical polymorphism could only evolve in ants if the basic division of labor system in their monomorphic ancestors was flexible and fault-tolerant. The need for this flexibility would explain why generalist workers are in the great majority even in the most polymorphic societies (Tofts and Franks 1992).

OTHER ORGANIZATIONAL PRINCIPLES IN TASK ALLOCATION

Along with foraging for work, there may of course be other principles that affect task allocation within ant societies, some of which share with foraging for work an element of positive feedback. First, there is learning, which may have effects independent of its interaction with foraging for work (see above). Thus, individuals that acquire skills and become faster at tasks they repeat will take tasks away from the less specialized or unemployed. Second, where the task-demand is too large, there may be recruitment of the unemployed by the employed. For example, foragers finding food may recruit unemployed workers in the nest (e.g. Traniello 1977). Third, where different physical castes have different intrinsic skills, each type of worker may avoid getting in the way of others depending on their location within the nest and the availability of suitable or unsuitable tasks in that region. For example, in *Pheidole pubiventris*, major workers are able to perform brood care (for which their large heads make them poorly suited), but usually do not because they avoid the smaller minor workers that typically tend the brood (and whose small size makes them more adept at doing so) (E.O. Wilson 1985b). Such *between-caste aversion* is a simple mechanism that tunes the colony's labor profile to changing needs, since it means that majors will undertake brood care if there are no individuals available better suited to this task. Comparable phenomena were described by Calabi and Traniello (1989a), Fresneau and Corbara (1990), and Theraulaz et al. (1991). Note that between-caste aversion is not inconsistent with the foraging for work algorithm. For example, it is conceivable that in addition to monitoring the number of tokens coming upstream or downstream, it might also be advantageous to a group of ants to monitor the abundance of operatives in adjacent worksites (rooms) before switching tasks.

SELF-ORGANIZATION AND NATURAL SELECTION

Finally we stress that the view, advocated here, that the division of labor is based on self-organization and simple algorithms is not in conflict with a gene selectionist viewpoint (Chapter 1). The foraging for work algorithm proposes a means by which a robust and flexible system of the division of labor may be generated within colonies at the proximate level. If colonies benefit from having such a system (by producing more sexuals), and the appropriate genetic variation exists, then the system can be naturally selected (cf. Section 2.3). Specifically, what is selected are the genes, expressed in the workers, for following the rules leading to behavior in conformity with the algorithm. By contrast, under a "hardwired" division of labor, what might be selected are genes for a hardwired relationship between age and task.

Page and Mitchell (1991) argued that the self-organization of insect societies challenges the view that the adaptive complexity of colony organization arises by natural selection. They suggested that colony organization is not a "direct" product of natural selection, but that selection operates on the parameters whose values result in the work-force acting as an adaptive, self-organized system (see also Kauffman 1991). However, this very argument means that Page and Mitchell's (1991) challenge to natural selection is not a challenge at all. In other words, there is nothing incompatible between the view that colonies are adaptively self-organized, and the view that they become so via natural selection in the way Page and Mitchell (1991) suggested. This point was also made by Reeve and Sherman (1993) in a critique of Page and Mitchell (1991). Reeve and Sherman (1993) added the valuable obser-vation that self-organized systems are not necessarily adaptive. Thus, one can imagine a colony that was self-organized in a manner that lead to a decrease in its output of sexuals. Therefore, another way of viewing the evolution of the division of labor is to regard natural selection as picking between alternative forms of self-organization (Reeve and Sherman 1993).

12.5 The Division of Labor: Empirical Studies

This section reviews case studies of the division of labor in ants, stress-ing where appropriate their bearing on the idea that division of labor stems from something akin to the foraging for work algorithm. First, however, we discuss some methodological issues.

A CRITIQUE OF THE USE OF BEHAVIORAL REPERTORIES AND ETHOGRAMS IN STUDYING THE DIVISION OF LABOR

The standard method used to study division of labor in ant species is to construct behavioral repertories from colonies in the laboratory. These are made by compiling an initial list of behaviors (Tables 12.1, 12.2), then recording which of them successive focal animals perform over a fixed period of time. Often both the frequency and the average duration of each behavior is noted. The total behavioral repertory of a species (the complete catalog of behaviors that it exhibits) can be calculated by plotting the number of new behaviors discovered as a function of sampling effort. Alternatively, in the Fagen–Goldman method, a theoretical frequency distribution may be fitted to the observed frequency distribution of behaviors, to calculate the number of behaviors likely to have been missed (Wilson and Fagen 1974; Fagen and Goldman 1977; Sudd and Franks 1987 p. 67). These methods have shown that ants perform few rare behaviors within their colonies, and hence that relatively little effort is required to record the whole repertoire. Behavioral repertoires can be constructed with or without identifying individual workers or knowing their age.

A related technique is the construction of ethograms. These document the frequency with which one type of behavior is followed by another. Ethograms can therefore be used to identify clusters of tasks. For example, the next most likely task of workers caring for eggs could be care of the queen or larvae. This would suggest that such workers are specialist nurses (e.g. Hölldobler and Wilson 1990 p. 300).

Initially, behavioral repertoires and ethograms seem to constitute powerful techniques. They are quick and quantitative and give rather

Table 12.1

Time Budgets of Foragers and Nest Workers in *Leptothorax acervorum*

Behavior	Percentage of Total Time Spent on Activity	
	Foragers	Nest Workers
Forage	40.9	8.7
Groom self	16.1	4.5
Rest	14.8	71.5
Feed worker	7.6	0
Interim movement (in nest)	6.4	8.9
Receive antennal contact from worker	5.3	1.1
Initiate antennal contact with worker	4.9	2.6
Groomed by worker	2.2	0.47
Feed larvae	0.7	0.03
Receive antennal contact from queen	0.4	0.001
Initiate antennal contact with queen	0.3	0.004
Receive food from worker	0.16	0.39
Groom larvae	0.10	0.71
Groom queen	0.07	0
Feed queen	0.07	0
Excrete	0.06	0
Antennal tipping	0.06	0
Groom worker	0	1.1

NOTE: The total observation times were 11.3 h for foragers and 11.5 h for nest workers. Reprinted from Franks et al. (1990b). Copyright © 1990, by kind permission of Elsevier Science Ltd.

Table 12.2

Relative Frequencies of Behavioral Acts by the Two Physical Worker Castes of the Ant
Pheidole dentata

Behavioral Act	Frequency	
	Minors	Majors
Self-grooming	0.18003	0.56373
Groom other adult		
Minor worker	0.04992	0
Major worker	0.00573	0
Alate or mother queen	0.01146	0
Brood care		
Carry or roll egg	0.01391	0
Lick egg	0.00245	0
Carry or roll larva	0.12357	0
Lick larva	0.09984	0.02941
Assist larval ecdysis	0.00409	0
Feed larva solid food	0.00573	0
Carry or roll pupa	0.03601	0
Lick pupa	0.01882	0
Assist eclosion of adult	0.00818	0
Regurgitate		
With larva	0.02128	0
With minor worker	0.03764	0.22059
With major worker	0.00573	0
With alate or mother queen	0.00327	0
Forage	0.12111	0.02941
Feed outside nest	0.04337	0.01471
Carry food particles inside nest	0.05237	0
Feed inside nest	0.05810	0.01471
Lick meconium	0.00573	0
Carry dead nestmate	0.01882	0.04902
Carry or drag live nestmate	0.00246	0
Eat dead nestmate	0.06383	0.07843
Handle nest material	0.00655	0
Totals	1.0	1.0

NOTE: The total number of acts recorded was 1,222 for minor workers and 204 for major workers. From E.O. Wilson (1976a Table 1). By permission of Springer-Verlag and the author; copyright © 1976 Springer-Verlag.

complete records. This facilitates comparisons within or between genera and species. Undoubtedly, the use of these techniques has done much to transform the study of ant behavior into a modern quantitative discipline (e.g. E.O. Wilson 1976a,c; Cole 1983b). However, they also involve several problems.

First, behavioral repertory and ethogram studies are based on a subjective assessment of the number of kinds of behavioral elements (or acts) present in a species. Jaisson et al. (1988) documented the difficulties that arise from this subjectivity. They showed that when different authors use the Fagen–Goldman method on the same species, they obtain substantially different behavioral repertories and ethograms. Seeley and Kolmes (1991) also describe some of the drawbacks of using ethograms in the study of honey bees.

Another difficulty is that ethograms are generally constructed serially for a succession of individuals. As a result almost all information is lost on the context in which social interactions take place. For example, one worker might give food to another, and it would be helpful to know which tasks both workers were undertaking before and after this event. Gordon (1985) questioned the value of ethograms for similar reasons. In addition, standard ethograms, behavioral repertories, and time budgets are often based on pooled data and are generally made over short periods relative to the lifespan of workers and the colony. As a result, they obscure the possibility that individuals learn which tasks to perform, acquire skills, and increase their personal efficiency. They also tend to overlook the rare but critical events associated with task switching. Moreover, ethograms tend to suppress data on individuality and may have helped to sustain the view that ants behave like preprogramed machine parts (Franks 1990). Because these techniques only yield information about age cohorts rather than the lifetime career of individuals, they have also tended to perpetuate the belief that division of labor is age-based (cf. Jaisson et al. 1988).

In addition, almost all detailed behavioral studies of the division of labor in social insects are undertaken in the laboratory. Often they are made on unusually small colonies, housed in nests with artificial internal geometries (Section 12.6), that lack any natural predation of older, foraging workers. The lack of natural mortality almost certainly distorts the age structure of laboratory colonies (Wehner et al. 1983; Schmid-Hempel and Schmid-Hempel 1984; Cole et al. 1994). Therefore, younger workers in these colonies probably experience little opportunity to demonstrate their flexibility. In sum, a large number of potential limitations need to be borne in mind when assessing empirical studies of the division of labor in ants.

Evidence of Flexibility in Individual
Workers and for Weak Correlations
Between Age and Task

Strong evidence for behavioral flexibility in individuals comes from the pioneering "sociotomy" experiments of Lenoir (1979a,b). He removed older workers from colonies of *Tapinoma erraticum* and observed that younger workers tended to become foragers earlier than normal (Lenoir 1979a). From subsequent experiments with *Lasius niger*, Lenoir (1979b) concluded that the behavioral plasticity of workers declines with age: older ants are largely specialists in foraging. Behavioral flexibility in these species therefore seemed to depend upon the flexibility of the younger, nurse ants, since they were able to show an accelerated behavioral development. Accelerated development was also demonstrated by McDonald and Topoff (1985) in *Novomessor albisetosus*. In control colonies, workers of this species became specialized foragers at about 68 days of age. Following removal of all the foragers, this transition occurred in only 19 days. The isolated foragers reverted to nursing behavior. Furthermore, McDonald and Topoff (1985) suggested that the presence of newly-emerged workers induces the shift to foraging behavior in young "domestic" workers. There are similar findings for honey bees (Huang and Robinson 1992). Significantly, this effect of the newly-emerged workers is exactly what one would predict according to the foraging for work algorithm (Tofts and Franks 1992; Tofts 1993; Franks and Tofts 1994). Large numbers of very young workers at the center of the nest might swamp the task demand there, leading to workers that were older on average moving to more peripheral stations in the production line.

Calabi (1988) and Gordon (1991) also emphasized the importance of individual flexibility and task switching in the division of labor in ants. But haphazard task switching or loose flexibility leads to anarchy rather than control, unless it is part of a suitable algorithm. The foraging for work algorithm shows how task switching and flexibility might enable a colony to function successfully in a changing environment (Section 12.4). Gordon (1989b) carried out field studies of the division of labor among *Pogonomyrmex* harvester ant workers. She marked workers performing different tasks and altered the amount of work they had to perform. It turned out that workers could very rapidly change tasks, and that changes in the numbers engaged in one activity altered the numbers engaged in other activities. Gordon (1989b) therefore suggested that, within a timescale of hours, one worker group receives information about events affecting other worker groups. These last two findings are consistent with the production line organization envisioned in the foraging for work algorithm. Numbers of workers moving from one workshop ("room") to another, as a result of changing task abundance and a low rate of task completion, might well have an effect that spreads like a ripple down the line.

Sendova-Franks and Franks (1993) studied individually-marked workers in *Leptothorax unifasciatus*. They found that workers which were at least one year old could specialize in a variety of tasks, from care of eggs and first-instar larvae, to queen care, nest guarding, and foraging. Callow workers could also be involved in all these tasks, including foraging. In addition, older workers were able to revert from foraging back to working in the nest. Clearly, workers did not undergo a steady progression through a set of tasks as they aged (Sendova-Franks and Franks 1993, 1994).

In a study of *Pheidole hortensis*, a species with two physical worker castes (majors and minors), Calabi et al. (1983) divided each caste into five categories, based on the degree of pigmentation. Since workers grow more pigmented with age, each category therefore represented an age class. Intriguingly, Calabi et al. (1983) found that the older classes maintained the level of performance of behavioral acts which character-ized the younger ones. For example, among the minor workers, the for-agers in the fifth, darkest class continued their nursing and grooming activities. Thus, the workers extended their behavioral repertoire with-out deleting behaviors that they exhibited earlier in their lives. Calabi et al. (1983) considered such a scheme "atypical" and thought it might have arisen from the presence of a surfeit of brood in the study colony. However, Jaisson et al. (1988) concluded that it is in fact much more typ-ical than atypical of ant social organization. They suggested that the greater behavioral variability of older individuals may result from their having had more and longer experience of items and tasks in their envi-ronment. This is again consistent with the foraging for work algorithm (Tofts 1993; Franks and Tofts 1994). The work of Calabi et al. (1983) also shows the limitations of a methodology that is not based on marking individuals (Jaisson et al. 1988). It can provide information about the effect of age on the average individual, but it does not reveal the variety of behavioral patterns shown by individuals in the same age class.

Further studies demonstrating behavioral variability in same-age workers include those of Calabi and Rosengaus (1988) on *Camponotus sericeiventris*, Corbara et al. (1989) on *Ectatomma ruidum*, and Dejean and Lachaud (1991) on *Odontomachus troglodytes*. In this last study, Dejean and Lachaud (1991) examined ten colonies that had been manipulated so that each contained one queen, 30 workers, and 30 lar-vae. They then studied the behavior of 91 workers from the time they emerged as adults until they were 7 weeks old. Workers of the same age

(A)

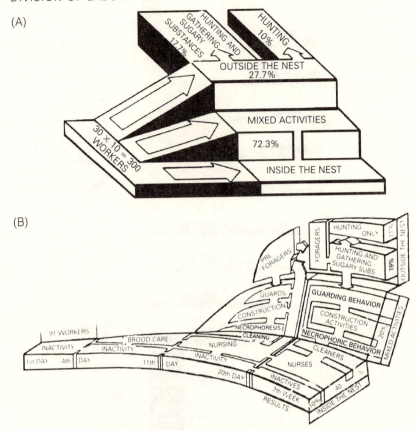

(B)

Figure 12.6   A. Distribution of 300 *Odontomachus troglodytes* adult workers among the three main behavioral categories (activities inside the nest, mixed activities, or activities outside the nest), one week after installation in artificial nests. B. Temporal polyethism in the distribution of 91 callow workers of *O. troglodytes* among the various behavioral subcastes during seven weeks of experimental observations. (From Dejean and Lachaud 1991; by permission)

often developed different subspecializations. For example, some of the nurses received far more honey droplets than others. Of greatest interest of all, some workers even remained in one role throughout their lives (Figure 12.6). These results resemble those of Sendova-Franks and Franks (1993) on *Leptothorax unifasciatus*. And, once more, all these findings are consistent with the foraging for work algorithm coupled with some specialization through task fixation. Furthermore, the fact that individuals of very different ages may be engaged in the same tasks is inconsistent with any meaningful definition of age polyethism (Franks and Tofts 1994).

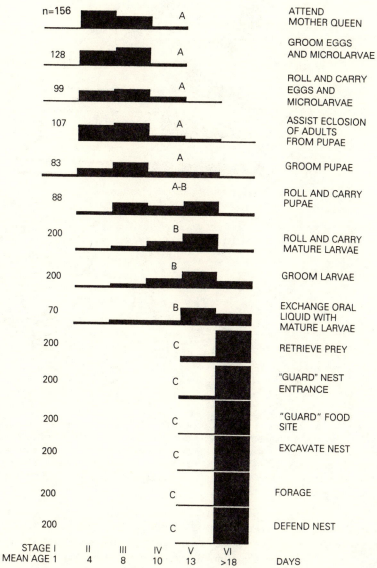

Figure 12.7 The proportions of workers of the six age groups attending to all of the principal tasks in colonies of *Pheidole dentata*, given in a series of histograms. The number of observed performances of each task, totalled through all of the age groups, is given on the left. The age groups (I–VI) and the average age of workers in each are given at the bottom. The histograms are classified into three groups (A, B, C) that are then identified as the temporal castes in a discretized caste system. (From E.O. Wilson 1976a, Figure 3. Copyright © 1976 Springer-Verlag; by permission of Springer-Verlag and the author)

STUDIES SHOWING STRONG CORRELATIONS
BETWEEN AGE AND TASK

E.O. Wilson (1976a) constructed ethograms for both the major and minor worker castes in *Pheidole dentata*. Since this is another species in which cuticle color darkens with age, Wilson was able to record how the principal tasks were distributed among different age groups of minor workers. He found that workers were organized into largely nonoverlapping age groups, each of which handled a distinct set of tasks (Figure 12.7). Wilson termed this a *discretized* caste system. One explanation for it is that workers of a certain age may tend to perform tasks that naturally occur in the same part of the nest, so promoting spatial efficiency (E.O. Wilson 1976a). For example, the queen tends to live in the deep center of the nest, where she produces eggs that later hatch into first-instar larvae. Thus queen care and nursery tasks occur together, and a likely result would be for these tasks to be performed by a single group of young workers (which, furthermore, emerged as adults in the neighborhood of the nursery).

However, these findings could have been partly artefactual. The results showed that, above the age of 18 days, workers performed hardly any brood care. Instead they either guarded or helped build the nest, or spent their time in external activities such as foraging (Figure 12.7). This seems strange, since worker ants typically live for weeks, or even months and years (Hölldobler and Wilson 1990 pp. 169–170). Furthermore, most studies of ant colonies reveal that over half the workers are usually inside the nest at any one time (e.g. Elmes and Wardlaw 1982b; Dejean and Lachaud 1991; Tschinkel 1993a). Therefore, unless *Pheidole dentata* workers are exceptionally short lived, or the colony produces them at a very high rate, it seems that only a tiny proportion of the workforce could actually be caring for the brood. So it is possible the workers in this study moved swiftly through a progression of tasks for some reason connected with their being in captivity, or because of some idiosyncrasy of the experimental colony. This is of course speculation, but what is not in doubt is the general need for longer-term observations of individually-marked workers in division of labor studies.

Tsuji (1990b) reported a division of labor associated with age in the thelytokous ant *Pristomyrmex pungens* (Section 10.7). Workers appeared to experience an irreversible transition between a period of reproduction and a period of post-reproductive activities outside the nest. Although Tsuji's (1990b) results therefore showed a strong correlation between age and task (doubtless partly due to selection on young, reproductive workers to avoid risky external tasks), it is possible that they too are compatible with the foraging for work algorithm. Thus, they would follow if the study colonies had rapidly growing populations, little unemployment among the workforce, and relatively little variation in task demand. For, under these conditions, one would expect age cohorts of workers to progress as a group from interior to exterior employment. In addition, it would be especially interesting to know if individual *P. pungens* workers could extend their period of reproduction when all other colony support tasks were being fulfilled. If so, this would show that they respond to task demand, as the foraging for work algorithm predicts, and not to their own physiological state.

WORKER POLYMORPHISM AND THE PRESENCE AND ABSENCE OF FLEXIBILITY

Behavioral repertory methods have been most successful when they have been used by a single investigator to compare the performance of different physical castes in the same ant colonies. Thus, these methods have revealed very clear differences between the behavior of major and minor *Pheidole* workers. Majors are always much more specialized than minors. Minors perform most of the brood care, nest maintenance, and standard foraging tasks, whereas majors use their disproportionately large heads and jaws to good effect as specialist soldiers or millers. For example, in *P. dentata* and *P. hortensis*, there was virtually no overlap between the sets of tasks performed by the two castes (E.O. Wilson 1976a; Calabi et al. 1983; Table 12.2).

On the other hand, Patel (1990) found that, despite their morphological specialization, *Pheidole morrisi* majors exhibited 25 types of behavior, which was only four fewer than the minors had. The study colony had a particularly high proportion of majors in the workforce (nearly 30%). In addition, in a between-species comparison of *Pheidole* ants, E.O. Wilson (1984) discovered a positive correlation between the proportion of majors within colonies and the fraction of all worker behaviors they undertook. These findings fit with the plasticity requirements of the foraging for work algorithm. Put simply, to achieve a reasonable level of plasticity, it is necessary for majors to show a greater diversity of behavior as they grow more numerous in the society.

Calabi et al. (1983) found that in *Pheidole hortensis*, in marked contrast to the case of *P. dentata* described above (E.O. Wilson 1976a), both castes had a *continuous* rather than a "discretized" system of division of labor. In other words, although worker age was correlated with employment (as earlier mentioned), age-groups of workers showed broad overlaps in their task repertory. The difference between *P. hortensis* and *P. dentata* has been attributed to differences in the scale of operations in these two species (Sudd and Franks 1987 p. 73). *P. hortensis* typically has smaller colonies. So each might be more like a cottage industry in which every individual has to be a flexible "jack of all trades." In *P. dentata*, by contrast, the entire factory is much larger and workers can move predictably from one set of tasks to another in a more rigid manner. This is because, in a large society, most tasks in the complete task repertory are likely to be present at any one time, whereas in a small society some may be absent through chance fluctuations. This hypothesis invoking the effects of scale may be true, but it remains untested, and it could be that there is considerable flexibility in both species. The foraging for work algorithm also suggests how, in colonies of different sizes, workers might have different rates of switching forwards and backwards (from task to task) along their production lines (Section 12.4).

There is little evidence that ant colonies with polymorphic workers can alter their physical caste ratios in response to changing needs. In *Camponotus* (*Colobopsis*) *impressus*, caste ratios changed both seasonally and during colony growth, but there appeared to be no adjustment to annual or spatial variation in resources, competitors, or predators (Walker and Stamps 1986). Moreover, when different proportions of soldiers were removed, the ability of *C. impressus* colonies to regulate their caste ratios turned out to be severely constrained by a low and constant production ratio of new soldiers to new minors. Walker and Stamps (1986) concluded that there are physiological and temporal constraints to caste ratio adjustment in ant colonies. Johnston and Wilson (1985) also provided evidence for a lack of plasticity in physical caste ratios in *Pheidole dentata*. A possible explanation for this could simply be that sufficient flexibility is already present in the division of labor. Since nonspecialized workers are always in the majority, their individual behavioral flexibility, coupled with the efficiency of task allocation provided by a system such as foraging for work, might mean that having fixed proportions of physical castes has little effect on colony efficiency.

Finally, although both *Leptothorax longispinosus* and *Paraponera clavata* have monomorphic workforces (with a unimodal size distribution), their division of labor is partly associated with size differences among the workers (Herbers and Cunningham 1983; Breed and Harrison 1988). Stuart and Page (1991) have also shown that, in other *Leptothorax* ants, separate genetic subgroups tended to be employed in different tasks. However, as with the honey bee work discussed in the previous section, all these observations could arise from the tendency of a system such as foraging for work to cause different phenotypes to be associated with different tasks (Franks and Tofts 1994).

12.6 New Approaches: Spatial Patterns and the Division of Labor

Among ants, there is a spectrum of foraging modes. For example, the highly individualistic workers of *Cataglyphis* (Wehner 1987) and many ponerine species (Hölldobler 1980; Beckers et al. 1989; Goss et al. 1989) literally go it alone once out of the nest, whereas army ant forays involve hundreds of thousands or even millions of workers (Franks 1989a; Section 10.8). Much attention has been paid to the temporal and spatial organization of these foraging systems (Hölldobler and Wilson 1990 pp. 378–388).

In the world outside the nest, workers forage for food resources whose distribution and abundance may be static or dynamic. Inside the nest, the "foraging environment" may be similarly complex. Workers must seek other adults, or brood of various stages of development, for grooming and food exchange, or perhaps find parts of the nest requiring maintenance and extension. So it seems reasonable to suppose that the ability of ants to explore space efficiently might also be used within the nest. This presupposes that different adult ants or brood stages are distributed predictably within the nest. (This is also an important assumption of the foraging for work algorithm.) Until recently, however, little quantitative research had been carried out on the internal social architecture of ants' nests.

Studies on *Leptothorax* species are beginning to show the extraordinary sophistication, again based on relative simplicity, that can occur in the spatial patterning of nursery functions within nests. This is particularly noteworthy, because members of this genus have some of the simplest nests of all ants. For example, *L. unifasciatus* often nests in extremely thin fissures in rocks. These are so narrow that the ants effectively live in a two dimensional environment. Nest building only involves arranging a simple, roughly circular wall of debris around the colony (Franks et al. 1992). Such species will readily nest between microscope slides in the laboratory, and are ideal for making photographic records of individual and colony activity (Franks et al. 1990b).

Franks and Sendova-Franks (1992) showed that *Leptothorax unifasciatus* colonies sort their brood into a sophisticated pattern. The ants make a single brood cluster composed of concentric rings each containing a different kind of brood item. Eggs and first-instar larvae are placed in the middle, with increasingly large larvae in progressively bigger rings. Prepupae (pharate pupae) and pupae are consistently placed between the outermost ring (containing the largest larvae) and the ring of larvae of medium size. Similarly, some form of brood sorting is, as far as is known, almost universal in ants, even if it only involves forming piles of same-age brood within a single chamber (e.g. R.W. Taylor 1978; Dejean and Lachaud 1991). In *L. unifasciatus*, the spatial patterning of brood is rapidly recreated when colonies emigrate to a new nest. Therefore the adult ants actively make the patterns, and they are not just "growth rings" resulting from the production of eggs by the queen at the center.

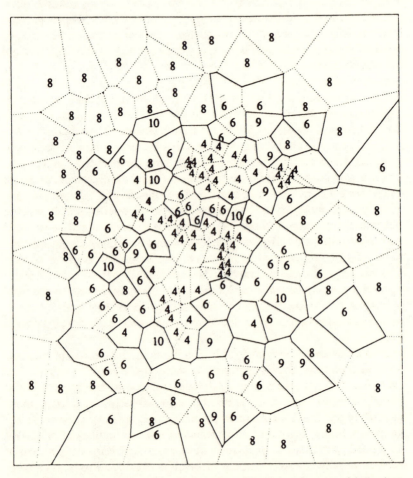

Figure 12.8 A map of the location of all the brood items in a flat nest of *Leptothorax unifasciatus*. The brood are sorted by type into concentric rings with eggs (labeled with the number 4) in the middle, medium larvae (6) around them, and the largest larvae (8) on the outside. Prepupae (9) and pupae (10) are in an intermediate position between the medium and large larvae. A Dirichlet tessellation designates a tile for each item (see text). (From Franks and Sendova-Franks 1992, Figure 2. Copyright © 1992 Springer-Verlag; by permission of Springer-Verlag and the authors)

In addition, in *Leptothorax unifasciatus*, each type of brood item is allocated a characteristic amount of space (Franks and Sendova-Franks 1992). This was shown by mapping the position of each item and then making a Dirichlet tessellation (Green and Sibson 1978) out of the map. This procedure allocates a polygonal tile to each item on the plane of the nest floor in such a way that the tile defines the space enclosing all points closest to that item (Figure 12.8). It was found that average tile area (excluding peripheral tiles that naturally have no outer boundaries) increased from the center of the cluster outwards. This is particularly intriguing because it means that tile area is not just a function of brood-item size. Thus, prepupae and pupae have tile areas intermediate in size between those of the medium and large larvae, and yet are among the largest brood items in the nest.

Franks and Sendova-Franks (1992) also showed that there is a significant positive correlation between the average tile area of each type of brood item and the item's predicted metabolic rate. They therefore suggested that the adult ants create a *domain of care* around each item proportional to its needs. This helps explain why the prepupae and pupae, which are large but require no feeding, have relatively small tile areas. The idea of a domain of care invokes the principle of a "domain of danger" (Hamilton 1971b, p. 296). In Hamilton's model of selfish herding behavior, prey cluster together to minimize personal risk associated with a predator that always moves towards the individual whose "tile area" (domain of danger) it happens to be inside. Now consider worker ants moving haphazardly about their dark nest. If they go to the nearest brood item whenever they have care to donate, they will always tend to apportion care according to the size of the tile area (domain of care), and hence appropriately to the requirements of each type of brood item.

By placing the most voracious brood items on the outside of the cluster, the ants may also be creating a hierarchy of care priorities. For example, nurse workers may first feed and groom the largest larvae (which are also the most accessible and valuable), and only move inwards repeating this automatic care-allocating process when these larvae are satiated. Stickland and Franks (1994), in testing the idea that the brood patterns in *Leptothorax* nests are associated with patterns of worker activity, showed that the larger brood do tend to be located in areas of greatest activity. Furthermore, experiments have shown that workers remain relatively faithful to certain areas of the nest, and to their relative positions, even after a disruptive emigration (Sendova-Franks and Franks 1993, 1994). This phenomenon, combined with the spatial patterning of brood, may serve to organize a flexible division of labor in which task fixation, learning, and skill acquisition all play a role, as the foraging for work algorithm supposes. Thus, it is possible that the spatial fidelity of workers is due to their learning particular roles and returning to familiar tasks at which they are most efficient. In any event, these studies confirm that the spatial arrangement of tasks and workers plays a central role in the division of labor in ants.

The mechanisms used by *Leptothorax* ants to sort their brood into concentric circles are not fully known. However, although the spatial arrangement of brood may be complex, the sorting mechanisms need not be. Deneubourg et al. (1991) developed a model showing that many individuals working independently can achieve brood sorting through self-organization (Figure 12.9). The model is based on simple rules that prescribe when to pick up or put down brood items as a function of the type and density of items already within the immediate neighborhood. These appear sufficient to enable ants to sort items by type into distinct clusters, as reported, for example, in *Odontomachus troglodytes* (Dejean and Lachaud 1991). These rules might also be employed by leptothoracines, but with an extra rule specifying that the size or perceived size of a brood item influences where it is placed.

RHYTHMIC ACTIVITY PATTERNS AND THE DIVISION OF LABOR

In *Leptothorax acervorum*, observations of the behavior of individual workers in the nest generate a fairly typical behavioral repertory for a small monomorphic myrmicine (Table 12.1). For example, on average, nest workers spend about 70% of their time resting. This dominant role of resting in the time budget of workers has been recorded many times in a wide variety of species (Hölldobler and Wilson 1990 pp. 342–343). However, as Franks and Bryant (1987) were first to show, resting periods of the majority of individuals in the nest may coincide, so that workers tend to be either inactive together or active together. Such is the synchronicity of the ants that frequently the whole workforce shows highly rhythmical behavior. In *L. acervorum*, periods of inactivity tend to be about 15 minutes long, and are followed by bursts of animation for 2 to 5 minutes. The outcome is a series of activity cycles with a period of about 20 minutes (Franks et al. 1990b; Figure 12.10). Cole (1991, 1992) reported similar activity rhythms in *L. allardycei*.

Almost all social interactions within the nest, including food exchange, grooming, and brood care, occur during the short, synchronized bursts of activity. Franks and Bryant (1987) therefore speculated that these serve for the exchange of timely and accurate information within the nest. In light of this, Hemerick et al. (1990) produced a model to explore whether the ants generate synchronized activity

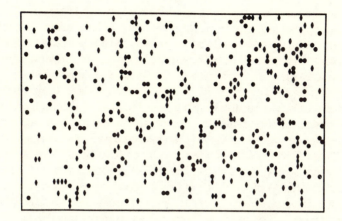

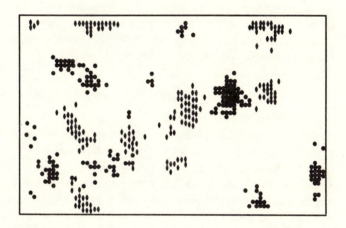

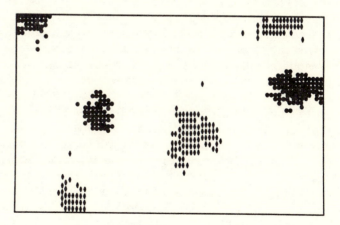

because they aim to exchange information about energy levels within the colony, while maximizing energy conservation. However, changing energy levels appear unlikely to be driving the cycles, first because the cycles seem insensitive to colony starvation, and second because energy levels change over too long a timescale (Tofts et al. 1992).

An alternative approach is to consider a mechanism enabling the ants to synchronize with one another, and then to consider the possible adaptive significance of activity rhythms. Goss and Deneubourg (1988) and Tofts et al. (1992) showed that synchronized waves of activity could be generated by an autocatalytic process. (In chemical systems, an auto-catalytic reaction is one that proceeds by positive feedback, because its products act as its own catalyst.) In the models of these authors, in-active ants that were resting after a bout of movement had an increasing probability of "waking up" or being woken as time passed. Periods of synchronized activity were generated spontaneously because waking ants tended to wake others (by running into them), so initiating a "chain reaction" in which yet more ants were woken. Tofts et al. (1992) also found that such self-organizing synchronicity is insensitive to the size of the worker population within the nest.

Activity rhythms could be adaptive for the colony by promoting an equitable distribution of workers over the necessary tasks. Assume, as a null hypothesis, that individual nurse workers become active at random relative to one another. When active, each worker looks for a brood item to groom or feed. Unless it leaves a mark on recently tended brood, some items may receive too much care and others too little. However, if there is synchronized activity, nurse workers will care for brood simultaneously, and each item being tended will automatically redirect other ants to untended brood. A model showed that brood would indeed be neglected very rarely when activity is synchronized, whereas brood risked death from lack of care if the ants were active at random (Hatcher et al. 1992).

Figure 12.9 (*facing page*)   A simulation of self-organized sorting of two types of brood item (represented by circles and ellipses respectively) based on simple local rules dictating the probabilities of ants picking up or putting down brood items in response to local densities of similar or dissimilar items. Starting with the top figure, small evenly spaced clusters of each type of object form and later merge into fewer larger clusters, yielding a high degree of sorting (bottom figure). (From Deneubourg et al. 1991; by permission of MIT Press)

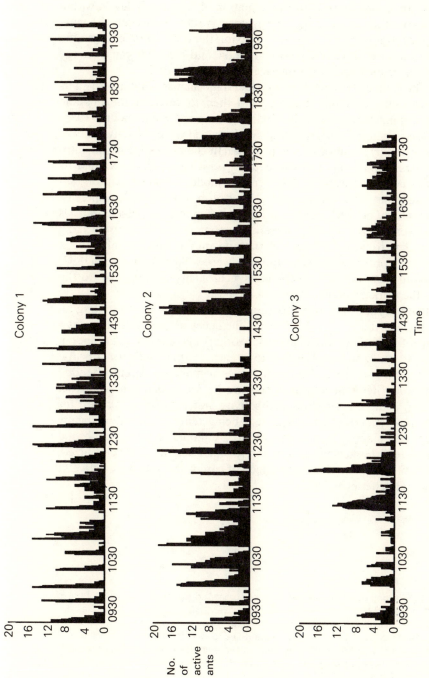

Figure 12.10 Activity time series for 20 workers in each of three colonies of *Leptothorax acervorum*, obtained with time-lapse video recording. (Reprinted from Franks et al. 1990b. Copyright © 1990; by kind permission of Elsevier Science Ltd)

12.7 Ant Colonies, Adaptive Redundancy, Complexity, and Organizations

Task allocation and role specialization among the members of animal societies can be seen as a continuation of the evolution of a division of labor that first began at the subcellular level (E.O. Wilson 1985a; Bonner 1988). The parallels at such different levels of biological organization are striking and potentially informative (cf. Sections 2.5, 2.6). For example, controversy still remains over the extent to which morphogenesis in multicellular organisms is controlled by hierarchical processes or localized self-organizing interactions (Goodwin and Cohen 1969; Wolpert 1974; J.D. Murray 1989). Increasingly, the importance of pattern formation through localized interactions is being emphasized in morphogenesis.

Consider the extreme example of the vertebrate brain. Instead of all the neurones and their connections being specified in the genetic code, it is much more likely that the genome specifies a far simpler set of growth rules for neurones. Thus, although the brain is ultimately the product of selfish genes, its structure and development are under decentralized control (e.g. Dawkins 1976). Moreover, it is believed that, as the brain starts to function, those neurones and synapses that participate in the function remain stable and become permanent, while the rest wither away (Bonner 1988; Edelman 1992).

We have suggested that genetic programming for flexibility is also at the heart of the division of labor in many ant colonies. Worker ants, especially young ones, may experiment with different social functions and temporarily delete those from their repertoire that are least effective in the current social context. Such event-driven programming might have the huge advantage, in both brain design and social design, of a combination of self-correction, automatic repair, flexibility, and robustness far beyond that possible with genetic predetermination. Analogies between brains and ant colonies have also been suggested by Meyer (1966), Hofstadter (1980), Minsky (1988), Franks (1989a), and Gordon et al. (1992).

A potentially powerful idea about the functional organization of brains is the suggestion that they have a high level of redundancy reminiscent of that seen in a holograph (Pribram 1976). A curious property of holographs is that, if one is smashed, each fragment can be used to reconstruct the whole image. The stimulus for comparing brains with holographs came from the finding that rats can run mazes they have learnt even after parts of their brains have been removed (Lashley 1950). This implied that certain aspects of memory are highly distributed within the structure of the brain. In other words, holography demonstrates, and the findings of Lashley (1950) suggest, that it is possible to create structures where the whole can be enfolded in the parts, so that each part somehow resembles the whole.

This holographic metaphor can be extended to the self-organizing ant colony. For one thing, the main "parts" of a colony (namely the workers, including different physical worker castes if present) certainly resemble each other to the extent that their differences are not based on dissimilar genetic information. In this sense they also resemble the queen, from which a whole colony may develop. In addition, some of the structural and behavioral patterns of workers do not depend on their numbers. Thus, T.R. Stickland and N.R. Franks (unpublished observations) found that a single *Leptothorax unifasciatus* worker could construct the concentric brood array typical of this species (Section 12.6). Similarly, the rhythmic activity patterns of *Leptothorax* ants appeared independent of colony size (Tofts et al. 1992; Section 12.6). Therefore, it seems that many ant colonies have such robust organizational algorithms that they can be reconstructed from just a few of their fragments. In fact, as earlier argued, this may be an essential feature, since ant colonies typically experience many sizes during their lifetimes.

Simon (1947) has suggested (and anyone with experience of them would agree) that human organizations can never be perfectly rational. People have to act on the basis of incomplete information about possible courses of action and their consequences. They can also only explore a limited number of alternatives relating to any given decision. Hence, in human societies, often the best that can be realistically achieved is a limited rationality involving "good-enough" decisions based on simple rules of thumb and limited investigation and information. In addition, it has been argued (Morgan 1986 pp. 84–95) that in human organization it is best openly to accept error, uncertainty, flexibility, and pluralism as inevitable features of life in complex and changing environments. In other words, we might do well to recognize the importance of inquiry-driven action.

As we have argued, these principles apply to a greater or lesser extent to self-organizing insect societies. For example, Morgan (1986 p. 98), considering human management systems, suggested that "Any system with an ability to self-organize must have an element of redundancy: a form of excess capacity which, appropriately designed and used, creates room for maneuver." Referring to the ideas of Emery (1967, 1969), he went on to suggest that the most useful form of such redundancy was a redundancy of functions and not just a redundancy of parts. Ants utilize both. Their worker populations combine a redundancy of parts (more workers than are needed at any one time) with a redundancy of functions (each worker can accomplish a wider range of tasks than it might be called upon to perform during its lifetime). In this way, spare parts (extra workers) are added to the system, and extra functions (potential skills) are added to all of the operating parts, so that each part is able to engage in a range of functions rather than just perform a single specialized activity. This again helps explain why most ant species have monomorphic, multi-skilled workers, and why in polymorphic worker populations the generalist caste always forms a great majority. In sum, much of relevance to a general theory of organizations can be learnt from the study of adaptive redundancy and self-organization in ant colonies (Franks 1995).

12.8 Conclusion

This chapter has argued that the assumption that the division of labor in ants and other social insects is based on a causal link between age and task has little justification either theoretically or empirically. Therefore, the term age polyethism should be dropped in favor of the more neutral temporal polyethism. This describes the process of task allocation changing over time, without implying that it is caused by ageing.

The foraging for work algorithm (Tofts and Franks 1992; Tofts 1993; Franks and Tofts 1994) is a self-organization model of the division of labor. According to this model, naive young workers perform the first jobs they encounter. Since similar jobs are likely to be in the same area (due to a high degree of spatial order within nests), workers are likely to repeat tasks and remain in that area. But if no task is available, or other individuals displace them, workers search for something else to do. Hence, the system as a whole pushes the unemployed towards zones where tasks are waiting. This automatic allocation of workers to tasks is the element of self-organization.

The foraging for work algorithm also predicts, as a side effect only, the existence of centrifugal polyethism involving a weak correlation between age and task. Therefore, future work should not, for example, take correlations between age and task as evidence for a causal link between the two. Instead, it should focus on long-term studies of large samples of individually-marked workers, and be sensitive to both their behavioral flexibility and spatial location.

The advantages of a self-organized system such as foraging for work are that it is decentralized, robust to perturbations, largely independent of colony size and age structure, and based on relatively few, simple behavioral rules. Such a system is more suitable for ant colonies in nature, at both the proximate and ultimate levels, than one based on complex, hardwired preprograming. Nevertheless, models show that foraging for work, along with other forms of self-organization, can generate complex patterns and behaviors. This is a general feature of self-organization, and one that is beginning to be recognized as underlying many types of complex biological system.

12.9 Summary

1. The division of labor among ant workers is a key feature of their efficient functioning and hence of their ecological success. Division of labor is a very general principle in biology and beyond. It is applicable, for example, to problems in both developmental biology and economics.

2. In many social insects, including ants, the tasks that workers perform tend to change as the workers grow older. However, the correlation between age and task is often weak, and a causal relation between the two has not been established. Moreover, detailed observations reveal that workers frequently show a great deal of flexibility. Individuals change the tasks they perform in response to changing task demand, irrespective of age. These findings call into question the idea that the division of labor is controlled by workers rigidly performing tasks according to their age. Nevertheless, this idea forms the basis of the theory that the age structure of a colony is a colony-level adaptation for an efficient division of labor (adaptive demography). Such a system also seems highly unsuited for dealing with any catastrophic loss of worker age-classes.

3. Another commonly observed pattern is that young workers tend to work in the nest as nurses, and old ones outside it as foragers (centrifugal polyethism). In species with worker reproduction, this could be because workers lay eggs in the safety of the nest when young, and increasingly expose themselves to risk by foraging as they senesce. However, this explanation does not address the detailed, proximate control of the division of labor.

4. An alternative approach to the division of labor proposes that it arises through a process of self-organization. A model of this process, the foraging for work algorithm, was presented by Tofts and Franks (1992), Tofts (1993), and Franks and Tofts (1994). It assumes that workers are ignorant of their age and that tasks in the nest are spatially ordered (for example, because the brood is kept in a consistent pattern). Young workers perform the first task they encounter, which is therefore usually nursing brood. Unemployed workers wander until they find a task. The model showed that, even with simple assumptions such as these, workers would be efficiently allocated to the available tasks. Furthermore, this self-organized division of labor was robust to

the presence of lazy workers, task-fixated workers, physically special-ized workers (physical castes), and absent age classes. Therefore, the foraging for work algorithm predicts both behaviorally flexible workers, and a division of labor that is flexible and fault-tolerant. Since the abili-ty to cope with environmental mishaps would be of great value to colonies in nature, these findings lend further support to the foraging for work algorithm.

5. The foraging for work algorithm generates centrifugal polyethism, and a weak correlation between age and task, as by-products. This is because workers tend to drift, with time, to tasks further and further from the center of the nest. The algorithm suggests that correlations between genotype and task may also be by-products, rather than evi-dence for the genetic determination of worker occupation.

6. The use of tabulations of worker behaviors and of their transition probabilities (behavioral repertories and ethograms) to study the divi-sion of labor may obscure essential individual variation and flexibility. Many previous division of labor studies are consistent with the assump-tions and predictions of the foraging for work algorithm.

7. Recent work on *Leptothorax* ants has shown that brood is posi-tioned in the nest in a highly predictable manner. This spatial pattern-ing is as assumed by foraging for work. Furthermore, it may promote the efficient care of brood, and itself be generated by a self-organization process. This could also be true of the characteristic short-term activity rhythms of *Leptothorax* ants.

8. Self-organization, with its ability to generate complex structure from simple processes, and its associated flexibility and robustness, may underlie not only the division of labor and other phenomena in ant colonies, but also the origin of complexity in many other types of com-plex system. Self-organization is entirely consistent with the evolution of adaptive complexity by natural selection.

References

Agosti, D., and E. Hauschteck-Jungen. 1987. Polymorphism of males in *Formica exsecta* Nyl. (Hym.: Formicidae). *Insectes Sociaux* 34: 280–290.

Alexander, R.D. 1974. The evolution of social behavior. *Annual Review of Ecology and Systematics* 5: 325–383.

Alexander, R.D. 1979. *Darwinism and Human Affairs*. University of Washington Press, Seattle.

Alexander, R.D., and G. Borgia. 1978. Group selection, altruism, and the levels of organization of life. *Annual Review of Ecology and Systematics* 9: 449–474.

Alexander, R.D., and P.W. Sherman. 1977. Local mate competition and parental investment in social insects. *Science* 196: 494–500.

Alexander, R.D., K.M. Noonan, and B.J. Crespi. 1991. The evolution of eusociality. In P.W. Sherman, J.U.M. Jarvis, and R.D. Alexander, eds., *The Biology of the Naked Mole-Rat*, pp. 3–44. Princeton University Press, Princeton, New Jersey.

Alloway, T.M., and M.G. Del Rio Pesado. 1983. Behavior of the slave-making ant, *Harpagoxenus americanus* (Emery), and its host species under "seminatural" laboratory conditions (Hymenoptera: Formicidae). *Psyche* 90: 425–436.

Alloway, T.M., A. Buschinger, M. Talbot, R. Stuart, and C. Thomas. 1982. Polygyny and polydomy in three North American species of the ant genus *Leptothorax* Mayr (Hymenoptera: Formicidae). *Psyche* 89: 249–274.

Andersson, M. 1984. The evolution of eusociality. *Annual Review of Ecology and Systematics* 15: 165–189.

Andersson, M. 1994. *Sexual Selection*. Princeton University Press, Princeton, New Jersey.

Antolin, M.F. 1992. Sex ratio variation in a parasitic wasp. II. Diallel cross. *Evolution* 46: 1511–1524.

Antolin, M.F. 1993. Genetics of biased sex ratios in subdivided populations: models, assumptions, and evidence. In D. Futuyma and J. Antonovics, eds., *Oxford Surveys in Evolutionary Biology*, Vol. 9, pp. 239–281. Oxford University Press, Oxford.

Aoki, K. 1981. Algebra of inclusive fitness. *Evolution* 35: 659–663.

Aoki, K. 1982. Additive polygenic formulation of Hamilton's model of kin selection. *Heredity* 49: 163–169.

Aoki, K., and M. Moody. 1981. One- and two-locus models of the origin of worker behavior in Hymenoptera. *Journal of Theoretical Biology* 89: 449–474.

Aoki, S. 1987. Evolution of sterile soldiers in aphids. In Y. Itô, J.L. Brown, and J. Kikkawa, eds., *Animal Societies: Theories and Facts*, pp. 53–65. Japan Scientific Societies Press, Tokyo.

Arnold, A.J., and K. Fristrup. 1982. The theory of evolution by natural selection: a hierarchical expansion. *Paleobiology* 8: 113–129.

Aron, S., L. Passera, and L. Keller. 1994. Queen-worker conflict over sex ratio: A comparison of primary and secondary sex ratios in the Argentine ant, *Iridomyrmex humilis*. *Journal of Evolutionary Biology* 7: 403–418.

Aron, S., E.L. Vargo, and L. Passera. 1995. Primary and secondary sex ratios in monogyne colonies of the fire ant, *Solenopsis invicta*. *Animal Behaviour* 49: 749–757.

Avilés, L. 1993. Interdemic selection and the sex ratio: a social spider perspective. *American Naturalist* 142: 320–345.

Axelrod, R. 1984. *The Evolution of Cooperation*. Basic Books Inc., New York.

Axelrod, R., and W.D. Hamilton. 1981. The evolution of cooperation. *Science* 211: 1390–1396.

Backus, V.L. 1993. Packaging of offspring by nests of the ant, *Leptothorax longispinosus*: parent-offspring conflict and queen-worker conflict. *Oecologia* 95: 283–289.

Backus, V.L., and J.M. Herbers. 1992. Sexual allocation ratios in forest ants: food limitation does not explain observed patterns. *Behavioral Ecology and Sociobiology* 30: 425–429.

Barnard, C. 1991. Kinship and social behaviour: the trouble with relatives. *Trends in Ecology and Evolution* 6: 310–312.

Baroni Urbani, C. 1968. Monogyny in ant societies. *Zoologischer Anzeiger* 181: 269–277.

Baroni Urbani, C. 1989a. Phylogeny and behavioural evolution in ants, with a discussion of the role of behaviour in evolutionary processes. *Ethology, Ecology and Evolution* 1: 137–168.

Baroni Urbani, C. 1989b. On a singular case of kin selection theory among Hymenoptera: justified fratricide and the natural way to avoid it. *Ethology, Ecology and Evolution* 1: 373–378.

Bartels, P.J. 1985. Field observations of multiple matings in *Lasius alienus* Foerster (Hymenoptera: Formicidae). *American Midland Naturalist* 113: 190–192.

Bartz, S.H. 1982. On the evolution of male workers in the Hymenoptera. *Behavioral Ecology and Sociobiology* 11: 223–228.

Bartz, S.H., and B. Hölldobler. 1982. Colony founding in *Myrmecocystus mimicus* Wheeler (Hymenoptera: Formicidae) and the evolution of foundress associations. *Behavioral Ecology and Sociobiology* 10: 137–147.

Bateson, P. 1982. Behavioural development and evolutionary processes. In King's College Sociobiology Group, ed., *Current Problems in Sociobiology*, pp. 133–151. Cambridge University Press, Cambridge.

Bateson, P. 1994. The dynamics of parent-offspring relationships in mammals. *Trends in Ecology and Evolution* 9: 399–403.

Beckers, R., S. Goss, J.-L. Deneubourg, and J.M. Pasteels. 1989. Colony size, communication and ant foraging strategy. *Psyche* 96: 239–256.

Begon, M., J.L. Harper, and C.R. Townsend. 1986. *Ecology: Individuals, Populations and Communities*. Blackwell Scientific Publications, Oxford.

Benford, F.A. 1978. Fisher's theory of the sex ratio applied to the social Hymenoptera. *Journal of Theoretical Biology* 72: 701–727.

Bennett, B. 1987a. Measures of relatedness. *Ethology* 74: 219–236.

Bennett, B. 1987b. Ecological differences between monogynous and polygynous sibling ant species (Hymenoptera: Formicidae). *Sociobiology* 13: 249–270.

Benton, T.G., and W.A. Foster. 1992. Altruistic housekeeping in a social aphid. *Proceedings of the Royal Society of London, Series B* 247: 199–202.

Bertram, B.C.R. 1982. Problems with altruism. In King's College Sociobiology Group, ed., *Current Problems in Sociobiology*, pp. 251–267. Cambridge University Press, Cambridge.

Beshers, S.N., and J.F.A. Traniello. 1994. The adaptiveness of worker demography in the attine ant *Trachymyrmex septentrionalis*. *Ecology* 75: 763–775.

Blick, J. 1977. Selection for traits which lower individual reproduction. *Journal of Theoretical Biology* 67: 597–601.

Blows, M.W., and M.P. Schwarz. 1991. Spatial distribution of a primitively social bee: does genetic population structure facilitate altruism? *Evolution* 45: 680–693.

Bodmer, W.F., and A.W.F. Edwards. 1960. Natural selection and the sex ratio. *Annals of Human Genetics* 24: 239–244.

Bolton, B. 1986. Apterous females and shift of dispersal strategy in the *Monomorium salomonis*-group (Hymenoptera: Formicidae). *Journal of Natural History* 20: 267–272.

Bolton, B. 1988. A new socially parasitic *Myrmica*, with a reassessment of the genus (Hymenoptera: Formicidae). *Systematic Entomology* 13: 1–11.

Bolton, B. 1990. Army ants reassessed: the phylogeny and classification of the doryline section (Hymenoptera, Formicidae). *Journal of Natural History* 24: 1339–1364.

Bonner, J.T. 1988. *The Evolution of Complexity by Means of Natural Selection*. Princeton University Press, Princeton, New Jersey.

Boomsma, J.J. 1988. Empirical analysis of sex allocation in ants: from descriptive surveys to population genetics. In G. de Jong, ed., *Population Genetics and Evolution*, pp. 42–51. Springer-Verlag, Berlin.

Boomsma, J.J. 1989. Sex-investment ratios in ants: has female bias been systematically overestimated? *American Naturalist* 133: 517–532.

Boomsma, J.J. 1990. How to capitalize on relative relatedness asymmetry? In G.K. Veeresh, B. Mallik, and C.A. Viraktamath, eds., *Social Insects and the Environment*, pp. 349–350. Oxford & IBH Publishing Co., New Delhi.

Boomsma, J.J. 1991. Adaptive colony sex ratios in primitively eusocial bees. *Trends in Ecology and Evolution* 6: 92–95.

Boomsma, J.J. 1993. Sex ratio variation in polygynous ants. In L. Keller, ed., *Queen Number and Sociality in Insects*, pp. 86–109. Oxford University Press, Oxford.

Boomsma, J.J., and G.C. Eickwort. 1993. Colony structure, provisioning and sex allocation in the sweat bee *Halictus ligatus* (Hymenoptera: Halictidae). *Biological Journal of the Linnean Society* 48: 355–377.

Boomsma, J.J., and A. Grafen. 1990. Intraspecific variation in ant sex ratios and the Trivers-Hare hypothesis. *Evolution* 44: 1026–1034.

Boomsma, J.J., and A. Grafen. 1991. Colony-level sex ratio selection in the eusocial Hymenoptera. *Journal of Evolutionary Biology* 3: 383–407.

Boomsma, J.J., and J.A. Isaaks. 1985. Energy investment and respiration in queens and males of *Lasius niger* (Hymenoptera: Formicidae). *Behavioral Ecology and Sociobiology* 18: 19–27.

Boomsma, J.J., and A. Leusink. 1981. Weather conditions during nuptial flights of four European ant species. *Oecologia* 50: 236–241.

Boomsma, J.J., G.A. Van der Lee, and T.M. Van der Have. 1982. On the production ecology of *Lasius niger* (Hymenoptera: Formicidae) in successive coastal dune valleys. *Journal of Animal Ecology* 51: 975–991.

Boomsma, J.J., A.H. Brouwer, and A.J. Van Loon. 1990. A new polygynous *Lasius* species (Hymenoptera: Formicidae) from Central Europe. II. Allozymatic confirmation of species status and social structure. *Insectes Sociaux* 37: 363–375.

Boorman, S.A., and P.R. Levitt. 1980. *The Genetics of Altruism.* Academic Press, New York.

Bourke, A.F.G. 1988a. Worker reproduction in the higher eusocial Hymenoptera. *Quarterly Review of Biology.* 63: 291–311.

Bourke, A.F.G. 1988b. Dominance orders, worker reproduction, and queen-worker conflict in the slave-making ant *Harpagoxenus sublaevis. Behavioral Ecology and Sociobiology* 23: 323–333.

Bourke, A.F.G. 1989. Comparative analysis of sex-investment ratios in slave-making ants. *Evolution* 43: 913–918.

Bourke, A.F.G. 1991. Queen behaviour, reproduction and egg cannibalism in multiple-queen colonies of the ant *Leptothorax acervorum. Animal Behaviour* 42: 295–310.

Bourke, A. 1992. Adaptive significance of avian helping behaviour. *Trends in Ecology and Evolution* 7: 102.

Bourke, A.F.G. 1993. Lack of experimental evidence for pheromonal inhibition of reproduction among queens in the ant *Leptothorax acervorum. Animal Behaviour* 45: 501–509.

Bourke, A.F.G. 1994a. Worker matricide in social bees and wasps. *Journal of Theoretical Biology* 167: 283–292.

Bourke, A.F.G. 1994b. Indiscriminate egg cannibalism and reproductive skew in a multiple-queen ant. *Proceedings of the Royal Society of London, Series B* 255: 55–59.

Bourke, A.F.G., and G.L. Chan. 1994. Split sex ratios in ants with multiple mating. *Trends in Ecology and Evolution* 9: 120–122.

Bourke, A.F.G., and N.R. Franks. 1991. Alternative adaptations, sympatric speciation and the evolution of parasitic, inquiline ants. *Biological Journal of the Linnean Society* 43: 157–178.

Bourke, A.F.G., and J. Heinze. 1994. The ecology of communal breeding: the case of multiple-queen leptothoracine ants. *Philosophical Transactions of the Royal Society, Series B* 345: 359–372.

Bourke, A.F.G., T.M. Van der Have, and N.R. Franks. 1988. Sex ratio determination and worker reproduction in the slave-making ant *Harpagoxenus sublaevis*. *Behavioral Ecology and Sociobiology* 23: 233–245.

Boyd, R., and P.J. Richerson. 1980. Effect of phenotypic variation on kin selection. *Proceedings of the National Academy of Sciences, U.S.A.* 77: 7506–7509.

Brandon, R.N. 1985. Adaptation explanations: are adaptations for the good of replicators or interactors? In D.J. Depew and B.H. Weber, eds., *Evolution at a Crossroads: The New Biology and the New Philosophy of Science*, pp. 81–96. MIT Press, Cambridge, Massachusetts.

Brandon, R.N. 1990. *Adaptation and Environment*. Princeton University Press, Princeton, New Jersey.

Brandon, R.N., and R.M. Burian (eds.) 1984. *Genes, Organisms, Populations: Controversies over the Units of Selection*. MIT Press, Cambridge, Massachusetts.

Breden, F. 1990. Partitioning of covariance as a method for studying kin selection. *Trends in Ecology and Evolution* 5: 224–228.

Breed, M.D., and B. Bennett. 1987. Kin recognition in highly eusocial insects. In D.J.C. Fletcher and C.D. Michener, eds., *Kin Recognition in Animals*, pp. 243–285. Wiley, Chichester, UK.

Breed, M.D., and J.M. Harrison. 1988. Worker size, ovary development and division of labor in the giant tropical ant, *Paraponera clavata* (Hymenoptera: Formicidae). *Journal of the Kansas Entomological Society* 61: 285–291.

Brian, M.V. 1955a. Food collection by a Scottish ant community. *Journal of Animal Ecology* 24: 336–351.

Brian, M.V. 1955b. Studies of caste differentiation in *Myrmica rubra* L. 3. Larval dormancy, winter size and vernalisation. *Insectes Sociaux* 2: 85–114.

Brian, M.V. 1956. The natural density of *Myrmica rubra* and associated ants in West Scotland. *Insectes Sociaux* 3: 473–487.

Brian, M.V. 1957. The growth and development of colonies of the ant *Myrmica*. *Insectes Sociaux* 4: 177–190.

Brian, M.V. 1969. Male production in the ant *Myrmica rubra* L. *Insectes Sociaux* 16: 249–268.

Brian, M.V. 1973a. Caste control through worker attack in the ant *Myrmica*. *Insectes Sociaux* 20: 87–102.

Brian, M.V. 1973b. Queen recognition by brood-rearing workers of the ant *Myrmica rubra* L. *Animal Behaviour* 21: 691–698.

Brian, M.V. 1975. Larval recognition by workers of the ant *Myrmica*. *Animal Behaviour* 23: 745–756.

Brian, M.V. 1979a. Caste differentiation and division of labor. In H.R. Hermann, ed., *Social Insects*, Vol. I, pp. 121–222. Academic Press, New York.

Brian, M.V. 1979b. Habitat differences in sexual production by two co-existent ants. *Journal of Animal Ecology* 48: 943–953.

Brian, M.V. 1980. Social control over sex and caste in bees, wasps and ants. *Biological Reviews* 55: 379–415.

Brian, M.V. 1981. Treatment of male larvae in ants of the genus *Myrmica*. *Insectes Sociaux* 28: 161–166.

Brian, M.V. 1983. *Social Insects: Ecology and Behavioural Biology*. Chapman and Hall, London.

Brian, M.V. 1986. The distribution, sociability and fecundity of queens in normal groups of the polygyne ant *Myrmica rubra* L. *Insectes Sociaux* 33: 118–131.

Brian, M.V. 1988. The behaviour and fecundity of queens of different ages in synthetic groups of *Myrmica rubra* L. with different worker populations. *Insectes Sociaux* 35: 153–166.

Brian, M.V. 1989. Social factors affecting queen fecundity in the ant *Myrmica rubra*. *Physiological Entomology* 14: 381–389.

Brian, M.V., and A.D. Brian. 1949. Observations on the taxonomy of the ants *Myrmica rubra* L. and *M. laevinodis* Nylander (Hymenoptera: Formicidae). *Transactions of the Royal Entomological Society of London* 100: 393–409.

Brian, M.V., and A.D. Brian. 1951. Insolation and ant population in the west of Scotland. *Transactions of the Royal Entomological Society of London* 102: 303–330.

Brian, M.V., and A.D. Brian. 1955. On the two forms macrogyna and microgyna of the ant *Myrmica rubra* L. *Evolution* 9: 280–290.

Brian, M.V., and C.A.H. Carr. 1960. The influence of the queen on brood rearing in ants of the genus *Myrmica*. *Journal of Insect Physiology* 5: 81–94.

Brian, M.V., and C. Rigby. 1978. The trophic eggs of *Myrmica rubra* L. *Insectes Sociaux* 25: 89–110.

Brian, M.V., R.T. Clarke, and R.M. Jones. 1981. A numerical model of an ant society. *Journal of Animal Ecology* 50: 387–405.

Briese, D.T. 1983. Different modes of reproductive behaviour (including a description of colony fission) in a species of *Chelaner* (Hymenoptera: Formicidae). *Insectes Sociaux* 30: 308–316.

Bristow, C.M., D. Cappaert, N.J. Campbell, and A. Heise. 1992. Nest structure and colony cycle of the Allegheny mound ant, *Formica exsectoides* Forel (Hymenoptera: Formicidae). *Insectes Sociaux* 39: 385–402.

Brockmann, H.J. 1984. The evolution of social behaviour in insects. In J.R. Krebs and N.B. Davies, eds., *Behavioural Ecology: An Evolutionary Approach*, 2nd edition, pp. 340–361. Blackwell, Oxford.

Brockmann, H.J. 1990. Primitive eusociality: comparisons between Hymenoptera and vertebrates. In G.K. Veeresh, B. Mallik, and C.A. Viraktamath, eds., *Social Insects and the Environment*, p. 77. Oxford & IBH Publishing Co., New Delhi.

Brockmann, H.J., and A. Grafen. 1989. Mate conflict and male behaviour in a solitary wasp, *Trypoxylon* (*Trypargilum*) *politum* (Hymenoptera: Sphecidae). *Animal Behaviour* 37: 232–255.

Brockmann, H.J., and A. Grafen. 1992. Sex ratios and life-history patterns of a solitary wasp, *Trypoxylon* (*Trypargilum*) *politum* (Hymenoptera: Sphecidae). *Behavioral Ecology and Sociobiology* 30: 7–27.

Bruckner, D. 1982. Different acceptance of foreign queens by colonies of *Leptothorax acervorum*. In M.D. Breed, C.D. Michener, and H.E. Evans, eds., *The Biology of Social Insects*, pp. 329–330. Westview Press, Boulder, Colorado.

Bull, J.J. 1983. *Evolution of Sex Determining Mechanisms*. Benjamin/Cummings, Menlo Park, California.

Bull, J.J., and E.L. Charnov. 1988. How fundamental are Fisherian sex ratios? In P.H. Harvey and L. Partridge, eds., *Oxford Surveys in Evolutionary Biology*, Vol. 5, pp. 96–135. Oxford University Press, Oxford.

Bulmer, M.G. 1981. Worker-queen conflict in annual social Hymenoptera. *Journal of Theoretical Biology* 93: 239–251.

Bulmer, M.G. 1983a. Sex ratio evolution in social Hymenoptera under worker control with behavioral dominance. *American Naturalist* 121: 899–902.

Bulmer, M.G. 1983b. Sex ratio theory in social insects with swarming. *Journal of Theoretical Biology* 100: 329–339.

Bulmer, M. 1986a. Sex ratios in geographically structured populations. *Trends in Ecology and Evolution* 1: 35–38.

Bulmer, M.G. 1986b. Sex ratio theory in geographically structured populations. *Heredity* 56: 69–73.

Bulmer, M.G., and P.D. Taylor. 1980. Dispersal and the sex ratio. *Nature* 284: 448–449.

Bulmer, M.G., and P.D. Taylor. 1981. Worker-queen conflict and sex ratio theory in social Hymenoptera. *Heredity* 47: 197–207.

Buschinger, A. 1966a. Untersuchungen an *Harpagoxenus sublaevis* Nyl. (Hym., Formicidae). I. Freilandbeobachtungen zu Verbreitung und Lebensweise. *Insectes Sociaux* 13: 5–16.

Buschinger, A. 1966b. Untersuchungen an *Harpagoxenus sublaevis* Nyl. (Hym., Formicidae). II. Haltung und Brutaufzucht. *Insectes Sociaux* 13: 311–322.

Buschinger, A. 1968a. Mono- und Polygynie bei Arten der Gattung *Leptothorax* Mayr (Hymenoptera, Formicidae). *Insectes Sociaux* 15: 217–225.

Buschinger, A. 1968b. Untersuchungen an *Harpagoxenus sublaevis* Nyl. (Hymenoptera, Formicidae). III. Kopula, Koloniegründung, Raubzüge. *Insectes Sociaux* 15: 89–104.

Buschinger, A. 1971. "Locksterzeln" und Kopula der sozialparasitischen Ameise *Leptothorax kutteri* Buschinger (Hym., Form.). *Zoologischer Anzeiger* 186: 242–248.

Buschinger, A. 1972. Kreuzung zweier sozialparasitischer Ameisenarten, *Doronomyrmex pacis* Kutter und *Leptothorax kutteri* Buschinger (Hym., Formicidae). *Zoologischer Anzeiger* 189: 169–179.

Buschinger, A. 1974a. Monogynie und Polygynie in Insektensozietäten. In G.H. Schmidt, ed., *Sozialpolymorphismus bei Insekten*, pp. 862–896. Wissenschaftliche Verlagsgesellschaft MBH, Stuttgart.

Buschinger, A. 1974b. Experimente und Beobachtungen zur Gründung und Entwicklung neuer Sozietäten der sklavenhaltenden Ameise *Harpagoxenus sublaevis* (Nyl). *Insectes Sociaux* 21: 381–406.

Buschinger, A. 1979. Functional monogyny in the American guest ant *Formicoxenus hirticornis* (Emery) (= *Leptothorax hirticornis*), (Hym., Form.). *Insectes Sociaux* 26: 61–68.

Buschinger, A. 1981. Biological and systematic relationships of social parasitic Leptothoracini from Europe and North America. In P.E. Howse and J.-L. Clément, eds., *Biosystematics of Social Insects*, pp. 211–222. Academic Press, London.

Buschinger, A. 1983. Sexual behavior and slave raiding of the dulotic ant,

Harpagoxenus sublaevis (Nyl.) under field conditions (Hym., Formicidae). *Insectes Sociaux* 30: 235–240.

Buschinger, A. 1987. Introduction to symposium on biosystematics of the ant tribe Leptothoracini. In J. Eder and H. Rembold, eds., *Chemistry and Biology of Social Insects*, pp. 27–28. Verlag J. Peperny, München.

Buschinger, A. 1989a. Evolution, speciation, and inbreeding in the parasitic ant genus *Epimyrma* (Hymenoptera, Formicidae). *Journal of Evolutionary Biology* 2: 265–283.

Buschinger, A. 1989b. Workerless *Epimyrma kraussei* Emery 1915, the first parasitic ant of Crete. *Psyche* 96: 69–74.

Buschinger, A. 1990a. Sympatric speciation and radiative evolution of socially parasitic ants - Heretic hypotheses and their factual background. *Zeitschrift für Zoologische Systematik und Evolutions-forschung* 28: 241–260.

Buschinger, A. 1990b. Regulation of worker and queen formation in ants with special reference to reproduction and colony development. In W. Engels, ed., *Social Insects: An Evolutionary Approach to Castes and Reproduction*, pp. 37–57. Springer-Verlag, Berlin.

Buschinger, A., and T.M. Alloway. 1979. Sexual behaviour in the slave-making ant, *Harpagoxenus canadensis* M.R. Smith, and sexual pheromone experiments with *H. canadensis*, *H. americanus* (Emery), and *H. sublaevis* (Nylander) (Hymenoptera; Formicidae). *Zeitschrift für Tierpsychologie* 49: 113–119.

Buschinger, A., and A. Francoeur. 1991. Queen polymorphism and functional monogyny in the ant, *Leptothorax sphagnicolus* Francoeur. *Psyche* 98: 119–133.

Buschinger, A., and E. Pfeifer. 1988. Effects of nutrition on brood production and slavery in ants (Hymenoptera, Formicidae). *Insectes Sociaux* 35: 61–69.

Buschinger, A., and U. Winter. 1976. Funktionelle Monogynie bei der Gastameise *Formicoxenus nitidulus* (Nyl.) (Hym., Form.). *Insectes Sociaux* 23: 549–558.

Buschinger, A., and U. Winter. 1978. Echte Arbeiterinnen, fertile Arbeiterinnen und sterile Wirtsweibchen in Völkern der dulotischen Ameise *Harpagoxenus sublaevis* (Nyl.) (Hym., Form.). *Insectes Sociaux* 25: 63–78.

Buschinger, A., and U. Winter. 1983. Population studies of the dulotic ant, *Epimyrma ravouxi*, and the degenerate slavemaker, *E. kraussei* (Hymenoptera: Formicidae). *Entomologia Generalis* 8: 251–266.

Buschinger, A., and U. Winter. 1985. Life history and male morphology of the workerless parasitic ant *Epimyrma corsica* (Hymenoptera: Formicidae). *Entomologia Generalis* 10: 65–75.

Buschinger, A., A. Francoeur, and K. Fischer. 1980. Functional monogyny, sexual behavior, and karyotype of the guest ant, *Leptothorax provancheri* Emery (Hymenoptera, Formicidae). *Psyche* 87: 1–12.

Buschinger, A., K. Fischer, H.-P. Guthy, K. Jessen, and U. Winter. 1986. Biosystematic revision of *Epimyrma kraussei*, *E. vandeli*, and *E. foreli* (Hymenoptera: Formicidae). *Psyche* 93: 253–276.

Buschinger, A., J. Heinze, K. Jessen, P. Douwes, and U. Winter. 1987. First European record of a queen ant carrying a mealybug during her mating flight. *Naturwissenschaften* 74: 139–140.

Buschinger, A., R.W. Klein, and U. Maschwitz. 1994. Colony structure of a bamboo-dwelling *Tetraponera* sp. (Hymenoptera: Formicidae: Pseudomyrmecinae) from Malaysia. *Insectes Sociaux* 41: 29–41.

Buss, L.W. 1987. *The Evolution of Individuality*. Princeton University Press, Princeton, New Jersey.

Cain, A.J. 1964. The perfection of animals. Reprinted in *Biological Journal of the Linnean Society* (1989) 36: 3–29.

Calabi, P. 1988. Behavioral flexibility in Hymenoptera: a re-examination of the concept of caste. In J.C. Trager, ed., *Advances in Myrmecology*, pp. 237–258. E.J. Brill, Leiden.

Calabi, P., and S.D. Porter. 1989. Worker longevity in the fire ant *Solenopsis invicta*: ergonomic considerations of correlations between temperature, size and metabolic rates. *Journal of Insect Physiology* 35: 643–649.

Calabi, P., and R. Rosengaus. 1988. Interindividual differences based on behavior transition probabilities in the ant *Camponotus sericeiventris*. In R.L. Jeanne, ed., *Interindividual Behavioral Variability in Social Insects*, pp. 61–90. Westview Press, Boulder, Colorado.

Calabi, P., and J.F.A. Traniello. 1989a. Behavioral flexibility in age castes of the ant *Pheidole dentata*. *Journal of Insect Behavior* 2: 663–677.

Calabi, P., and J.F.A. Traniello. 1989b. Social organization in the ant *Pheidole dentata*: Physical and temporal caste ratios lack ecological correlates. *Behavioral Ecology and Sociobiology* 24: 69–78.

Calabi, P., J.F.A. Traniello, and M.H. Werner. 1983. Age polyethism: its occurrence in the ant *Pheidole hortensis*, and some general considerations. *Psyche* 90: 395–412.

Camazine, S. 1991. Self-organizing pattern formation on the combs of honey bee colonies. *Behavioral Ecology and Sociobiology* 28: 61–76.

Camazine, S., J. Sneyd, M.J. Jenkins, and J.D. Murray. 1990. A mathematical model of self-organized pattern formation on the combs of honeybee colonies. *Journal of Theoretical Biology* 147: 553–571.

Cameron, S.A. 1993. Multiple origins of advanced eusociality in bees inferred from mitochondrial DNA sequences. *Proceedings of the National Academy of Sciences, U.S.A.* 90: 8687–8691.

Cammaerts-Tricot, M.-C. 1974. Production and perception of attractive pheromones by differently aged workers of *Myrmica rubra* (Hymenoptera Formicidae). *Insectes Sociaux* 21: 235–247.

Carlin, N.F. 1988. Species, kin and other forms of recognition in the brood discrimination behavior of ants. In J.C. Trager, ed., *Advances in Myrmecology*, pp. 267–295. E.J. Brill, Leiden.

Carlin, N.F. 1989. Discrimination between and within colonies of social insects: two null hypotheses. *Netherlands Journal of Zoology* 39: 86–100.

Carlin, N.F., and P.C. Frumhoff. 1990. Nepotism in the honey bee. *Nature* 346: 706–707.

Carlin, N.F., H.K. Reeve, and S.P. Cover. 1993. Kin discrimination and division of labour among matrilines in the polygynous carpenter ant, *Camponotus planatus*. In L. Keller, ed., *Queen Number and Sociality in Insects*, pp. 362–401. Oxford University Press, Oxford.

Carpenter, J.M. 1989. Testing scenarios: wasp social behavior. *Cladistics* 5: 131–144.

Chan, G.L., and A.F.G. Bourke. 1994. Split sex ratios in a multiple-queen ant population. *Proceedings of the Royal Society of London, Series B* 258: 261–266.

Chapman. J.A. 1969. Winged ants return after removal from a summit swarming site. *Annals of the Entomological Society of America* 62: 1256–1259.

Charlesworth, B. 1978. Some models of the evolution of altruistic behaviour between siblings. *Journal of Theoretical Biology* 72: 297–319.

Charlesworth, B. 1980. Models of kin selection. In H. Markl, ed., *Evolution of Social Behavior: Hypotheses and Empirical Tests*, pp. 11–26. Verlag Chemie, Weinheim.

Charlesworth, B., and E.L. Charnov. 1981. Kin selection in age-structured populations. *Journal of Theoretical Biology* 88: 103–119.

Charnov, E.L. 1977. An elementary treatment of the genetical theory of kin-selection. *Journal of Theoretical Biology* 66: 541–550.

Charnov, E.L. 1978a. Evolution of eusocial behavior: offspring choice or parental parasitism? *Journal of Theoretical Biology* 75: 451–465.

Charnov, E.L. 1978b. Sex-ratio selection in eusocial Hymenoptera. *American Naturalist* 112: 317–326.

Charnov, E.L. 1982. *The Theory of Sex Allocation*. Princeton University Press, Princeton, New Jersey.

Charnov, E.L. 1993. *Life History Invariants*. Oxford University Press, Oxford.

Charnov, E.L., and J.P. Finerty. 1980. Vole population cycles: a case for kin-selection? *Oecologia* 45: 1–2.

Charnov, E.L., and W.M. Schaffer. 1973. Life-history consequences of natural selection: Cole's result revisited. *American Naturalist* 107: 791–793.

Cherix, D. 1980. Note préliminaire sur la structure, la phénologie et le régime alimentaire d'une super-colonie de *Formica lugubris* Zett. *Insectes Sociaux* 27: 226–236.

Cherix, D., D. Chautems, D.J.C. Fletcher, W. Fortelius, G. Gris, L. Keller, L. Passera, R. Rosengren, E.L. Vargo, and F. Walter. 1991. Alternative reproductive strategies in *Formica lugubris* Zett. (Hymenoptera Formicidae). *Ethology, Ecology and Evolution, Special Issue* 1: 61–66.

Cheverud, J.M. 1985. A quantitative genetic model of altruistic selection. *Behavioral Ecology and Sociobiology* 16: 239–243.

Choe, J.C. 1988. Worker reproduction and social evolution in ants (Hymenoptera: Formicidae). In J.C. Trager, ed., *Advances in Myrmecology*, pp. 163–187. E.J. Brill, Leiden.

Clark, A.B. 1978. Sex ratio and local resource competition in a prosimian primate. *Science* 201: 163–165.

Clutton-Brock, T.H. 1991. *The Evolution of Parental Care*. Princeton University Press, Princeton, New Jersey.

Clutton-Brock, T., and C. Godfray. 1991. Parental investment. In J.R. Krebs and N.B. Davies, eds., *Behavioural Ecology: An Evolutionary Approach*, 3rd edition, pp. 234–262. Blackwell, Oxford.

Clutton-Brock, T.H., and G.A. Parker. 1992. Potential reproductive rates and the operation of sexual selection. *Quarterly Review of Biology* 67: 437–456.

Clutton-Brock, T.H., and A.C.J. Vincent. 1991. Sexual selection and the potential reproductive rates of males and females. *Nature* 351: 58–60.

Cole, B.J. 1981. Dominance hierarchies in *Leptothorax* ants. *Science* 212: 83–84.

Cole, B.J. 1983a. Multiple mating and the evolution of social behavior in the Hymenoptera. *Behavioral Ecology and Sociobiology* 12: 191–201.

Cole, B.J. 1983b. Assembly of mangrove ant communities: colonization abilities. *Journal of Animal Ecology* 52: 349–355.

Cole, B.J. 1984. Colony efficiency and the reproductivity effect in *Leptothorax allardycei* (Mann). *Insectes Sociaux* 31: 403–407.

Cole, B.J. 1986. The social behavior of *Leptothorax allardycei* (Hymenoptera, Formicidae): time budgets and the evolution of worker reproduction. *Behavioral Ecology and Sociobiology* 18: 165–173.

Cole, B. 1988. Individual differences in social insect behavior: movement and space use in *Leptothorax allardycei*. In R.L. Jeanne, ed., *Interindividual Behavioral Variability in Social Insects*, pp. 113–146. Westview Press, Boulder, Colorado.

Cole, B.J. 1991. Short-term activity cycles in ants: generation of periodicity by worker interaction. *American Naturalist* 137: 244–259.

Cole, B.J. 1992. Short-term activity cycles in ants: age-related changes in tempo and colony synchrony. *Behavioral Ecology and Sociobiology* 31: 181–187.

Cole, B.J., J. McDowell, and D. Cheshire. 1994. Demography of the worker caste of *Leptothorax allardycei* (Hymenoptera: Formicidae). *Annals of the Entomological Society of America* 87: 562–565.

Cole, L.C. 1954. The population consequences of life history phenomena. *Quarterly Review of Biology* 29: 103–137.

Collingwood, C.A. 1958. The ants of the genus *Myrmica* in Britain. *Proceedings of the Royal Entomological Society of London, Series A* 33: 65–75.

Collingwood, C.A. 1979. *The Formicidae (Hymenoptera) of Fennoscandia and Denmark*. Scandinavian Science Press Ltd, Klampenborg.

Colwell, R.K. 1981. Group selection is implicated in the evolution of female-biased sex ratios. *Nature* 290: 401–404.

Cook, J.M. 1993. Sex determination in the Hymenoptera: a review of models and evidence. *Heredity* 71: 421–435.

Corbara, B., J.-P. Lachaud, and D. Fresneau. 1989. Individual variability, social structure and division of labour in the ponerine ant *Ectatomma ruidum* Roger (Hymenoptera, Formicidae). *Ethology* 82: 89–100.

Cosmides, L.M., and J. Tooby. 1981. Cytoplasmic inheritance and intragenomic conflict. *Journal of Theoretical Biology* 89: 83–129.

Craig, R. 1979. Parental manipulation, kin selection, and the evolution of altruism. *Evolution* 33: 319–334.

Craig, R. 1980a. Sex ratio changes and the evolution of eusociality in the Hymenoptera: simulation and games theory studies. *Journal of Theoretical Biology* 87: 55–70.

Craig, R. 1980b. Sex investment ratios in social Hymenoptera. *American Naturalist* 116: 311–323.

Craig, R. 1982a. Evolution of eusociality by kin selection: the effect of inbreeding between siblings. *Journal of Theoretical Biology* 94: 119–128.

Craig, R. 1982b. Evolution of male workers in the Hymenoptera. *Journal of Theoretical Biology* 94: 95–105.

Craig, R. 1983a. Familial selection and the evolution of social behaviour re-examined. *Journal of Theoretical Biology* 103: 287–294.

Craig, R. 1983b. Subfertility and the evolution of eusociality by kin selection. *Journal of Theoretical Biology* 100: 379–397.

Craig, R., and R.H. Crozier. 1978. No evidence for role of heterozygosity in ant caste determination. *Isozyme Bulletin* 11: 66–67.

Creel, S. 1990. How to measure inclusive fitness. *Proceedings of the Royal Society of London, Series B* 241: 229–231.

Crespi, B.J. 1992a. Eusociality in Australian gall thrips. *Nature* 359: 724–726.

Crespi, B.J. 1992b. Cannibalism and trophic eggs in subsocial and eusocial insects. In M.A. Elgar and B.J. Crespi, eds., *Cannibalism: Ecology and Evolution among Diverse Taxa*, pp. 176–213. Oxford University Press, Oxford.

Crespi, B.J. 1994. Three conditions for the evolution of eusociality: Are they sufficient? *Insectes Sociaux* 41: 395–400.

Crespi, B.J., and D. Yanega. 1995. The definition of eusociality. *Behavioral Ecology* 6: 109–115.

Cronin, H. 1991. *The Ant and the Peacock*. Cambridge University Press, Cambridge.

Crosland, M.W.J., and R.H. Crozier. 1986. *Myrmecia pilosula*, an ant with only one pair of chromosomes. *Science* 231: 1278.

Crow, J.F. 1979. Genes that violate Mendel's rules. *Scientific American* 240(2): 104–113.

Crow, J.F. 1991. Why is Mendelian segregation so exact? *BioEssays* 13: 305–312.

Crozier, R.H. 1970. Coefficients of relationship and the identity of genes by descent in the Hymenoptera. *American Naturalist* 104: 216–217.

Crozier, R.H. 1974. Allozyme analysis of reproductive strategy in the ant *Aphaenogaster rudis*. *Isozyme Bulletin* 7: 18.

Crozier, R.H. 1975. *Animal Cytogenetics. Vol. 3: Insecta 7. Hymenoptera.* Gebrüder Borntraeger, Berlin.

Crozier, R.H. 1977. Evolutionary genetics of the Hymenoptera. *Annual Review of Entomology* 22: 263–288.

Crozier, R.H. 1979. Genetics of sociality. In H.R. Hermann, ed., *Social Insects*, Vol. I, pp. 223–286. Academic Press, New York.

Crozier, R.H. 1980. Genetical structure of social insect populations. In H. Markl, ed., *Evolution of Social Behavior: Hypotheses and Empirical Tests*, pp. 129–145. Verlag Chemie, Weinheim.

Crozier, R.H. 1982. On insects and insects: twists and turns in our understanding of the evolution of eusociality. In M.D. Breed, C.D. Michener, and H.E. Evans, eds., *The Biology of Social Insects*, pp. 4–9. Westview Press, Boulder, Colorado.

Crozier, R.H. 1989. Insect sociobiology. *Science* 245: 313–314.

Crozier, R.H. 1992. The genetic evolution of flexible strategies. *American Naturalist* 139: 218–223.

Crozier, R.H., and D. Brückner. 1981. Sperm clumping and the population genetics of Hymenoptera. *American Naturalist* 117: 561–563.

Crozier, R.H., and P.C. Consul. 1976. Conditions for genetic polymorphism in social Hymenoptera under selection at the colony level. *Theoretical Population Biology* 10: 1–9.

Crozier, R.H., and R.E. Page. 1985. On being the right size: male contributions and multiple mating in social Hymenoptera. *Behavioral Ecology and Sociobiology* 18: 105–115.

Crozier, R.H., and P. Pamilo. 1980. Asymmetry in relatedness: who is related to whom? *Nature* 283: 604.

Crozier, R.H., and P. Pamilo. 1986. Relatedness within and between colonies of a queenless ant species of the genus *Rhytidoponera* (Hymenoptera: Formicidae). *Entomologia Generalis* 11: 113–117.

Crozier, R.H., and P. Pamilo. 1993. Sex allocation in social insects: problems in prediction and estimation. In D.L. Wrensch and M.A. Ebbert, eds., *Evolution and Diversity of Sex Ratio in Insects and Mites*, pp. 369–383. Chapman and Hall, New York.

Crozier, R.H., P. Pamilo, and Y.C. Crozier. 1984. Relatedness and microgeographic genetic variation in *Rhytidoponera mayri*, an Australian arid-zone ant. *Behavioral Ecology and Sociobiology* 15: 143–150.

Darwin, C. 1859. *On the Origin of Species*. John Murray, London. (Facsimile of 1st Edition, Harvard University Press, Cambridge, Massachusetts, 1964).

Davidson, D.W. 1982. Sexual selection in harvester ants (Hymenoptera: Formicidae: *Pogonomyrmex*). *Behavioral Ecology and Sociobiology* 10: 245–250.

Dawkins, R. 1976. *The Selfish Gene*. 1st edition. Oxford University Press, Oxford.

Dawkins, R. 1978. Replicator selection and the extended phenotype. *Zeitschrift für Tierpsychologie* 47: 61–76.

Dawkins, R. 1979. Twelve misunderstandings of kin selection. *Zeitschrift für Tierpsychologie* 51: 184–200.

Dawkins, R. 1982a. *The Extended Phenotype*. W.H. Freeman and Co., Oxford.

Dawkins, R. 1982b. Replicators and vehicles. In King's College Sociobiology Group, ed., *Current Problems in Sociobiology*, pp. 45–64. Cambridge University Press, Cambridge.

Dawkins, R. 1983. Universal Darwinism. In D.S. Bendall, ed., *Evolution from Molecules to Men*, pp. 403–425. Cambridge University Press, Cambridge.

Dawkins, R. 1986. *The Blind Watchmaker*. Longman, Harlow, U.K.

Dawkins, R. 1989a. *The Selfish Gene*. 2nd edition. Oxford University Press, Oxford.

Dawkins, R. 1989b. The evolution of evolvability. In C.G. Langton, ed., *Artificial Life*, pp. 201–220. Addison-Wesley, Redwood City, California.

Dawkins, R. 1990. Parasites, desiderata lists and the paradox of the organism. *Parasitology* 100: S 63–S 73.

Dawkins, R., and J.R. Krebs. 1979. Arms races between and within species. *Proceedings of the Royal Society of London, Series B* 205: 489–511.

Dejean, A., and J.-P. Lachaud. 1991. Polyethism in the ponerine ant

Odontomachus troglodytes: interaction of age and interindividual variability. *Sociobiology* 18: 177–196.

Deneubourg, J.-L., and S. Goss. 1989. Collective patterns and decision-making. *Ethology, Ecology and Evolution* 1: 295–311.

Deneubourg, J.-L., S. Goss, J.M. Pasteels, D. Fresneau, and J.-P. Lachaud. 1987. Self-organization mechanisms in ant societies (II): Learning in foraging and division of labor. In J.M. Pasteels and J.-L. Deneubourg, eds., *From Individual to Collective Behavior in Social Insects*, pp. 177–196. Birkhäuser Verlag, Basel.

Deneubourg, J.-L., S. Goss, N. Franks, and J.M. Pasteels. 1989. The blind leading the blind: modeling chemically mediated army ant raid patterns. *Journal of Insect Behavior* 2: 719–725.

Deneubourg, J.-L., S. Goss, N. Franks, A. Sendova-Franks, C. Detrain, and L. Chrétien. 1991. The dynamics of collective sorting: robot-like ants and ant-like robots. In J.-A. Meyer and S.W. Wilson, eds., *From Animals to Animats: Proceedings of the First International Conference on Simulation of Adaptive Behavior*, pp. 356–363. MIT Press, Cambridge, Massachusetts.

Denlinger, D.L. 1985. Hormonal control of diapause. In G.A. Kerkut and L.I. Gilbert, eds., *Comprehensive Insect Physiology, Biochemistry and Pharmacology, Vol. 8, Endocrinology II*, pp. 353–412. Pergamon Press, Oxford.

Dlussky, G.M., and A.N. Kupianskaya. 1972. Consumption of protein food and growth of *Myrmica* colonies. *Ekologia Polska* 20: 73–82.

Donisthorpe, H.St.J.K. 1927. *British Ants. Their Life-History and Classification.* 2nd edition. George Routledge and Sons Ltd, London.

Doolittle, W.F., and C. Sapienza. 1980. Selfish genes, the phenotype paradigm and genome evolution. *Nature* 284: 601–603.

Douwes, P., K. Jessen, and A. Buschinger. 1988. *Epimyrma adlerzi* sp. n. (Hymenoptera: Formicidae) from Greece: morphology and life history. *Entomologica Scandinavica* 19: 239–249.

Douwes, P., L. Sivusaari, M. Niklasson, and B. Stille. 1987. Relatedness among queens in polygynous nests of the ant *Leptothorax acervorum*. *Genetica* 75: 23–29.

Dover, G.A. 1988. Evolving the improbable. *Trends in Ecology and Evolution* 3: 81–84.

Dugatkin, L.A., and H.K. Reeve. 1994. Behavioral ecology and levels of selection: dissolving the group selection controversy. *Advances in the Study of Behavior* 23: 101–133.

Dugatkin, L.A., M. Mesterton-Gibbons, and A.I. Houston. 1992. Beyond the prisoner's dilemma: toward models to discriminate among mechanisms of cooperation in nature. *Trends in Ecology and Evolution* 7: 202–205.

Dugatkin, L.A., D.S. Wilson, L. Farrand, and R.T. Wilkens. 1994. Altruism, tit for tat and "outlaw" genes. *Evolutionary Ecology* 8: 431–437.

Dumpert, K. 1981. *The Social Biology of Ants.* Translated by C. Johnson. Pitman, Boston.

Ebbert, M.A. 1993. Endosymbiotic sex ratio distorters in insects and mites. In D.L. Wrensch and M.A. Ebbert, eds., *Evolution and Diversity of Sex Ratio in Insects and Mites*, pp. 150–191. Chapman and Hall, New York.

Eberhard, W.G. 1980. Evolutionary consequences of intracellular organelle competition. *Quarterly Review of Biology* 55: 231–249.

Eberhard, W.G. 1990. Evolution in bacterial plasmids and levels of selection. *Quarterly Review of Biology* 65: 3–22.

Edelman, G.M. 1992. *Bright Air, Brilliant Fire: On the Matter of the Mind.* Basic Books, New York.

Edwards, J.P. 1987. Caste regulation in the pharaoh's ant *Monomorium pharaonis*: the influence of queens on the production of new sexual forms. *Physiological Entomology* 12: 31–39.

Edwards, J.P. 1991. Caste regulation in the pharaoh's ant *Monomorium pharaonis*: recognition and cannibalism of sexual brood by workers. *Physiological Entomology* 16: 263–271.

Eickwort, G.C. 1981. Presocial insects. In H.R. Hermann, ed., *Social Insects*, Vol. II, pp. 199–280. Academic Press, New York.

Eldredge, N. 1985. *Unfinished Synthesis. Biological Hierarchies and Modern Evolutionary Thought.* Oxford University Press, New York.

Elmes, G.W. 1973a. Observations on the density of queens in natural colonies of *Myrmica rubra* L. (Hymenoptera: Formicidae). *Journal of Animal Ecology* 42: 761–771.

Elmes, G.W. 1973b. Miniature queens of the ant *Myrmica rubra* L. (Hymenoptera, Formicidae). *The Entomologist* 106: 133–136.

Elmes, G.W. 1974a. Colony populations of *Myrmica sulcinodis* Nyl. (Hym. Formicidae). *Oecologia* 15: 337–343.

Elmes, G.W. 1974b. The effect of colony population on caste size in three species of *Myrmica* (Hymenoptera, Formicidae). *Insectes Sociaux* 21: 213–229.

Elmes, G.W. 1976. Some observations on the microgyne form of *Myrmica rubra* L. (Hymenoptera, Formicidae). *Insectes Sociaux* 23: 3–21.

Elmes, G.W. 1978. A morphometric comparison of three closely related species of *Myrmica* (Formicidae) including a new species from England. *Systematic Entomology* 3: 131–145.

Elmes, G.W. 1980. Queen numbers in colonies of ants of the genus *Myrmica*. *Insectes Sociaux* 27: 43–60.

Elmes, G.W. 1982. The phenology of five species of *Myrmica* (Hym. Formicidae) from South Dorset, England. *Insectes Sociaux* 29: 548–559.

Elmes, G.W. 1983. Some experimental observations on the parasitic *Myrmica hirsuta* Elmes. *Insectes Sociaux* 30: 221–234.

Elmes, G.W. 1987a. Temporal variation in colony populations of the ant *Myrmica sulcinodis*. I. Changes in queen number, worker number and spring production. *Journal of Animal Ecology* 56: 559–571.

Elmes, G.W. 1987b. Temporal variation in colony populations of the ant *Myrmica sulcinodis*. II. Sexual production and sex ratios. *Journal of Animal Ecology* 56: 573–583.

Elmes, G.W. 1989. The effect of multiple queens in small groups of *Myrmica rubra* L. *Actes des Colloques Insectes Sociaux* 5: 137–144.

Elmes, G.W. 1991. The social biology of *Myrmica* ants. *Actes des Colloques Insectes Sociaux* 7: 17–34.

Elmes, G.W., and M.V. Brian. 1991. The importance of the egg-mass to the activity of normal queens and microgynes of *Myrmica rubra* L. (Hym., Formicidae). *Insectes Sociaux* 38: 51–62.

Elmes, G.W., and R.T. Clarke. 1981. A biometric investigation of variation of workers of *Myrmica ruginodis* Nylander (Formicidae). In P.E. Howse and J.-L. Clément, eds., *Biosystematics of Social Insects*, pp. 121–140. Academic Press, London.

Elmes, G.W., and L. Keller. 1993. Distribution and ecology of queen number in ants of the genus *Myrmica*. In L. Keller, ed., *Queen Number and Sociality in Insects*, pp. 294–307. Oxford University Press, Oxford.

Elmes. G.W., and J. Petal. 1990. Queen number as an adaptable trait: evidence from wild populations of two red ant species (genus *Myrmica*). *Journal of Animal Ecology* 59: 675–690.

Elmes, G.W., and J.C. Wardlaw. 1982a. Variations in populations of *Myrmica sabuleti* and *M. scabrinodis* (Formicidae: Hymenoptera) living in Southern England. *Pedobiologia* 23: 90–97.

Elmes, G.W., and J.C. Wardlaw. 1982b. A population study of the ants *Myrmica sabuleti* and *Myrmica scabrinodis*, living at two sites in the south of England. I. A comparison of colony populations. *Journal of Animal Ecology* 51: 651–664.

Elmes, G.W., and J.C. Wardlaw. 1982c. A population study of the ants *Myrmica sabuleti* and *Myrmica scabrinodis* living at two sites in the south of England. II. Effect of above-nest vegetation. *Journal of Animal Ecology* 51: 665–680.

Elmes, G.W., and J.C. Wardlaw. 1983. A comparison of the effect of a queen upon the development of large hibernated larvae of six species of the genus *Myrmica* (Hym. Formicidae). *Insectes Sociaux* 30: 134–148.

Elton, E.T.G. 1991. Labial gland disease in the genus *Formica* (Formicidae, Hymenoptera). *Insectes Sociaux* 38: 91–93.

Emery, F.E. 1967. The next thirty years: concepts, methods and anticipations. *Human Relations* 20: 199–237.

Emery, F.E. (ed.) 1969. *Systems Thinking. Selected Readings.* Penguin, Harmondsworth, U.K.

Emlen, S.T. 1982. The evolution of helping. I. An ecological constraints model. *American Naturalist* 119: 29–39.

Emlen, S.T. 1991. Evolution of cooperative breeding in birds and mammals. In J.R. Krebs and N.B. Davies, eds., *Behavioural Ecology: An Evolutionary Approach*, 3rd edition, pp. 301–337. Blackwell, Oxford.

Emlen, S.T., and L.W. Oring. 1977. Ecology, sexual selection, and the evolution of mating systems. *Science* 197: 215–223.

Endler, J.A. 1986. *Natural Selection in the Wild.* Princeton University Press, Princeton, New Jersey.

Enquist, M., and O. Leimar. 1990. The evolution of fatal fighting. *Animal Behaviour* 39: 1–9.

Eshel, I., and M.W. Feldman. 1991. The handicap principle in parent-offspring conflict: comparison of optimality and population-genetic analyses. *American Naturalist* 137: 167–185.

Estoup, A., M. Solignac, and J.-M. Cornuet. 1994. Precise assessment of the number of patrilines and of genetic relatedness in honeybee colonies. *Proceedings of the Royal Society of London, Series B* 258: 1–7.

Evans, H.E. 1977. Extrinsic versus intrinsic factors in the evolution of insect eusociality. *BioScience* 27: 613–617.

Evans, J.D. 1993. Parentage analyses in ant colonies using simple sequence repeat loci. *Molecular Ecology* 2: 393–397.

Evesham, E.J.M. 1984. Queen distribution movements and interactions in a semi-natural nest of the ant *Myrmica rubra* L. *Insectes Sociaux* 31: 5–19.

Fagen, R.M., and R.N. Goldman. 1977. Behavioural catalogue analysis methods. *Animal Behaviour* 25: 261–274.

Fahrig, L., D.P. Coffin, W.K. Lauenroth, and H.H. Shugart. 1994. The advantage of long-distance clonal spreading in highly disturbed habitats. *Evolutionary Ecology* 8: 172–187.

Faulkes, C.G., D.H. Abbott, C.E. Liddell, L.M. George, and J.U.M. Jarvis. 1991. Hormonal and behavioral aspects of reproductive suppression in female naked mole-rats. In P.W. Sherman, J.U.M. Jarvis, and R.D. Alexander, eds., *The Biology of the Naked Mole-Rat*, pp. 426–445. Princeton University Press, Princeton, New Jersey.

Feldman, M.W., and I. Eshel. 1982. On the theory of parent-offspring conflict: a two-locus genetic model. *American Naturalist* 119: 285–292.

Field, J., and W.A. Foster. 1995. Nest co-occupation in the digger wasp *Cerceris arenaria* – cooperation or usurpation? *Animal Behaviour*: in press.

Fisher, R.A. 1930. *The Genetical Theory of Natural Selection*. 1st edition. Clarendon Press, Oxford.

Fisher, R.A. 1958. *The Genetical Theory of Natural Selection*. 2nd edition. Dover, New York.

Fisher, R.M. 1992. Sex ratios in bumble bee social parasites: support for queen-worker conflict theory? (Hymenoptera: Apidae). *Sociobiology* 20: 205–217.

Fisher, R.M. 1993. How important is the sting in insect social evolution? *Ethology, Ecology and Evolution* 5: 157–168.

Fletcher, D.J.C. 1983. Three newly-discovered polygynous populations of the fire ant, *Solenopsis invicta*, and their significance. *Journal of the Georgia Entomological Society* 18: 538–543.

Fletcher, D.J.C., and M.S. Blum. 1981. Pheromonal control of dealation and oogenesis in virgin queen fire ants. *Science* 212: 73–75.

Fletcher, D.J.C., and M.S. Blum. 1983. Regulation of queen number by workers in colonies of social insects. *Science* 219: 312–314.

Fletcher, D.J.C., and K.G. Ross. 1985. Regulation of reproduction in eusocial Hymenoptera. *Annual Review of Entomology* 30: 319–343.

Fletcher, D.J.C., M.S. Blum, T.V. Whitt, and N. Temple. 1980. Monogyny and polygyny in the fire ant, *Solenopsis invicta. Annals of the Entomological Society of America* 73: 658–661.

Fluri, P., M. Lüscher, H. Wille, and L. Gerig. 1982. Changes in weight of the pharyngeal gland and haemolymph titres of juvenile hormone, protein and vitellogenin in worker honey bees. *Journal of Insect Physiology* 28: 61–68.

Forsyth, A. 1980. Worker control of queen density in Hymenopteran societies. *American Naturalist* 116: 895–898.

Forsyth, A. 1981. Sex ratio and parental investment in an ant population. *Evolution* 35: 1252–1253.

Fortelius, W. 1987. Different patterns of female behaviour in mono- and poly-domous *Formica* populations. In J. Eder and H. Rembold, eds., *Chemistry and Biology of Social Insects*, pp. 293–294. Verlag J. Peperny, München.

Fortelius, W., P. Pamilo, R. Rosengren, and L. Sundström. 1987. Male size dimorphism and alternative reproductive tactics in *Formica exsecta* ants (Hymenoptera, Formicidae). *Annales Zoologici Fennici* 24: 45–54.

Fortelius, W., R. Rosengren, D. Cherix, and D. Chautems. 1993. Queen recruit-ment in a highly polygynous supercolony of *Formica lugubris* (Hymenoptera, Formicidae). *Oikos* 67: 193–199.

Francoeur, A. 1986. Deux nouvelles fourmis néarctiques: *Leptothorax retractus* et *L. sphagnicolus* (Formicidae, Hymenoptera). *Canadian Entomologist* 118: 1151–1164.

Francoeur, A., R. Loiselle, and A. Buschinger. 1985. Biosystématique de la tribu Leptothoracini (Formicidae, Hymenoptera). 1. Le genre *Formicoxenus* dans la région holarctique. *Naturaliste Canadien* 112: 343–403.

Frank, S.A. 1986. Hierarchical selection theory and sex ratios. I. General solu-tions for structured populations. *Theoretical Population Biology* 29: 312–342.

Frank, S.A. 1987a. Individual and population sex allocation patterns. *Theoretical Population Biology* 31: 47–74.

Frank, S.A. 1987b. Variable sex ratio among colonies of ants. *Behavioral Ecology and Sociobiology* 20: 195–201.

Frank, S.A. 1990. Sex allocation theory for birds and mammals. *Annual Review of Ecology and Systematics* 21: 13–55.

Frank, S.A. 1994. Kin selection and virulence in the evolution of protocells and parasites. *Proceedings of the Royal Society of London, Series B* 258: 153–161.

Frank, S.A., and B.J. Crespi. 1989. Synergism between sib-rearing and sex ratio in Hymenoptera. *Behavioral Ecology and Sociobiology* 24: 155–162.

Franks, N. 1982a. Ecology and population regulation in the army ant *Eciton burchelli*. In E.G. Leigh, A.S. Rand, and D.W. Windsor, eds., *The Ecology of a Tropical Forest: Seasonal Rhythms and Long-term Changes*, pp. 389–395. Smithsonian Institute Press, Washington, D.C.

Franks, N.R. 1982b. A new method for censusing animal populations: the num-ber of *Eciton burchelli* army ant colonies on Barro Colorado Island, Panama. *Oecologia* 52: 266–268.

Franks, N.R. 1985. Reproduction, foraging efficiency and worker polymorphism in army ants. In M. Lindauer and B. Hölldobler, eds., *Experimental Behavioral Ecology and Sociobiology*, pp. 91–107. Gustav Fischer Verlag, Stuttgart.

Franks, N.R. 1989a. Army ants: a collective intelligence. *American Scientist* 77: 138–145.

Franks, N.R. 1989b. Thermoregulation in army ant bivouacs. *Physiological Entomology* 14: 397–404.

Franks, N.R. 1990. Considering the ants. *Science* 248: 897–898.

Franks, N.R. 1995. Collective intelligence and limited rationality in the organi-zation of ant colonies and human societies. *Jahrbuch des Wissen-*

schaftskollegs 1993/1994, Berlin (*1993/1994 Year Book of the Institute of Advanced Study, Berlin*). Nicolai-Verlag, Berlin. In press.

Franks, N.R., and W.H. Bossert. 1983. The influence of swarm raiding army ants on the patchiness and diversity of a tropical leaf litter ant community. In S.L. Sutton, T.C. Whitmore, and A.C. Chadwick, eds., *Tropical Rain Forest: Ecology and Management (Special Publication No. 2 of the British Ecological Society)*, pp. 151–163. Blackwell, Oxford.

Franks, N., and A. Bourke. 1988. Slaves of circumstance. *New Scientist* 119 (1627): 45–49.

Franks, N., and S. Bryant. 1987. Rhythmical patterns of activity within the nests of ants. In J. Eder and H. Rembold, eds., *Chemistry and Biology of Social Insects*, pp. 122–123. Verlag J. Peperny, München.

Franks, N.R., and C.R. Fletcher. 1983. Spatial patterns in army ant foraging and migration: *Eciton burchelli* on Barro Colorado Island, Panama. *Behavioral Ecology and Sociobiology* 12: 261–270.

Franks, N.R., and B. Hölldobler. 1987. Sexual competition during colony reproduction in army ants. *Biological Journal of the Linnean Society* 30: 229–243.

Franks, N.R., and P.J. Norris. 1987. Constraints on the division of labour in ants: D'Arcy Thompson's cartesian transformations applied to worker polymorphism. In J.M. Pasteels and J.-L. Deneubourg, eds., *From Individual to Collective Behavior in Social Insects*, pp. 253–270. Birkhäuser Verlag, Basel.

Franks, N.R., and L.W. Partridge. 1993. Lanchester battles and the evolution of combat in ants. *Animal Behaviour* 45: 197–199.

Franks, N.R., and E. Scovell. 1983. Dominance and reproductive success among slave-making worker ants. *Nature* 304: 724–725.

Franks, N.R., and A.B. Sendova-Franks. 1992. Brood sorting by ants: distributing the workload over the work-surface. *Behavioral Ecology and Sociobiology* 30: 109–123.

Franks, N.R., and C. Tofts. 1994. Foraging for work: how tasks allocate workers. *Animal Behaviour* 48: 470–472.

Franks, N.R., B. Ireland, and A.F.G. Bourke. 1990a. Conflicts, social economics and life history strategies in ants. *Behavioral Ecology and Sociobiology* 27: 175–181.

Franks, N.R., S. Bryant, R. Griffiths, and L. Hemerik. 1990b. Synchronization of the behaviour within nests of the ant *Leptothorax acervorum* (Fabricius) – I. Discovering the phenomenon and its relation to the level of starvation. *Bulletin of Mathematical Biology* 52: 597–612.

Franks, N.R., A.B. Sendova-Franks, M. Sendova-Vassileva, and L. Vassilev. 1991a. Nuptial flights and calling behaviour in the ant *Leptothorax acervorum* (Fabr.). *Insectes Sociaux* 38: 327–330.

Franks, N.R., N. Gomez, S. Goss, and J.-L. Deneubourg. 1991b. The blind leading the blind in army ant raid patterns: testing a model of self-organization (Hymenoptera: Formicidae). *Journal of Insect Behavior* 4: 583–607.

Franks, N.R., A. Wilby, B.W. Silverman, and C. Tofts. 1992. Self-organizing nest construction in ants: sophisticated building by blind bulldozing. *Animal Behaviour* 44: 357–375.

Free, J.B. 1965. The allocation of duties among worker honeybees. *Symposia of the Zoological Society of London* 14: 39–59.

Fresneau, D., and B. Corbara. 1990. Spatial organization in the nest during colony ontogeny in the ponerine ant *Pachycondyla* (*Neoponera*) *apicalis*. In G.K. Veeresh, B. Mallik, and C.A. Viraktamath, eds., *Social Insects and the Environment*, p. 387. Oxford and IBH Publishing Co., New Delhi.

Frumhoff, P.C., and J. Baker. 1988. A genetic component to division of labour within honey bee colonies. *Nature* 333: 358–361.

Frumhoff, P.C., and S. Schneider. 1987. The social consequences of honey bee polyandry: the effects of kinship on worker interactions within colonies. *Animal Behaviour* 35: 255–262.

Frumhoff, P.C., and P.S. Ward. 1992. Individual-level selection, colony-level selection, and the association between polygyny and worker monomorphism in ants. *American Naturalist* 139: 559–590.

Fuchs, S., and V. Schade. 1994. Lower performance in honeybee colonies of uniform paternity. *Apidologie* 25: 155–168.

Fukuda, H. 1983. The relationship between work efficiency and population size in a honeybee colony. *Researches in Population Ecology* 25: 249–263.

Fukumoto, Y., T. Abe, and A. Taki. 1989. A novel form of colony organization in the "queenless" ant *Diacamma rugosum*. *Physiology and Ecology Japan* 26: 55–61.

Furey, R.E. 1992. Division of labour can be morphological and/or temporal: a reply to Tsuji. *Animal Behaviour* 44: 571.

Gadagkar, R. 1985a. Evolution of insect sociality – A review of some attempts to test modern theories. *Proceedings of the Indian Academy of Sciences (Animal Sciences)* 94: 309–324.

Gadagkar, R. 1985b. Kin recognition in social insects and other animals – A review of recent findings and a consideration of their relevance for the theory of kin selection. *Proceedings of the Indian Academy of Sciences (Animal Sciences)* 94: 587–621.

Gadagkar, R. 1990a. The haplodiploidy threshold and social evolution. *Current Science* 59: 374–375.

Gadagkar, R. 1990b. Origin and evolution of eusociality: a perspective from studying primitively eusocial wasps. *Journal of Genetics* 69: 113–125.

Gadagkar, R. 1990c. Evolution of eusociality: the advantage of assured fitness returns. *Philosophical Transactions of the Royal Society of London, Series B* 329: 17–25.

Gadagkar, R. 1991a. On testing the role of genetic asymmetries created by haplodiploidy in the evolution of eusociality in the Hymenoptera. *Journal of Genetics* 70: 1–31.

Gadagkar, R. 1991b. Demographic predisposition to the evolution of eusociality: a hierarchy of models. *Proceedings of the National Academy of Sciences, U.S.A.* 88: 10993–10997.

Gadagkar, R. 1991c. *Belonogaster*, *Mischocyttarus*, *Parapolybia*, and independent-founding *Ropalidia*. In K.G. Ross and R.W. Matthews, eds., *The Social Biology of Wasps*, pp. 149–190. Comstock, Ithaca, New York.

Gadagkar, R. 1992. Disease and social evolution. *Current Science* 63: 285–286.

Gadagkar, R. 1993. Can animals be spiteful? *Trends in Ecology and Evolution* 8: 232–234.

Gadagkar, R. 1994. Why the definition of eusociality is not helpful to understand its evolution and what should we do about it. *Oikos* 70: 485–488.

Gayley, T. 1993. Genetics of kin selection: the role of behavioral inclusive fitness. *American Naturalist* 141: 928–953.

Gerber, H.S., and E.C. Klostermeyer. 1970. Sex control by bees: a voluntary act of egg fertilization during oviposition. *Science* 167: 82–84.

Getz, W.M. 1991. The honey bee as a model kin recognition system. In P.G. Hepper, ed., *Kin Recognition*, pp. 358–412. Cambridge University Press, Cambridge.

Getz, W.M., D. Brückner, and T.R. Parisian. 1982. Kin structure and the swarming behavior of the honey bee *Apis mellifera*. *Behavioral Ecology and Sociobiology* 10: 265–270.

Ghiselin, M.T. 1974. *The Economy of Nature and the Evolution of Sex*. University of California Press, Berkeley.

Glancey, B.M., C.H. Craig, C.E. Stringer, and P.M. Bishop. 1973. Multiple fertile queens in colonies of the imported fire ant, *Solenopsis invicta*. *Journal of the Georgia Entomological Society* 8: 237–238.

Glancey, B.M., C.E. Stringer, C.H. Craig, and P.M. Bishop. 1975. An extraordinary case of polygyny in the red imported fire ant. *Annals of the Entomological Society of America* 68: 922.

Gliddon, C.J., and P.-H. Gouyon. 1989. The units of selection. *Trends in Ecology and Evolution* 4: 204–208.

Godfray, H.C.J. 1990. The causes and consequences of constrained sex allocation in haplodiploid animals. *Journal of Evolutionary Biology* 3: 3–17.

Godfray, H.C.J. 1991. Signalling of need by offspring to their parents. *Nature* 352: 328–330.

Godfray, H.C.J. 1994. *Parasitoids: Behavioral and Evolutionary Ecology*. Princeton University Press, Princeton, New Jersey.

Godfray, H.C.J., and A. Grafen. 1988. Unmatedness and the evolution of eusociality. *American Naturalist* 131: 303–305.

Godfray, H.C.J., and G.A. Parker. 1991. Clutch size, fecundity and parent-offspring conflict. *Philosophical Transactions of the Royal Society of London, Series B* 332: 67–79.

Godfray, H.C.J., and G.A. Parker. 1992. Sibling competition, parent-offspring conflict and clutch size. *Animal Behaviour* 43: 473–490.

Goodnight, C.J., J.M. Schwartz, and L. Stevens. 1992. Contextual analysis of models of group selection, soft selection, hard selection, and the evolution of altruism. *American Naturalist* 140: 743–761.

Goodnight, K.F. 1992. The effect of stochastic variation on kin selection in a budding-viscous population. *American Naturalist* 140: 1028–1040.

Goodwin, B.C., and M.H. Cohen. 1969. A phase-shift model for the spatial and temporal organization of developing systems. *Journal of Theoretical Biology* 25: 49–107.

Gordon, D.M. 1985. Do we need more ethograms? *Zeitschrift für Tierpsychologie* 68: 340–342.

Gordon, D.M. 1989a. Caste and change in social insects. In P.H. Harvey and L. Partridge, eds., *Oxford Surveys in Evolutionary Biology*, Vol. 6, pp. 55–72. Oxford University Press, Oxford.

Gordon, D.M. 1989b. Dynamics of task switching in harvester ants. *Animal Behaviour* 38: 194–204.

Gordon, D.M. 1991. Behavioral flexibility and the foraging ecology of seed-eating ants. *American Naturalist* 138: 379–411.

Gordon, D.M., B.C. Goodwin, and L.E.H. Trainor. 1992. A parallel distributed model of the behaviour of ant colonies. *Journal of Theoretical Biology* 156: 293–307.

Goss, S., and J.-L. Deneubourg. 1988. Autocatalysis as a source of synchronised rhythmical activity in social insects. *Insectes Sociaux* 35: 310–315.

Goss, S., D. Fresneau, J.-L. Deneubourg, J.-P. Lachaud, and J. Valenzuela-Gonzalez. 1989. Individual foraging in the ant *Pachycondyla apicalis*. *Oecologia* 80: 65–69.

Gotwald, W.H. 1982. Army ants. In H.R. Hermann, ed., *Social Insects*, Vol. IV, pp. 157–254. Academic Press, New York.

Gould, S.J. 1980. Is a new and general theory of evolution emerging? *Paleobiology* 6: 119–130.

Gould, S.J. 1982. Darwinism and the expansion of evolutionary theory. *Science* 216: 380–387.

Gould, S.J. 1983a. Caring groups and selfish genes. In S.J. Gould, *The Panda's Thumb*, pp. 72–78. Penguin, Harmondsworth, U.K.

Gould, S.J. 1983b. Irrelevance, submission, and partnership: the changing role of palaeontology in Darwin's three centennials, and a modest proposal for macroevolution. In D.S. Bendall, ed., *Evolution from Molecules to Men*, pp. 347–366. Cambridge University Press, Cambridge.

Gould, S.J. 1984. What happens to bodies if genes act for themselves? In S.J. Gould, *Hen's Teeth and Horse's Toes*, pp. 166–176. Penguin, Harmondsworth, U.K.

Gould, S.J. 1992. The confusion over evolution. *The New York Review of Books* 39(19): 47–54 (19 November 1992).

Gould, S.J., and R.C. Lewontin. 1979. The spandrels of San Marco and the Panglossian paradigm: a critique of the adaptationist programme. *Proceedings of the Royal Society of London, Series B* 205: 581–598.

Grafen, A. 1980. Models of *r* and *d*. *Nature* 284: 494–495.

Grafen, A. 1982. How not to measure inclusive fitness. *Nature* 298: 425–426.

Grafen, A. 1984. Natural selection, kin selection and group selection. In J.R. Krebs and N.B. Davies, eds., *Behavioural Ecology: An Evolutionary Approach*, 2nd edition, pp. 62–84. Blackwell, Oxford.

Grafen, A. 1985. A geometric view of relatedness. In R. Dawkins and M. Ridley, eds., *Oxford Surveys in Evolutionary Biology*, Vol. 2, pp. 28–89. Oxford University Press, Oxford.

Grafen, A. 1986. Split sex ratios and the evolutionary origins of eusociality. *Journal of Theoretical Biology* 122: 95–121.

Grafen, A. 1988. On the uses of data on lifetime reproductive success. In

T.H. Clutton-Brock, ed., *Reproductive Success*, pp. 454–471. University of Chicago Press, Chicago.

Grafen, A. 1990a. Do animals really recognize kin? *Animal Behaviour* 39: 42–54.

Grafen, A. 1990b. Biological signals as handicaps. *Journal of Theoretical Biology* 144: 517–546.

Grafen, A. 1991. Modelling in behavioural ecology. In J.R. Krebs and N.B. Davies, eds., *Behavioural Ecology: An Evolutionary Approach*, 3rd edition, pp. 5–31. Blackwell, Oxford.

Grant, V. 1963. *The Origin of Adaptations*. Columbia University Press, New York.

Green, P.J., and R. Sibson. 1978. Computing Dirichlet tessellations in the plane. *Computer Journal* 21: 168–173.

Greenberg, L., D.J.C. Fletcher, and S.B. Vinson. 1985. Differences in worker size and mound distribution in monogynous and polygynous colonies of the fire ant, *Solenopsis invicta* Buren. *Journal of the Kansas Entomological Society* 58: 9–18.

Hagen, R.H., D.R. Smith, and S.W. Rissing. 1988. Genetic relatedness among co-foundresses of two desert ants, *Veromessor pergandei* and *Acromyrmex versicolor* (Hymenoptera: Formicidae). *Psyche* 95: 191–201.

Haig, D. 1987. Kin conflict in seed plants. *Trends in Ecology and Evolution* 2: 337–340.

Haig, D. 1992a. Genomic imprinting and the theory of parent-offspring conflict. *Developmental Biology* 3: 153–160.

Haig, D. 1992b. Intragenomic conflict and the evolution of eusociality. *Journal of Theoretical Biology* 156: 401–403.

Haig, D., and A. Grafen. 1991. Genetic scrambling as a defence against meiotic drive. *Journal of Theoretical Biology* 153: 531–558.

Haldane, J.B.S. 1932. *The Causes of Evolution*. Longmans, London.

Haldane, J.B.S. 1955. Population genetics. In M.L. Johnson, M. Abercrombie, and G.E. Fogg, eds., *New Biology* No. 18, pp. 34–51. Penguin, Harmondsworth, U.K.

Halliday, R.B. 1983. Social organization of meat ants *Iridomyrmex purpureus* analysed by gel electrophoresis of enzymes. *Insectes Sociaux* 30: 45–56.

Hamaguchi, K., Y. Itô, and O. Takenaka. 1993. GT dinucleotide repeat polymorphisms in a polygynous ant, *Leptothorax spinosior* and their use for measurement of relatedness. *Naturwissenschaften* 80: 179–181.

Hamilton, W.D. 1963. The evolution of altruistic behavior. *American Naturalist* 97: 354–356.

Hamilton, W.D. 1964a. The genetical evolution of social behaviour. I. *Journal of Theoretical Biology* 7: 1–16.

Hamilton, W.D. 1964b. The genetical evolution of social behaviour. II. *Journal of Theoretical Biology* 7: 17–52.

Hamilton, W.D. 1966. The moulding of senescence by natural selection. *Journal of Theoretical Biology* 12: 12–45.

Hamilton, W.D. 1967. Extraordinary sex ratios. *Science* 156: 477–488.

Hamilton, W.D. 1970. Selfish and spiteful behaviour in an evolutionary model. *Nature* 228: 1218–1220.

Hamilton, W.D. 1971a. Selection of selfish and altruistic behavior in some extreme models. In J.F. Eisenberg and W.S. Dillon, eds., *Man and Beast: Comparative Social Behavior*, pp. 57–91. Smithsonian Institution Press, Washington, D.C.

Hamilton, W.D. 1971b. Geometry for the selfish herd. *Journal of Theoretical Biology* 31: 295–311.

Hamilton, W.D. 1971c. Addendum: sex ratio and social coefficients of relationship under male haploidy. In G.C. Williams, ed., *Group Selection*, pp. 87–89. Aldine Atherton, Chicago.

Hamilton, W.D. 1972. Altruism and related phenomena, mainly in social insects. *Annual Review of Ecology and Systematics* 3: 193–232.

Hamilton, W.D. 1975a. Innate social aptitudes of man: an approach from evolutionary genetics. In R. Fox, ed., *Biosocial Anthropology*, pp. 133–155. Malaby Press, London.

Hamilton, W.D. 1975b. Gamblers since life began: barnacles, aphids, elms. *Quarterly Review of Biology* 50: 175–180.

Hamilton, W.D. 1979. Wingless and fighting males in fig wasps and other insects. In M.S. Blum and N.A. Blum, eds., *Sexual Selection and Reproductive Competition in Insects*, pp. 167–220. Academic Press, New York.

Hamilton, W.D. 1987a. Discriminating nepotism: expectable, common, overlooked. In D.J.C. Fletcher and C.D. Michener, eds., *Kin Recognition in Animals*, pp. 417–437. Wiley, Chichester, U.K.

Hamilton, W.D. 1987b. Kinship, recognition, disease, and intelligence: constraints of social evolution. In Y. Itô, J.L. Brown, and J. Kikkawa, eds., *Animal Societies: Theories and Facts*, pp. 81–102. Japan Scientific Societies Press, Tokyo.

Hamilton, W.D., and R.M. May. 1977. Dispersal in stable habitats. *Nature* 269: 578–581.

Hansell, M. 1987. Nest building as a facilitating and limiting factor in the evolution of eusociality in the Hymenoptera. In P.H. Harvey and L. Partridge, eds., *Oxford Surveys in Evolutionary Biology*, Vol. 4, pp. 155–181. Oxford University Press, Oxford.

Harborne, J.B. 1988. *Introduction to Ecological Biochemistry*. 3rd edition. Academic Press, London.

Harcourt, A.H. 1991. Sperm competition and the evolution of nonfertilizing sperm in mammals. *Evolution* 45: 314–328.

Hardy, I.C.W. 1992. Non-binomial sex allocation and brood sex ratio variances in the parasitoid Hymenoptera. *Oikos* 65: 143–158.

Harpending, H.C. 1979. The population genetics of interactions. *American Naturalist* 113: 622–630.

Harper, J.L. 1980. Plant demography and ecological theory. *Oikos* 35: 244–253.

Harper, J.L., and A.D. Bell. 1979. The population dynamics of growth form in organisms with modular construction. In R.M. Anderson, B.D. Turner, and L.R. Taylor, eds., *Population Dynamics*, pp. 29–52. Blackwell, Oxford.

Harvey, P.H. 1985. Intrademic group selection and the sex ratio. In R.M. Sibly and R.H. Smith, eds., *Behavioural Ecology: Ecological Consequences of Adaptive Behaviour*, pp. 59–73. Blackwell, Oxford.

Harvey, P.H., and M.D. Pagel. 1991. *The Comparative Method in Evolutionary Biology.* Oxford University Press, Oxford.

Hasegawa, E. 1993. Nest defense and early production of the major workers in the dimorphic ant *Colobopsis nipponicus* (Wheeler) (Hymenoptera: Formicidae). *Behavioral Ecology and Sociobiology* 33: 73–77.

Hasegawa, E. 1994. Sex allocation in the ant *Colobopsis nipponicus* (Wheeler). I. Population sex ratio. *Evolution* 48: 1121–1129.

Hasegawa, E., and T. Yamaguchi. 1994. Population structure, local mate competition and sex allocation patterns in the ant *Messor aciculatus*. In A. Lenoir, G. Arnold, and M. Lepage, eds., *Les Insectes Sociaux*, p. 77. Université Paris Nord, Villetaneuse.

Haskins, C.P. 1970. Researches in the biology and social behavior of primitive ants. In L.R. Aronson, E. Tobach, D.S. Lehrman, and J.S. Rosenblatt, eds., *Development and Evolution of Behavior*, pp. 355–388. W.H. Freeman, San Francisco.

Haskins, C.P. 1978. Sexual calling behavior in highly primitive ants. *Psyche* 85: 407–415.

Hatcher, M.J., C. Tofts, and N.R. Franks. 1992. Mutual exclusion as a mechanism for information exchange within ant nests. *Naturwissenschaften* 79: 32–34.

Headley, A.E. 1943. Population studies of two species of ants, *Leptothorax longispinosus* Roger and *Leptothorax curvispinosus* Mayr. *Annals of the Entomological Society of America* 36: 743–753.

Heinze, J. 1989. Alternative dispersal strategies in a North American ant. *Naturwissenschaften* 76: 477–478.

Heinze, J. 1992. Ecological correlates of functional monogyny and queen dominance in leptothoracine ants. In J. Billen, ed., *Biology and Evolution of Social Insects*, pp. 25–33. Leuven University Press, Leuven, Belgium.

Heinze, J. 1993a. Queen-queen interactions in polygynous ants. In L. Keller, ed., *Queen Number and Sociality in Insects*, pp. 334–361. Oxford University Press, Oxford.

Heinze, J. 1993b. Habitat structure, dispersal strategies and queen number in two boreal *Leptothorax* ants. *Oecologia* 96: 32–39.

Heinze, J. 1993c. Life histories of subarctic ants. *Arctic* 46: 354–358.

Heinze, J., and A. Buschinger. 1987. Queen polymorphism in a non-parasitic *Leptothorax* species (Hymenoptera, Formicidae). *Insectes Sociaux* 34: 28–43.

Heinze, J., and A. Buschinger. 1988. Polygyny and functional monogyny in *Leptothorax* ants (Hymenoptera: Formicidae). *Psyche* 95: 309–325.

Heinze, J., and A. Buschinger. 1989. Queen polymorphism in *Leptothorax* spec. A: its genetic and ecological background (Hymenoptera: Formicidae). *Insectes Sociaux* 36: 139–155.

Heinze, J., and B. Hölldobler. 1993a. Queen polymorphism in an Australian weaver ant, *Polyrhachis* cf. *doddi. Psyche* 100: 83–92.

Heinze, J., and B. Hölldobler. 1993b. Fighting for a harem of queens: physiology of reproduction in *Cardiocondyla* male ants. *Proceedings of the National Academy of Sciences, U.S.A.* 90: 8412–8414.

Heinze, J., and B. Hölldobler. 1994. Ants in the cold. *Memorabilia Zoologica* 48: 99–108.

Heinze, J., and D. Ortius. 1991. Social organization of *Leptothorax acervorum* from Alaska (Hymenoptera: Formicidae). *Psyche* 98: 227–240.

Heinze, J., and T.A. Smith. 1990. Dominance and fertility in a functionally monogynous ant. *Behavioral Ecology and Sociobiology* 27: 1–10.

Heinze, J., B. Hölldobler, and S.P. Cover. 1992a. Queen polymorphism in the North American harvester ant, *Ephebomyrmex imberbiculus*. *Insectes Sociaux* 39: 267–273.

Heinze, J., N. Lipski, and B. Hölldobler. 1992b. Reproductive competition in colonies of the ant *Leptothorax gredleri*. *Ethology* 90: 265–278.

Heinze, J., S. Kühnholz, K. Schilder, and B. Hölldobler. 1993a. Behavior of ergatoid males in the ant, *Cardiocondyla nuda*. *Insectes Sociaux* 40: 273–282.

Heinze, J., T. Gübitz, C. Errard, A. Lenoir, and B. Hölldobler. 1993b. Reproductive competition and colony fragmentation in the guest-ant, *Formicoxenus provancheri*. *Experientia* 49: 814–816.

Heinze, J., B. Hölldobler, and C. Peeters. 1994. Conflict and cooperation in ant societies. *Naturwissenschaften* 81: 489–497.

Heinze, J., N. Lipski, B. Hölldobler, and A.F.G. Bourke. 1995. Geographical variation in the social and genetic structure of the ant, *Leptothorax acervorum*. *Zoology* 98: 127–135.

Helms, K.R. 1994. Sexual size dimorphism and sex ratios in bees and wasps. *American Naturalist* 143: 418–434.

Hemerik, L., N.F. Britton, and N.R. Franks. 1990. Synchronization of the behaviour within nests of the ant *Leptothorax acervorum* (Fabricius) – II. Modelling the phenomenon and predictions from the model. *Bulletin of Mathematical Biology* 52: 613–628.

Henderson, G., and R.L. Jeanne. 1992. Population biology and foraging ecology of prairie ants in southern Wisconsin (Hymenoptera: Formicidae). *Journal of the Kansas Entomological Society* 65: 16–29.

Herbers, J.M. 1979. The evolution of sex-ratio strategies in Hymenopteran societies. *American Naturalist* 114: 818–834.

Herbers, J.M. 1982. Queen number and colony ergonomics in *Leptothorax longispinosus*. In M.D. Breed, C.D. Michener, and H.E. Evans, eds., *The Biology of Social Insects*, pp. 238–242. Westview Press, Boulder, Colorado.

Herbers, J.M. 1984. Queen-worker conflict and eusocial evolution in a polygynous ant species. *Evolution* 38: 631–643.

Herbers, J.M. 1986a. Ecological genetics of queen number in *Leptothorax longispinosus* (Hymenoptera: Formicidae). *Entomologia Generalis* 11: 119–123.

Herbers, J.M. 1986b. Effects of ecological parameters on queen number in *Leptothorax longispinosus* (Hymenoptera: Formicidae). *Journal of the Kansas Entomological Society* 59: 675–686.

Herbers, J.M. 1986c. Nest site limitation and facultative polygyny in the ant *Leptothorax longispinosus*. *Behavioral Ecology and Sociobiology* 19: 115–122.

Herbers, J.M. 1989. Community structure in north temperate ants: temporal and spatial variation. *Oecologia* 81: 201–211.

Herbers, J.M. 1990. Reproductive investment and allocation ratios for the ant

Leptothorax longispinosus: sorting out the variation. *American Naturalist* 136: 178–208.

Herbers, J.M. 1991. The population biology of *Tapinoma minutum* (Hymenoptera: Formicidae) in Australia. *Insectes Sociaux* 38: 195–204.

Herbers, J.M. 1993. Ecological determinants of queen number in ants. In L. Keller, ed., *Queen Number and Sociality in Insects*, pp. 262–293. Oxford University Press, Oxford.

Herbers, J.M., and M. Cunningham. 1983. Social organization in *Leptothorax longispinosus* Mayr. *Animal Behaviour* 31: 759–771.

Herbers, J.M., and S. Grieco. 1994. Population structure of *Leptothorax ambiguus*, a facultatively polygynous and polydomous ant species. *Journal of Evolutionary Biology* 7: 581–598.

Herbers, J.M., and R.J. Stuart. 1990. Relatedness and queen number in *Leptothorax longispinosus*. In G.K. Veeresh, B. Mallik, and C.A. Viraktamath, eds., *Social Insects and the Environment*, pp. 258–259. Oxford & IBH Publishing Co., New Delhi.

Herre, E.A. 1985. Sex ratio adjustment in fig wasps. *Science* 228: 896–898.

Herre, E.A., E.G. Leigh, and E.A. Fischer. 1987. Sex allocation in animals. In S.C. Stearns, ed., *The Evolution of Sex and its Consequences*, pp. 219–244. Birkhäuser Verlag, Basel.

Higashi, M., N. Yamamura, T. Abe, and T.P. Burns. 1991. Why don't all termite species have a sterile worker caste? *Proceedings of the Royal Society of London, Series B* 246: 25–29.

Higashi, S. 1983. Polygyny and nuptial flight of *Formica (Formica) yessensis* Forel at Ishikari Coast, Hokkaido, Japan. *Insectes Sociaux* 30: 287–297.

Higashi, S., and K. Yamauchi. 1979. Influence of a supercolonial ant *Formica (Formica) yessensis* Forel on the distribution of other ants in Ishikari coast. *Japanese Journal of Ecology* 29: 257–264.

Higashi, S., F. Ito, N. Sugiura, and K. Ohkawara. 1994. Worker's age regulates the linear dominance hierarchy in the queenless ponerine ant, *Pachycondyla sublaevis* (Hymenoptera: Formicidae). *Animal Behaviour* 47: 179–184.

Hillesheim, E., N. Koeniger, and R.F.A. Moritz. 1989. Colony performance in honeybees (*Apis mellifera capensis* Esch.) depends on the proportion of subordinate and dominant workers. *Behavioral Ecology and Sociobiology* 24: 291–296.

Hofstadter, D.R. 1980. *Gödel, Escher, Bach: An Eternal Golden Braid*. Penguin, Harmondsworth, U.K.

Hölldobler, B. 1976. The behavioral ecology of mating in harvester ants (Hymenoptera: Formicidae: *Pogonomyrmex*). *Behavioral Ecology and Sociobiology* 1: 405–423.

Hölldobler, B. 1980. Canopy orientation: a new kind of orientation in ants. *Science* 210: 86–88.

Hölldobler, B. 1982. Communication, raiding behavior and prey storage in *Cerapachys* (Hymenoptera: Formicidae). *Psyche* 89: 3–23.

Hölldobler, B. 1983. Territorial behavior in the green tree ant (*Oecophylla smaragdina*). *Biotropica* 15: 241–250.

Hölldobler, B., and S.H. Bartz. 1985. Sociobiology of reproduction in ants. In B.

Hölldobler and M. Lindauer, eds., *Experimental Behavioral Ecology and Sociobiology*, pp. 237–257. Gustav Fischer Verlag, Stuttgart.

Hölldobler, B., and N.F. Carlin. 1985. Colony founding, queen dominance and oligogyny in the Australian meat ant *Iridomyrmex purpureus*. *Behavioral Ecology and Sociobiology* 18: 45–58.

Hölldobler, B., and N.F. Carlin. 1989. Colony founding, queen control and worker reproduction in the ant *Aphaenogaster* (= *Novomessor*) *cockerelli* (Hymenoptera: Formicidae). *Psyche* 96: 131–151.

Hölldobler, B., and C.P. Haskins. 1977. Sexual calling behavior in primitive ants. *Science* 195: 793–794.

Hölldobler, B., and C.J. Lumsden. 1980. Territorial strategies in ants. *Science* 210: 732–739.

Hölldobler, B., and U. Maschwitz. 1965. Der Hochzeitsschwarm der Rossameise *Camponotus herculeanus* L. (Hym. Formicidae). *Zeitschrift für Vergleichende Physiologie* 50: 551–568.

Hölldobler, B., and C.D. Michener. 1980. Mechanisms of identification and discrimination in social Hymenoptera. In H. Markl, ed., *Evolution of Social Behavior: Hypotheses and Empirical Tests*, pp. 35–57. Verlag Chemie, Weinheim.

Hölldobler, B., and E.O. Wilson. 1977. The number of queens: an important trait in ant evolution. *Naturwissenschaften* 64: 8–15.

Hölldobler, B., and E.O. Wilson. 1983. Queen control in colonies of weaver ants (Hymenoptera: Formicidae). *Annals of the Entomological Society of America* 76: 235–238.

Hölldobler, B., and E.O. Wilson. 1990. *The Ants*. Springer Verlag, Berlin.

Horn, H.S., and D.I. Rubenstein. 1984. Behavioural adaptations and life history. In J.R. Krebs and N.B. Davies, eds., *Behavioural Ecology: An Evolutionary Approach*, 2nd edition, pp. 279–298. Blackwell, Oxford.

Houston, A., P. Schmid-Hempel, and A. Kacelnik. 1988. Foraging strategy, worker mortality, and the growth of the colony in social insects. *American Naturalist* 131: 107–114.

Howard, D.F., and W.R. Tschinkel. 1981. The flow of food in colonies of the fire ant, *Solenopsis invicta*: a multifactorial study. *Physiological Entomology* 6: 297–306.

Huang, Z.-Y., and G.E. Robinson. 1992. Honeybee colony integration: Worker-worker interactions mediate hormonally regulated plasticity in division of labor. *Proceedings of the National Academy of Sciences, U.S.A.* 89: 11726–11729.

Hughes, C.R., D.C. Queller, J.E. Strassmann, and S.K. Davis. 1993a. Relatedness and altruism in *Polistes* wasps. *Behavioral Ecology* 4: 128–137.

Hughes, C.R., D.C. Queller, J.E. Strassmann, C.R. Solís, J.A. Negrón-Sotomayor, and K.R. Gastreich. 1993b. The maintenance of high genetic relatedness in multi-queen colonies of social wasps. In L. Keller, ed., *Queen Number and Sociality in Insects*, pp. 153–170. Oxford University Press, Oxford.

Hull, D.L. 1980. Individuality and selection. *Annual Review of Ecology and Systematics* 11: 311–332.

Hurst, G.D.D., L.D. Hurst, and R.A. Johnstone. 1992. Intranuclear conflict and its role in evolution. *Trends in Ecology and Evolution* 7: 373–378.

Hurst, L.D. 1991. The incidences and evolution of cytoplasmic male killers. *Proceedings of the Royal Society of London, Series B* 244: 91–99.

Hurst, L.D. 1992. Intragenomic conflict as an evolutionary force. *Proceedings of the Royal Society of London, Series B* 248: 135–140.

Hurst, L.D. 1993a. The incidences, mechanisms and evolution of cytoplasmic sex ratio distorters in animals. *Biological Reviews* 68: 121–194.

Hurst, L.D. 1993b. Drunken walk of the diploid. *Nature* 365: 206–207.

Hurst, L.D., and A. Pomiankowski. 1991. Causes of sex ratio bias may account for unisexual sterility in hybrids: a new explanation of Haldane's rule and related phenomena. *Genetics* 128: 841–858.

Hurst, L.D., and F. Vollrath. 1992. Sex-ratio adjustment in solitary and social spiders. *Trends in Ecology and Evolution* 7: 326–327.

Huxley, J.S. 1932. *Problems of Relative Growth*. Dial Press, New York.

Ito, F. 1990. Functional monogyny of *Leptothorax acervorum* in northern Japan. *Psyche* 97: 203–211.

Ito, F. 1993a. Functional monogyny and dominance hierarchy in the queenless ponerine ant *Pachycondyla* (= *Bothroponera*) sp. in West Java, Indonesia (Hymenoptera, Formicidae, Ponerinae). *Ethology* 95: 126–140.

Ito, F. 1993b. Social organization in a primitive ponerine ant: queenless reproduction, dominance hierarchy and functional polygyny in *Amblyopone* sp. (*reclinata* group) (Hymenoptera: Formicidae: Ponerinae). *Journal of Natural History* 27: 1315–1324.

Ito, F., and S. Higashi. 1991. A linear dominance hierarchy regulating reproduction and polyethism of the queenless ant *Pachycondyla sublaevis*. *Naturwissenschaften* 78: 80–82.

Ito, F., and K. Ohkawara. 1994. Spermatheca size differentiation between queens and workers in primitive ants. Relationship with reproductive structure of colonies. *Naturwissenschaften* 81: 138–140.

Itô, Y. 1989. The evolutionary biology of sterile soldiers in aphids. *Trends in Ecology and Evolution* 4: 69–73.

Itô, Y. 1993. *Behaviour and Social Evolution of Wasps: The Communal Aggregation Hypothesis*. Oxford University Press, Oxford.

Itow, T., K. Kobayashi, M. Kubota, K. Ogata, H.T. Imai, and R.H. Crozier. 1984. The reproductive cycle of the queenless ant *Pristomyrmex pungens*. *Insectes Sociaux* 31: 87–102.

Iwasa, Y. 1981. Role of sex ratio in the evolution of eusociality in haplodiploid social insects. *Journal of Theoretical Biology* 93: 125–142.

Jaisson, P. 1975. L'imprégnation dans l'ontogenèse des comportements de soins aux cocons chez la jeune fourmi rousse (*Formica polyctena* Först.). *Behaviour* 52: 1–37.

Jaisson, P. 1985. Social behaviour. In G.A. Kerkut and L.I. Gilbert, eds., *Comprehensive Insect Physiology, Biochemistry and Pharmacology, Vol. 9, Behaviour*, pp. 673–694. Pergamon Press, Oxford.

Jaisson, P. 1991. Kinship and fellowship in ants and social wasps. In P.G. Hepper, ed., *Kin Recognition*, pp. 60–93. Cambridge University Press, Cambridge.

Jaisson, P., D. Fresneau, and J.-P. Lachaud. 1988. Individual traits of social behavior in ants. In R.L. Jeanne, ed., *Interindividual Behavioral Variability in Social Insects*, pp. 1–51. Westview Press, Boulder, Colorado.

Janzen, D.H. 1973. Evolution of polygynous obligate acacia-ants in western Mexico. *Journal of Animal Ecology* 42: 727–750.

Jarvis, J.U.M., M.J. O'Riain, N.C. Bennett, and P.W. Sherman. 1994. Mammalian eusociality: a family affair. *Trends in Ecology and Evolution* 9: 47–51.

Johnston, A.B., and E.O. Wilson. 1985. Correlations of variation in the major/minor ratio of the ant, *Pheidole dentata* (Hymenoptera: Formicidae). *Annals of the Entomological Society of America* 78: 8–11.

Joshi, N.V., and R. Gadagkar. 1985. Evolution of sex ratios in social Hymenoptera: kin selection, local mate competition, polyandry and kin recognition. *Journal of Genetics* 64: 41–58.

Jouvenaz, D.P., D.P. Wojcik, and R.K. Vander Meer. 1989. First observation of polygyny in fire ants, *Solenopsis* spp., in South America. *Psyche* 96: 161–165.

Karlin, S., and S. Lessard. 1986. *Theoretical Studies on Sex Ratio Evolution*. Princeton University Press, Princeton, New Jersey.

Kasuya, E. 1982. Factors governing the evolution of eusociality through kin selection. *Researches on Population Ecology* 24: 174–192.

Kauffman, S.A. 1991. Antichaos and adaptation. *Scientific American* 265: 64–70.

Kaufmann, B., J.J. Boomsma, L. Passera, and K.N. Petersen. 1992. Relatedness and inbreeding in a French population of the unicolonial ant *Iridomyrmex humilis* (Mayr). *Insectes Sociaux* 39: 195–200.

Keeler, K.H. 1988. Colony survivorship in *Pogonomyrmex occidentalis*, western harvester ant, in western Nebraska. *The Southwestern Naturalist* 33: 480–482.

Keeler, K.H. 1993. Fifteen years of colony dynamics in *Pogonomyrmex occidentalis*, the western harvester ant, in western Nebraska. *The Southwestern Naturalist* 38: 286–289.

Keller, L. 1988. Evolutionary implications of polygyny in the Argentine ant, *Iridomyrmex humilis* (Mayr) (Hymenoptera: Formicidae): an experimental study. *Animal Behaviour* 36: 159–165.

Keller, L. 1991. Queen number, mode of colony founding, and queen reproductive success in ants (Hymenoptera Formicidae). *Ethology, Ecology and Evolution* 3: 307–316.

Keller, L. (ed.) 1993a. *Queen Number and Sociality in Insects*. Oxford University Press, Oxford.

Keller, L. 1993b. The assessment of reproductive success of queens in ants and other social insects. *Oikos* 67: 177–180.

Keller, L. 1995a. Sterility by deleterious alleles and the evolution of sociality. *Journal of Theoretical Biology:* in press.

Keller, L. 1995b. Parasites, worker polymorphism, and queen number in social insects. *American Naturalist* 145: 842–847.

Keller, L., and P. Nonacs. 1993. The role of queen pheromones in social insects: queen control or queen signal? *Animal Behaviour* 45: 787–794.

Keller, L., and L. Passera. 1989a. Size and fat content of gynes in relation to the mode of colony founding in ants (Hymenoptera; Formicidae). *Oecologia* 80: 236–240.

Keller, L., and L. Passera. 1989b. Influence of the number of queens on nest-mate recognition and attractiveness of queens to workers in the Argentine ant, *Iridomyrmex humilis* (Mayr). *Animal Behaviour* 37: 733–740.

Keller, L., and L. Passera. 1990. Fecundity of ant queens in relation to their age and the mode of colony founding. *Insectes Sociaux* 37: 116–130.

Keller, L., and L. Passera. 1992. Mating system, optimal number of matings, and sperm transfer in the Argentine ant *Iridomyrmex humilis*. *Behavioral Ecology and Sociobiology* 31: 359–366.

Keller, L., and L. Passera. 1993. Incest avoidance, fluctuating asymmetry, and the consequences of inbreeding in *Iridomyrmex humilis*, an ant with multiple queen colonies. *Behavioral Ecology and Sociobiology* 33: 191–199.

Keller, L., and H.K. Reeve. 1994a. Partitioning of reproduction in animal societies. *Trends in Ecology and Evolution* 9: 98–102.

Keller, L., and H.K. Reeve. 1994b. Genetic variability, queen number, and polyandry in social Hymenoptera. *Evolution* 48: 694–704.

Keller, L., and K.G. Ross. 1993a. Phenotypic basis of reproductive success in a social insect: genetic and social determinants. *Science* 260: 1107–1110.

Keller, L., and K.G. Ross. 1993b. Phenotypic plasticity and "cultural transmission" of alternative social organizations in the fire ant *Solenopsis invicta*. *Behavioral Ecology and Sociobiology* 33: 121–129.

Keller, L., and E.L. Vargo. 1993. Reproductive structure and reproductive roles in colonies of eusocial insects. In L. Keller, ed., *Queen Number and Sociality in Insects*, pp. 16–44. Oxford University Press, Oxford.

Keller, L., L. Passera, and J.-P. Suzzoni. 1989. Queen execution in the Argentine ant, *Iridomyrmex humilis*. *Physiological Entomology* 14: 157–163.

Kelly, J.K. 1992a. Kin selection in density regulated populations. *Journal of Theoretical Biology* 157: 447–461.

Kelly, J.K. 1992b. Restricted migration and the evolution of altruism. *Evolution* 46: 1492–1495.

Kelly, J.K. 1994. The effect of scale dependent processes on kin selection: mating and density regulation. *Theoretical Population Biology* 46: 32–57.

Kent, D.S., and J.A. Simpson. 1992. Eusociality in the beetle *Austroplatypus incompertus* (Coleoptera: Curculionidae). *Naturwissenschaften* 79: 86–87.

Kinomura, K., and K. Yamauchi. 1987. Fighting and mating behaviors of dimorphic males in the ant *Cardiocondyla wroughtoni*. *Journal of Ethology* 5: 75–81.

Klein, R.W., D. Kovac, A. Schellerich, and U. Maschwitz. 1992. Mealybug-carrying by swarming queens of a Southeast Asian bamboo-inhabiting ant. *Naturwissenschaften* 79: 422–423.

Koenig, W.D., F.A. Pitelka, W.J. Carmen, R.L. Mumme, and M.T. Stanback. 1992. The evolution of delayed dispersal in cooperative breeders. *Quarterly Review of Biology* 67: 111–150.

Kolman, W.A. 1960. The mechanism of natural selection for the sex ratio. *American Naturalist* 94: 373–377.

Kozlowski, J., and S.C. Stearns. 1989. Hypotheses for the production of excess zygotes: models of bet-hedging and selective abortion. *Evolution* 43: 1369–1377.

Krebs, J.R., and N.B. Davies. 1987. *An Introduction to Behavioural Ecology,* 2nd edition. Blackwell, Oxford.

Krebs, J.R., and N.B. Davies. 1993. *An Introduction to Behavioural Ecology,* 3rd edition. Blackwell, Oxford.

Krebs, R.A., and S.W. Rissing. 1991. Preference for larger foundress associations in the desert ant *Messor pergandei. Animal Behaviour* 41: 361–363.

Kukuk, P.F., G.C. Eickwort, M. Raveret-Richter, B. Alexander, R. Gibson, R.A. Morse, and F. Ratnieks. 1989. Importance of the sting in the evolution of sociality in the Hymenoptera. *Annals of the Entomological Society of America* 82: 1–5.

Kurland, J.A. 1980. Kin selection theory: a review and selective bibliography. *Ethology and Sociobiology* 1: 255–274.

Lachaud, J.-P., and D. Fresneau. 1987. Social regulation in ponerine ants. In J.M. Pasteels and J.-L. Deneubourg, eds., *From Individual to Collective Behavior in Social Insects,* pp. 197–217. Birkhäuser Verlag, Basel.

Lack, D. 1966. *Population Studies of Birds.* Clarendon Press, Oxford.

Laidlaw, H.H., and R.E. Page. 1984. Polyandry in honey bees (*Apis mellifera* L.): sperm utilization and intracolony genetic relationships. *Genetics* 108: 985–997.

Lashley, K.S. 1950. In search of the engram. In J.F. Danielli and R. Brown, eds., *Physiological Mechanisms in Animal Behaviour (Symposia of the Society for Experimental Biology, No. 4),* pp. 454–482. Cambridge University Press, Cambridge.

Le Masne, G. 1956. La signification des reproducteurs aptères chez la fourmi *Ponera eduardi* Forel. *Insectes Sociaux* 3: 239–259.

Leigh, E.G. 1971. *Adaptation and Diversity.* Freeman, Cooper and Co., San Francisco, California.

Leigh, E.G. 1977. How does selection reconcile individual advantage with the good of the group? *Proceedings of the National Academy of Sciences, U.S.A.* 74: 4542–4546.

Leigh, E.G. 1991. Genes, bees and ecosystems: the evolution of a common interest among individuals. *Trends in Ecology and Evolution* 6: 257–262.

Lenoir, A. 1979a. Feeding behaviour in young societies of the ant *Tapinoma erraticum* L.: trophallaxis and polyethism. *Insectes Sociaux* 26: 19–37.

Lenoir, A. 1979b. Le comportement alimentaire et la division du travail chez la fourmi *Lasius niger. Bulletin Biologique de la France et de la Belgique* 113: 79–314.

Lenoir, A., and H. Cagniant. 1986. Role of worker thelytoky in colonies of the ant *Cataglyphis cursor* (Hymenoptera: Formicidae). *Entomologia Generalis* 11: 153–157.

Lenoir, A., L. Querard, N. Pondicq, and F. Berton. 1988. Reproduction and dispersal in the ant *Cataglyphis cursor* (Hymenoptera, Formicidae). *Psyche* 95: 21–44.

Léon, J.A. 1985. Germination strategies. In P.J. Greenwood, P.H. Harvey, and M. Slatkin, eds., *Evolution: Essays in Honour of John Maynard Smith,* pp. 129–143. Cambridge University Press, Cambridge.

Lessard, S. 1992. Relatedness and inclusive fitness with inbreeding. *Theoretical Population Biology* 42: 284–307.

Levings, S.C., and N.R. Franks. 1982. Patterns of nest dispersion in a tropical ground ant community. *Ecology* 63: 338–344.

Levings, S.C., and J.F.A. Traniello. 1981. Territoriality, nest dispersion, and community structure in ants. *Psyche* 88: 265–319.

Lewis, T. 1975. Colony size, density and distribution of the leaf-cutting ant, *Acromyrmex octospinosus* (Reich) in cultivated fields. *Transactions of the Royal Entomological Society of London* 127: 51–64.

Lewontin, R.C. 1965. Selection for colonizing ability. In H.G. Baker and G.L. Stebbins, eds., *The Genetics of Colonizing Species*, pp. 77–94. Academic Press, New York.

Lewontin, R.C. 1970. The units of selection. *Annual Review of Ecology and Systematics* 1: 1–18.

Lewontin, R.C. 1974. *The Genetic Basis of Evolutionary Change*. Columbia University Press, New York.

Ligon, J.D. 1991. Co-operation and reciprocity in birds and mammals. In P.G. Hepper, ed., *Kin Recognition*, pp. 30–59. Cambridge University Press, Cambridge.

Lin, N., and C.D. Michener. 1972. Evolution of sociality in insects. *Quarterly Review of Biology* 47: 131–159.

Lipski, N., J. Heinze, and B. Hölldobler. 1992. Social organization of three European *Leptothorax* species (Hymenoptera, Formicidae). In J. Billen, ed., *Biology and Evolution of Social Insects*, pp. 287–290. Leuven University Press, Leuven, Belgium.

Lloyd, E.A., and S.J. Gould. 1993. Species selection on variability. *Proceedings of the National Academy of Sciences, U.S.A.* 90: 595–599.

Lloyd, J.E. 1986. Firefly communication and deception: "Oh, what a tangled web". In R.W. Mitchell and N.S. Thompson, eds., *Deception: Perspectives on Human and Nonhuman Deceit*, pp. 113–128. SUNY Press, Albany, New York.

Lofgren, C.S., W.A. Banks, and B.M. Glancey. 1975. Biology and control of imported fire ants. *Annual Review of Entomology* 20: 1–30.

López, F., J.M. Serrano, and F.J. Acosta. 1994. Parallels between the foraging strategies of ants and plants. *Trends in Ecology and Evolution* 9: 150–153.

Lumsden, C.J. 1982. The social regulation of physical caste: the superorganism revived. *Journal of Theoretical Biology* 95: 749–781.

Luther, A. 1987. Production of sexuals in relation to nest-condition and habitat quality in *Formica aquilonia* Yarrow (Formicidae, Hymenoptera). In J. Eder and H. Rembold, eds., *Chemistry and Biology of Social Insects*, pp. 303–304. Verlag J. Peperny, München.

Lyttle, T.W. 1991. Segregation distorters. *Annual Review of Genetics* 25: 511–557.

MacArthur, R.H., and E.O. Wilson. 1967. *The Theory of Island Biogeography*. Princeton University Press, Princeton, New Jersey.

Macevicz, S. 1979. Some consequences of Fisher's sex ratio principle for social Hymenoptera that reproduce by colony fission. *American Naturalist* 113: 363–371.

MacKay, W.P. 1983. Stratification of workers in harvester ant nests (Hymenoptera: Formicidae). *Journal of the Kansas Entomological Society* 56: 538–542.

Mackie, G.O. 1986. From aggregates to integrates: physiological aspects of modularity in colonial animals. *Philosophical Transactions of the Royal Society of London, Series B* 313: 175–196.

Macnair, M.R. 1978. An ESS for the sex ratio in animals, with particular reference to the social Hymenoptera. *Journal of Theoretical Biology* 70: 449–459.

Macnair, M.R., and G.A. Parker. 1978. Models of parent-offspring conflict. II. Promiscuity. *Animal Behaviour* 26: 111–122.

Macnair, M.R., and G.A. Parker. 1979. Models of parent-offspring conflict. III. Intra-brood conflict. *Animal Behaviour* 27: 1202–1209.

Madsen, T., R. Shine, J. Loman, and T. Håkansson. 1992. Why do female adders copulate so frequently? *Nature* 355: 440–441.

Marchal, P. 1917. La fourmi d'Argentine (*Iridomyrmex humilis* Mayr). *Bulletin de la Société d'Etude et de Vulgarisation de Zoologie et d'Agriculture de Bordeaux* 16: 23–26.

Margulis, L. 1981. *Symbiosis in Cell Evolution*. W.H. Freeman, San Francisco.

Marikovsky, P.I. 1961. Material on sexual biology of the ant *Formica rufa* L. *Insectes Sociaux* 8: 23–30.

Markin, G.P. 1970. The seasonal life cycle of the Argentine ant, *Iridomyrmex humilis* (Hymenoptera, Formicidae), in southern California. *Annals of the Entomological Society of America* 63: 1238–1242.

Markin, G.P., J.H. Dillier, S.O. Hill, M.S. Blum, and H.R. Hermann. 1971. Nuptial flight and flight ranges of the imported fire ant, *Solenopsis saevissima richteri* (Hymenoptera: Formicidae). *Journal of the Georgia Entomological Society* 6: 145–156.

Markin, G.P., J.H. Dillier, and H.L. Collins. 1973. Growth and development of colonies of the red imported fire ant, *Solenopsis invicta*. *Annals of the Entomological Society of America* 66: 803–808.

Matessi, C., and I. Eshel. 1992. Sex ratio in the social Hymenoptera: a population-genetics study of long-term evolution. *American Naturalist* 139: 276–312.

Matessi, C., and S.D. Jayakar. 1976. Conditions for the evolution of altruism under Darwinian selection. *Theoretical Population Biology* 9: 360–387.

Maynard Smith, J. 1964. Group selection and kin selection. *Nature* 201: 1145–1147.

Maynard Smith, J. 1976. Group selection. *Quarterly Review of Biology* 51: 277–283.

Maynard Smith, J. 1982a. The evolution of social behaviour – a classification of models. In King's College Sociobiology Group, ed., *Current Problems in Sociobiology*, pp. 29–44. Cambridge University Press, Cambridge.

Maynard Smith, J. 1982b. *Evolution and the Theory of Games*. Cambridge University Press, Cambridge.

Maynard Smith, J. 1983. Models of evolution. *Proceedings of the Royal Society of London, Series B* 219: 315–325.

Maynard Smith, J. 1984. The ecology of sex. In J.R. Krebs and N.B. Davies, eds., *Behavioural Ecology: An Evolutionary Approach*, 2nd edition, pp. 201–221. Blackwell, Oxford.

Maynard Smith, J. 1989. *Evolutionary Genetics*. Oxford University Press, Oxford.

Mayr, E. 1983. How to carry out the adaptationist program? *American Naturalist* 121: 324–334.

McDonald, P., and H. Topoff. 1985. Social regulation of behavioral development in the ant, *Novomessor albisetosus* (Mayr). *Journal of Comparative Psychology* 99: 3–14.

McIver, J.D. 1991. Dispersed central place foraging in Australian meat ants. *Insectes Sociaux* 38: 129–137.

Medawar, P.B. 1957. *The Uniqueness of the Individual*. Methuen, London.

Medeiros, F.N.S., L.E. Lopes, P.R.S. Moutinho, P.S. Oliveira, and B. Hölldobler. 1992. Functional polygyny, agonistic interactions and reproductive dominance in the neotropical ant *Odontomachus chelifer* (Hymenoptera, Formicidae, Ponerinae). *Ethology* 91: 134–146.

Mercier, B., L. Passera, and J.-P. Suzzoni. 1985a. Etude de la polygynie chez la fourmi *Plagiolepis pygmaea* Latr. (Hym. Formicidae). I. La fécondité des reines en condition expérimentale monogyne. *Insectes Sociaux* 32: 335–348.

Mercier, B., L. Passera, and J.-P. Suzzoni. 1985b. Etude de la polygynie chez la fourmi *Plagiolepis pygmaea* Latr. (Hym., Formicidae). II. La fécondité des reines en condition expérimentale polygyne. *Insectes Sociaux* 32: 349–362.

Mesterton-Gibbons, M., and L.A. Dugatkin. 1992. Cooperation among unrelated individuals: evolutionary factors. *Quarterly Review of Biology* 67: 267–281.

Metcalf, R.A. 1980a. Measuring fitness in social insects. In H. Markl, ed., *Evolution of Social Behavior: Hypotheses and Empirical Tests*, pp. 81–95. Verlag Chemie, Weinheim.

Metcalf, R.A. 1980b. Sex ratios, parent–offspring conflict, and local competition for mates in the social wasps *Polistes metricus* and *Polistes variatus*. *American Naturalist* 116: 642–654.

Metcalf, R.A., and G.S. Whitt. 1977. Intra-nest relatedness in the social wasp *Polistes metricus*: a genetic analysis. *Behavioral Ecology and Sociobiology* 2: 339–351.

Metcalf, R.A., J.A. Stamps, and V.V. Krishnan. 1979. Parent–offspring conflict that is not limited by degree of kinship. *Journal of Theoretical Biology* 76: 99–107.

Meyer, J. 1966. Essai d'application de certains modèles cybernétiques à la coordination chez les insectes sociaux. *Insectes Sociaux* 13: 127–138.

Michener, C.D. 1964. Reproductive efficiency in relation to colony size in Hymenopterous societies. *Insectes Sociaux* 11: 317–341.

Michener, C.D. 1969. Comparative social behavior of bees. *Annual Review of Entomology* 14: 299–342.

Michener, C.D. 1974. *The Social Behavior of the Bees*. Harvard University Press, Cambridge, Massachusetts.

Michener, C.D. 1985. From solitary to eusocial: need there be a series of intervening species? In B. Hölldobler and M. Lindauer, eds., *Experimental Behavioral Ecology and Sociobiology*, pp. 293–305. Gustav Fischer Verlag, Stuttgart.

Michener, C.D., and D.J. Brothers. 1974. Were workers of eusocial Hymenoptera initially altruistic or oppressed? *Proceedings of the National*

Academy of Sciences, U.S.A. 71: 671–674.

Michod, R. 1979. Genetical aspects of kin selection: effects of inbreeding. *Journal of Theoretical Biology* 81: 223–233.

Michod, R.E. 1982. The theory of kin selection. *Annual Review of Ecology and Systematics* 13: 23–55.

Michod, R.E. 1993. Inbreeding and the evolution of social behavior. In N.W. Thornhill, ed., *The Natural History of Inbreeding and Outbreeding*, pp. 74–96. University of Chicago Press, Chicago.

Michod, R.E., and W.W. Anderson. 1979. Measures of genetic relationship and the concept of inclusive fitness. *American Naturalist* 114: 637–647.

Michod, R.E., and W.D. Hamilton. 1980. Coefficients of relatedness in sociobiology. *Nature* 288: 694–697.

Milinski, M. 1978. Kin selection and reproductive value. *Zeitschrift für Tierpsychologie* 47: 328–329.

Milner, R. 1989. *Communication and Concurrency.* Prentice Hall, New York.

Minsky, M. 1988. *The Society of Mind.* Pan Books Ltd, London.

Mintzer, A.C. 1982. Copulatory behavior and mate selection in the harvester ant, *Pogonomyrmex californicus* (Hymenoptera: Formicidae). *Annals of the Entomological Society of America* 75: 323–326.

Mintzer, A.C. 1987. Primary polygyny in the ant *Atta texana*: number and weight of females and colony foundation success in the laboratory. *Insectes Sociaux* 34: 108–117.

Mintzer, A., and S.B. Vinson. 1985. Cooperative colony foundation by females of the leafcutting ant *Atta texana* in the laboratory. *Journal of the New York Entomological Society* 93: 1047–1051.

Mizutani, A. 1981. On the two forms of the ant *Myrmica ruginodis* Nylander (Hymenoptera, Formicidae) from Sapporo and its vicinity, Japan. *Japanese Journal of Ecology* 31: 131–137.

Mock, D.W., and L.S. Forbes. 1992. Parent-offspring conflict: a case of arrested development. *Trends in Ecology and Evolution* 7: 409–413.

Moore, T., and D. Haig. 1991. Genomic imprinting in mammalian development: a parental tug-of-war. *Trends in Genetics* 7: 45–49.

Morgan, G. 1986. *Images of Organization.* Sage Publications, Newbury Park, California.

Mori, A., P. d'Ettorre, and F. Le Moli. 1994. Mating and post-mating behaviour of the European amazon ant, *Polyergus rufescens* (Hymenoptera, Fomicidae). *Bolletino di Zoologia* 61: 203–206

Moritz, R.F.A. 1985. The effects of multiple mating on the worker-queen conflict in *Apis mellifera* L. *Behavioral Ecology and Sociobiology* 16: 375–377.

Moritz, R.F.A. 1986a. Two parthenogenetical strategies of laying workers in populations of the honeybee, *Apis mellifera* (Hymenoptera: Apidae) *Entomologia Generalis* 11: 159–164.

Moritz, R.F.A. 1986b. Intracolonial worker relationship and sperm competition in the honeybee (*Apis mellifera* L.). *Experientia* 42: 445–448.

Moritz, R.F.A. 1989. Colony level and within colony level selection in honeybees. A two allele population model for *Apis mellifera capensis. Behavioral Ecology and Sociobiology* 25: 437–444.

Moritz, R.F.A. 1991. Kin recognition in honeybees: experimental artefact or biological reality? In L.J. Goodman and R.C. Fisher, eds., *The Behaviour and Physiology of Bees*, pp. 48–59. C.A.B. International, Wallingford.

Moritz, R.F.A., and E. Hillesheim. 1990. Trophallaxis and genetic variance of kin recognition in honey bees, *Apis mellifera* L. *Animal Behaviour* 40: 641–647.

Moritz, R.F.A., and E.E. Southwick. 1992. *Bees as Superorganisms: An Evolutionary Reality*. Springer-Verlag, Berlin.

Mueller, U.G. 1991. Haplodiploidy and the evolution of facultative sex ratios in a primitively eusocial bee. *Science* 254: 442–444.

Mueller, U.G., G.C. Eickwort, and C.F. Aquadro. 1994. DNA fingerprinting analysis of parent-offspring conflict in a bee. *Proceedings of the National Academy of Sciences, U.S.A.* 91: 5143–5147.

Mumme, R.L., W.D. Koenig, and F.L.W. Ratnieks. 1989. Helping behaviour, reproductive value, and the future component of indirect fitness. *Animal Behaviour* 38: 331–343.

Munger, J.C. 1992. Reproductive potential of colonies of desert harvester ants (*Pogonomyrmex desertorum*): effects of predation and food. *Oecologia* 90: 276–282.

Murray, J.D. 1989. *Mathematical Biology*. Springer-Verlag, Berlin.

Murray, M.G. 1989. Environmental constraints on fighting in flightless male fig wasps. *Animal Behaviour* 38: 186–193.

Myles, T.G. 1988. Resource inheritance in social evolution from termites to man. In C.N. Slobodchikoff, ed., *The Ecology of Social Behavior*, pp. 379–423. Academic Press, San Diego.

Myles, T.G., and W.L. Nutting. 1988. Termite eusocial evolution: a re-examination of Bartz's hypothesis and assumptions. *Quarterly Review of Biology* 63: 1–23.

Nee, S. 1989. Does Hamilton's rule describe the evolution of reciprocal altruism? *Journal of Theoretical Biology* 141: 81–91.

Newell, W., and T.C. Barber. 1913. *The Argentine Ant*. (*Bureau of Entomology Bulletin* No. 122). United States Department of Agriculture, Washington, D.C.

Nickerson, J.C., H.L. Cromroy, W.H. Whitcomb, and J.A. Cornell. 1975. Colony organization and queen numbers in two species of *Conomyrma*. *Annals of the Entomological Society of America* 68: 1083–1085.

Nicolis, G., and I. Prigogine. 1977. *Self-Organization in Nonequilibrium Systems*. Wiley, New York.

Nijhout, H.F., and D.E. Wheeler. 1982. Juvenile hormone and the physiological basis of insect polymorphisms. *Quarterly Review of Biology* 57: 109–133.

Noirot, C., and J.M. Pasteels. 1987. Ontogenetic development and evolution of the worker caste in termites. *Experientia* 43: 851–860.

Noirot, C., and J.M. Pasteels. 1988. The worker caste is polyphyletic in termites. *Sociobiology* 14: 15–20.

Nonacs, P. 1986a. Ant reproductive strategies and sex allocation theory. *Quarterly Review of Biology* 61: 1–21.

Nonacs, P. 1986b. Sex-ratio determination within colonies of ants. *Evolution* 40: 199–204.

Nonacs, P. 1988. Queen number in colonies of social Hymenoptera as a kin-selected adaptation. *Evolution* 42: 566–580.

Nonacs, P. 1989. Competition and kin discrimination in colony founding by social Hymenoptera. *Evolutionary Ecology* 3: 221–235.

Nonacs, P. 1990. Size and kinship affect success of co-founding *Lasius pallitarsis* queens. *Psyche* 97: 217–228.

Nonacs, P. 1991a. Alloparental care and eusocial evolution: the limits of Queller's head-start advantage. *Oikos* 61: 122–125.

Nonacs, P. 1991b. Less growth with more food: How insect-prey availability changes colony demographics in the ant, *Camponotus floridanus*. *Journal of Insect Physiology* 37: 891–898.

Nonacs, P. 1992. Queen condition and alate density affect pleometrosis in the ant *Lasius pallitarsis*. *Insectes Sociaux* 39: 3–13.

Nonacs, P. 1993a. The effects of polygyny and colony life history on optimal sex investment. In L. Keller, ed., *Queen Number and Sociality in Insects*, pp. 110–131. Oxford University Press, Oxford.

Nonacs, P. 1993b. Male parentage and sexual deception in the social Hymenoptera. In D.L. Wrensch and M.A. Ebbert, eds., *Evolution and Diversity of Sex Ratio in Insects and Mites*, pp. 384–401. Chapman and Hall, New York.

Nonacs, P. 1993c. The economics of brood raiding and nest consolidation during ant colony founding. *Evolutionary Ecology* 7: 625–633.

Nonacs, P., and N.F. Carlin. 1990. When can ants discriminate the sex of brood? A new aspect of queen-worker conflict. *Proceedings of the National Academy of Sciences, U.S.A.* 87: 9670–9673.

Nonacs, P., and J.E. Tobin. 1992. Selfish larvae: development and the evolution of parasitic behavior in the Hymenoptera. *Evolution* 46: 1605–1620.

Noonan, K.M. 1978. Sex ratio of parental investment in colonies of the social wasp *Polistes fuscatus*. *Science* 199: 1354–1356.

Nunney, L. 1985a. Group selection, altruism, and structured-deme models. *American Naturalist* 126: 212–230.

Nunney, L. 1985b. Female-biased sex ratios: individual or group selection? *Evolution* 39: 349–361.

Ohkawara, K., F. Ito, and S. Higashi. 1993. Production and reproductive function of intercastes in *Myrmecina graminicola nipponica* colonies (Hymenoptera: Formicidae). *Insectes Sociaux* 40: 1–10.

Ohta, T. 1992. The meaning of natural selection revisited at the molecular level. *Trends in Ecology and Evolution* 7: 311–312.

Oliveira, P.S., and B. Hölldobler. 1990. Dominance orders in the ponerine ant *Pachycondyla apicalis* (Hymenoptera, Formicidae). *Behavioral Ecology and Sociobiology* 27: 385–393.

Oliveira, P.S., and B. Hölldobler. 1991. Agonistic interactions and reproductive dominance in *Pachycondyla obscuricornis* (Hymenoptera: Formicidae). *Psyche* 98: 215–225.

O'Neill, K.M. 1994. The male mating strategy of the ant *Formica subpolita* Mayr (Hymenoptera: Formicidae): swarming, mating, and predation risk. *Psyche* 101: 93–108.

Orgel, L.E., and F.H.C. Crick. 1980. Selfish DNA: the ultimate parasite. *Nature* 284: 604–607.

Orlove, M.J. 1975. A model of kin selection not invoking coefficients of relationship. *Journal of Theoretical Biology* 49: 289–310.

Orlove, M.J. 1981. A model of the sex ratio in flying ants with an unexpected equilibrium and an expected ESS. *Journal of Theoretical Biology* 93: 523–532.

Orzack, S.H., and J. Gladstone. 1994. Quantitative genetics of sex ratio traits in the parasitic wasp, *Nasonia vitripennis*. *Genetics* 137: 211–220.

Oster, G.F., and E.O. Wilson. 1978. *Caste and Ecology in the Social Insects*. Princeton University Press, Princeton, New Jersey.

Oster, G., I. Eshel, and D. Cohen. 1977. Worker-queen conflict and the evolution of social insects. *Theoretical Population Biology* 12: 49–85.

Otto, D. 1958. Über die Arbeitsteilung im Staate von *Formica rufa rufo-pratensis minor* Gössw. und ihre verhaltensphysiologischen Grundlagen: ein Beitrag zur Biologie der Roten Waldameise. *Wissenschaftliche Abhandlungen der Deutschen Akademie der Landwirtschaftswissenschaften zu Berlin* 30: 1–169.

Owen, R.E. 1986. Colony-level selection in the social insects: single locus additive and nonadditive models. *Theoretical Population Biology* 29: 198–234.

Owen, R.E. 1989. The genetics of colony-level selection. In M.D. Breed and R.E. Page, eds., *The Genetics of Social Evolution*, pp. 31–59. Westview Press, Boulder, Colorado.

Owen, R.E., F.H. Rodd, and R.C. Plowright. 1980. Sex ratios in bumble bee colonies: complications due to orphaning? *Behavioral Ecology and Sociobiology* 7: 287–291.

Packer, L. 1990. Solitary and eusocial nests in a population of *Augochlorella striata* (Provancher) (Hymenoptera; Halictidae) at the northern edge of its range. *Behavioral Ecology and Sociobiology* 27: 339–344.

Packer, L., and R.E. Owen. 1994. Relatedness and sex ratio in a primitively eusocial halictine bee. *Behavioral Ecology and Sociobiology* 34: 1–10.

Page, R.E. 1980. The evolution of multiple mating behavior by honey bee queens (*Apis mellifera* L.). *Genetics* 96: 263–273.

Page, R.E. 1986. Sperm utilization in social insects. *Annual Review of Entomology* 31: 297–320.

Page, R.E., and R.A. Metcalf. 1982. Multiple mating, sperm utilization, and social evolution. *American Naturalist* 119: 263–281.

Page, R.E., and R.A. Metcalf. 1984. A population investment sex ratio for the honey bee (*Apis mellifera* L.). *American Naturalist* 124: 680–702.

Page, R.E., and S.D. Mitchell. 1991. Self organization and adaptation in insect societies. In A. Fine, M. Forbes, and L. Wessels, eds., *PSA 1990*, Vol. 2, pp. 289–298. Philosophy of Science Association, East Lansing.

Page, R.E., G.E. Robinson, and M.K. Fondrk. 1989a. Genetic specialists, kin recognition and nepotism in honey-bee colonies. *Nature* 338: 576–579.

Page, R.E., G.E. Robinson, N.W. Calderone, and W.C. Rothenbuhler. 1989b. Genetic structure, division of labor, and the evolution of insect societies. In M.D. Breed and R.E. Page, eds., *The Genetics of Social Evolution*, pp. 15–29. Westview Press, Boulder, Colorado.

Page, R.E., M.K. Fondrk, and G.E. Robinson. 1993. Selectable components of sex allocation in colonies of the honeybee (*Apis mellifera* L.). *Behavioral Ecology* 4: 239–245.

Pamilo, P. 1981. Genetic organization of *Formica sanguinea* populations. *Behavioral Ecology and Sociobiology* 9: 45–50.

Pamilo, P. 1982a. Genetic evolution of sex ratios in eusocial Hymenoptera: allele frequency simulations. *American Naturalist* 119: 638–656.

Pamilo, P. 1982b. Multiple mating in *Formica* ants. *Hereditas* 97: 37–45.

Pamilo, P. 1982c. Genetic population structure in polygynous *Formica* ants. *Heredity* 48: 95–106.

Pamilo, P. 1983. Genetic differentiation within subdivided populations of *Formica* ants. *Evolution* 37: 1010–1022.

Pamilo, P. 1984. Genetic relatedness and evolution of insect sociality. *Behavioral Ecology and Sociobiology* 15: 241–248.

Pamilo, P. 1985. Effect of inbreeding on genetic relatedness. *Hereditas* 103: 195–200.

Pamilo, P. 1987. Sex ratios and the evolution of eusociality in the Hymenoptera. *Journal of Genetics* 66: 111–122.

Pamilo, P. 1989. Estimating relatedness in social groups. *Trends in Ecology and Evolution* 4: 353–355.

Pamilo, P. 1990a. Comparison of relatedness estimators. *Evolution* 44: 1378–1382.

Pamilo, P. 1990b. Sex allocation and queen-worker conflict in polygynous ants. *Behavioral Ecology and Sociobiology* 27: 31–36.

Pamilo, P. 1991a. Evolution of the sterile caste. *Journal of Theoretical Biology* 149: 75–95.

Pamilo, P. 1991b. Evolution of colony characteristics in social insects. I. Sex allocation. *American Naturalist* 137: 83–107.

Pamilo, P. 1991c. Evolution of colony characteristics in social insects. II. Number of reproductive individuals. *American Naturalist* 138: 412–433.

Pamilo, P. 1991d. Life span of queens in the ant *Formica exsecta*. *Insectes Sociaux* 38: 111–119.

Pamilo, P. 1993. Polyandry and allele frequency differences between the sexes in the ant *Formica aquilonia*. *Heredity* 70: 472–480.

Pamilo, P., and R.H. Crozier. 1982. Measuring genetic relatedness in natural populations: methodology. *Theoretical Population Biology* 21: 171–193.

Pamilo, P., and R. Rosengren. 1983. Sex ratio strategies in *Formica* ants. *Oikos* 40: 24–35.

Pamilo, P., and R. Rosengren. 1984. Evolution of nesting strategies of ants: genetic evidence from different population types of *Formica* ants. *Biological Journal of the Linnean Society* 21: 331–348.

Pamilo, P., and P. Seppä. 1994. Reproductive competition and conflicts in colonies of the ant *Formica sanguinea*. *Animal Behaviour* 48: 1201–1206.

Pamilo, P., and S.-L. Varvio-Aho. 1979. Genetic structure of nests in the ant *Formica sanguinea*. *Behavioral Ecology and Sociobiology* 6: 91–98.

Pamilo, P., R. Rosengren, K. Vepsäläinen, S.-L. Varvio-Aho, and B. Pisarski. 1978. Population genetics of *Formica* ants. I. Patterns of enzyme gene variation. *Hereditas* 89: 233–248.

Pamilo, P., D. Chautems, and D. Cherix. 1992. Genetic differentiation of disjunct populations of the ants *Formica aquilonia* and *Formica lugubris* in Europe. *Insectes Sociaux* 39: 15–29.

Pamilo, P., L. Sundström, W. Fortelius, and R. Rosengren. 1994. Diploid males and colony-level selection in *Formica* ants. *Ethology, Ecology and Evolution* 6: 221–235.

Parker, G.A. 1970. Sperm competition and its evolutionary consequences in the insects. *Biological Reviews* 45: 525–567.

Parker, G.A. 1985. Models of parent-offspring conflict. V. Effects of the behaviour of the two parents. *Animal Behaviour* 33: 519–533.

Parker, G.A. 1989. Hamilton's rule and conditionality. *Ethology, Ecology and Evolution* 1: 195–211.

Parker, G.A., and M.R. Macnair. 1978. Models of parent-offspring conflict. I. Monogamy. *Animal Behaviour* 26: 97–110.

Parker, G.A., and M.R. Macnair. 1979. Models of parent-offspring conflict. IV. Suppression: evolutionary retaliation by the parent. *Animal Behaviour* 27: 1210–1235.

Parker, G.A., and J. Maynard Smith. 1990. Optimality theory in evolutionary biology. *Nature* 348: 27–33.

Passera, L. 1980a. La fonction inhibitrice des reines de la fourmi *Plagiolepis pygmaea* Latr.: rôle des phéromones. *Insectes Sociaux* 27: 212–225.

Passera, L. 1980b. La ponte d'oeufs préorientés chez la fourmi *Pheidole pallidula* (Nyl.) (Hymenoptera - Formicidae). *Insectes Sociaux* 27: 79–95.

Passera, L. 1984. *L'Organisation Sociale des Fourmis*. Privat, Toulouse.

Passera, L. 1994. Characteristics of tramp species. In D.F. Williams, ed., *Exotic Ants. Biology, Impact, and Control of Introduced Species*, pp. 23–43. Westview Press, Boulder, Colorado.

Passera, L., and S. Aron. 1993a. Factors controlling dealation and egg laying in virgin queens of the Argentine ant *Linepithema humile* (Mayr) (= *Iridomyrmex humilis*). *Psyche* 100: 51–63.

Passera, L., and S. Aron. 1993b. Social control over the survival and selection of winged virgin queens in an ant without nuptial flight: *Iridomyrmex humilis*. *Ethology* 93: 225–235.

Passera, L., and L. Keller. 1987. Energy investment during the differentiation of sexuals and workers in the Argentine ant *Iridomyrmex humilis* (Mayr). *Mitteilungen der Schweizerischen Entomologischen Gesellschaft* 60: 249–260.

Passera, L., and L. Keller. 1990. Loss of mating flight and shift in the pattern of carbohydrate storage in sexuals of ants (Hymenoptera; Formicidae). *Journal of Comparative Physiology B* 160: 207–211.

Passera, L., and L. Keller. 1992. The period of sexual maturation and the age at mating in *Iridomyrmex humilis*, an ant with intranidal mating. *Journal of Zoology* 228: 141–153.

Passera, L., and L. Keller. 1994. Mate availability and male dispersal in the Argentine ant *Linepithema humile* (Mayr) (= *Iridomyrmex humilis*). *Animal Behaviour* 48: 361–369.

Passera, L., L. Keller, and J.-P. Suzzoni. 1987. Male differentiation in the Argentine ant *Iridomyrmex humilis* (Mayr). In J. Eder and H. Rembold, eds.,

Chemistry and Biology of Social Insects, pp. 297–298. Verlag J. Peperny, München.

Passera, L., L. Keller, and J.-P. Suzzoni. 1988a. Control of brood male production in the Argentine ant *Iridomyrmex humilis* (Mayr). *Insectes Sociaux* 35: 19–33.

Passera, L., L. Keller, and J.-P. Suzzoni. 1988b. Queen replacement in dequeened colonies of the Argentine ant *Iridomyrmex humilis* (Mayr). *Psyche* 95: 59–65.

Passera, L., E.L. Vargo, and L. Keller. 1991. Le nombre de reines chez les fourmis et sa conséquence sur l'organisation sociale. *Anneé Biologique* 30: 137–173.

Patel, A.D. 1990. An unusually broad behavioral repertory for a major worker in a dimorphic ant species: *Pheidole morrisi* (Hymenoptera, Formicidae). *Psyche* 97: 181–191.

Pearson, B. 1981. The electrophoretic determination of *Myrmica rubra* microgynes as a social parasite: possible significance in the evolution of ant social parasites. In P.E. Howse and J.-L. Clément, eds., *Biosystematics of Social Insects*, pp. 75–84. Academic Press, London.

Pearson, B. 1982. Relatedness of normal queens (macrogynes) in nests of the polygynous ant *Myrmica rubra* Latreille. *Evolution* 36: 107–112.

Pearson, B., and A.R. Child. 1980. The distribution of an esterase polymorphism in macrogynes and microgynes of *Myrmica rubra* Latreille. *Evolution* 34: 105–109.

Pearson, B., and A.F. Raybould. 1993. The effects of sampling and nest structure on relatedness in *Myrmica rubra* populations (Hymenoptera: Formicidae). *Sociobiology* 21: 209–216.

Pearson, B., A.F. Raybould, and R.T. Clarke. 1995. Breeding behaviour, relatedness and sex-investment ratios in *Leptothorax tuberum* Fabricius. *Entomologia Experimentalis et Applicata:* in press.

Pedersen, J.S., and J.J. Boomsma. 1994. Colony-level variation in relatedness structure of the polygynous ant, *Myrmica sulcinodis*. *Hereditas* 121: 221–222.

Peeters, C. 1990. Monogyny and polygyny in ponerine ants without queens. In G.K. Veeresh, B. Mallik, and C.A. Viraktamath, eds., *Social Insects and the Environment*, pp. 234–235. Oxford & IBH Publishing Co., New Delhi.

Peeters, C. 1991a. The occurrence of sexual reproduction among ant workers. *Biological Journal of the Linnean Society* 44: 141–152.

Peeters, C.P. 1991b. Ergatoid queens and intercastes in ants: two distinct adult forms which look morphologically intermediate between workers and winged queens. *Insectes Sociaux* 38: 1–15.

Peeters, C. 1993. Monogyny and polygyny in ponerine ants with or without queens. In L. Keller, ed., *Queen Number and Sociality in Insects*, pp. 234–261. Oxford University Press, Oxford.

Peeters, C., and A.N. Andersen. 1989. Cooperation between dealate queens during colony foundation in the green tree ant, *Oecophylla smaragdina*. *Psyche* 96: 39–44.

Peeters, C., and J.P.J. Billen. 1991. A novel exocrine gland inside the thoracic appendages ("gemmae") of the queenless ant *Diacamma australe*. *Experientia* 47: 229–231.

Peeters, C., and S. Higashi. 1989. Reproductive dominance controlled by mutilation in the queenless ant *Diacamma australe. Naturwissenschaften* 76: 177–180.

Peeters, C., and K. Tsuji. 1993. Reproductive conflict among ant workers in *Diacamma* sp. from Japan: dominance and oviposition in the absence of the gamergate. *Insectes Sociaux* 40: 119–136.

Peeters, C., J. Billen, and B. Hölldobler. 1992. Alternative dominance mechanisms regulating monogyny in the queenless ant genus *Diacamma. Naturwissenschaften* 79: 572–573.

Peeters, C., B. Hölldobler, M. Moffett, and T.M. Musthak Ali. 1994. "Wallpapering" and elaborate nest architecture in the ponerine ant *Harpegnathos saltator. Insectes Sociaux* 41: 211–218.

Perrins, C. 1964. Survival of young swifts in relation to brood-size. *Nature* 201: 1147–1148.

Péru, L., L. Plateaux, A. Buschinger, P. Douwes, A. Perramon, and J.C. Quentin. 1990. New records of *Leptothorax* ants with cysticercoids of the cestode, *Choanotaenia unicoronata*, and the rearing of the tapeworm in quails. *Spixiana* 13: 223–225.

Pisarski, B. 1981. Intraspecific variations in ants of the genus *Formica* L. In P.E. Howse and J.-L. Clément, eds., *Biosystematics of Social Insects*, pp. 17–25. Academic Press, London.

Plateaux, L. 1981. The *pallens* morph of the ant *Leptothorax nylanderi*: description, formal genetics, and study of populations. In P.E. Howse and J.-L. Clément, eds., *Biosystematics of Social Insects*, pp. 63–74. Academic Press, London.

Pollock, G.B. 1983. Population viscosity and kin selection. *American Naturalist* 122: 817–829.

Pollock, G.B. 1989. Suspending disbelief - of Wynne-Edwards and his reception. *Journal of Evolutionary Biology* 2: 205–221.

Pollock, G.B., and S.W. Rissing. 1988. Social competition under mandatory group life. In C.N. Slobodchikoff, ed., *The Ecology of Social Behavior*, pp. 315–334. Academic Press, San Diego.

Porter, S.D. 1989. Effects of diet on the growth of laboratory fire ant colonies (Hymenoptera: Formicidae). *Journal of the Kansas Entomological Society* 62: 288–291.

Porter, S.D. 1991. Origins of new queens in polygyne red imported fire ant colonies (Hymenoptera: Formicidae). *Journal of Entomological Science* 26: 474–478.

Porter, S.D., and C.D. Jorgensen. 1988. Longevity of harvester ant colonies in southern Idaho. *Journal of Range Management* 41: 104–107.

Porter, S.D., and D.A. Savignano. 1990. Invasion of polygyne fire ants decimates native ants and disrupts arthropod community. *Ecology* 71: 2095–2106.

Porter, S.D., B. Van Eimeren, and L.E. Gilbert. 1988. Invasion of red imported fire ants (Hymenoptera: Formicidae): microgeography of competitive replacement. *Annals of the Entomological Society of America* 81: 913–918.

Pribram, K.H. 1976. Problems concerning the structure of consciousness. In G.G. Globus, G. Maxwell, and I. Savodnik, eds., *Consciousness and the Brain*, pp. 297–313. Plenum Press, New York.

Price, G.R. 1970. Selection and covariance. *Nature* 227: 520–521.

Price, G.R. 1972. Extension of covariance selection mathematics. *Annals of Human Genetics* 35: 485–490.

Provost, E., and P. Cerdan. 1990. Experimental polygyny and colony closure in the ant *Messor barbarus* (L.) (Hym., Formicidae). *Behaviour* 115: 114–126.

Queller, D.C. 1984. Kin selection and frequency dependence: a game theoretic approach. *Biological Journal of the Linnean Society* 23: 133–143.

Queller, D.C. 1985. Kinship, reciprocity and synergism in the evolution of social behaviour. *Nature* 318: 366–367.

Queller, D.C. 1989a. Inclusive fitness in a nutshell. In P.H. Harvey and L. Partridge, eds., *Oxford Surveys in Evolutionary Biology*, Vol. 6, pp. 73–109. Oxford University Press, Oxford.

Queller, D.C. 1989b. The evolution of eusociality: reproductive head starts of workers. *Proceedings of the National Academy of Sciences, U.S.A.* 86: 3224–3226.

Queller, D.C. 1992a. A general model for kin selection. *Evolution* 46: 376–380.

Queller, D.C. 1992b. Quantitative genetics, inclusive fitness, and group selection. *American Naturalist* 139: 540–558.

Queller, D.C. 1992c. Does population viscosity promote kin selection? *Trends in Ecology and Evolution* 7: 322–324.

Queller, D.C. 1993a. Genetic relatedness and its components in polygynous colonies of social insects. In L. Keller, ed., *Queen Number and Sociality in Insects*, pp. 132–152. Oxford University Press, Oxford.

Queller, D.C. 1993b. Worker control of sex ratios and selection for extreme multiple mating by queens. *American Naturalist* 142: 346–351.

Queller, D.C. 1994a. Genetic relatedness in viscous populations. *Evolutionary Ecology* 8: 70–73.

Queller, D.C. 1994b. Extended parental care and the origin of eusociality. *Proceedings of the Royal Society of London, Series B* 256: 105–111.

Queller, D.C., and K.F. Goodnight. 1989. Estimating relatedness using genetic markers. *Evolution* 43: 258–275.

Queller, D.C., and J.E. Strassmann. 1988. Reproductive success and group nesting in the paper wasp, *Polistes annularis*. In T.H. Clutton-Brock, ed., *Reproductive Success*, pp. 76–96. University of Chicago Press, Chicago.

Queller, D.C., and J.E. Strassmann. 1989. Measuring inclusive fitness in social wasps. In M.D. Breed and R.E. Page, eds., *The Genetics of Social Evolution*, pp. 103–122. Westview Press, Boulder, Colorado.

Queller, D.C., C.R. Hughes, and J.E. Strassmann. 1990. Wasps fail to make distinctions. *Nature* 344: 388.

Queller, D.C., J.E. Strassmann, C.R. Solís, C.R. Hughes, and D.M. DeLoach. 1993. A selfish strategy of social insect workers that promotes social cohesion. *Nature* 365: 639–641.

Ratnieks, F.L.W. 1988. Reproductive harmony via mutual policing by workers in eusocial Hymenoptera. *American Naturalist* 132: 217–236.

Ratnieks, F.L.W. 1990a. Assessment of queen mating frequency by workers in social Hymenoptera. *Journal of Theoretical Biology* 142: 87–93.

Ratnieks, F.L.W. 1990b. The evolution of polyandry by queens in social

Hymenoptera: the significance of the timing of removal of diploid males. *Behavioral Ecology and Sociobiology* 26: 343–348.

Ratnieks, F.L.W. 1991a. Facultative sex allocation biasing by workers in social Hymenoptera. *Evolution* 45: 281–292.

Ratnieks, F.L.W. 1991b. The evolution of genetic odor-cue diversity in social Hymenoptera. *American Naturalist* 137: 202–226.

Ratnieks, F.L.W. 1993. Egg-laying, egg-removal, and ovary development by workers in queenright honey bee colonies. *Behavioral Ecology and Sociobiology* 32: 191–198.

Ratnieks, F.L.W., and J.J. Boomsma. 1995. Facultative sex allocation by workers and the evolution of polyandry by queens in social Hymenoptera. *American Naturalist* 145: 969–993.

Ratnieks, F.L.W., and D.G. Miller. 1993. Division of honey bee drones during swarming. *Animal Behaviour* 46: 803–805.

Ratnieks, F.L.W., and H.K. Reeve. 1991. The evolution of queen-rearing nepotism in social Hymenoptera: Effects of discrimination costs in swarming species. *Journal of Evolutionary Biology* 4: 93–115.

Ratnieks, F.L.W., and H.K. Reeve. 1992. Conflict in single-queen Hymenopteran societies: the structure of conflict and processes that reduce conflict in advanced eusocial species. *Journal of Theoretical Biology* 158: 33–65.

Ratnieks, F.L.W., and P.K. Visscher. 1989. Worker policing in the honeybee. *Nature* 342: 796–797.

Rayner, A.D.M., and N.R. Franks. 1987. Evolutionary and ecological parallels between ants and fungi. *Trends in Ecology and Evolution* 2: 127–133.

Reeve, H.K. 1991. *Polistes*. In K.G. Ross and R.W. Matthews, eds., *The Social Biology of Wasps*, pp. 99–148. Comstock, Ithaca, New York.

Reeve, H.K. 1992. Queen activation of lazy workers in colonies of the eusocial naked mole-rat. *Nature* 358: 147–149.

Reeve, H.K. 1993. Haplodiploidy, eusociality and absence of male parental and alloparental care in Hymenoptera: a unifying genetic hypothesis distinct from kin selection theory. *Philosophical Transactions of the Royal Society of London, Series B* 342: 335–352.

Reeve, H.K., and L. Keller. 1995. Partitioning of reproduction in mother-daughter versus sibling associations: a test of optimal skew theory. *American Naturalist* 145: 119–132.

Reeve, H.K., and F.L.W. Ratnieks. 1993. Queen-queen conflicts in polygynous societies: mutual tolerance and reproductive skew. In L. Keller, ed., *Queen Number and Sociality in Insects*, pp. 45–85. Oxford University Press, Oxford.

Reeve, H.K., and P.W. Sherman. 1991. Intracolonial aggression and nepotism by the breeding female naked mole-rat. In P.W. Sherman, J.U.M. Jarvis, and R.D. Alexander, eds., *The Biology of the Naked Mole-Rat*, pp. 337–357. Princeton University Press, Princeton, New Jersey.

Reeve, H.K., and P.W. Sherman. 1993. Adaptation and the goals of evolutionary research. *Quarterly Review of Biology* 68: 1–32.

Rettenmeyer, C.W. 1963. Behavioral studies of army ants. *University of Kansas Science Bulletin* 44: 281–465.

Rettenmeyer, C.W., and J.F. Watkins. 1978. Polygyny and monogyny in army ants (Hymenoptera: Formicidae). *Journal of the Kansas Entomological Society* 51: 581–591.

Richards, O.W., and R.G. Davies. 1977. *Imms' General Textbook of Entomology, 10th edition., Vol. 1, Structure, Physiology and Development.* Chapman and Hall, London.

Ridley, M. 1985. *The Problems of Evolution.* Oxford University Press, Oxford.

Ridley, M. 1988. Mating frequency and fecundity in insects. *Biological Reviews* 63: 509–549.

Ridley, M., and R. Dawkins. 1981. The natural selection of altruism. In J.P Rushton and R.M. Sorrentino, eds., *Altruism and Helping Behavior,* pp. 19–39. Lawrence Erlbaum Associates, Hillsdale, New Jersey.

Rissing, S.W., and G.B. Pollock. 1986. Social interaction among pleometrotic queens of *Veromessor pergandei* (Hymenoptera: Formicidae) during colony foundation. *Animal Behaviour* 34: 226–233.

Rissing, S.W., and G.B. Pollock. 1987. Queen aggression, pleometrotic advantage and brood raiding in the ant *Veromessor pergandei* (Hymenoptera: Formicidae). *Animal Behaviour* 35: 975–981.

Rissing, S.W., and G.B. Pollock. 1988. Pleometrosis and polygyny in ants. In R.L. Jeanne, ed., *Interindividual Behavioral Variability in Social Insects,* pp. 179–222. Westview Press, Boulder, Colorado.

Rissing, S.W., and G.B. Pollock. 1991. An experimental analysis of pleometrotic advantage in the desert seed-harvester ant *Messor pergandei* (Hymenoptera; Formicidae). *Insectes Sociaux* 38: 205–211.

Rissing, S.W., G.B. Pollock, M.R. Higgins, R.H. Hagen, and D.R. Smith. 1989. Foraging specialization without relatedness or dominance among co-founding ant queens. *Nature* 338: 420–422.

Robertson, H.G., and M. Villet. 1989. Mating behaviour in three species of myrmicine ants (Hymenoptera: Formicidae). *Journal of Natural History* 23: 767–773.

Robinson, G.E. 1985. Effects of a juvenile hormone analogue on honey bee foraging behaviour and alarm pheromone production. *Journal of Insect Physiology* 31: 277–282.

Robinson, G.E. 1992. Regulation of division of labor in insect societies. *Annual Review of Entomology* 37: 637–665.

Robinson, G.E., and R.E. Page. 1988. Genetic determination of guarding and undertaking in honey-bee colonies. *Nature* 333: 356–358.

Robinson, G.E., and R.E. Page. 1989. Genetic determination of nectar foraging, pollen foraging, and nest-site scouting in honey bee colonies. *Behavioral Ecology and Sociobiology* 24: 317–323.

Robinson, G.E., R.E. Page, C. Strambi, and A. Strambi. 1989. Hormonal and genetic control of behavioral integration in honey bee colonies. *Science* 246: 109–112.

Robinson, G.E., R.E. Page, and Z.-Y. Huang. 1994. Temporal polyethism in social insects is a developmental process. *Animal Behaviour* 48: 467–469.

Rosengren, R. 1977. Foraging strategy of wood ants (*Formica rufa* group). II. Nocturnal orientation and diel periodicity. *Acta Zoologica Fennica* 150: 1–30.

Rosengren, R., and P. Pamilo. 1983. The evolution of polygyny and polydomy in mound-building *Formica* ants. *Acta Entomologica Fennica* 42: 65–77.

Rosengren, R., and P. Pamilo. 1986. Sex ratio strategy as related to queen number, dispersal behaviour and habitat quality in *Formica* ants (Hymenoptera: Formicidae). *Entomologia Generalis* 11: 139–151.

Rosengren, R., and L. Sundström. 1991. The interaction between red wood ants, *Cinara* aphids, and pines. A ghost of mutualism past? In C.R. Huxley and D.F. Cutler, eds., *Ant-Plant Interactions*, pp. 80–91. Oxford University Press, Oxford.

Rosengren, R., K. Vepsäläinen, and H. Wuorenrinne. 1979. Distribution, nest densities, and ecological significance of wood ants (the *Formica rufa* group) in Finland. *Bulletin SROP (Section Régionale Ouest Paléarctique) de l'Organisation Internationale de Lutte Biologique contre les Animaux et les Plantes Nuisibles, Union Internationale des Sciences Biologiques* II-3: 181–213.

Rosengren, R., D. Cherix, and P. Pamilo. 1985. Insular ecology of the red wood ant *Formica truncorum* Fabr., I. Polydomous nesting, population size and foraging. *Mitteilungen der Schweizerischen Entomologischen Gesellschaft* 58: 147–175.

Rosengren, R., W. Fortelius, K. Lindström, and A. Luther. 1987. Phenology and causation of nest heating and thermoregulation in red wood ants of the *Formica rufa* group studied in coniferous forest habitats in southern Finland. *Annales Zoologici Fennici* 24: 147–155.

Rosengren, R., L. Sundström, and W. Fortelius. 1993. Monogyny and polygyny in *Formica* ants: the result of alternative dispersal tactics. In L. Keller, ed., *Queen Number and Sociality in Insects*, pp. 308–333. Oxford University Press, Oxford.

Ross, K.G. 1986. Kin selection and the problem of sperm utilization in social insects. *Nature* 323: 798–800.

Ross, K.G. 1988a. Population and colony-level genetic studies of ants. In J.C. Trager, ed., *Advances in Myrmecology*, pp. 189–215. E.J. Brill, Leiden.

Ross, K.G. 1988b. Differential reproduction in multiple-queen colonies of the fire ant *Solenopsis invicta* (Hymenoptera: Formicidae). *Behavioral Ecology and Sociobiology* 23: 341–355.

Ross, K.G. 1989. Reproductive and social structure in polygynous fire ant colonies. In M.D. Breed and R.E. Page, eds., *The Genetics of Social Evolution*, pp. 149–162. Westview Press, Boulder, Colorado.

Ross, K.G. 1990. Breeding systems and kin selection in social Hymenoptera. In G.K. Veeresh, B. Mallik, and C.A. Viraktamath, eds., *Social Insects and the Environment*, pp. 347–348. Oxford & IBH Publishing Co., New Delhi.

Ross, K.G. 1992. Strong selection on a gene that influences reproductive competition in a social insect. *Nature* 355: 347–349.

Ross, K.G. 1993. The breeding system of the fire ant *Solenopsis invicta*: effects on colony genetic structure. *American Naturalist* 141: 554–576.

Ross, K.G., and J.M. Carpenter. 1991a. Population genetic structure, relatedness, and breeding systems. In K.G. Ross and R.W. Matthews, eds., *The Social Biology of Wasps*, pp. 451–479. Comstock, Ithaca, New York.

Ross, K.G., and J.M. Carpenter. 1991b. Phylogenetic analysis and the evolution of queen number in eusocial Hymenoptera. *Journal of Evolutionary Biology* 4: 117–130.

Ross, K.G., and D.J.C. Fletcher. 1985a. Comparative study of genetic and social structure in two forms of the fire ant *Solenopsis invicta* (Hymenoptera: Formicidae). *Behavioral Ecology and Sociobiology* 17: 349–356.

Ross, K.G., and D.J.C. Fletcher. 1985b. Genetic origin of male diploidy in the fire ant, *Solenopsis invicta* (Hymenoptera: Formicidae), and its evolutionary significance. *Evolution* 39: 888–903.

Ross, K.G., and D.J.C. Fletcher. 1986. Diploid male production - a significant colony mortality factor in the fire ant *Solenopsis invicta* (Hymenoptera: Formicidae). *Behavioral Ecology and Sociobiology* 19: 283–291.

Ross, K.G., and L. Keller. 1995. Joint influence of gene flow and selection on a reproductively important genetic polymorphism in the fire ant *Solenopsis invicta*. *American Naturalist*: in press.

Ross, K.G., and R.W. Matthews. 1989. Population genetic structure and social evolution in the sphecid wasp *Microstigmus comes*. *American Naturalist* 134: 574–598.

Ross, K.G., and R.W. Matthews (eds.) 1991. *The Social Biology of Wasps*. Comstock, Ithaca, New York.

Ross, K.G., and D.D. Shoemaker. 1993. An unusual pattern of gene flow between the two social forms of the fire ant *Solenopsis invicta*. *Evolution* 47: 1595–1605.

Ross, K.G., R.K. Vander Meer, D.J.C. Fletcher, and E.L. Vargo. 1987a. Biochemical phenotypic and genetic studies of two introduced fire ants and their hybrid (Hymenoptera: Formicidae). *Evolution* 41: 280–293.

Ross, K.G., E.L. Vargo, and D.J.C. Fletcher. 1987b. Comparative biochemical genetics of three fire ant species in North America, with special reference to the two social forms of *Solenopsis invicta* (Hymenoptera: Formicidae). *Evolution* 41: 979–990.

Ross, K.G., E.L. Vargo, and D.J.C. Fletcher. 1988. Colony genetic structure and queen mating frequency in fire ants of the subgenus *Solenopsis* (Hymenoptera: Formicidae). *Biological Journal of the Linnean Society* 34: 105–117.

Ross, K.G., E.L. Vargo, L. Keller, and J.C. Trager. 1993. Effect of a founder event on variation in the genetic sex-determining system of the fire ant *Solenopsis invicta*. *Genetics* 135: 843–854.

Rust, R.W. 1988. Nuptial flights and mating behavior in the harvester ant, *Pogonomyrmex salinus* Olsen (Hymenoptera: Formicidae). *Journal of the Kansas Entomological Society* 61: 492–494.

Rutz, W., L. Gerig, H. Wille, and M. Lüscher. 1976. The function of juvenile hormone in adult worker honeybees, *Apis mellifera*. *Journal of Insect Physiology* 22: 1485–1491.

Saito, Y. 1994. Is sterility by deleterious recessives an origin of inequalities in the evolution of eusociality? *Journal of Theoretical Biology* 166: 113–115.

Samson, D.A., and K.S. Werk. 1986. Size-dependent effects in the analysis of reproductive effort in plants. *American Naturalist* 127: 667–680.

Sanchez-Peña, S.R., A. Buschinger, and R.A. Humber. 1993. *Myrmicinosporidium durum*, an enigmatic fungal parasite of ants. *Journal of Invertebrate Pathology* 61: 90–96.

Satoh, T. 1989. Comparisons between two apparently distinct forms of *Camponotus nawai* Ito (Hymenoptera: Formicidae). *Insectes Sociaux* 36: 277–292.

Satoh, T. 1991. Behavioral differences of queens in monogynous and polygynous nests of the *Camponotus nawai* complex (Hymenoptera: Formicidae). *Insectes Sociaux* 38: 37–44.

Savolainen, R., and K. Vepsäläinen. 1988. A competition hierarchy among boreal ants: impact on resource partitioning and community structure. *Oikos* 51: 135–155.

Savolainen, R., and K. Vepsäläinen. 1989. Niche differentiation of ant species within territories of the wood ant *Formica polyctena*. *Oikos* 56: 3–16.

Schmid-Hempel, P. 1990. Reproductive competition and the evolution of work load in social insects. *American Naturalist* 135: 501–526.

Schmid-Hempel, P. 1991. The ergonomics of worker behavior in social Hymenoptera. *Advances in the Study of Behavior* 20: 87–134.

Schmid-Hempel, P. 1994. Infection and colony variability in social insects. *Philosophical Transactions of the Royal Society of London, Series B* 346: 313–321.

Schmid-Hempel, P., and R. Schmid-Hempel. 1984. Life duration and turnover of foragers in the ant *Cataglyphis bicolor* (Hymenoptera, Formicidae). *Insectes Sociaux* 31: 345–360.

Schneirla, T.C. 1943. The nature of ant learning, II: The intermediate stage of segmental maze adjustment. *Journal of Comparative Psychology* 35: 149–176.

Schneirla, T.C. 1949. Army-ant life and behavior under dry season conditions. 3. The course of reproduction and colony behavior. *Bulletin of the American Museum of Natural History* 94: 1–82.

Schneirla, T.C. 1971. *Army Ants, a Study in Social Organization.* W.H. Freeman, San Francisco.

Schwarz, M.P. 1988. Local resource enhancement and sex ratios in a primitively social bee. *Nature* 331: 346–348.

Seeley, T.D. 1982. Adaptive significance of the age polyethism schedule in honeybee colonies. *Behavioral Ecology and Sociobiology* 11: 287–293.

Seeley, T.D. 1985. *Honeybee Ecology. A Study of Adaptation in Social Life.* Princeton University Press, Princeton, New Jersey.

Seeley, T.D. 1989. The honey bee colony as a superorganism. *American Scientist* 77: 546–553.

Seeley, T.D., and S.A. Kolmes. 1991. Age polyethism for hive duties in honey bees – illusion or reality? *Ethology* 87: 284–297.

Seger, J. 1981. Kinship and covariance. *Journal of Theoretical Biology* 91: 191–213.

Seger, J. 1983. Partial bivoltinism may cause alternating sex-ratio biases that favour eusociality. *Nature* 301: 59–62.

Seger, J. 1989. Who are the drone police? *Nature* 342: 741–742.

Seger, J. 1991. Cooperation and conflict in social insects. In J.R. Krebs and N.B. Davies, eds., *Behavioural Ecology: An Evolutionary Approach*, 3rd edition, pp. 338–373. Blackwell, Oxford.

Seger, J. 1993. Opportunities and pitfalls in co-operative reproduction. In L. Keller, ed., *Queen Number and Sociality in Insects*, pp. 1–15. Oxford University Press, Oxford.

Seger, J., and H.J. Brockmann. 1987. What is bet-hedging? In P.H. Harvey and L. Partridge, eds., *Oxford Surveys in Evolutionary Biology*, Vol. 4, pp. 182–211. Oxford University Press, Oxford.

Seger, J., and E.L. Charnov. 1988. Benevolent sisterhood. *Nature* 331: 303.

Sendova-Franks, A., and N.R. Franks. 1993. Task allocation in ant colonies within variable environments (A study of temporal polyethism: experimental). *Bulletin of Mathematical Biology* 55: 75–96.

Sendova-Franks, A.B., and N.R. Franks. 1994. Social resilience in individual worker ants and its role in division of labour. *Proceedings of the Royal Society of London, Series B* 256: 305–309.

Seppä, P. 1992. Genetic relatedness of worker nestmates in *Myrmica ruginodis* (Hymenoptera: Formicidae) populations. *Behavioral Ecology and Sociobiology* 30: 253–260.

Seppä, P. 1994. Sociogenetic organization of the ants *Myrmica ruginodis* and *Myrmica lobicornis*: number, relatedness and longevity of reproducing individuals. *Journal of Evolutionary Biology* 7: 71–95.

Shattuck, S.O. 1992. Review of the dolichoderine ant genus *Iridomyrmex* Mayr with descriptions of three new genera (Hymenoptera: Formicidae). *Journal of the Australian Entomological Society* 31: 13–18.

Shaw, R.F., and J.D. Mohler. 1953. The selective significance of the sex ratio. *American Naturalist* 87: 337–342.

Sherman, P.W. 1979. Insect chromosome numbers and eusociality. *American Naturalist* 113: 925–935.

Sherman, P.W., and J.S. Shellman-Reeve. 1994. Ant sex ratios. *Nature* 370: 257.

Sherman, P.W., T.D. Seeley, and H.K. Reeve. 1988. Parasites, pathogens, and polyandry in social Hymenoptera. *American Naturalist* 131: 602–610.

Sherman, P.W., J.U.M. Jarvis, and R.D. Alexander. (eds.) 1991. *The Biology of the Naked Mole-Rat*. Princeton University Press, Princeton, New Jersey.

Sherman, P.W., E.A. Lacey, H.K. Reeve, and L. Keller. 1995. The eusociality continuum. *Behavioral Ecology:* 6: 102–108.

Shub, D.A. 1994. Bacterial altruism? *Current Biology* 4: 555–556.

Shykoff, J.A., and P. Schmid-Hempel. 1991a. Parasites delay worker reproduction in bumblebees: consequences for eusociality. *Behavioral Ecology* 2: 242–248.

Shykoff, J.A., and P. Schmid-Hempel. 1991b. Genetic relatedness and eusociality: parasite-mediated selection on the genetic composition of groups. *Behavioral Ecology and Sociobiology* 28: 371–376.

Shykoff, J.A., and P. Schmid-Hempel. 1991c. Parasites and the advantage of genetic variability within social insect colonies. *Proceedings of the Royal Society of London, Series B* 243: 55–58.

Sibly, R.M. 1994. An allelocentric analysis of Hamilton's rule for overlapping generations. *Journal of Theoretical Biology* 167: 301–305.

Sibly, R., and P. Calow. 1986. Why breeding earlier is always worthwhile. *Journal of Theoretical Biology* 123: 311–319.

Sibly, R., P. Calow, and R.H. Smith. 1988. Optimal size of seasonal breeders. *Journal of Theoretical Biology* 133: 13–21.

Simon, H.A. 1947. *Administrative Behavior: A Study of Decision-making Processes in Administrative Organization*. Macmillan, New York.

Skaife, S.H. 1955. The Argentine ant, *Iridomyrmex humilis* Mayr. *Transactions of the Royal Society of South Africa* 34: 355–377.

Skinner, G.J. 1980. The feeding habits of the wood-ant, *Formica rufa* (Hymenoptera: Formicidae), in limestone woodland in north-west England. *Journal of Animal Ecology* 49: 417–433.

Smeeton, L. 1981. The source of males in *Myrmica rubra* L. (Hym. Formicidae). *Insectes Sociaux* 28: 263–278.

Smeeton, L. 1982a. The effect of larvae on the production of reproductive eggs by workers of *Myrmica rubra* L. (Hym. Formicidae). *Insectes Sociaux* 29: 455–464.

Smeeton, L. 1982b. The effect of age on the production of reproductive eggs by workers of *Myrmica rubra* L. (Hym. Formicidae). *Insectes Sociaux* 29: 465–474.

Smeeton, L. 1982c. The effects of the sizes of colony worker and food store on the production of reproductive eggs by workers of *Myrmica rubra* L. (Hym. Formicidae). *Insectes Sociaux* 29: 475–484.

Smith, A. 1776. *The Wealth of Nations, Books I–III*. Reprinted 1986 (A. Skinner, ed.). Penguin, Harmondsworth, U.K.

Smith, J.A., and W.D. Ross (eds.) 1910. *The Works of Aristotle, Vol. IV, Historia Animalium*. Translated by D'Arcy Wentworth Thompson. Clarendon Press, Oxford.

Smith, R.H., and M.R. Shaw. 1980. Haplodiploid sex ratios and the mutation rate. *Nature* 287: 728–729.

Snyder, L.E. 1992. The genetics of social behavior in a polygynous ant. *Naturwissenschaften* 79: 525–527.

Snyder, L.E. 1993. Non-random behavioural interactions among genetic subgroups in a polygynous ant. *Animal Behaviour* 46: 431–439.

Snyder, L.E., and J.M. Herbers. 1991. Polydomy and sexual allocation ratios in the ant *Myrmica punctiventris*. *Behavioral Ecology and Sociobiology* 28: 409–415.

Sober, E. 1984a. *The Nature of Selection*. MIT Press, Cambridge, Massachusetts.

Sober, E. (ed.) 1984b. *Conceptual Issues in Evolutionary Biology. An Anthology*. MIT Press, Cambridge, Massachusetts.

Sommeijer, M.J., and J.W. Van Veen. 1990. The polygyny of *Myrmica rubra*: selective oophagy and trophallaxis as mechanisms of reproductive dominance. *Entomologia Experimentalis et Applicata* 56: 229–239.

Sommer, K., and B. Hölldobler. 1992a. Pleometrosis in *Lasius niger*. In J. Billen, ed., *Biology and Evolution of Social Insects*, pp. 47–50. Leuven University Press, Leuven, Belgium.

Sommer, K., and B. Hölldobler. 1992b. Coexistence and dominance among queens and mated workers in the ant *Pachycondyla tridentata*. *Naturwissenschaften* 79: 470–472.

Sorensen, A.A., T.M. Busch, and S.B. Vinson. 1983. Factors affecting brood cannibalism in laboratory colonies of the imported fire ant, *Solenopsis invicta* Buren (Hymenoptera: Formicidae). *Journal of the Kansas Entomological Society* 56: 140–150.

Sorensen, A.A., T.M. Busch, and S.B. Vinson. 1985. Trophallaxis by temporal subcastes in the fire ant, *Solenopsis invicta*, in response to honey. *Physiological Entomology* 10: 105–111.

Southwick, E.E. 1991. The colony as a thermoregulating superorganism. In L.J. Goodman and R.C. Fisher, eds., *The Behaviour and Physiology of Bees*, pp. 28–47. C.A.B. International, Wallingford, U.K.

Stamps, J.A., R.A. Metcalf, and V.V. Krishnan. 1978. A genetic analysis of parent-offspring conflict. *Behavioral Ecology and Sociobiology* 3: 369–392.

Stanley, S.M. 1979. *Macroevolution. Pattern and Process*. W.H. Freeman and Co., San Francisco.

Stark, R.E. 1992. Cooperative nesting in the multivoltine large carpenter bee *Xylocopa sulcatipes* Maa (Apoidea: Anthophoridae): do helpers gain or lose to solitary females? *Ethology* 91: 301–310.

Starr, C.K. 1979. Origin and evolution of insect sociality: a review of modern theory. In H.R. Hermann, ed., *Social Insects*, Vol. I, pp. 35–79. Academic Press, New York.

Starr, C.K. 1984. Sperm competition, kinship, and sociality in the aculeate Hymenoptera. In R.L. Smith, ed., *Sperm Competition and the Evolution of Animal Mating Systems*, pp. 427–464. Academic Press, Orlando.

Starr, C.K. 1985a. What if workers in social Hymenoptera *were* males? *Journal of Theoretical Biology* 117: 11–18.

Starr, C.K. 1985b. Enabling mechanisms in the origin of sociality in the Hymenoptera – the sting's the thing. *Annals of the Entomological Society of America* 78: 836–840.

Starr, C.K. 1989. In reply, is the sting the thing? *Annals of the Entomological Society of America* 82: 6–8.

Stearns, S.C. 1983. The influence of size and phylogeny on patterns of covariation among life-history traits in the mammals. *Oikos* 41: 173–187.

Stearns, S.C. 1992. *The Evolution of Life Histories*. Oxford University Press, Oxford.

Stearns, S.C., and R.E. Crandall. 1981. Bet-hedging and persistence as adaptations of colonizers. In G.G.E. Scudder and J.L. Reveal, eds., *Evolution Today*, pp. 371–383. Hunt Institute, Philadelphia, Pennsylvania.

Stearns, S.C., and J.C. Koella. 1986. The evolution of phenotypic plasticity in life-history traits: predictions of reaction norms for age and size at maturity. *Evolution* 40: 893–913.

Stern, D.L. 1994. A phylogenetic analysis of soldier evolution in the aphid family Hormaphididae. *Proceedings of the Royal Society of London, Series B* 256: 203–209.

Stickland, T.R., and N.R. Franks. 1994. Computer image analysis provides new observations of ant behaviour patterns. *Proceedings of the Royal Society of London, Series B* 257: 279–286.

Stille, M., and B. Stille. 1992. Intra- and inter-nest variation in mitochondrial DNA in the polygynous ant *Leptothorax acervorum* (Hymenoptera; Formicidae). *Insectes Sociaux* 39: 335–340.

Stille, M., and B. Stille. 1993. Intrapopulation nestclusters of maternal mtDNA lineages in the polygynous ant *Leptothorax acervorum* (Hymenoptera: Formicidae). *Insect Molecular Biology* 1: 117–121.

Stille, M., B. Stille, and P. Douwes. 1991. Polygyny, relatedness and nest founding in the polygynous myrmicine ant *Leptothorax acervorum* (Hymenoptera; Formicidae). *Behavioral Ecology and Sociobiology* 28: 91–96.

Strassmann, J.E. 1984. Female-biased sex ratios in social insects lacking morphological castes. *Evolution* 38: 256–266.

Strassmann, J.E. 1989. Altruism and relatedness at colony foundation in social insects. *Trends in Ecology and Evolution* 4: 371–374.

Strassmann, J.E., and D.C. Queller. 1989. Ecological determinants of social evolution. In M.D. Breed and R.E. Page, eds., *The Genetics of Social Evolution*, pp. 81–101. Westview Press, Boulder, Colorado.

Strassmann, J.E., D.C. Queller, C.R. Solís, and C.R. Hughes. 1991. Relatedness and queen number in the Neotropical wasp, *Parachartergus colobopterus*. *Animal Behaviour* 42: 461–470.

Stuart, R.J. 1991. Nestmate recognition in leptothoracine ants: testing for effects of queen number, colony size and species of intruder. *Animal Behaviour* 42: 277–284.

Stuart, R.J., and R.E. Page. 1991. Genetic component to division of labor among workers of a leptothoracine ant. *Naturwissenschaften* 78: 375–377.

Stuart, R.J., A. Francoeur, and R. Loiselle. 1987. Lethal fighting among dimorphic males of the ant, *Cardiocondyla wroughtonii*. *Naturwissenschaften* 74: 548–549.

Stuart, R.J., L. Gresham-Bissett, and T.M. Alloway. 1993. Queen adoption in the polygynous and polydomous ant, *Leptothorax curvispinosus*. *Behavioral Ecology* 4: 276–281.

Stubblefield, J.W., and E.L. Charnov. 1986. Some conceptual issues in the origin of eusociality. *Heredity* 57: 181–187.

Stubblefield, J.W., and J. Seger. 1994. Sexual dimorphism in the Hymenoptera. In R.V. Short and E. Balaban, eds., *The Differences between the Sexes*, pp. 71–103. Cambridge University Press, Cambridge.

Sturtevant, A.H. 1938. Essays on evolution. II. On the effects of selection on social insects. *Quarterly Review of Biology* 13: 74–76.

Sudd, J.H. 1967. *An Introduction to the Behaviour of Ants*. Edward Arnold (Publishers) Ltd, London.

Sudd, J.H., and N.R. Franks. 1987. *The Behavioural Ecology of Ants*. Blackie, Glasgow.

Sundström, L. 1993a. Genetic population structure and sociogenetic organisation in *Formica truncorum* (Hymenoptera; Formicidae). *Behavioral Ecology and Sociobiology* 33: 345–354.

Sundström, L. 1993b. Intraspecific variation in sociogenetic organisation and worker-queen conflict in the ant *Formica truncorum*. Ph.D. thesis, University of Helsinki.

Sundström, L. 1994a. Sex ratio bias, relatedness asymmetry and queen mating frequency in ants. *Nature* 367: 266–268.

Sundström, L. 1994b. Ant sex ratios. *Nature* 370: 257–258.

Sundström, L. 1995a. Sex allocation and colony maintenance in monogyne and polygyne colonies of *Formica truncorum* (Hymenoptera: Formicidae); The impact of kinship and mating structure. *American Naturalist:* in press

Sundström, L. 1995b. Dispersal polymorphism and physiological condition of males and females in the ant, *Formica truncorum*. *Behavioral Ecology* 6: 132–139.

Suzuki, T. 1986. Production schedules of males and reproductive females, investment sex ratios, and worker-queen conflict in paper wasps. *American Naturalist* 128: 366–378.

Talbot, M. 1948. A comparison of two ants of the genus *Formica*. *Ecology* 29: 316–325.

Talbot, M. 1957. Population studies of the slave-making ant *Leptothorax duloticus* and its slave, *Leptothorax curvispinosus*. *Ecology* 38: 449–456.

Taylor, P.D. 1981a. Sex ratio compensation in ant populations. *Evolution* 35: 1250–1251.

Taylor, P.D. 1981b. Intra-sex and inter-sex sibling interactions as sex ratio determinants. *Nature* 291: 64–66.

Taylor, P.D. 1985. A general mathematical model for sex allocation. *Journal of Theoretical Biology* 112: 799–818.

Taylor, P.D. 1988. Inclusive fitness models with two sexes. *Theoretical Population Biology* 34: 145–168.

Taylor, P.D. 1989. Evolutionary stability in one-parameter models under weak selection. *Theoretical Population Biology* 36: 125–143.

Taylor, P.D. 1990. Allele-frequency change in a class-structured population. *American Naturalist* 135: 95–106.

Taylor, P.D. 1992a. Altruism in viscous populations - an inclusive fitness model. *Evolutionary Ecology* 6: 352–356.

Taylor, P.D. 1992b. Inclusive fitness in a homogeneous environment. *Proceedings of the Royal Society of London, Series B* 249: 299–302.

Taylor, P.D. 1993. Female-biased sex ratios under local mate competition: an experimental confirmation. *Evolutionary Ecology* 7: 306–308.

Taylor, P.D., and M.G. Bulmer. 1980. Local mate competition and the sex ratio. *Journal of Theoretical Biology* 86: 409–419.

Taylor, P.D., and A. Sauer. 1980. The selective advantage of sex-ratio homeostasis. *American Naturalist* 116: 305–310.

Taylor, R.W. 1978. *Nothomyrmecia macrops:* A living-fossil ant rediscovered. *Science* 201: 979–985.

Terayama, M., and T. Satoh. 1990. A new species of the genus *Camponotus* from Japan, with notes on two known forms of the subgenus *Myrmamblys* (Hymenoptera, Formicidae). *Japanese Journal of Entomology* 58: 405–414.

Theraulaz, G., S. Goss, J. Gervet, and J.-L. Deneubourg. 1991. Task differentiation in *Polistes* wasp colonies: a model for self-organizing groups of robots. In J.-A. Meyer and S.W. Wilson, eds., *From Animals to Animats: Proceedings of the First International Conference on Simulation of Adaptive Behavior*, pp. 346–355. MIT Press, Cambridge, Massachusetts.

Thornhill, R., and J. Alcock. 1983. *The Evolution of Insect Mating Systems.* Harvard University Press, Cambridge, Massachusetts.

Thurber, D.K., M.C. Belk, H.L. Black, C.D. Jorgensen, S.P. Hubbell, and R.B. Foster. 1993. Dispersion and mortality of colonies of the tropical ant *Paraponera clavata. Biotropica* 25: 215–221.

Tinaut, A., and J. Heinze. 1992. Wing reduction in ant queens from arid habitats. *Naturwissenschaften* 79: 84–85.

Tofts, C. 1993. Algorithms for task allocation in ants (A study of temporal polyethism: theory). *Bulletin of Mathematical Biology* 55: 891–918.

Tofts, C., and N.R. Franks. 1992. Doing the right thing: ants, honeybees and naked mole-rats. *Trends in Ecology and Evolution* 7: 346–349.

Tofts, C., M. Hatcher, and N.R. Franks. 1992. The autosynchronization of the ant *Leptothorax acervorum* (Fabricius): Theory, testability and experiment. *Journal of Theoretical Biology* 157: 71–82.

Topoff, H., and J. Mirenda. 1978. Precocial behaviour of callow workers of the army ant *Neivamyrmex nigrescens*: importance of stimulation by adults during mass recruitment. *Animal Behaviour* 26: 698–706.

Topoff, H., K. Lawson, and P. Richards. 1972. Trail following and its development in the Neotropical army ant genus *Eciton* (Hymenoptera: Formicidae: Dorylinae). *Psyche* 79: 357–364.

Traniello, J.F.A. 1977. Recruitment behavior, orientation, and the organization of foraging in the carpenter ant *Camponotus pennsylvanicus* DeGeer (Hymenoptera: Formicidae). *Behavioral Ecology and Sociobiology* 2: 61–79.

Traniello, J.F.A., and S.C. Levings. 1986. Intra- and intercolony patterns of nest dispersion in the ant *Lasius neoniger*: correlations with territoriality and foraging ecology. *Oecologia* 69: 413–419.

Trivers, R.L. 1971. The evolution of reciprocal altruism. *Quarterly Review of Biology* 46: 35–57.

Trivers, R.L. 1974. Parent-offspring conflict. *American Zoologist* 14: 249–264.

Trivers, R. 1985. *Social Evolution.* Benjamin/Cummings, Menlo Park, California.

Trivers, R.L., and H. Hare. 1976. Haplodiploidy and the evolution of the social insects. *Science* 191: 249–263.

Trivers, R.L., and D.E. Willard. 1973. Natural selection of parental ability to vary the sex ratio of offspring. *Science* 179: 90–92.

Tschinkel, W.R. 1987a. Fire ant queen longevity and age: estimation by sperm depletion. *Annals of the Entomological Society of America* 80: 263–266.

Tschinkel, W.R. 1987b. The fire ant, *Solenopsis invicta*, as a successful "weed". In J. Eder and H. Rembold, eds., *Chemistry and Biology of Social Insects*, pp. 585–588. Verlag J. Peperny, München.

Tschinkel, W.R. 1987c. Relationship between ovariole number and spermathecal sperm count in ant queens: a new allometry. *Annals of the Entomological Society of America* 80: 208–211.

Tschinkel, W.R. 1988a. Social control of egg-laying rate in queens of the fire ant, *Solenopsis invicta. Physiological Entomology* 13: 327–350.

Tschinkel, W.R. 1988b. Colony growth and the ontogeny of worker polymorphism in the fire ant, *Solenopsis invicta. Behavioral Ecology and Sociobiology* 22: 103–115.

Tschinkel, W.R. 1991. Insect sociometry, a field in search of data. *Insectes Sociaux* 38: 77–82.

Tschinkel, W.R. 1992a. Brood raiding in the fire ant, *Solenopsis invicta* (Hymenoptera: Formicidae): laboratory and field observations. *Annals of the Entomological Society of America* 85: 638–646.

Tschinkel, W.R. 1992b. Brood raiding and the population dynamics of founding and incipient colonies of the fire ant, *Solenopsis invicta*. *Ecological Entomology* 17: 179–188.

Tschinkel, W.R. 1993a. Sociometry and sociogenesis of colonies of the fire ant *Solenopsis invicta* during one annual cycle. *Ecological Monographs* 63: 425–457.

Tschinkel, W.R. 1993b. Resource allocation, brood production and cannibalism during colony founding in the fire ant, *Solenopsis invicta*. *Behavioral Ecology and Sociobiology* 33: 209–223.

Tschinkel, W.R., and D.F. Howard. 1978. Queen replacement in orphaned colonies of the fire ant, *Solenopsis invicta*. *Behavioral Ecology and Sociobiology* 3: 297–310.

Tschinkel, W.R., and D.F. Howard. 1983. Colony founding by pleometrosis in the fire ant, *Solenopsis invicta*. *Behavioral Ecology and Sociobiology* 12: 103–113.

Tschinkel, W.R., and S.D. Porter. 1988. Efficiency of sperm use in queens of the fire ant, *Solenopsis invicta* (Hymenoptera: Formicidae). *Annals of the Entomological Society of America* 81: 777–781.

Tsuchida, K. 1994. Genetic relatedness and the breeding structure of the Japanese paper wasp, *Polistes jadwigae*. *Ethology, Ecology and Evolution* 6: 237–242.

Tsuji, K. 1988a. Nest relocations in the Japanese queenless ant, *Pristomyrmex pungens* Mayr (Hymenoptera: Formicidae). *Insectes Sociaux* 35: 321–340.

Tsuji, K. 1988b. Obligate parthenogenesis and reproductive division of labor in the Japanese queenless ant *Pristomyrmex pungens*. Comparison of intranidal and extranidal workers. *Behavioral Ecology and Sociobiology* 23: 247–255.

Tsuji, K. 1990a. Kin recognition in *Pristomyrmex pungens* (Hymenoptera: Formicidae): asymmetrical change in acceptance and rejection due to odour transfer. *Animal Behaviour* 40: 306–312.

Tsuji, K. 1990b. Reproductive division of labour related to age in the Japanese queenless ant, *Pristomyrmex pungens*. *Animal Behaviour* 39: 843–849.

Tsuji, K. 1992. Sterility for life: applying the concept of eusociality. *Animal Behaviour* 44: 572–573.

Tsuji, K. 1994. Inter-colonial selection for the maintenance of cooperative breeding in the ant *Pristomyrmex pungens*: a laboratory experiment. *Behavioral Ecology and Sociobiology* 35: 109–113.

Tsuji, K., and K. Yamauchi. 1994. Colony level sex allocation in a polygynous and polydomous ant. *Behavioral Ecology and Sociobiology* 34: 157–167.

Uyenoyama, M.K. 1984. Inbreeding and the evolution of altruism under kin selection: effects on relatedness and group structure. *Evolution* 38: 778–795.

Uyenoyama, M.K., and B.O. Bengtsson. 1979. Towards a genetic theory for the evolution of the sex ratio. *Genetics* 93: 721–736.

Uyenoyama, M.K., and B.O. Bengtsson. 1981. Towards a genetic theory for the evolution of the sex ratio. II. Haplodiploid and diploid models with sibling and parental control of the brood sex ratio and brood size. *Theoretical Population Biology* 20: 57–79.

Uyenoyama, M.K., and B.O. Bengtsson. 1982. Towards a genetic theory for the evolution of the sex ratio. III. Parental and sibling control of brood investment ratio under partial sib-mating. *Theoretical Population Biology* 22: 43–68.

Uyenoyama, M., and M.W. Feldman. 1980. Theories of kin and group selection: a population genetics perspective. *Theoretical Population Biology* 17: 380–414.

Uyenoyama, M.K., and M.W. Feldman. 1981. On relatedness and adaptive topography in kin selection. *Theoretical Population Biology* 19: 87–123.

Van der Blom, J. 1991. Social regulation of egg-laying by queenless honeybee workers (*Apis mellifera* L.). *Behavioral Ecology and Sociobiology* 29: 341–346.

Van der Blom, J. 1993a. 'Division of labour' in honeybees: a revision. In M. Sommeijer and J. Van der Blom, eds., *Proceedings of the Section Experimental and Applied Entomology of the Netherlands Entomological Society*, Vol. 4, pp. 91–96. Nederlandse Entomologische Vereniging, Amsterdam.

Van der Blom, J. 1993b. Individual differentiation in behaviour of honey bee workers (*Apis mellifera* L.). *Insectes Sociaux* 40: 345–361.

Van der Have, T.M., J.J. Boomsma, and S.B.J. Menken. 1988. Sex-investment ratios and relatedness in the monogynous ant *Lasius niger* (L.). *Evolution* 42: 160–172.

Vander Meer, R.K., M.S. Obin, and L. Morel. 1990. Nestmate recognition in fire ants: monogyne and polygyne populations. In R.K. Vander Meer, K. Jaffe, and A. Cedeno, eds., *Applied Myrmecology: A World Perspective*, pp. 322–328. Westview Press, Boulder, Colorado.

Vander Meer, R.K., L. Morel, and C.S. Lofgren. 1992. A comparison of queen oviposition rates from monogyne and polygyne fire ant, *Solenopsis invicta*, colonies. *Physiological Entomology* 17: 384–390.

Van Loon, A.J., J.J. Boomsma, and A. Andrasfalvy. 1990. A new polygynous *Lasius* species (Hymenoptera; Formicidae) from Central Europe. I. Description and general biology. *Insectes Sociaux* 37: 348–362.

Van Noordwijk, A.J., and G. de Jong. 1986. Acquisition and allocation of resources: their influence on variation in life history tactics. *American Naturalist* 128: 137–142.

Vargo, E.L. 1988. Effect of pleometrosis and colony size on the production of sexuals in monogyne colonies of the fire ant *Solenopsis invicta*. In J.C. Trager, ed., *Advances in Myrmecology*, pp. 217–225. E.J. Brill, Leiden.

Vargo, E.L. 1990. Social control of reproduction in fire ant colonies. In R.K. Vander Meer, K. Jaffe, and A. Cedeno, eds., *Applied Myrmecology: A World Perspective*, pp. 158–172. Westview Press, Boulder, Colorado.

Vargo, E.L. 1992. Mutual pheromonal inhibition among queens in polygyne colonies of the fire ant *Solenopsis invicta*. *Behavioral Ecology and Sociobiology* 31: 205–210.

Vargo, E.L. 1993. Colony reproductive structure in a polygyne population of *Solenopsis geminata* (Hymenoptera: Formicidae). *Annals of the Entomological Society of America* 86: 441–449.

Vargo, E.L., and D.J.C. Fletcher. 1986a. Evidence of pheromonal queen control over the production of male and female sexuals in the fire ant, *Solenopsis invicta. Journal of Comparative Physiology A* 159: 741–749.

Vargo, E.L., and D.J.C. Fletcher. 1986b. Queen number and the production of sexuals in the fire ant, *Solenopsis invicta* (Hymenoptera: Formicidae). *Behavioral Ecology and Sociobiology* 19: 41–47.

Vargo, E.L., and D.J.C. Fletcher. 1987. Effect of queen number on the production of sexuals in natural populations of the fire ant, *Solenopsis invicta. Physiological Entomology* 12: 109–116.

Vargo, E.L., and D.J.C. Fletcher. 1989. On the relationship between queen number and fecundity in polygyne colonies of the fire ant *Solenopsis invicta. Physiological Entomology* 14: 223–232.

Vargo, E.L., and L. Passera. 1991. Pheromonal and behavioral queen control over the production of gynes in the Argentine ant *Iridomyrmex humilis* (Mayr). *Behavioral Ecology and Sociobiology* 28: 161–169.

Vargo, E.L., and L. Passera. 1992. Gyne development in the Argentine ant *Iridomyrmex humilis:* role of over-wintering and queen control. *Physiological Entomology* 17: 193–201.

Vargo, E.L., and S.D. Porter. 1989. Colony reproduction by budding in the polygyne form of *Solenopsis invicta* (Hymenoptera: Formicidae). *Annals of the Entomological Society of America* 82: 307–313.

Vargo, E.L., and S.D. Porter. 1993. Reproduction by virgin queen fire ants in queenless colonies: comparative study of three taxa (*Solenopsis richteri,* hybrid *S. invicta/richteri, S. geminata*) (Hymenoptera: Formicidae). *Insectes Sociaux* 40: 283–293.

Vargo, E.L., and K.G. Ross. 1989. Differential viability of eggs laid by queens in polygyne colonies of the fire ant, *Solenopsis invicta. Journal of Insect Physiology* 35: 587–593.

Vehrencamp, S.L. 1979. The roles of individual, kin, and group selection in the evolution of sociality. In P. Marler and J.G. Vandenbergh, eds., *Handbook of Behavioral Neurobiology. Vol. 3. Social Behavior and Communication*, pp. 351–394. Plenum Press, New York.

Vehrencamp, S.L. 1983a. Optimal degree of skew in cooperative societies. *American Zoologist* 23: 327–335.

Vehrencamp, S.L. 1983b. A model for the evolution of despotic versus egalitarian societies. *Animal Behaviour* 31: 667–682.

Venkataraman, A.B., V.B. Swarnalatha, P. Nair, and R. Gadagkar. 1988. The mechanism of nestmate discrimination in the tropical social wasp *Ropalidia marginata* and its implications for the evolution of sociality. *Behavioral Ecology and Sociobiology* 23: 271–279.

Vepsäläinen, K., and R. Savolainen. 1990. The effect of interference by formicine ants on the foraging of *Myrmica. Journal of Animal Ecology* 59: 643–654.

Verner, J. 1965. Selection for sex ratio. *American Naturalist* 99: 419–421.

Villet, M. 1989. A syndrome leading to ergatoid queens in ponerine ants (Hymenoptera: Formicidae). *Journal of Natural History* 23: 825–832.

Villet, M.H. 1991. Colony foundation in *Plectroctena mandibularis* F. Smith, and the evolution of ergatoid queens in *Plectroctena* (Hymenoptera: Formicidae). *Journal of Natural History* 25: 979–983.

Visscher, P.K. 1989. A quantitative study of worker reproduction in honey bee colonies. *Behavioral Ecology and Sociobiology* 25: 247–254.

Visscher, P.K. 1993. A theoretical analysis of individual interests and intra-colony conflict during swarming of honey bee colonies. *Journal of Theoretical Biology* 165: 191–212.

Visscher, P.K., and R. Dukas. 1995. Honey bees recognize development of nest-mates' ovaries. *Animal Behaviour* 49: 542–544.

Vrba, E.S. 1989. Levels of selection and sorting with special reference to the species level. In P.H. Harvey and L. Partridge, eds., *Oxford Surveys in Evolutionary Biology*, Vol. 6, pp. 111–168. Oxford University Press, Oxford.

Waage, J.K. 1979. Dual function of the damselfly penis: sperm removal and transfer. *Science* 203: 916–918.

Wade, M.J. 1978a. Kin selection: a classical approach and a general solution. *Proceedings of the National Academy of Sciences, U.S.A.* 75: 6154–6158.

Wade, M.J. 1978b. A critical review of the models of group selection. *Quarterly Review of Biology* 53: 101–114.

Wade, M.J. 1979. The evolution of social interactions by family selection. *American Naturalist* 113: 399–417.

Wade, M.J. 1980. Kin selection: its components. *Science* 210: 665–667.

Wade, M.J. 1982a. Evolution of interference competition by individual, family, and group selection. *Proceedings of the National Academy of Sciences, U.S.A.* 79: 3575–3578.

Wade, M.J. 1982b. The effect of multiple inseminations on the evolution of social behaviors in diploid and haplo-diploid organisms. *Journal of Theoretical Biology* 95: 351–368.

Wade, M.J. 1985a. Soft selection, hard selection, kin selection, and group selection. *American Naturalist* 125: 61–73.

Wade, M.J. 1985b. The influence of multiple inseminations and multiple foundresses on social evolution. *Journal of Theoretical Biology* 112: 109–121.

Wade, M.J., and F. Breden. 1981. Effect of inbreeding on the evolution of altruistic behavior by kin selection. *Evolution* 35: 844–858.

Walker, J., and J. Stamps. 1986. A test of optimal caste ratio theory using the ant *Camponotus (Colobopsis) impressus*. *Ecology* 67: 1052–1062.

Waloff, N. 1957. The effect of the number of queens of the ant *Lasius flavus* (Fab.) (Hym., Formicidae) on their survival and on the rate of development of the first brood. *Insectes Sociaux* 4: 391–408.

Walter, F., D.J.C. Fletcher, D. Chautems, D. Cherix, L. Keller, W. Francke, W. Fortelius, R. Rosengren, and E.L. Vargo. 1993. Identification of the sex pheromone of an ant, *Formica lugubris* (Hymenoptera, Formicidae). *Naturwissenschaften* 80: 30–34.

Ward, P.S. 1983a. Genetic relatedness and colony organization in a species complex of ponerine ants. I. Phenotypic and genotypic composition of colonies. *Behavioral Ecology and Sociobiology* 12: 285–299.

Ward, P.S. 1983b. Genetic relatedness and colony organization in a species complex of ponerine ants. II. Patterns of sex ratio investment. *Behavioral Ecology and Sociobiology* 12: 301–307.

Ward, P.S. 1989. Genetic and social changes associated with ant speciation. In M.D. Breed and R.E. Page, eds., *The Genetics of Social Evolution*, pp. 123–148. Westview Press, Boulder, Colorado.

Watkins, J.F. 1976. *The Identification and Distribution of New World Army Ants (Dorylinae: Formicidae)*. The Markham Press Fund of Baylor University Press, Waco, Texas.

Watkinson, A.R., and J. White. 1986. Some life-history consequences of modular construction in plants. *Philosophical Transactions of the Royal Society of London, Series B* 313: 31–51.

Watson, J.A.L., B.M. Okot-Kotber, and C. Noirot. (eds.) 1985. *Caste Differentiation in Social Insects*. Pergamon Press, Oxford.

Wehner, R. 1987. Spatial organization of foraging behavior in individually searching desert ants, *Cataglyphis* (Sahara Desert) and *Ocymyrmex* (Namib Desert). In J.M. Pasteels and J.-L. Deneubourg, eds., *From Individual to Collective Behavior in Social Insects*, pp. 15–42. Birkhäuser Verlag, Basel.

Wehner, R., R.D. Harkness, and P. Schmid-Hempel. 1983. *Foraging strategies in individually searching ants*, Cataglyphis bicolor *(Hymenoptera: Formicidae)*. (*Information Processing in Animals*, Vol. 1). Gustav Fischer Verlag, Stuttgart.

Wenzel, J.W., and J. Pickering. 1991. Cooperative foraging, productivity, and the central limit theorem. *Proceedings of the National Academy of Sciences, U.S.A.* 88: 36–38.

Werren, J.H., U. Nur, and C.-I. Wu. 1988. Selfish genetic elements. *Trends in Ecology and Evolution* 3: 297–302.

Wesson, L.G. 1940. An experimental study on caste determination in ants. *Psyche* 47: 105–111.

West-Eberhard, M.J. 1975. The evolution of social behavior by kin selection. *Quarterly Review of Biology* 50: 1–33.

West-Eberhard, M.J. 1978a. Polygyny and the evolution of social behavior in wasps. *Journal of the Kansas Entomological Society* 51: 832–856.

West-Eberhard, M.J. 1978b. Temporary queens in *Metapolybia* wasps: nonreproductive helpers without altruism? *Science* 200: 441–443.

West-Eberhard, M.J. 1979. Sexual selection, social competition, and evolution. *Proceedings of the American Philosophical Society* 123: 222–234.

West-Eberhard, M.J. 1981. Intragroup selection and the evolution of insect societies. In R.D. Alexander and D.W. Tinkle, eds., *Natural Selection and Social Behavior*, pp. 3–17. Chiron Press, New York.

West-Eberhard, M.J. 1987a. The epigenetical origins of insect sociality. In J. Eder and H. Rembold, eds., *Chemistry and Biology of Social Insects*, pp. 369–372. Verlag J. Peperny, München.

West-Eberhard, M.J. 1987b. Flexible strategy and social evolution. In Y. Itô, J.L. Brown, and J. Kikkawa, eds., *Animal Societies: Theories and Facts*, pp. 35–51. Japan Scientific Societies Press, Tokyo.

West-Eberhard, M.J. 1988. Phenotypic plasticity and "genetic" theories of insect sociality. In G. Greenberg and E. Tobach, eds., *Evolution of Social Behavior and Integrative Levels*, pp. 123–133. Lawrence Erlbaum, Hillsdale, New Jersey.

West-Eberhard, M.J. 1989. Phenotypic plasticity and the origins of diversity. *Annual Review of Ecology and Systematics* 20: 249–278.

West-Eberhard, M.J. 1992. Genetics, epigenetics, and flexibility: a reply to Crozier. *American Naturalist* 139: 224–226.

Westoby, M. 1994. Adaptive thinking and medicine. *Trends in Ecology and Evolution* 9: 1–2.

Wheeler, D.E. 1986. Developmental and physiological determinants of caste in social Hymenoptera: evolutionary implications. *American Naturalist* 128: 13–34.

Wheeler, D.E. 1991. The developmental basis of worker caste polymorphism in ants. *American Naturalist* 138: 1218–1238.

Wheeler, D.E., and P.H. Krutzsch. 1992. Internal reproductive system in adult males of the genus *Camponotus* (Hymenoptera: Formicidae: Formicinae). *Journal of Morphology* 211: 307–317.

Wheeler, D.E., and H.F. Nijhout. 1984. Soldier determination in *Pheidole bicarinata*: inhibition by adult soldiers. *Journal of Insect Physiology* 30: 127–135.

Wheeler, W.M. 1910. *Ants: Their Structure, Development and Behavior.* Columbia University Press, New York.

Wheeler, W.M. 1911. The ant-colony as an organism. *Journal of Morphology* 22: 307–325.

Wickler, W. 1976. Evolution-oriented ethology, kin selection, and altruistic parasites. *Zeitschrift für Tierpsychologie* 42: 206–214.

Willer, D.E., and D.J.C. Fletcher. 1986. Differences in inhibitory capability among queens of the ant *Solenopsis invicta*. *Physiological Entomology* 11: 475–482.

Williams, B.J. 1981. A critical review of models in sociobiology. *Annual Review of Anthropology* 10: 163–192.

Williams, G.C. 1957. Pleiotropy, natural selection, and the evolution of senescence. *Evolution* 11: 398–411.

Williams, G.C. 1966. *Adaptation and Natural Selection.* Princeton University Press, Princeton, New Jersey.

Williams, G.C. (ed.) 1971. *Group Selection.* Aldine Atherton, Chicago.

Williams, G.C. 1979. The question of adaptive sex ratio in outcrossed vertebrates. *Proceedings of the Royal Society of London, Series B* 205: 567–580.

Williams, G.C. 1985. A defense of reductionism in evolutionary biology. In R. Dawkins and M. Ridley, eds., *Oxford Surveys in Evolutionary Biology*, Vol. 2, pp. 1–27. Oxford University Press, Oxford.

Williams, G.C. 1992. *Natural Selection: Domains, Levels, and Challenges.* Oxford University Press, New York.

Williams, G.C., and D.C. Williams. 1957. Natural selection of individually harmful social adaptations among sibs with special reference to social insects. *Evolution* 11: 32–39.

Willis, E.O. 1967. The behavior of bicolored antbirds. *University of California Publications in Zoology* 79: 1–127.

Wilson, D.S. 1975. A theory of group selection. *Proceedings of the National Academy of Sciences, U.S.A.* 72: 143–146.

Wilson, D.S. 1976. Evolution on the level of communities. *Science* 192: 1358–1360.

Wilson, D.S. 1977. Structured demes and the evolution of group-advantageous traits. *American Naturalist* 111: 157–185.

Wilson, D.S. 1979. Structured demes and trait-group variation. *American Naturalist* 113: 606–610.

Wilson, D.S. 1980. *The Natural Selection of Populations and Communities.* Benjamin/Cummings, Menlo Park, California.

Wilson, D.S. 1983. The group selection controversy: history and current status. *Annual Review of Ecology and Systematics* 14: 159–187.

Wilson, D.S. 1990. Weak altruism, strong group selection. *Oikos* 59: 135–140.

Wilson, D.S., and R.K. Colwell. 1981. Evolution of sex ratio in structured demes. *Evolution* 35: 882–897.

Wilson, D.S., and E. Sober. 1989. Reviving the superorganism. *Journal of Theoretical Biology* 136: 337–356.

Wilson, D.S., G.B. Pollock, and L.A. Dugatkin. 1992. Can altruism evolve in purely viscous populations? *Evolutionary Ecology* 6: 331–341.

Wilson, E.O. 1953. The origin and evolution of polymorphism in ants. *Quarterly Review of Biology* 28: 136–156.

Wilson, E.O. 1963. Social modifications related to rareness in ant species. *Evolution* 17: 249–253.

Wilson, E.O. 1966. Behaviour of social insects. In P.T. Haskell, ed., *Insect Behaviour.* Symposia of the Royal Entomological Society of London No. 3, pp. 81–96. Royal Entomological Society, London.

Wilson, E.O. 1968. The ergonomics of caste in the social insects. *American Naturalist* 102: 41–66.

Wilson, E.O. 1971. *The Insect Societies.* Harvard University Press, Cambridge, Massachusetts.

Wilson, E.O. 1974a. Aversive behavior and competition within colonies of the ant *Leptothorax curvispinosus. Annals of the Entomological Society of America* 67: 777–780.

Wilson, E.O. 1974b. The population consequences of polygyny in the ant *Leptothorax curvispinosus. Annals of the Entomological Society of America* 67: 781–786.

Wilson, E.O. 1975. *Sociobiology. The New Synthesis.* Harvard University Press, Cambridge, Massachusetts.

Wilson, E.O. 1976a. Behavioral discretization and the number of castes in an ant species. *Behavioral Ecology and Sociobiology* 1: 141–154.

Wilson, E.O. 1976b. The central problems of sociobiology. In R.M. May, ed., *Theoretical Ecology: Principles and Applications,* 1st edition, pp. 205–217. Blackwell, Oxford.

Wilson, E.O. 1976c. A social ethogram of the neotropical arboreal ant *Zacryptocerus varians* (Fr. Smith). *Animal Behaviour* 24: 354–363.

Wilson, E.O. 1978. Division of labor in fire ants based on physical castes (Hymenoptera: Formicidae: *Solenopsis*). *Journal of the Kansas Entomological Society* 51: 615–636.

Wilson, E.O. 1980a. Caste and division of labor in leaf-cutter ants (Hymenoptera: Formicidae: *Atta*). I. The overall pattern in *A. sexdens*. *Behavioral Ecology and Sociobiology* 7: 143–156.

Wilson, E.O. 1980b. Caste and division of labor in leaf-cutter ants (Hymenoptera: Formicidae: *Atta*). II. The ergonomic optimization of leaf cutting. *Behavioural Ecology and Sociobiology* 7: 157–165.

Wilson, E.O. 1984. The relation between caste ratios and division of labor in the genus *Pheidole* (Hymenoptera: Formicidae). *Behavioral Ecology and Sociobiology* 16: 89–98.

Wilson, E.O. 1985a. The sociogenesis of insect colonies. *Science* 228: 1489–1495.

Wilson, E.O. 1985b. Between-caste aversion as a basis for division of labor in the ant *Pheidole pubiventris* (Hymenoptera: Formicidae). *Behavioral Ecology and Sociobiology* 17: 35–37.

Wilson, E.O. 1985c. The principles of caste evolution. In B. Hölldobler and M. Lindauer, eds., *Experimental Behavioral Ecology and Sociobiology*, pp. 307–324. Gustav Fischer Verlag, Stuttgart.

Wilson, E.O. 1987. Causes of ecological success: the case of the ants. *Journal of Animal Ecology* 56: 1–9.

Wilson, E.O. 1990. *Success and Dominance in Ecosystems: the Case of the Social Insects.* Ecology Institute, Oldendorf/Luhe, Germany.

Wilson, E.O., and T. Eisner. 1957. Quantitative studies of liquid food transmission in ants. *Insectes Sociaux* 4: 157–166.

Wilson, E.O., and R.M. Fagen. 1974. On the estimation of total behavioral repertories in ants. *Journal of the New York Entomological Society* 82: 106–112.

Wilson, E.O., and B. Hölldobler. 1980. Sex differences in cooperative silk-spinning by weaver ant larvae. *Proceedings of the National Academy of Sciences, U.S.A.* 77: 2343–2347.

Wilson, E.O., and B. Hölldobler. 1988. Dense heterarchies and mass communication as the basis of organization in ant colonies. *Trends in Ecology and Evolution* 3: 65–68.

Wimsatt, W. 1980. Reductionist research strategies and their biases in the units of selection controversy. Reprinted in E. Sober (ed.) 1984. *Conceptual Issues in Evolutionary Biology. An Anthology*, pp. 142–183. MIT Press, Cambridge, Massachusetts.

Winter, U., and A. Buschinger. 1983. The reproductive biology of a slavemaker ant, *Epimyrma ravouxi*, and a degenerate slavemaker, *E. kraussei* (Hymenoptera: Formicidae). *Entomologia Generalis* 9: 1–15.

Winter, U., and A. Buschinger. 1986. Genetically mediated queen polymorphism and caste determination in the slave-making ant, *Harpagoxenus sublaevis* (Hymenoptera: Formicidae). *Entomologia Generalis* 11: 125–137.

Wolpert, L. 1974. *The Development of Pattern and Form in Animals.* (Oxford Biology Readers No. 51), Oxford University Press, London.

Woyciechowski, M. 1990. Do honey bee, *Apis mellifera* L., workers favour sibling eggs and larvae in queen rearing? *Animal Behaviour* 39: 1220–1222.

Woyciechowski, M., and A. Lomnicki. 1987. Multiple mating of queens and the sterility of workers among eusocial Hymenoptera. *Journal of Theoretical Biology* 128: 317–327.

Woyciechowski, M., L. Kabat, and E. Król. 1994a. The function of the mating sign in honey bees, *Apis mellifera* L.: new evidence. *Animal Behaviour* 47: 733–735.

Woyciechowski, M., E. Król, E. Figurny, M. Stachowicz, and M. Tracz. 1994b. Genetic diversity of workers and infection by the parasite *Nosema apis* in honey bee colonies (*Apis mellifera*). In A. Lenoir, G. Arnold, and M. Lepage, eds., *Les Insectes Sociaux*, p. 347. Université Paris Nord, Villetaneuse.

Woyke, J. 1963. What happens to diploid drone larvae in a honeybee colony. *Journal of Apicultural Research* 2: 73–75.

Wrensch, D.L., and M.A. Ebbert (eds.) 1993. *Evolution and Diversity of Sex Ratio in Insects and Mites*. Chapman and Hall, New York.

Wright, S. 1931. Evolution in Mendelian populations. *Genetics* 16: 97–159. Reprinted in S. Wright. 1986. *Evolution: Selected Papers* (W.B. Provine, ed.), pp. 98–160. University of Chicago Press, Chicago.

Wright, S. 1945. Tempo and mode in evolution: a critical review. *Ecology* 26: 415–419.

Wright, S. 1980. Genic and organismic selection. *Evolution* 34: 825–843.

Wynne-Edwards, V.C. 1962. *Animal Dispersion in Relation to Social Behaviour*. Oliver and Boyd, Edinburgh.

Wynne-Edwards, V.C. 1963. Intergroup selection in the evolution of social systems. *Nature* 200: 623–626.

Wynne-Edwards, V.C. 1964a. Response to Maynard Smith (1964). *Nature* 201: 1147.

Wynne-Edwards, V.C. 1964b. Response to Perrins (1964). *Nature* 201: 1148–1149.

Wynne-Edwards, V.C. 1986. *Evolution through Group Selection*. Blackwell, Oxford.

Wynne-Edwards, V.C. 1993. A rationale for group selection. *Journal of Theoretical Biology* 162: 1–22.

Yamaguchi, T. 1992. Interspecific interference for nest sites between *Leptothorax congruus* and *Monomorium intrudens*. *Insectes Sociaux* 39: 117–127.

Yamaguchi, Y. 1985. Sex ratios of an aphid subject to local mate competition with variable maternal condition. *Nature* 318: 460–462.

Yamamura, N., and M. Higashi. 1992. An evolutionary theory of conflict resolution between relatives: altruism, manipulation, compromise. *Evolution* 46: 1236–1239.

Yamauchi, K., and N. Kawase. 1992. Pheromonal manipulation of workers by a fighting male to kill his rival males in the ant *Cardiocondyla wroughtonii*. *Naturwissenschaften* 79: 274–276.

Yamauchi, K., K. Kinomura, and S. Miyake. 1981. Sociobiological studies of the polygynic ant *Lasius sakagamii*. I. General features of its polydomous system. *Insectes Sociaux* 28: 279–296.

Yamauchi, K., K. Kinomura, and S. Miyake. 1982. Sociobiological studies of the polygynic ant *Lasius sakagamii*. II. Production of colony members. *Insectes Sociaux* 29: 164–174.

Yamauchi, K., T. Furukawa, K. Kinomura, H. Takamine, and K. Tsuji. 1991.

Secondary polygyny by inbred wingless sexuals in the dolichoderine ant *Technomyrmex albipes. Behavioral Ecology and Sociobiology* 29: 313–319.

Yamauchi, K., W. Czechowski, and B. Pisarski. 1994. Multiple mating and queen adoption in the wood ant, *Formica polyctena* Foerst. (Hymenoptera, Formicidae). *Memorabilia Zoologica* 48: 267–278.

Yanega, D. 1988. Social plasticity and early-diapausing females in a primitively social bee. *Proceedings of the National Academy of Sciences, U.S.A.* 85: 4374–4377.

Yanega, D. 1989. Caste determination and differential diapause within the first brood of *Halictus rubicundus* in New York (Hymenoptera: Halictidae). *Behavioral Ecology and Sociobiology* 24: 97–107.

Yokoyama, S., and J. Felsenstein. 1978. A model of kin selection for an altruistic trait considered as a quantitative character. *Proceedings of the National Academy of Sciences, U.S.A.* 75: 420–422.

Author Index

Subject Index

Taxonomic Index